Tectonics

Eldridge M. Moores
Robert J. Twiss
University of California at Davis

W. H. Freeman and Company
New York

To our wives, Judy and Helen,
and our children, Geneva, Brian and Kathryn, and Ian and Andrew.

Cover image: Landsat TM image of an area in northern Pakistan showing the southern margin of part of the active fold and thrust belt that marks the Eurasian-Indian continent-continent collision. North is approximately at the top. The Indus River flows from the northeast corner to the south center of the image. The prominent curved range in the southern part of the image is the Salt Range. The area northeast of the Salt Range is the Potwar Plateau, a highland with elevations up to 1500 m, underlain by Neogene sediments. The patterned ground within the C of the word "TECTONICS" is the city of Kohat and surroundings. The light-colored region at the top center is the southern margin of the late Neogene Peshawar basin. The entire area from the Salt Range north involves folds and thrusts in a Paleozoic-Cenozoic sedimentary sequence above a décollement in Cambrian salt. The area is highly active seismically. *(Image courtesy of GEOPIC®, Earth Satellite Corporation)*

Library of Congress Cataloging-in-Publication Data

Moores, Eldridge M., 1938–
    Tectonics/Eldridge M. Moores, Robert J. Twiss.
       p.   cm.
    Includes bibliographical references and index.
    ISBN 0-7167-2437-5
    1. Geology, Structural.   I. Twiss, Robert J.   II. Title.
    QE601.M65   1995
    551.8—dc20

                           95-10975
                           CIP

© 1995 by W. H. Freeman and Company

No part of this book may be reproduced by any mechanical, photographic, or electronic process, or in the form of a phonographic recording, nor may it be stored in a retrieval system, transmitted, or otherwise copied for public or private use, without written permission from the publisher.

Printed in the United States of America

First printing 1995, VB

Contents

Preface

IN THIS BOOK and in its companion *Structural Geology,* we have endeavored to present tectonics and structural geology as a unified subject that is divided only by the scale of the deformational features in the Earth's crust that we choose to consider. The continuous variation in scale of structures, however, means that there is really no clear boundary between tectonics and structural geology. Because our goal is to understand the processes by which deformation of the Earth's crust occurs and to interpret the structures we observe in terms of those processes, we need to integrate our understanding of the deformation that occurs at all scales. In *Tectonics* we deal with features at scales ranging from the regional to the global. In *Structural Geology* we consider scales from the submicroscopic to the regional. Both books emphasize the value of integrating information over all available scales. We firmly believe that the work of tectonicists and structural geologists is complementary and each group needs to pay attention to the work of the other.

This book is divided into three main parts. In Part I, we present first an overview, and we discuss techniques of studying structural geology and tectonics. In Part II, we review the principal tectonic features of the Earth and then we set out the basic principles of plate tectonics, discussing the geometry and kinematic principles of plate motions and plate reconstructions, and describing the processes and characteristic structures associated with active rift, transform, and subduction boundaries, and with triple junctions and collision zones. In an "Interlude," we discuss the nature of the scientific method, especially as it relates to geology, how scientific revolutions come about, and the history of development of plate tectonics as an example of a scientific revolution. In Part III, we discuss the nature of orogenic belts, the field of neotectonics, and several case studies of selected orogenic systems. We conclude the book with a brief introduction to the tectonics of the terrestrial planets and Earth's Moon.

Our organization is based on the time-honored strategy for much geologic investigation: using the present as a key to the past. We use recent and on-going tectonic processes to infer the tectonics of the ocean basins. These are relatively young areas of the Earth's crust whose history is relatively simple. We infer the history from the pattern of linear magnetic anomaly stripes that record the history of creation of the ocean floor for roughly the past 180 million years. Because the record of tectonic activity on the continents reaches much further back in time than it does in the ocean basins, we also consider how these tectonic processes interact with continental crust and what we can observe as the consequences of this interaction. We consider the effects on continents both of processes at plate margins and in regions of crustal collision. These observations then provide the basis for our understanding of neotectonic processes and of older orogenic belts and the tectonic processes that occurred in the geologic past. In particular, we find that the plate tectonic model is adequate for understanding orogenic belts as old as Neoproterozoic, but that for more ancient deformation, evidence suggests differences that provide clues about the evolution of tectonic processes in the Earth.

In addition to using present tectonic activity on the Earth as a key to the past, we can also compare tectonic activity on Earth with that on other terrestrial bodies in the solar system. To the extent that our models of the causes and evolution of tectonic activity on Earth are correct, the models should be adaptable to the specific conditions that occur in other planets of comparable composition and structure, and observation of these planets thus begins to provide us with a broader realm within which to test our models and to build better ones.

We are acutely aware of the fact that tectonics is a fast-moving field in which many new ideas and observations are continuously appearing in print. Any tectonics textbook such as this one is in danger of rapidly becoming obsolete. We have tried, however, to emphasize the basic principles of the field—principles that we hope have achieved some permanence, at least for a time. Further, we have tried to illustrate the methodology by which those principles can be applied to the interpretation and understanding of geologic observation.

In this work we also have tried to maintain a clear distinction between what we know from observing the Earth and what we infer from interpretation. Hence we emphasize throughout the book the use of models as a means of understanding our observations and putting them in context. In principle, the observations should remain reliable, although the models may change. In fact, however, it is an interesting characteristic of scientific research that the choice of which observations should be made is governed in no small part by the models that the observer has in mind at the time of the observation. Maintaining a clear distinction between observation and interpretation is nevertheless important, because models are always limited by simplifying assumptions, and it is often by challenging such assumptions that we open doorways to new and improved models that advance our understanding and that generate the need for new observations. With this approach, therefore, we hope to present the subject less as a fixed body of information than as an on-going and open-ended field of study in which assumptions can and should be challenged and interpretations changed to further our understanding.

Tectonics and its companion *Structural Geology* originally were conceived as one volume. They were separated into two volumes only when the amount of material seemed too much for a single book. In order to make each book a stand-alone product, some duplication of text became unavoidable. Thus Chapters 1, 2, 3, and 10 in this book duplicate, in varying degrees, the material in Chapters 1, 2, 21, and 22 of *Structural Geology*.

Much of the material in this book has been the subject of a separate quarter-course in tectonics, which has been taught at the University of California, Davis, for over two decades. There is more, however, than can be covered in a single quarter. Our typical quarter course treats principally the information in Chapters 3–10. Our experience is that to cover all the material in the book would require more than one quarter, probably at least a semester or, with supplementary reading from the scientific literature, even two quarters.

We welcome comments on the book from professors who teach with the book, from professional geologists who use it as a resource, and also from students. Let us know what parts of the book you don't like, what parts are not clear, and even what parts you do like, so that we can keep striving to make it more effectively meet your needs.

Acknowledgments

In the course of writing this book, we have benefited immensely from the help and suggestions of our own students and of a number of colleagues who have read parts or all of the manuscript in its various stages of revision. The constructive comments of Roy Dokka, Roy Kligfield, William MacDonald, Stephen Marshak, Peter Mattson, Cris Mawer, Sharon Mosher, Carol Simpson, and Doug Walker all helped us to improve the manuscript. We are particularly grateful to Carol Simpson, Sharon Mosher, and William MacDonald for their exceptionally detailed and helpful comments. We thank our own students for their patience with early photocopy versions of the manuscript and for the help they have given us by their reactions and questions about the manuscript. We are also grateful to all those authors who have helped by allowing us to use their published diagrams, and to those whose research has provided the basis for much of our discussion.

We wish to express our appreciation to the editorial and the production staff at W. H. Freeman and Company, including the late John Staples, who initiated the projected. Our thanks also go to Holly Hodder, Acqusitions Editor, and Kay Ueno, Project Editor, who have diplomatically kept us going, to Larry Marcus, Assistant Photo Editor, whose research efforts found us our spectacular cover image, and last, but not least, to John Hatzakis, Designer, Sheila Anderson, Production Coordinator, Bill Page, Illustration Coordinator, who coordinated the work of the illustrators Ian Worpole and Network Graphics, and Sheridan Sellers, EPC Manager, who prepared page makeup.

Finally, we wish to thank John R. Everett of Earth Satellite Corporation for providing us with the cover image.

September 1995

Eldridge M. Moores
Robert J. Twiss

PART I

Introduction

TECTONICS and structural geology are closely related in both their subject matter and their approach to the study of Earth's evolution. Although traditionally taught as courses that are distinct from the other branches of geological study such as petrology, paleontology, and geophysics, tectonics and structural geology have large areas of interdependence and overlap with these other fields. The overall goal of geology is to understand the evolution of a single planet, after all, and the pieces of the jigsaw puzzle must ultimately fit together. The practice and study of tectonics and structural geology require a familiarity with the scientific method and with a variety of standard techniques of data acquisition and analysis. Thus, an overview of tectonics and structural geology and a discussion of some of these basic techniques are the topics of Part I.

In Chapter 1, after defining what we mean by tectonics and structural geology, we discuss the use of models in scientific investigation. The model deductive approach is common to most of the sciences today, and it provides a fundamental theme that is the basis for much of the book's organization. We then present an overview of the fields of structural geology and tectonics and their relationship to the study of Earth processes and to the large-scale view of planetary bodies.

In Chapter 2, we present brief summaries of geophysical techniques that provide our primary sources of data on the structure of the deep crust and upper mantle. These techniques enable us to view the third dimension of the solid Earth that is so central to the study of tectonics. At least a rudimentary understanding of the meaning and significance of geophysical data is important, but we encourage students to go beyond these rather cursory discussions and take separate courses in geophysics.

In Chapter 3, we give a brief introduction to the major tectonic features of the Earth, as developed in continents and ocean basins.

We wish to point out that Part I of this book is essentially the same as Chapter 1, and parts of Chapter 2 and Chapter 21 in our companion book, *Structural Geology*. The material in Chapters 1 and 2 applies equally to both structural geology and tectonics. The material in Chapter 3 of this book introduced selected aspects of tectonics in *Structural Geology*. Here we provide a more detailed discussion of tectonics.

CHAPTER

1 Overview

What Are Tectonics and Structural Geology?

Earth is a dynamic planet, and the evidence of its dynamism is all around us. Earthquakes and volcanic eruptions regularly jar many parts of the world. Rocks exposed at the Earth's surface reveal a continuous history of such activity, and many have been uplifted from much deeper levels in the crust where they were broken, bent, and contorted. Geological processes take place in slow motion, however, compared with the time scale of a human lifetime or even of human history. The "continual" eruption of a volcano can mean that it erupts once in one or more human lifetimes. The "continual" shifting and grinding along a fault in the crust means that a major earthquake might occur once every 50 to 150 years. At the almost imperceptible rate of a few millimeters per year, high mountain ranges can be uplifted in the geologically short span of only a million years. A million years, however, is already more than 100 times longer than the whole of recorded history. These processes have been going on for hundreds or thousands of millions of years, leaving a record of constant dynamic activity in the Earth's crust.

Tectonics and structural geology[1] are both concerned with the reconstruction of these inexorable motions that have shaped the evolution of Earth's outer layers. For example, the Earth's crust breaks along faults and the two pieces slide past each other. Sections of continental crust break apart as oceans open, or they collide as oceans close. These events bend and break rocks in the shallow crust and cause puttylike flowing of rocks at depth. Mountain ranges are uplifted and subsequently eroded, exposing the deeper levels of the crust. The breaking, bending, and flowing of rocks all produce permanent structures such as fractures, faults, and folds that we can use as clues to reconstruct the deformation that produced them. Even at much smaller scales, the preferred alignment of platy and elongate mineral grains in the rocks and the submicroscopic imperfections in crystalline structure help to tell the tale of the deformation.

Tectonics and structural geology both deal with motion and deformation in the Earth's crust and

---

[1] These terms come from similar etymological roots. Structure comes from the Latin *struere*, to build, and tectonics from the Greek *tektos*, builder.

upper mantle. They differ in that tectonics is predominantly the study of the history of motion and deformation on a regional to global scale, whereas structural geology is predominantly the study of deformation in rocks at a scale ranging from submicroscopic to regional. The two realms of study are interdependent, and at the regional scale, they overlap considerably. Our understanding of the history of large-scale motions is constrained by observations of the deformation that has occurred in the rocks. Conversely, the origins of deformation can be understood in the context of the history of large-scale motions.

Both tectonics and structural geology have developed rapidly since the 1960s. Tectonics has undergone a revolution based largely on the formulation of the theory of plate tectonics. This theory now provides the framework for the study of almost all large-scale motions and deformations affecting the Earth's crust and mantle. Field-based studies of structures and the history of regional deformation have taken on new meaning when interpreted in the context of plate tectonic theory.

Geophysical data have become increasingly important to both tectonics and structure, as numerous diagrams throughout this book attest. Seismic, magnetic, and gravity data provide information on the geometry of large-scale structures at depth, which adds the critical third spatial dimension to our observations. Studies of magnetism and paleomagnetism as well as seismology provide data on the past and present geometry and motions of plates that are essential for reconstructing global tectonic patterns. Heat flow and geothermal data provide information on temperature distribution within the Earth, which bears upon possible patterns of convective flow in the Earth's mantle.

Tectonics also depends on other branches of geology. Petrology and geochemistry provide data on the temperatures, pressures, and ages of deformation and metamorphism, which permit accurate interpretation of deformation and its tectonic significance. Sedimentology and paleontology are also important in reconstructing the patterns and ages of tectonic events.

1.2 Tectonics, Structure, and the Use of Models

All field studies in structural geology and many in tectonics rely upon observations of deformed rocks at the Earth's surface. Other tectonic studies rely upon regional-scale studies of magnetic or gravity field variations or seismicity. Field tectonic studies generally begin with observations of features at outcrop scale; that is, a scale of a few millimeters to several meters. They then may proceed "downscale" to observations made at the microscopic or even electron-microscopic level of microns[2] or "upscale" to regional observations on a scale of hundreds to thousands of kilometers. At the largest scale, observations are generally based on a compilation of observations from smaller scales. *None of these observations alone provides a complete view of all structural and tectonic processes.* Our understanding increases with the integration of observations of the Earth at all scales. In addition to direct observation of the Earth, we also use the results of laboratory experiments and mathematical calculations to aid our interpretations.

Our first task in trying to unravel the tectonic history of an area or region is to observe and record, carefully and systematically, the patterns revealed in the structures of the rock, including such features as lithologic contacts, fractures, faults, folds, and preferred orientations of mineral grains, or the patterns revealed in geophysical surveys. In general, this is a process of determining the **geometry** of the features. We must try to answer such questions as, where are the structures located in the rocks? What are their characteristics? How are they oriented in space and with respect to one another? How many times in the past have the rocks been deformed? Which structures belong to which episodes of deformation? These and similar questions define the initial phase of any structural or tectonic investigation.

In some circumstances, determining the geometry of structures in the rock is an end in itself. For example, it may be important in locating economic deposits. To understand the processes that occur in the Earth, however, we need an explanation of the geometry. Ultimately we want to know the **kinematics** of formation of the features, that is, the motions that have occurred in producing them. Such motions, when integrated over large areas, form the basis for inferring ancient tectonic motions. Beyond the kinematics, we want to understand the **mechanics** both of the formation of small-scale structures and of large-scale plate motions, as well as how the motions at these different scales are connected. Therefore, we want to understand the forces that were applied, how they were applied, and how the rocks reacted to those forces.

---

[2] One micron is one micrometer, or one-millionth (10^{-6}) meter, or one-thousandth (10^{-3}) millimeter.

Much of our understanding progresses by making conceptual **models** of how tectonic features form and then testing predictions derived from the models against observation. **Geometric models** are three-dimensional interpretations of the distribution and orientation of features within the Earth, based on mapping, geophysical data, and any other information we have. We often present such models as geologic maps and as vertical cross sections along a particular line through an area.

Kinematic models describe a specific history of motion that could have carried the system from the undeformed to the deformed state or from one configuration to another, or that could have produced the geophysical pattern in question. These models are not concerned with why or how the motion occurred or with what the physical properties of the system were. The model of plate tectonics is a good example of a kinematic model. The constraints on the accuracy of such models derive from comparing the geometries of the motion and deformation observed in the Earth with those deduced from the model.

Mechanical models are based on our understanding of the basic laws of continuum mechanics—such as the conservation of mass, linear momentum, angular momentum, and energy—and on our understanding of how rocks behave in response to applied forces. This latter information comes largely from laboratory experiments in which rocks are deformed under conditions that reproduce as nearly as possible the conditions within the Earth. Using mechanical models, we can calculate the theoretical deformation of a body of rock that is subjected to a given set of physical conditions such as forces, temperatures, and pressures. A model of the driving forces of plate tectonics based on the mechanics of convection in the mantle, for example, is a mechanical model. Such models represent a deeper level of analysis than kinematic models, since the motions of the model are not assumed but must be a consequence of the physical and mechanical properties of the system. Thus, the models are constrained not only by the geometry of the deformation, but also by the physical conditions and mechanical properties of the rocks when they deformed. We use geometric, kinematic, and mechanical models to help us understand deformation on both the small scale and the global scale.

Even though we may be able to invent some model whose properties resemble observations of part of the Earth, such a model is not necessarily a good one. It is important to recognize that predictions based on a model tell us *only* about the properties and characteristics of the *model*, not about the actual conditions in the Earth. The relevance of a model for understanding the Earth, therefore, must always be tested.

Observations guide the formulation of models. The models in turn provide predictions that can be compared with reality, thereby stimulating new observations of the real world. Comparisons of a model's predictions with observations of the Earth constitute tests of the model's relevance. New observations that confirm the predictions support the model, and to that extent we accept it as a reasonable representation of the processes occurring in the Earth. If our observations are inconsistent with the predictions, we must refine the old model or reject it and devise a new one. Our understanding of structural and tectonic processes improves gradually through a continual repetition of the acts of making observations, making models based on those observations, making predictions from the models, testing the models with new observations, and changing the models. This process, in fact, is common to all science.

1.3 The Interior of the Earth and Other Terrestrial Planets

Although most tectonic processes discussed in this book are observed either at the surface or within the outermost layer of the Earth (the crust), the motions and deformations must reflect deeper interior processes. Moreover, Earth is not alone in the solar system. In this age of space exploration, the models we make of the dynamic processes in the Earth are relevant to our understanding of other planets.

According to current models, the Earth is divided into three approximately concentric shells including, from the center outward, the core, the mantle, and the crust (Fig. 1.1). The core is composed of very dense material believed to be predominantly an iron-nickel alloy. It includes a solid inner core and a liquid outer core. Surrounding the core is the mantle, a thick shell much less dense than the core and composed primarily of solid magnesium-iron silicates. The crust is a thin layer of relatively low density minerals that surrounds the mantle. It is composed of igneous rocks of granitic to basaltic composition, sediments and sedimentary rocks, and the metamorphic equivalents of these rocks, predominantly sodium, potassium, and calcium aluminosilicates.

The temperature of the Earth increases with depth at a gradient of approximately 20°C to 30°C per kilometer in the crust and upper mantle and at considerably smaller gradients deeper within the Earth. Several sources of heat account for this

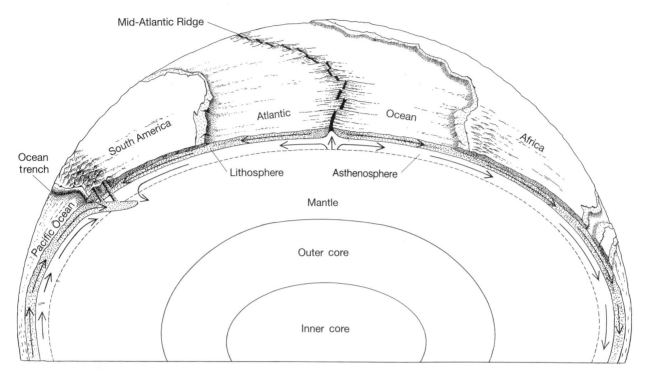

Figure 1.1 Diagrammatic cross section of the Earth showing the inner and outer core, the mantle, the lithosphere, and the crust. The spreading centers and subduction zones of plate tectonics are also indicated. *(After Wyllie, 1975)*

increase of temperature with depth: residual heat trapped during the original accretion of the Earth approximately 4500 million years ago; heat produced continually by the spontaneous decay of radioactive elements within the Earth; latent heat of crystallization from slow solidification of the liquid outer core; and heat produced by the dissipation of tidal energy resulting from the gravitational interaction among the Earth, Moon, and Sun.

The increase of temperature with increasing depth results in a flow of heat energy toward the surface that may involve the convective cycling of material both within the liquid core and within the solid mantle. Convection in the core may release heat to the lower mantle that is then carried toward the surface in a separate mantle convection system. The top of the mantle is a relatively cold, strong boundary layer called the **lithosphere,** which includes the crust and upper mantle to a depth of about 100 km under ocean basins and possibly up to 200 or 300 km under continents. Heat escapes through the lithosphere to the surface largely by conduction and by transport in upwardly migrating fluids such as igneous melts.

The study of the motion and deformation of the lithosphere, as revealed mostly in its outer 20 or 30 km, is the primary focus of tectonics. The task of understanding Earth's dynamics can be put in perspective by recognizing that what we have available to study is a thin rind of relatively cold material that probably rides passively atop mantle convection currents.

The other terrestrial planets apparently possess a concentric structure similar to that of the Earth, in that they they all have a crust, mantle, and core, as discussed in Part IV. Each planet exhibits different tectonic patterns, however, that reflect its unique structure and history.

1.4 Characteristics of the Earth's Crust and Plate Tectonics

Earth's crust is broadly divided into continental crust of approximately granitic composition and oceanic crust of roughly basaltic composition. Land—that part of Earth's surface above sea level—is principally continental, the exceptions being islands in the oceans. At present, 29.22 percent of the Earth's surface is land, and 70.78 percent is oceans and seas. Continental crust makes up 34.7 percent of Earth's total area, and it underlies most of the land area as well as the continental shelves and continental regions covered by shallow seas, such as Hudson's Bay and the North Sea. Oceanic crust makes up the other 65.3 percent of Earth's surface (Fig. 1.2).

The distribution of Earth's surface elevation is strongly bimodal. Most of the continental surface lies within a few hundred meters of sea level, and most of the ocean floor lies approximately 5 km below the sea surface. This distribution is evident from the two types of **hypsometric[3] diagram** shown in Figure 1.3. The cumulative plot (Fig. 1.3*A*) shows the total percentage of surface above a given elevation; the specific plot, or histogram (Fig. 1.3*B*), shows the percentage of the surface within a given elevation interval. The difference in elevation between the continental surface and the ocean floor results from a number of factors, such as the thickness and density differences between the continental

and oceanic crusts; tectonic activity; erosion; sea level; and the ultimate strength of continental rocks, which determines their ability to maintain an unsupported slope above oceanic crust.

The characteristics of Earth's crust are largely the direct or indirect result of motions of the lithosphere. The **theory of plate tectonics** describes these motions and accounts for most observable tectonic activity in the Earth, as well as the tectonic history recorded in the ocean basins. The theory holds that the Earth's lithosphere is divided at present into seven major and several minor **plates** that are in motion with respect to one another (see Fig. 1.2) and that the motion of each plate is, to a first approximation, a rigid-body motion. The term *rigid-body motion* does not imply that the plates are in fact rigid and undeformable. It simply means that to a first approximation, the plate tectonic process does not impose significant deformation on the plate interiors. Most deformation of the plates is concentrated in belts tens to hundreds of kilometers wide along the plate boundaries. In a few regions, however, deformation extends deep into plate interiors.

---

[3] Derived from the Greek *hypsos*, elevation, and *metron*, measure.

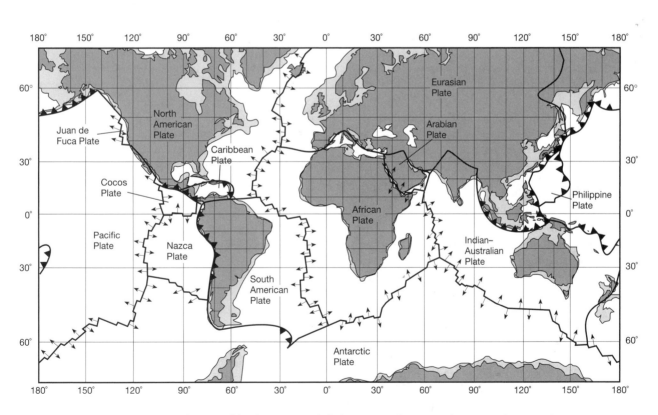

Figure 1.2 Distribution of land, continental shelves, ocean basins, and tectonic plates on the surface of the Earth. *(After Uyeda, 1978)*

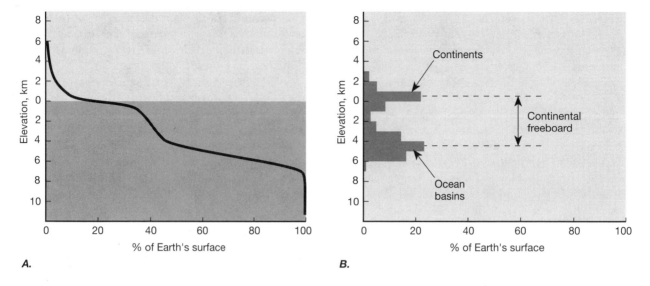

Figure 1.3 Distribution of topographic elevations on the Earth. *A.* Cumulative curve showing the percentage of the Earth's surface that is above a particular elevation. *B.* Histogram showing the percentage of the Earth's surface within elevation intervals of 1 km. The mean elevation difference (MED) is the difference between the average elevations of the continents and the ocean basins.

The types of boundaries between the plates include divergent boundaries, where plates move away from each other; convergent or consuming boundaries (also called subduction zones), where plates move toward each other and one descends into the mantle beneath the other; and conservative or transform fault boundaries, where plates move horizontally past each other, without creation or destruction of lithosphere.

The most direct evidence for plate tectonic processes and seafloor spreading comes from the oceanic crust, where divergent motion at mid-ocean ridges adds new material to lithospheric plates. As indicated in Figure 1.4, however, the maximum age of the oceanic crust limits this evidence to only the last 190 million years (m.y.); that is, *the last four percent,* of Earth's history. Any evidence of plate tectonic processes for the preceding 96 percent of Earth's history must come from the continental crust, which contains a much longer record of Earth's geological activity. Therefore, we must learn to understand the large-scale tectonic significance of deformation in the continental crust so that we can see further back into the history of Earth's dynamic activity.

In the geologic record, highly deformed continental rocks tend to be concentrated in long, linear belts comparable to the belts of deformation associated with present plate boundaries. This observation suggests that belts of deformation in the continental crust record the existence and location of former plate boundaries. If this hypothesis is correct, and if we can learn the structural characteristics of deformation corresponding to the different types of plate boundaries, we can use these structures in ancient continental rocks to infer the pattern and processes of tectonic activity. In this sense, the plate tectonic model has united the disciplines of structural geology and tectonics and made them interdependent.

In the belts of deformation along the plate boundaries, the relative motions of adjacent plates largely determine the orientation and intensity of deforming forces. Consequently, the style of deformation varies depending upon whether the boundary is divergent, convergent, or conservative. Differences between oceanic and continental crust also affect the nature of deformation along plate boundaries.

At divergent boundaries, oceanic crust is produced by the partial melting of upwelling mantle material to form basaltic magma. Igneous intrusion and extrusion of these basalts produce the new oceanic crust. The relative motion of the plates creates structures in the crust that accommodate stretching, such as systems of normal faults near the surface and ductile thinning at deeper levels.

Where divergent plate boundaries develop within continents, the associated stretching and thinning lower the mean elevation, ultimately resulting in flooding of the surface by the sea. Such structures often underlie the wide continental shelves (see Fig. 1.2).

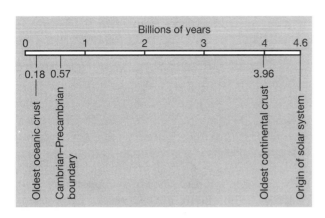

Figure 1.4 Time line showing events in Earth's history and the ages of the oldest oceanic and continental crusts.

In subduction zones at convergent boundaries, oceanic crust is recycled into the mantle. The plate plunges back into the interior of the Earth. Sediments on the downgoing plate are partly scraped off, and partial melting of the downgoing plate produces characteristic volcanic arcs on the overriding plate. Structures at the plate boundary are predominantly systems of thrust faults. Along the volcanic arc, however, normal faults are common. If a continent is involved with either the overriding or downgoing plate at a subduction zone, there is often shortening and thickening of its crust produced by characteristic systems of thrust faults.

The structures that form at conservative or transform fault boundaries are typically systems of strike-slip faults or, at deeper levels, zones of ductile deformation having a subhorizontal direction of displacement.

A variety of secondary structures can develop in any of these tectonic environments, so the presence of any particular structure per se is not necessarily indicative of the type of boundary at which it developed. The genesis of structures can only be inferred through a careful study of the regional pattern of the structures and their associations.

Tectonic processes also influence other Earth processes. For example, the number, size, and geographic position of continents have varied in the past as a result of plate tectonic processes. As plate tectonics changed the shape and distribution of continents and ocean basins, the patterns of oceanic and atmospheric circulation changed accordingly. The resulting changes in environmental conditions in turn affected the patterns of sedimentary environments, as revealed by studies of stratigraphy and sedimentology, as well as the patterns of natural selection and evolution, as revealed by paleontological studies.

Because plate boundaries are sites of major thermal anomalies in the crust and upper mantle, these areas control the occurrence and distribution of igneous and metamorphic rocks studied in "hard rock" petrology. Similarly, the formation, concentration, and preservation of mineral deposits are profoundly affected by structures and their tectonic environments, as well as by the thermal anomalies at plate boundaries. As the Earth's resources become depleted, increasingly sophisticated and subtle exploration strategies are required to find and develop new mineral deposits. Tectonics and structural geology are assuming a crucial role in the search for metal and hydrocarbon deposits.

Plate boundaries are also the sites of most hazardous earth activity, such as earthquakes and volcanic eruptions. Recognition of this fact has given new meaning to the statement by the historian Will Durant that "Civilization exists by geologic consent, subject to change without notice." For example, the volcanic eruptions of Santorini, Greece (ca 2400 years BP [before present]), Tamboro, Indonesia (1815), Krakatau, Indonesia (1883), Mt. St. Helens, U.S. (1980), Nevado del Ruiz, Columbia (1986), Unzen, Japan (1992), Pinatubo, Philippines (1992–94); and the earthquakes of Chile (1960), Alaska (1964), Mexico City (1985), Armenia (1988), Kobe, Japan (1995), and Colombia (1995) occurred along convergent boundaries. The earthquakes of San Francisco (1906), Loma Prieta (1989), and Northridge (1994) occurred along the San Andreas fault system, a transform fault boundary between the Pacific and American plates. Even the damaging earthquakes of New Madrid, Illinois (1811–1812) and Charleston, South Carolina (1886), which occurred in areas not presently at plate boundaries, probably were caused by movement on ancient faults formed at a plate boundary in the geologic past. Thus, the study of plate tectonics contributes in a very real way to an understanding and possible mitigation of such hazardous events in an increasingly densely populated Earth.

1.5 Summary and Preview

In a sense, the study of deformation processes on the Earth is an exercise in detective work. As with all other branches of geology, our evidence is usually incomplete, and we must use all available paths of investigation to limit the uncertainties. Thus, we study modern processes to help us understand the results of past deformations. We use indirect geophysical observation to detect structures beneath the surface of the Earth where we cannot see them. We make observations on all scales from the submicroscopic to the regional and try to integrate them into a unified model. We perform laboratory experiments in which the behavior of rocks can be studied under conditions that at least partially reproduce those found in the Earth, and we use mechanical modeling in which we apply the principles of continuum mechanics to calculate the expected behavior of rocks under various conditions.

At the level of this book and its companion, *Structural Geology* (Twiss and Moores, 1992), we cannot hope to cover all these aspects of tectonics and structure in detail. We do aim, however, to provide a thorough basis for field observation of geologic structures and their tectonic significance and to introduce the various paths of investigation that can add valuable data to our observations and lead to a deeper understanding of tectonic and structural processes. We also hope to instill an appreciation for the interdependence and essential unity of the disciplines of tectonics and structural geology.

In our book *Structural Geology*, we have covered the basic topics of structural geology, including a discussion of the structures typically associated with brittle and ductile deformation and the characteristics and mechanisms of brittle and ductile deformation in rocks. In this book, *Tectonics,* we examine in more detail the tectonic processes that ultimately are the origin of the deformation recorded by such structures. In Part I, Introduction, we summarize the major tectonic features of the Earth and introduce the geophysical techniques that have become indispensable in tectonic inquiry. In Part II, Plate Tectonics, we examine modern or active tectonic processes and describe the structures and associa-

tions of structures that are observed to develop in response to these processes. We follow Part II with an interlude in which we discuss the scientific method, the evolution of scientific inquiry, and a possible model for the normal progress of science and scientific revolutions, using the plate tectonics revolution as an example. In Part III, Historical Tectonics, we use the tectonic models that we construct from observations of recent tectonic processes to examine systems of structures in ancient orogenic belts that are exposed mostly in major mountain ranges of the world. By studying these structures and applying our understanding of how they form, we try to reconstruct the geometry, kinematics, and mechanics of their formation and finally to integrate this information into a tectonic interpretation consistent with the tectonic models. In this manner, we can push our reconstructions of the Earth's tectonic history further and further back into the past, where the geologic record becomes increasingly fragmentary and obscure. We conclude the book with a brief overview of our current understanding of the tectonics of the nearby terrestrial bodies, the Moon, Mercury, Venus, and Mars, as a comparison with the tectonics of the Earth.

The current theory of plate tectonics does not solve all the problems of Earth's tectonic evolution. Tectonic processes, for example, need not have remained the same throughout Earth's entire history. One challenge of modern tectonic study is to analyze ancient deformation to see whether models based on modern tectonics are able to account for the observed structural patterns and associations. Deficiencies in the modern tectonic model may indicate the need for different models to account for patterns in different stages of Earth's evolution.

Additional Readings

Bailey, E. B. 1968. *Tectonic Essays, Mainly Alpine.* London: Oxford.

Bally, A. W. 1980. Basins and subsidence: A summary. In *Dynamics of Plate Interiors.* A. W. Bally, P. L. Bender, T. R. McGetchin, and R. I. Walcott, eds. Geodynamics Series, vol. 1. Washington, D.C.: American Geophysical Union; Boulder, Colo.: Geological Society of America.

Goguel, Jean. 1962. *Tectonics.* From the French edition of 1952. H. E. Thalmann, trans. San Francisco: W. H. Freeman and Company.

Siever, R., ed. 1983. *The Dynamic Earth.* Special issue of *Scientific American.* September.

Twiss, R. J., and E. M. Moores. 1992. *Structural Geology.* New York: W. H. Freeman and Company.

Uyeda, S. 1978. *The New View of the Earth.* San Francisco: W. H. Freeman and Company.

Wyllie, P. J. 1975. The Earth's mantle. *Scientific American.* March. Reprinted in *Continents Adrift and Continents Aground.* J. T. Wilson, ed. 1976. New York: W. H. Freeman and Company.

CHAPTER

2 Geophysical Techniques in Tectonics

2.1 Introduction

Geologic maps provide the basis for a detailed understanding of the structure and tectonics of an area or region. Such a map, however, is only a two-dimensional representation of the three-dimensional structures. Mapping rocks that are exposed at the surface provides good information about the three-dimensional structure near the surface, but it cannot reveal the structure of areas covered by alluvium, deep soils, vegetation, or water (lakes, seas, or oceans). Nor can surface mapping provide information about the structure at great depth. Information about the shapes of major faults at depth, the presence of magma chambers at depth, the location of the crust-mantle boundary, or the thickness and nature of the lithosphere and the lower mantle can come only from the interpretation of geophysical measurements, especially seismic, gravity, and magnetic measurements. We will review briefly the application of these aspects of geophysics to large-scale structure and tectonics because they have become so essential and because a student of tectonics must be aware, at least, of the techniques and their limitations. Adequate coverage of these topics, however, would require a separate book, and we encourage students to take appropriate courses in geophysics.

Geologists often combine information from surface geology with geophysical data to produce cross-sections that portray our interpretation of the three-dimensional geometry of structures at depth. To show an undistorted view of this geometry, we must take the vertical scale equal to the horizontal scale. In some cases, however, it is useful to show features using **vertical exaggeration,** for which the vertical scale is larger than the horizontal scale. The relatively small changes in topography and in stratigraphic thicknesses over large distances can be shown clearly only with vertical exaggeration. As a result, vertically exaggerated cross sections are common in marine geology and stratigraphy.

The habitual use of vertical exaggeration to portray certain features, however, gives a false picture of those features. Figure 2.1, for example, shows cross sections of a subduction zone and adjacent volcanic island arc. The true scale section is shown in Figure 2.1*A*. Note that the topographic variation is almost impossible to see! A vertically exaggerated section is shown in Figure 2.1*B*. To emphasize the variations in topography, the surface slopes become extremely exaggerated and not at all representative of actual slopes. The effect of vertical exaggeration is just as dramatic on the dip of planar features, as can be seen by comparing the angle of the subducting slab in Figures 2.1*A* and 2.1*B*. Most published informal cross sections give an even more confusing view by combining vertically exaggerated topography with no vertical exaggeration below the surface, as shown in Figure 2.1C.

Another effect of vertical exaggeration is to cause beds of the same thickness but different dip to appear to have different thicknesses. Figure 2.1*B* shows that the subducted lithospheric slab is apparently thinner where it is dipping than where it is horizontal, an effect caused simply by the exaggeration of the vertical dimension.

Thus, vertical exaggeration distorts the true geometry and changes the way that features appear. Because much of structural geology involves visualization of the true shape of features in three dimensions, the use of vertical exaggeration with structural cross sections should be avoided whenever possible.

2.2 Seismic Studies

Seismic waves are oscillations of elastic deformation that propagate away from a source. Waves from large sources such as major earthquakes or nuclear explosions can be detected all around the world. Small explosions are often used as sources to investigate structure at a more local scale. Geophysical studies that use seismic waves from earthquakes differ from those that use human-made (artificial) sources, such as explosions or vibrations. This difference arises largely because the location and exact time of an earthquake are unknowns for which we try to solve, whereas we can know precisely the location and time of an artificial source. Although both sources of seismic energy are used to investigate the Earth's internal structure, the shallow structure is usually investigated by reflection seismology (see below), using artificial sources of seismic energy.

Body waves, which can travel anywhere through a solid body, are of two kinds: compressional (P)

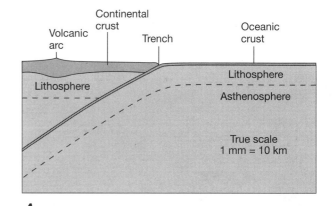

A.

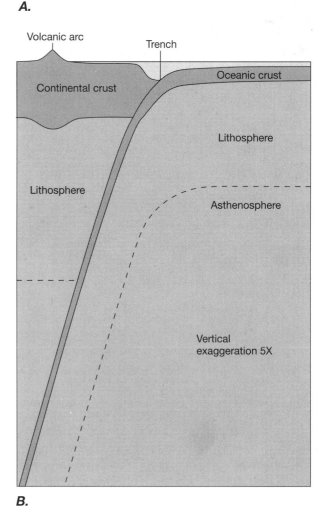

B.

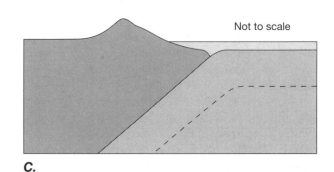

C.

Figure 2.1 The effect of vertical exaggeration on cross sections. *A.* True-scale cross section of a subduction zone in which the horizontal and vertical scales are equal. Note that topographic variations are almost lost. *B.* A cross section of the same subduction zone as in (*A*) shown at a vertical exaggeration of 5×; that is, the vertical scale is five times the horizontal scale. Notice the exaggeration of the dips of the subducting slab. *C.* Schematic cross section that combines vertical exaggeration for the topography with no vertical exaggeration for structures below the surface. This mixing of vertical scales precludes the construction of an accurate cross section.

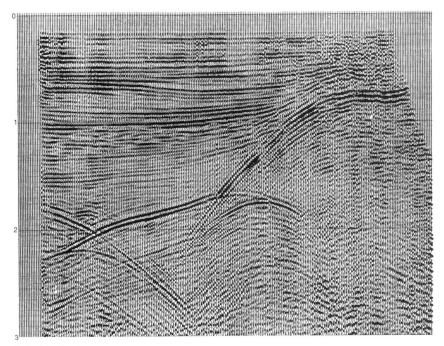

A. Unmigrated section

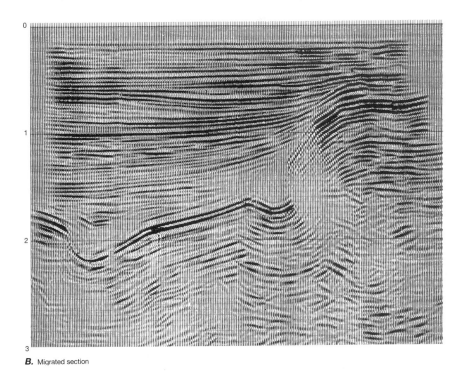

B. Migrated section

Figure 2.2 Seismic reflection profiles. Individual seismic records are the wavy vertical lines plotted side by side along the distance axis. The vertical axis is the two-way travel time. Peaks in each record are shaded black to show reflectors that can be traced from one record to the next. Good horizontal reflectors are particularly evident below about 1.7 s in the left half of the profile. *A.* Unmigrated seismic profile. *B.* Migrated seismic profile. *(From Lindseth, 1982)*

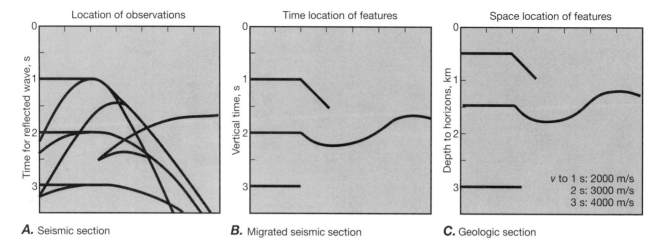

A. Seismic section   **B.** Migrated seismic section   **C.** Geologic section

Figure 2.3 Diagrams illustrating effects of migration. *A.* An unmigrated seismic section with multiple intersecting curved reflections. *B.* The same section as in (*A*) after migration. The ambiguities and artifacts of the unmigrated section are all removed. *C.* The corresponding geologic section. Note that the depth scale is different from the two-way travel-time scale because seismic velocity varies with depth. *(After Sheriff, 1978)*

large-scale tectonic movements of the plates. Figure 2.4 shows the characteristic radiation patterns for the three common types of faulting: normal, thrust, and strike-slip.

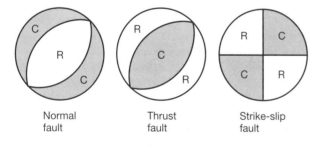

Normal fault   Thrust fault   Strike-slip fault

Figure 2.4 Equal-area projections showing the radiation pattern of compression (C) and rarefaction (R) first motions for the three main types of faults. The fault on which the earthquake occurs is assumed to be at the center of the plotting sphere. All orientations that plot within a given sector are the orientations of rays that have the indicated first motion when they leave the source. Planes separating the sectors are nodal planes, one of which must be the fault plane. The sense of slip on the fault is into the compression sectors.

2.3   Analysis of Gravity Anomalies

Gravity measurements are perhaps the second most common geophysical technique (after seismic techniques) used in tectonics. A gravity anomaly[1] is the difference between a measured value of the acceleration of gravity, to which certain corrections are applied, and the reference value for the corresponding location. The reference value is determined from an internationally accepted formula that gives the gravitational field for an elliptically symmetric Earth. Because gravity anomalies arise from differences in the density of rocks, the goal in tectonics and structural geology is to relate these density differences to structural features. If no density contrasts exist, then the structure can have no effect on the gravitational field, and gravity anomalies cannot aid in the interpretation of that structure.

We interpret the structure at depth by matching the gravity anomaly profile observed along a linear traverse with the anomaly profile calculated from an assumed model of the structure. The model is adjusted until the model anomaly profile shows a satisfac-

---

[1] From the Greek *anomalia*, irregularity or unevenness.

Box 2.2 Stacking of Seismic Records

Figure 2.2.1 illustrates the principle involved in the common depth point stacking of seismic records. If explosions are detonated at shot points S_1, S_2, S_3, and S_4, reflections from the same point P on a horizontal subsurface boundary will be received, respectively, at geophones G_1, G_2, G_3, and G_4. The same is true for all other horizontal reflectors below the surface point p and P. The corresponding shot points and geophones (S_i and G_i) are equidistant from the point p. If the travel times are corrected for the difference in lengths of the ray paths, the records can be added together, or stacked. The time-corrected signals from the reflections at P reinforce one another, and the signals from random noise tend to cancel out, thereby increasing the signal-to-noise ratio. The result is an enhanced seismogram showing the reflections as they would appear if the shot point and receiver were both at p.

Figure 2.2.1 The principle of stacking of seismic records. Explosions are set off at show points S_1, S_2, S_3, and S_4. The reflections of rays from the same point P on a reflector at depth are received for each of the explosions, respectively, at geophones G_1, G_2, G_3, and G_4. Adjusting the arrival times of these reflections for the different ray path lengths allows the four records to be added together, which enhances the signal from the reflection at P and cancels out random noise. The result is an increase in the signal-to-noise ratio.

In practice, data are gathered from a large linear array of shot points and geophones. Each geophone records many reflections from different depths, and the stacking is done by computer to produce enhanced seismograms at each point in the profile. Figure 2.2A is an example of a stacked seismic profile. Horizontal reflectors are abundant in the shallow portion of the profile.

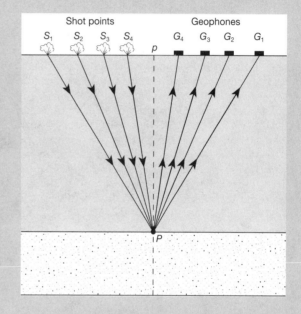

tory fit to the observed anomaly profile. Although the final adjusted model can never be unique, as explained in more detail below, it is constrained by surface mapping and sometimes by seismic data.

To calculate an anomaly, the measured value must be corrected to the same reference as used for the standard field. All measurements are therefore corrected to sea level as a common reference level. This altitude correction, called the **free-air correction,** results in an increase in most land-based values but leaves surface observations at sea unchanged. If this is the only correction applied, the calculated anomaly is called a **free-air anomaly.**

In addition, the **Bouguer correction** is usually applied. It is assumed that between sea level and the altitude of the measurement is a uniform layer of continental crustal rock that represents an excess of mass

piled on the surface. This assumed excess gravitational attraction is subtracted from land-based measurements. At sea, it is assumed that all depths of water represent a deficiency of mass because water is less dense than rock. This assumed deficiency in gravitational attraction is added to sea-based measurements. The **Bouguer anomaly** results from the application of both the free-air and Bouguer corrections.

The gravitational effects of local topography, however, differ measurably from those of a uniform layer. Thus, a refinement of the simple Bouguer anomaly, called the **complete Bouguer anomaly,** requires a **terrain correction** to account for the effects of local topography.

In effect, the Bouguer anomaly compares the mass of existing rocks at depth with the mass of standard continental crust whose elevation is at sea

Box 2.3   Migration of Seismic Records

Although horizontal or very shallowly dipping reflectors are common in undeformed sedimentary basins (shallow parts of Fig. 2.2), much of the structure of interest in structural and tectonic investigations is far more complex (deeper parts of Fig. 2.2). Beds with significant and variable dips, and discontinuous beds—truncated, for example, by a fault—are common. Such structures give rise to distortions and artifacts in seismic profiles (see Fig. 2.2A) that must be corrected by migration.

We describe the principle of migration using an example for which the seismic velocity of the material is constant and the source and detector are at the same point p (Fig. 2.3.1A). A particular reflection that apparently plots at point P below the detector could come from a boundary that is tangent to any point on a circular arc of constant two-way travel time having radius pP around p; for example, from P'. On two adjacent reflection seismograms (Fig. 2.3.1B), a reflection apparently plots at P_1 beneath p_1 and at P_2 beneath p_2. Therefore, it would appear that the reflector had the dip of the line P_1P_2. In fact, however, the reflector must be the common tangent to the two constant-travel-time arcs of radius p_1P_1 about p_1 and p_2P_2 about p_2. So the reflections must come

from points P_1' and P_2'. Thus, an uncorrected profile shows erroneous locations and dips for dipping reflectors, and the reflection points P_1 and P_2 must be migrated along their respective constant-travel-time arcs to the correct locations at P_1' and P_2'. The higher the true dip, the greater the distortion. Vertical reflectors plot on unmigrated seismic profiles as an alignment of reflections having a 45° dip.

If the seismic source and the detector are not at the same point, the arc of constant two-way travel time becomes an ellipse, and if the velocity is not constant, the arc is distorted further. These are complications that must be accounted for in any analysis of a real seismic record, although the principle remains the same.

Another problem develops if a reflector is discontinuous. The end of the reflector acts as a diffraction point, which takes energy from any angle of incidence and radiates it in all directions as if the point were a new source (point D in Fig. 2.3.2). The signals are recorded by the nearby detectors—for example, at p_1, p_2, and p_3—then plotted along a parabolic arc on an uncorrected seismic profile (dotted line, Fig. 2.3.2; see also Figs. 2.2A and 2.3A), because the two-way travel times for the signal increase as the distance of the detector

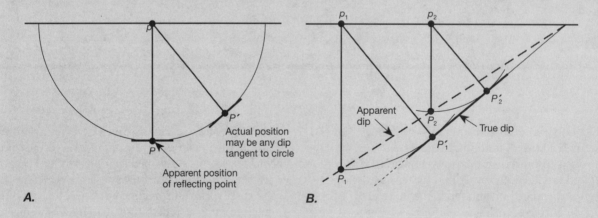

Figure 2.3.1 The migration of seismic signals corrects the seismic records to give the true location and dip of reflectors. In this example, seismic velocity is considered constant, and the shot point and receiver are both located at the same point. A. A reflection received at p appears on the seismic record at a two-way travel time that plots at P. In fact, the signal could come from any reflector, such as P', that is tangent to the semicircular arc of radius pP around p. That arc is the locus of constant two-way travel time. B. Reflections detected at p_1 and p_2 plot vertically below each point at P_1 and P_2, respectively, giving the reflector the apparent dip and location of the line P_1P_2. The true location and dip of the reflector, however, must be given by the line $P_1'P_2'$, which is the common tangent to the constant two-way travel-time arcs about p_1 and p_2, respectively. Note that P_1' is the actual location of the reflector below p_2.

from the end of the reflector increases. The locations of all possible diffraction points that could generate the signal recorded at a given receiver, however, must lie along a constant two-way travel-time arc about that receiver (dashed arcs, Fig. 2.3.2). The true location of the diffraction point is the common intersection of the arcs constructed for several detectors. Thus, migrating each signal along its arc to the common point identifies the true location of the diffraction point D.

In practice, the process of migration consists of taking each individual signal on a reflection seismogram, migrating it along its arc of constant two-way travel time, and adding that signal to any other seismogram intersected by that arc at the point of intersection. The resulting seismic profile is then a series of seismograms each of which consists of the original record to which is added all the signals that migrate to that record. With this procedure, reflecting boundaries appear as coherent traces of signals across the section in their correct locations, and diffracted signals sum together at the location of the diffraction point. The other additions to the various seismograms tend to cancel one another out and do not produce coherent patterns on the seismic profile.

Determining the constant two-way travel-time arc, of course, requires a determination of the velocity structure. The amount of computation required to migrate every signal in a profile, such as in Figure 2.2, to every seismogram intersected by its constant-travel-time arc is prodigious; as a practical means of analysis, it can be handled only by a computer.

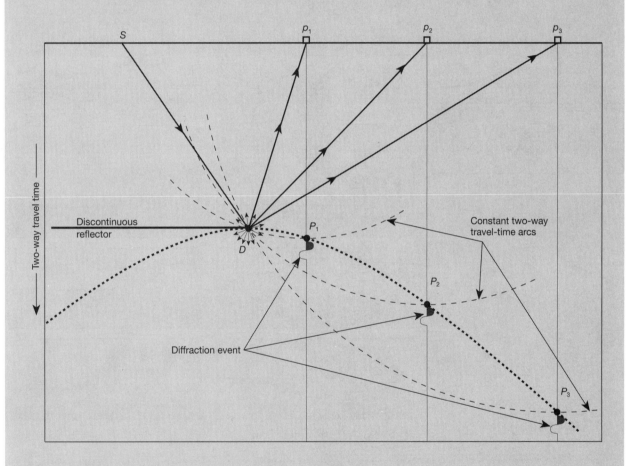

Figure 2.3.2 The end of a discontinuous reflector (D) acts as a diffraction point and radiates seismic energy in all directions for any angle of incidence. The farther the receiver is from D, the later the diffracted ray arrives. Thus, the diffracted energy arrives at receivers p_1, p_2, and p_3 at times that fall along a parabolic arc at P_1, P_2, and P_3, respectively. The three constant-travel-time arcs constructed about the three receivers with radii p_1P_1, p_2P_2, and p_3P_3, respectively, must intersect at the location of the diffraction point. Thus, to reconstruct its true location, each signal must be migrated along its constant-travel-time arc to the common point at D.

level. Generally, Bouguer anomalies are strongly negative over areas of high topography, indicating that there is a mass deficiency below sea level compared to standard continental crust; they are strongly positive over ocean basins, indicating that there is a mass excess below the ocean bottom compared to standard continental crust.

The area under a gravity anomaly profile provides a unique measure of the total excess or deficiency of mass at depth, and the shape of the profile constrains the possible distribution of the anomalous mass. The interpretation of mass distribution is not unique, however, because a given anomaly profile can be produced by a wide range of density differences and distributions. For example, Figure 2.5A shows three symmetric bodies of the same density, each of

which produces the same symmetric gravity anomaly. Figures 2.5B and 2.5C illustrate how the faulting of different density distributions affects the gravity anomaly profile. If a low-density layer overlies a thick, higher density layer and the structure is faulted (Fig. 2.5B), the gravity anomaly profile is asymmetric, but the different geometries of faulting have only a minor effect on the anomaly shape. If the denser material is in a relatively thin layer (Fig. 2.5C), the gravity anomaly profile is again asymmetric, although the shape is different from that in Figure 2.5B, and the effect of different fault geometry is significant. Thus, although the anomaly shape provides constraints on the possible structure, gravity models should be based on additional structural and geophysical information if they are to be reliable.

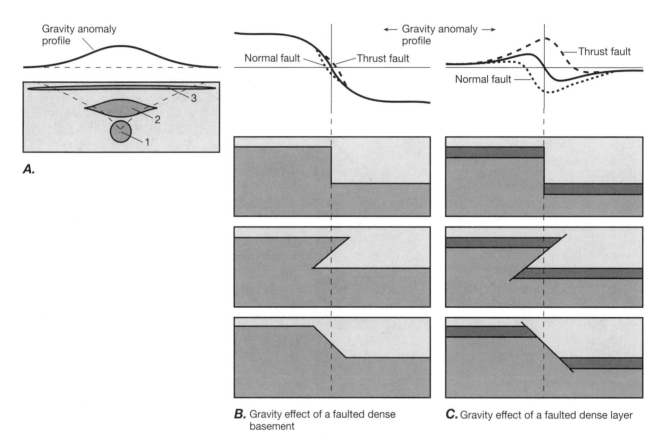

Figure 2.5 Illustration of the ambiguity and nonuniqueness inherent in the interpretation of gravity anomalies. *A.* Three symmetrical bodies of the same density, each of which can produce the observed symmetrical anomaly. *B.* The effect on gravity anomalies of vertical, thrust, and normal displacement of a dense basement overlain by less dense strata. The asymmetry of the anomaly reflects that of the underlying structure, but the distinction among the three different structures is almost negligible. *C.* The effect on gravity anomalies of vertical, thrust, and normal displacement of a dense layer within less dense layers. The three structures produce markedly different gravity anomalies. *(After Sheriff, 1978)*

2.4 Geomagnetic Studies

A magnetic field is described at any point in the field by a vector, which has both magnitude and direction. For the Earth's magnetic field, the magnitude is specified by the magnitudes of the horizontal and vertical components of the field. The orientation is specified by the declination and inclination, which are essentially the trend and plunge of the field line, although the inclination also includes the polarity, which defines whether the magnetic vector points up or down. Studies of Earth's magnetic field include the study of magnetic anomalies and of paleomagnetism.

Magnetic anomalies are measurements of the variation of Earth's magnetic field relative to some locally defined reference. There is no international standard reference field from which anomalies are measured, because the Earth's magnetic field is not constant through time and varies significantly even on the scale of a human lifetime.

Regional maps of magnetic anomalies are made using both aerial and surface measurements. Continental magnetic anomaly maps are used principally to infer the presence of rock types and structures that are covered by other rocks, sediments, or water. In some cases, the presence of particular rock types at depth may be inferred on the basis of characteristic patterns on a magnetic anomaly map. For example, the extension of rocks of the Canadian Shield beneath thrust faults of the Canadian Rocky Mountains can be inferred from the extension of the shield magnetic pattern beneath the thrust front.

Marine magnetic surveys have resulted in the well-known maps of the symmetrical patterns of magnetic anomalies that have been so fundamental to the development of plate tectonic theory. When correlated with the magnetic reversal time scale, these maps can be interpreted to give the age of ocean basins (see, for example, Fig. 3.1).

Magnetic anomalies, like gravity anomalies, can be used to infer the structure at depth, except that the magnetic anomalies are caused by differences in the magnetic properties of the rocks rather than their densities. The modeling of magnetic information is also more complex because a given anomaly in total field intensity can result either from differences in the intensity of magnetization of the rocks or from different orientations of the magnetic vector. Models of structure based on magnetic anomalies suffer from the same nonuniqueness as models based on gravity anomalies, and for similar reasons.

Rocks can become magnetized in a direction parallel to the ambient field by a variety of processes, including crystallization, cooling, sedimentation, and chemical reaction in the Earth's magnetic field. Such magnetization can be preserved in the rocks even if they are rotated to new orientations. In studies of **paleomagnetism,** the orientation of the magnetic field preserved in rocks is measured and compared with the orientation of the present field. If the original horizontal plane in the sample is known, these measurements can be interpreted to indicate the declination and inclination of the Earth's field at the time of magnetization. Gently dipping, unaltered sediments and volcanic rocks provide the most reliable paleomagnetic measurements, but deformed or metamorphosed rocks and plutonic rocks can also be useful. Rocks that have been tilted since magnetization are generally assumed to have tilted about a horizontal axis, so they are restored to the original horizontal by rotation about an axis parallel to the strike of the bedding.

The Earth's magnetic field is approximately symmetrical about the axis of rotation, so the declination of the field lines indicates the location of the Earth's axis of rotation, and the inclination of the field lines varies systematically with latitude from vertically down at the North Pole to horizontal at the equator to vertically up at the South Pole. Because this relationship is assumed to have been constant throughout geologic time, the paleodeclination determined for the sample, corrected for any tilting that may have occurred subsequent to magnetization, indicates the amount of rotation a rock has undergone about a vertical axis, and the paleoinclination, similarly corrected, indicates the latitude at which the sample was magnetized. Such measurements, therefore, can define the rotations about a vertical axis and the changes in latitude that have resulted from the large-scale tectonic motions to which the rocks have been subjected since magnetization. They cannot provide information, however, on the changes in longitude associated with these motions.

Plotting the apparent paleomagnetic pole positions for different time periods in a particular region provides an approximate indication of the movement of that area with respect to the Earth's geographic pole. These results are usually presented in the form of apparent polar wander maps, such as the map of paleopole positions for North America and Europe during the Phanerozoic (Fig. 2.6A). If the continents are restored to their relative positions before the opening of the Atlantic Ocean, the apparent polar wander paths coincide approximately from the Silurian through the Triassic, indicating the period of time over which the continents were joined (Fig. 2.6B).

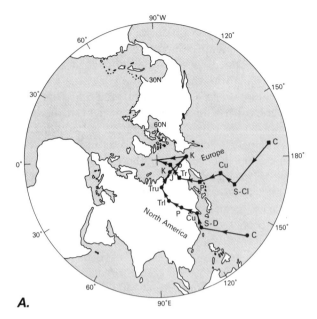

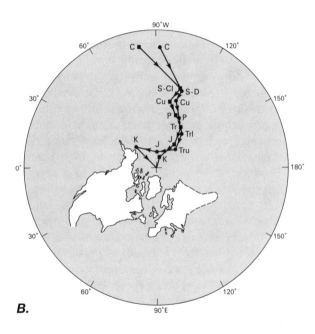

Figure 2.6 Apparent polar wander path (APW) for North America (circles) and Europe (squares). (C) Cambrian; (S) Silurian; (D) Devonian; (Cl, Cu) lower and upper Carboniferous; (P) Permian; (Tr, Trl, Tru) Triassic, lower and upper; (K) Cretaceous. *A.* Polar wander paths for the continents in their present positions. *B.* Polar wander paths for the continents before the opening of the Atlantic Ocean. The coincidence of the paths from Silurian through Triassic indicates the period of time during which the continents were joined. *(After M. W. McElhinny, 1973; also see Press and Siever, 1986)*

Heat is a measure of the internal energy of the atoms and molecules of a material. It includes the kinetic energy of translational and rotational motions of the atoms and molecules, as well as the potential energy associated with interatomic forces. Temperature is an arbitrary numerical scale that is proportional to the average translational kinetic energy of the molecules of a material. Energy is transferred from regions of high temperature to regions of low temperature; that is, from regions where the average translational kinetic energy is high to regions where it is low. This transfer of energy is described as a flux of heat.

In the nineteenth century, it was recognized from observations in caves and mines that the temperature of the Earth increases with depth. Because heat flows down a temperature gradient—that is, from higher to lower temperatures—that observation can only mean that heat is being lost from the Earth's interior.

This fact raises a host of questions that bear on our fundamental understanding of the composition and processes of the Earth's interior. How is the heat transferred? Where does it come from? How is it generated? Is the Earth heating up or cooling off? What constraints are implied about internal processes in the Earth? What is the thermal structure of the Earth's interior? What has been the thermal history of the planet? We cannot begin to address all these questions here, but we summarize briefly the basic principles upon which the analysis of heat flow is based. We discuss the tectonic implications of heat flow in the next chapter (see Box 3.1).

Heat Transfer

Heat is transferred from place to place by three main mechanisms: conduction, radiation, and convection.

Conduction is the process by which heat is transferred through a material by molecular collision. If molecules having a higher vibrational kinetic energy (higher temperature) collide with those having a lower vibrational kinetic energy (lower temperature), some energy is transferred from the higher to the lower energy molecules. Conduction accounts for the warming of the handle of a spoon, for example, when the opposite end is immersed in hot coffee. The basic equation that describes conductive transport of heat is known as Fourier's law:

$$q_x = -k \, (dT/dx) \qquad (2.1)$$

It says that the heat flux in the x-direction q_x is proportional to the magnitude of the temperature change across unit distance in the x-direction (dT/dx). The constant of proportionality k is the thermal conductivity, which is a characteristic of any particular material. Thus, if we measure the temperature difference ($T_2 - T_1$) between two points a known distance ($x_2 - x_1$) apart, and we measure the thermal conductivity k of the material, we can determine the heat flux:

$$q_x = -k\,(T_2 - T_1)/(x_2 - x_1) \qquad (2.2)$$

The minus sign means that heat is transferred in the positive x-direction from x_1 to x_2 if the temperature at x_2 is less than the temperature at x_1 ($T_2 < T_1$). In other words, heat flows from higher to lower temperatures.

Heat can also be transferred by radiation. Radiative transfer occurs when the internal energy at one place in a material is converted to electromagnetic radiation that is then absorbed by the material at another place and converted back to internal energy. The radiation, called infrared radiation, is similar to light but occurs in the electromagnetic spectrum at wavelengths just longer than visible light. Although our eyes are not sensitive to that wavelength, our skin can sense the heat radiated, for example, by the sun, a fire, or the heating element on a stove. The higher the temperature, the more internal energy is converted to radiation. Thus, a point at a lower temperature will absorb more radiation from a point at a higher temperature than it radiates back to that point, and the result is a net radiative transfer of heat down a temperature gradient; that is, from higher to lower temperatures. For radiative transfer to occur in the Earth, the material must be at least somewhat transparent to infrared wavelengths. Although transparency increases with temperature, this type of heat transfer is important only on a small scale, and its effects are usually incorporated into the definition of the thermal conductivity k in Equation (2.1).

Convection is the process by which heat is transferred by the motion of the material itself. The motion is driven by the differences in density associated with differences in temperature. Higher temperature material has a lower density because of thermal expansion. In a gravitational field such as the Earth's, lower density material tends to rise relative to higher density material, carrying with it heat from depth toward the surface. At the surface, the material cools off, largely by conduction, and the cooler, denser material then tends to sink back down into the Earth's interior, where it warms up again. This mechanism of heat transfer can be very efficient, and in the Earth it is the fundamental mechanism driving all tectonic activity including plate tectonics, orogeny, volcanism, and earthquakes.

The calculation of the heat flux associated with convection involves the equations describing fluid flow. There are so many factors that govern flow in the solid mantle that it is still not thoroughly understood. Thus, an accurate calculation of the heat transfer provided by convection is not yet possible. Nevertheless, the observed heat flux at the surface is a boundary condition that must be satisfied by any model of convection in the mantle. Despite the difficulty of making such models, the efficiency of convection as a heat transfer mechanism allows us to make some simplifying assumptions that are useful in understanding the Earth's heat budget and temperature structure, as discussed below.

Heat Flow Measurement

The measurement of heat flow from the Earth relies on the process of thermal conduction. Typically, we measure both the temperature difference between two points a known distance apart and the thermal conductivity of the rock to determine the heat flux (see Eq. 2.2). Measurements are made in deep boreholes or in the sediments on the ocean floor. Care must be taken, however, to eliminate perturbations of the ambient temperature.

In continental crust, perturbations can be caused by the fluctuations of surface temperature. Temperature fluctuations due to climatic variations such as the ice ages can affect the temperature profile to depths of nearly 300 m. These fluctuations are themselves of great interest to scientists studying the history of climatic variations. For the study of the heat flux from the Earth's interior, however, holes deeper than this are necessary to obtain temperature data unaffected by these fluctuations. The circulation of drilling mud can affect the temperatures in a drill hole, as can the circulation of groundwater in the hole after drilling. Thus, the temperatures in a drill hole can take one to two years to equilibrate with the ambient geothermal gradient once drilling is complete. As a result, such measurements are by no means trivial to make.

On the ocean floor, the temperatures are much more stable because polar ice has provided dense bottom water of relatively constant temperature for over a million years at least. Thus, the temperature gradient is generally measured by driving a probe

Box 2.4 First-Motion Radiation Pattern from a Faulting Event

The first-motion radiation pattern from a fault-slip event can be accounted for by the two-dimensional model shown in Figure 2.4.1. The undeformed state is shown in Figure 2.4.1A and is represented by two squares drawn on opposite sides of an east-west line that represents the future location of a fault. Gradual prefaulting deformation of the rock (Fig. 2.4.1B) elastically deforms the squares into parallelograms, shortening the NW-oriented dimensions of the squares (such as AD and CF) and lengthening the NE-oriented dimensions (such as BC and DE). N-S and E-W dimensions remain unchanged.

An earthquake occurs when cohesion on the fault plane is lost and sudden slip returns each parallelogram separately to its undeformed condition (Fig. 2.4.1C). During faulting, the outer points A, B, E, and F remain stationary while the points on the fault, C and D, separate into the respective pairs C_N and D_N and C_S and D_S.

In this process, the NW-oriented dimensions suddenly become longer (for example, D_N moves away from A, and C_S moves away from F), creating a rarefaction for the first motion. The NE-oriented dimensions, however, suddenly become shorter (for example, C_N moves closer to B and D_S moves closer to E), creating a compression for the first motion. Again the N-S and E-W dimensions remain unchanged. Thus, compressive first motions radiate outward in the NE and SW quadrants, and rarefaction first motions radiate outward in the NW and SE quadrants. The quadrants are separated by nodal planes, which are the fault plane and the plane normal to it, since dimensions do not change in these directions during faulting, and the amplitude of the seismic wave is therefore zero.

The first-motion radiation pattern of compressions and rarefactions thus allows the fault plane and the nodal plane to be identified. It also indicates the sense of slip on either plane that would generate that pattern, because slip would have to be toward the quadrant of compressional first motions on either side of either nodal plane. The actual fault plane sometimes can be identified from geologic considerations or by studying the location of aftershocks that occur along the fault plane.

Figure 2.4.1 A two-dimensional model for the mechanism of first-motion radiation patterns. A. Undeformed state represented by squares on either side of a future fault. B. Deformed state before faulting. N-S and E-W dimensions of the squares are unchanged, but NE-SW dimensions are lengthened (such as BC and DE) and NW-SE dimensions are shortened (such as AD and CF). C. Faulted state. Sudden slip on the fault generates an earthquake. N-S and E-W dimensions of the squares are still unchanged, but NE-SW dimensions are suddenly shortened (for example, BC_N and D_SE), and NW-SE dimensions are suddenly lengthened (for example, AD_N and C_SF). Thus, first motions are compressions for rays leaving the source in the quadrants marked C and rarefactions for rays leaving the source in the quadrants marked **R**. The fault plane and the plane normal to it are nodal planes along which no change in dimension occurs and the amplitude of the first motion is therefore zero.

about 3 m long into the seafloor sediment. The probe is equipped with temperature-measuring thermistors along its length and with a heater. The temperature profile is first determined, and then the heater is activated and the thermal conductivity is measured by monitoring the resulting change in temperature. In some parts of the oceanic crust, however, there is ample evidence that a significant amount of heat is transported by convecting pore water or water circulating through cracks and along faults. This transfer of heat is not accounted for by the standard measurement techniques.

Temperature gradients in the Earth's crust are typically 20°C/km to 30°C/km. Thermal conductivities are on the order of 2 to 3 W m^{-1} K^{-1}, leading to typical heat flows of 40 to 90 mW/m^2 or about 1 to 2 hfu (1 heat flow unit = 41.84 mW/m^2).

Heat Sources

Is the Earth steadily cooling off, or could it be heating up? If the Earth is steadily losing heat, where is the heat coming from? One source of internal heat is primordial heat, which is the heat left over from the

The same principle works in three dimensions, and the nodal planes can be identified from first-motion studies using a worldwide system of seismometers that, in effect, forms a three-dimensional array surrounding the earthquake. It is a source of potential confusion that the axis of maximum shortening, usually assumed to be the axis of maximum compressive stress, lies in the quadrant of rarefactional first motions, and the axis of maximum lengthening, usually identified with the axis of minimum compressive stress, lies in the quadrant of compressional first motions.

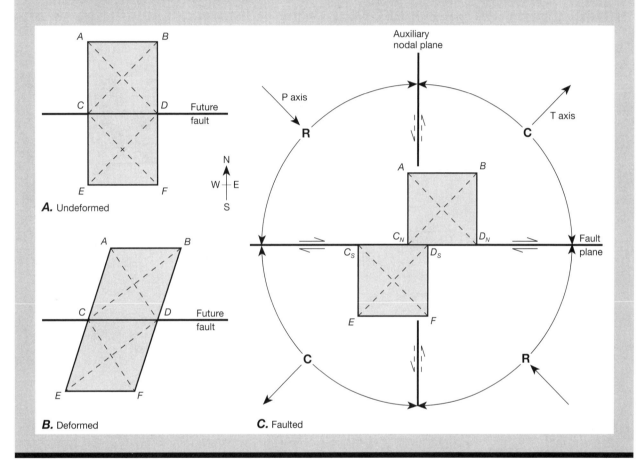

A. Undeformed

B. Deformed

C. Faulted

original accretion of the Earth from the planetary nebula. A less obvious but more significant source is the heat generated by the decay of radioactive elements in the Earth, the most important of which are uranium, thorium, and potassium. During the partial melting of rocks, these elements tend to be fractionated into the melt phase. Because most of the Earth's crust must have originated as a partial melt (granitic melts for the continental crust, basaltic melts for the oceanic crust), the radioactive elements are highly concentrated in the crust relative to the mantle. In granitic crust, concentrations are roughly an order of magnitude higher than in basaltic crust, and in basaltic crust they are roughly a factor of four higher than in unfractionated mantle rocks.

Other sources of heat in the Earth include the crystallization of the liquid core. The released latent heat of crystallization may be the energy source that is driving convection in the core and providing a source of heat to the base of the mantle. The dissipation of mechanical work during deformation is another source of heat that could be significant locally, but it contributes minimally to the overall heat budget.

Additional Readings

Fowler, C. M. R. 1990. *The Solid Earth: An Introduction to Global Geophysics.* Cambridge (England); New York: Cambridge University Press.

Gurnis, M. 1992. Long-term controls on eustatic and epeirogenic motions by mantle convection. *GSA Today* 2(7):141.

Lindseth, R. O. 1982. *Digital Processing of Geophysical Data. A Review.* Continuing Education Program, Society of Exploitation Geophysicists. Calgary, Alberta, Canada: Teknica Resource Development Ltd.

Sheriff, R. E. 1978. *A First Course in Geophysical Exploration and Interpretation.* Boston: International Human Resources Development Corporation.

Stacey, F. D. 1969. Physics of the earth. In space science text series. New York: Wiley.

Turcotte, D. L., and G. Schubert. 1982. *Geodynamics: Applications of Continuum Physics to Geological Problems.* New York: Wiley.

PART II focuses in detail on plate tectonics of the Earth and its geologic consequences. Chapter 3 presents a review of the types and distribution of tectonic features around the Earth. In Chapter 4, we examine the geometry, kinematics, and mechanisms of modern plate tectonics. The rigid-body motions that we use to describe plate movements are only first approximations, and the plates in fact can undergo large amounts of deformation, especially along the boundaries where they interact with one another. The geologic record of the plate interactions is preserved in the structures of the plate boundaries, and understanding these structures provides a key to unlocking the geologic history of the Earth. Thus, we look at modern boundaries where plate motions are well understood and where we can associate the various structures with the plate tectonic environment in which they develop. We want to know what kind of rocks, structures, and processes characterize the various plate tectonic environments currently observed on Earth today.

In Chapter 5, we examine divergent or constructional boundaries (also called spreading centers) in oceanic lithosphere, which occur where two plates separate from each other and new lithosphere is created. This type of boundary typifies mid-ocean ridges and major continental rift zones, where crustal extension and thinning is accommodated predominantly by normal faulting and/or emplacement of igneous rocks.

In Chapter 6, we examine conservative or transform fault boundaries where one plate slides horizontally past another and lithosphere is neither created nor destroyed. These boundaries are characterized by strike-slip faulting.

In Chapter 7, we discuss convergent or consuming margins, also called subduction zones, where one plate slides beneath another and lithosphere is recycled back into the mantle. These regions typically exhibit thrust faults that accommodate crustal shortening and thickening.

In Chapter 8, we briefly discuss the complex geology of triple junctions. These are areas where three plates come together and interact.

In Chapter 9, we examine cases where a convergent or subduction-zone boundary evolves into a collisional zone (such as between two continents or between a continent and an island arc) and where the steady-state recycling of lithosphere is ultimately terminated, possibly causing the relocation of the subduction zone.

CHAPTER

3

Principal Tectonic Features of the Earth

3.1 Introduction

Earth's major tectonic features and the types of structures they exhibit provide the keys to understanding the large-scale dynamic processes of the Earth and how they have evolved through time. In this chapter, we discuss these tectonic features, considering first the ocean basins and then the continental crust. Figure 3.1 summarizes the age of the oceans and the broad outlines of continental geology. Although the oceans occupy by far the majority of Earth's surface area, oceanic crust is substantially younger than continental crust. Relatively young oceanic crust near active spreading centers occupies a larger area than the oldest crust, most of which has disappeared down subduction zones (see Fig. 3.1). Thus, most oceanic crust has a relatively simple history reflecting only the most recent events of the Earth's tectonic evolution.

Continental crust, on the other hand, ranges in age from 0 to at least 3.96 billion years and displays a more complex geology (see Fig. 1.4). Thus the geologic age, which we use to characterize the oceanic crust, is not as useful a basis for describing the tectonic features of continents. Instead we distinguish Pre-

cambrian shields, interior lowlands, orogenic belts, and continental rifts and margins.

Although oceanic crust makes up much more of the Earth's surface, the continental regions are better exposed and more accessible, and thus they are easier to study. Moreover, information about the tectonic processes that preceded the formation of the oldest oceanic crust can come only from continental crust because of its greater range of ages. Interpretation of many ancient continental features, however, is hampered by the effects of a long and complex history during which the oldest records may have been obscured, destroyed, or buried.

3.2 Ocean Basins

Vast areas of the ocean bottom are flat or nearly so. The oceanic crust underlying these areas is remarkably uniform both in thickness and composition. The thickness ranges from 3 to 10 km, with an average of 5 km, and thus the oceanic crust is substantially thinner than the continental crust, which averages about 35 km. Oceanic crust consists predominantly of igneous rocks of basaltic composition.

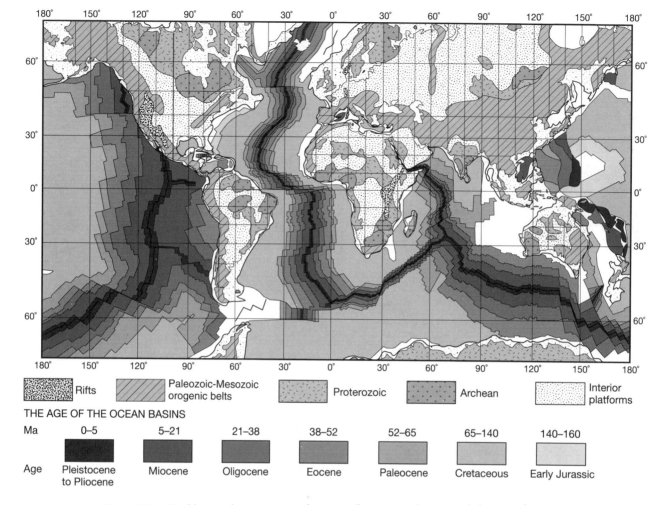

| Rifts | Paleozoic-Mesozoic orogenic belts | Proterozoic | Archean | Interior platforms |

THE AGE OF THE OCEAN BASINS

Ma	0–5	5–21	21–38	38–52	52–65	65–140	140–160
Age	Pleistocene to Pliocene	Miocene	Oligocene	Eocene	Paleocene	Cretaceous	Early Jurassic

Figure 3.1 World map showing major features of continental crust and the age of oceanic crust. Continental features include Precambrian shields with Archean and Proterozoic areas, interior lowlands, orogenic belts, rifts, and margins. (*After Anonymous, 1950; Stanley, 1986*)

These nearly flat areas of the ocean bottom include the ridges[1] and the abyssal plains. Scattered throughout the ocean basins are plateaus of anomalously thick crust, island arc-trench systems, and aseismic ridges of relatively thick crust (Fig. 3.2).

Gravity measurements over the oceans indicate that generally the free-air anomaly is nearly zero (see Sect. 2.3). Thus, for the most part, ocean basins are in isostatic equilibrium, and differences in elevation reflect differences in the density or thickness of the underlying crust and/or mantle.

An average layered model for the oceanic crust, partly shown in Figure 3.3, is based on P-wave seis-mic velocity (V_P) measurements. The lithologic interpretation of these layers results from direct sampling of the oceanic crust and from comparison with on-land exposures of rock sequences thought to represent old oceanic crust.

The uppermost layer, layer 1 (not shown in Figure 3.3), has a V_P of 3 to 5 km/s and is interpreted as unconsolidated sediment of pelagic, hemipelagic, or turbiditic origin.[2] Layer 2—commonly subdivided

---

[1] Although the mid-ocean ridge system represents one of the most important topographic and tectonic features on Earth, the average slope of its flanks is generally less than 1° or 2°.

[2] Pelagic sediments are derived from the settling of suspended material throughout the ocean water column. The material comes either from wind-borne dust from land or from shells of microscopic animals and plants. Hemipelagic sediments contain significant amounts of continental or volcanic material. Turbidite is a sediment deposited by sediment-laden bottom currents generally derived from a continent or island source.

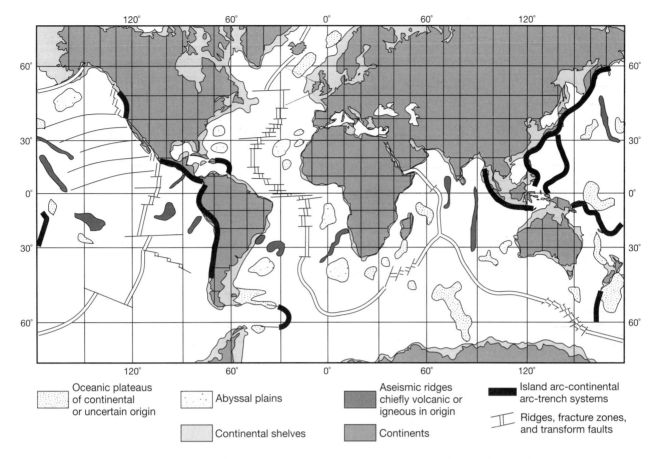

Oceanic plateaus of continental or uncertain origin

Continental shelves

Abyssal plains

Aseismic ridges chiefly volcanic or igneous in origin

Continents

Island arc-continental arc-trench systems

Ridges, fracture zones, and transform faults

Figure 3.2 World map showing major oceanic features: ridges, transform faults and fracture zones, oceanic plateaus, aseismic ridges, and island or continental arc-trench systems. *(After Bally, 1980; Uyeda, 1978)*

into layers 2A, 2B, and 2C—has a V_P ranging from 5 to 6 km/s and is interpreted as predominantly submarine basaltic extrusive and shallow intrusive rocks. Layer 2A has a relatively low velocity that increases rapidly with depth, layer 2B has a relatively constant velocity, and layer 2C again displays a rapid increase of seismic velocity with depth. Layer 3, with common subdivisions 3A and 3B and V_P ranging from 6 to 7.5 km/s, is thought to represent mafic-ultramafic[3] plutonic rocks and/or serpentinized

[3] *Mafic* and *ultramafic* are terms used to indicate the composition of igneous rocks. A mafic rock has one or more Fe–Mg–bearing minerals, such as amphibole, pyroxene, or olivine. An ultramafic rock consists chiefly of Fe–Mg–bearing minerals. Rocks of basaltic or gabbroic composition are mafic, whereas peridotites and serpentinites are ultramafic.

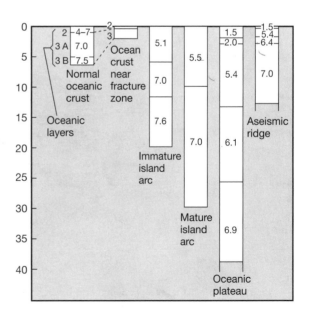

Figure 3.3 Seismic-velocity-layer models of typical oceanic crust and other oceanic features. P-wave velocities are indicated for the various layers.

mantle peridotite. Layer 3 subdivisions possibly reflect varying quantities of olivine in plutonic rocks.

For descriptive purposes, we divide the principal features of oceanic crust into those characteristic of plate margins and those characteristic of plate interiors.

Features of Oceanic Plate Margins

Divergent plate margins are topographically high regions characteristically in the middle of the ocean basins (except for those in the eastern Pacific Ocean and northwestern Indian Ocean). These **mid-oceanic ridges** form a continuous, world-girdling topographic swell that is approximately 40,000 km long, rises 2.5 km high above the abyssal floors of the ocean basins on either side, and extends to widths of 1000 to 3000 km. Structures on mid-ocean ridges are predominantly active normal faults, as revealed by morphology and first-motion studies of earthquakes (Fig. 3.4). The faulting is consistent with extension perpendicular to the trend of the ridge and parallel to the inferred relative plate motion.

Transform fault boundaries in the oceans are the seismically active portions of **fracture zones**—great rectilinear fracture systems within the oceanic crust. They are characterized by pronounced topographic relief, sharp ridge and trough topography, steeply dipping faults, and deformed oceanic rocks (Fig. 3.5). They range in length up to 10,000 km. Although they typically are fairly narrow features, some transform faults are up to 100 km or more wide. The seismically inactive portions of these fracture zones represent fossil transform faults. First-motion studies of the earthquakes along the active transform fault portions indicate strike-slip faults with characteristic horizontal relative motion *opposite* the sense of the apparent offset of the ridge crest. The oceanic crust near fracture zones and transform faults tends to be thinner than average (see Fig. 3.3B).

Convergent plate margins in the oceans exhibit chains of volcanic islands accompanied by parallel trenches, which are the deepest parts of the ocean

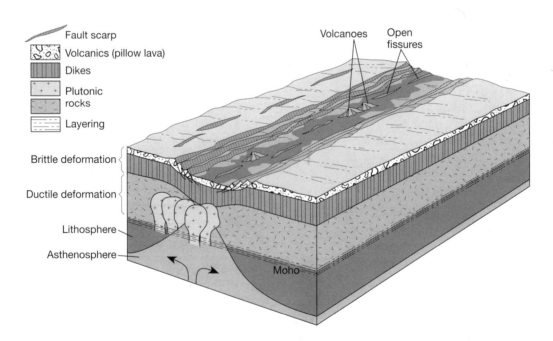

Figure 3.4 Schematic block diagram illustrating principal features of a mid-oceanic divergent plate margin. Extensional (normal fault) structures at the surface pass downward into a zone of magmatic intrusion and ductile stretching. The lithosphere thickens away from the plate margin. Not to scale.

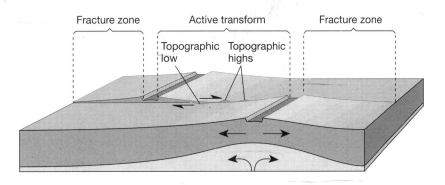

Figure 3.5 Schematic block diagram illustrating a conservative, or transform fault, boundary in oceanic crust offsetting a divergent margin (ridge). Structures of each offset portion of the ridge are as in Figure 3.4. Not to scale.

basins. Generally, these **island arc–deep-sea trench** pairs are arrayed in a series of arcs that join at cusps and extend for thousands of kilometers. The volcanic islands are spaced approximately 80 km apart and rise above submerged ridges that tend to be a few hundred kilometers wide. Trenches are up to 12 km deep and approximately 100 km wide. Systems of active thrust faults characterize the landward side of trenches, whereas active normal faults are typical of island arcs and regions behind the arcs (Fig. 3.6).

Trenches are associated with pronounced negative Bouguer gravity anomalies, indicating a marked mass deficiency below the seafloor.

The crust of island arc regions averages 25 km in thickness, considerably thicker than normal oceanic crust. It is rather variable, however, thinning abruptly to oceanic thicknesses on either side of the arc axis (see Figs. 3.3C and 3.3D). Younger, immature arcs tend to have a thinner crust (see Fig. 3.3C) than older, mature ones (see Fig. 3.3D).

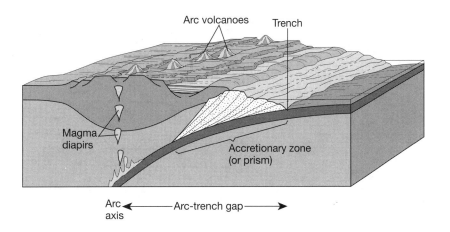

Figure 3.6 Schematic block diagram illustrating principal features in an intraoceanic convergent plate margin, or subduction zone. One plate descends beneath another along a marginal zone of thrust faults. Partial melting of downgoing crust produces blobs of magma that rise and become volcanoes. Not to scale.

Features of Oceanic Plate Interiors

Away from plate margins, the deepest regions of the ocean are vast areas of very flat ocean floor, the **abyssal plains.** These plains represent areas of normal oceanic crust covered by pelagic and turbidite sediments.

Broad elevated regions, or **oceanic plateaus,** have a variety of origins. Some are apparently continental rocks, others are inactive volcanic arcs, and the origin of still others is unclear. They range in area from a few hundred to many thousands of square kilometers and stand 1 to 4 km above the ocean floor. Crustal thickness is generally of continental rather than oceanic dimensions (see Fig. 3.3E).

Linear ridges characterized by high elevation, anomalously thick oceanic crust, and a general lack of associated seismic activity are called **aseismic ridges.** Their lack of seismic activity and more limited dimensions, as illustrated in Figure 3.2, set them apart from the mid-ocean ridges. In most cases, they represent linear constructional ridges formed by chains of basaltic volcanoes. The Hawaiian Islands–Emperor Seamount chain, extending northwest from Hawaii to Midway Island and thence north to the Kamchatka Trench (see Fig. 3.2), forms the most famous example of this type of crustal feature. The crustal thickness of aseismic ridges is considerably greater than that of normal oceanic crust and is comparable to that of island arcs (see Fig. 3.3F).

3.3 Structure of the Continental Crust

Continental crust is thicker and less dense than oceanic crust, has lower seismic velocity, and is more complex in structure because it is older and has experienced a longer tectonic history. Figure 3.7 shows an idealized cross section of North America, which provides an example of many features typical of continental crust.

The average thickness of continental crust is about 35 km. There is considerable deviation from that average, however, depending upon location and tectonic setting. The crust tends to be thickest (up to 70 km or more) under mountainous regions, about average under sedimentary platforms, and relatively thin under rifts such as the Basin and Range province, Precambrian shields, and marginal areas. The crustal seismic velocity tends to increase with depth, but velocity inversions are reported in some regions, such as the Basin and Range province.

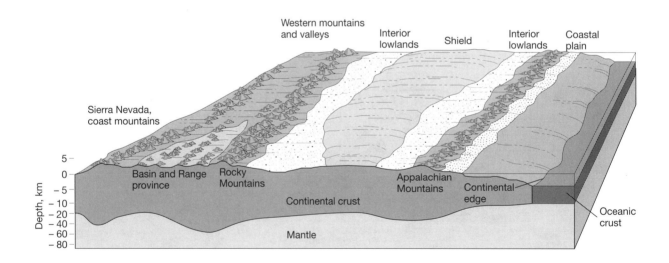

Figure 3.7 Generalized block diagram of the North American continent showing variations in crustal thickness in diverse tectonic provinces. Note the change in vertical exaggeration at about 10 km.

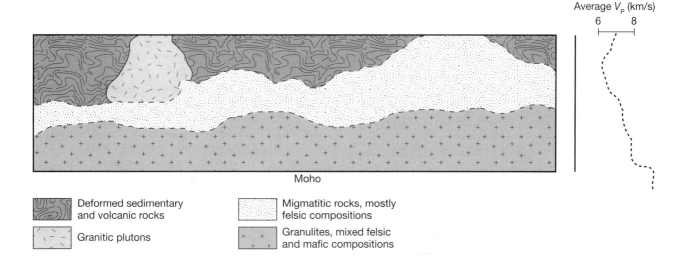

Average V_p (km/s)

6   8

Moho

	Deformed sedimentary and volcanic rocks
	Granitic plutons
	Migmatitic rocks, mostly felsic compositions
	Granulites, mixed felsic and mafic compositions

Figure 3.8 Crustal model showing lateral and vertical inhomogeneities, to account for observations in deeply eroded regions and for observed variations in seismic velocity. The diagram at right shows idealized average P-wave velocities (V_P) through the crust. *(After Smithson et al, 1977)*

Clues about the properties of the continental crust at depth come from direct examination of rocks formed at depth and exposed by uplift and erosion and from geophysical sounding of the crust. Exposures of shallow crustal rocks show that faulting, folding, metamorphism, and igneous intrusion are characteristic of orogenic activity. High-grade metamorphic rocks of the lower crust may be exposed in very deeply eroded terranes. These rocks contain very complex structures, such as the folded and refolded interlayers of pyroxene granulite and granitic gneiss from Greenland. Such structures are relatively common in many Precambrian shields and in deeply eroded central zones of Phanerozoic orogenic belts. It seems reasonable to infer that they are common in the lower crust everywhere.

Correlation of field data with seismic velocity structure of the continental crust suggests considerable lateral and vertical heterogeneity in rock types. Nevertheless, the crust is characterized by a general increase in seismic velocity with increasing depth. The actual nature of the deepest rocks is obscure because we can only see those rocks in regions where they have been uplifted and exposed at the surface, and such regions may not be representative of the entire crust. Figure 3.8 shows one petrologic and

seismic model that is not, however, necessarily representative of all continental crust. Beneath the sedimentary cover, the upper levels of the crust consist predominantly of metasedimentary and metavolcanic rocks intruded in places by granitic rocks. Middle levels of the crust may include extensive volumes of migmatite, a silicic rock that formed by partial melting during metamorphism. The lower levels of the crust consist of highly folded rocks commonly metamorphosed to the granulite facies,[4] and intruded by mafic and silicic plutonic rocks. There is a general trend of increasing seismic velocity with increasing depth in the crust.

Detailed exploration of the continental crust is taking place around the world, and we can expect much change in our knowledge and concepts in the years ahead.

---

[4] The term *metamorphic facies* refers to a distinctive asemblage of metamorphic minerals that is characteristic of a certain range of pressures and temperatures. The granulite facies of metamorphism is characterized by an assemblage of garnet, pyroxene, and feldspar. It usually indicates metamorphism at high temperature (above 650°C) and high pressure (above 500 MPa).

3.4 Precambrian Shields

All continents exhibit large areas where Precambrian rocks greater than 600 million years (m.y.) old are exposed at the surface (see Fig. 3.1). These regions generally form topographically rolling uplands that stand higher than the surrounding lowlands, giving rise to the name Precambrian *shield*.

We subdivide Precambrian shields into Archean and Proterozoic[5] terranes based on the age of the rocks. This subdivision has tectonic significance of worldwide utility. Most Archean rocks are greater than 2500 Ma old, and Proterozoic rocks range in age from approximately 2500 to 550 Ma. Archean regions display evidence of greater crustal instability or mobility than Proterozoic regions. The tectonic distinction is not universal or abrupt, however, and the transition from one tectonic style to the other varies by a few hundred million years or so from place to place.

Archean Terranes

Rocks in Archean terranes are divisible on the basis of their metamorphic grade into **high-grade gneissic regions**, exhibiting amphibolite or granulite metamorphic facies,[6] and **greenstone belts**, characterized by rocks at greenschist or lower grades of metamorphism. Both types are characteristically intruded by younger granitic plutons. The part of the Kalahari craton of southern Africa that is shown in Figure 3.9A displays the typical division into greenstone belt, gneiss, and granitic rocks (Fig. 3.10B).

High-grade gneisses form the bulk of Archean regions. They consist mostly of quartzofeldspathic gneisses derived by metamorphism of felsic igneous rocks, but they also contain subordinate metasedimentary rocks, including metamorphosed quartzites, volcanogenic sediments, iron formations, and carbonate rocks (Fig. 3.9C). Deformed mafic-ultramafic complexes make up the rest of the gneissic regions.

The high-grade gneissic regions are complexly mixed on a scale of tens to hundreds of kilometers with lower grade greenstone belts (Fig. 3.9B) that contain mafic to silicic volcanic rocks and shallow intrusive bodies, volcanogenic sediments of similar composition, and subordinate flows and shallow sills of olivine-rich magmas (Fig. 3.9D).

Three tectonic and structural features are common to all Archean terranes. First, most rocks are highly deformed and display more than one generation of folds (see Figs. 3.9B and 3.9D). The most obvious structural features are upright folds; less obvious are refolded low-angle faults and recumbent folds. Figure 3.9D shows a complex pattern of folding in the Barberton Mountains of Swaziland and South Africa; the overall pattern of the belt is reminiscent of a type 2 interference structure (for example, see Twiss and Moores, *Structural Geology*, Fig. 12.31B, p. 257). Figure 3.10 shows a map and cross section of a complex region of type 1 and 3 interference folds (Twiss and Moores, *Structural Geology*, Figs. 12.31A and 12.31C, p. 257) in a mafic-ultramafic complex (the Fiskenaesset complex) and surrounding gneiss and amphibolite in southwestern Greenland.

Second, the contacts between greenstone belts and high-grade gneissic areas are complex. In some places, the contacts are shear zones that mask the original relationship. Elsewhere, greenstone rocks are deposited on older gneissic basement. In still other areas, gneissic granitic rocks intrude rocks of the greenstone belt.

Third, the sedimentary rock types fall into one of two broad associations: either they are immature volcanogenic sediments that are characteristic of the greenstone belts and parts of the gneissic terranes, or they are a quartzite–carbonate–iron formation assemblage associated in many areas with multiply deformed mafic-ultramafic layered igneous complexes and found only in gneissic terranes (see Fig. 3.9C).

The study of Archean tectonics is a relatively young field because numerous large-scale detailed maps and sufficiently precise radiometric dating techniques have been available only since about 1975. The worldwide presence of these characteristic sedimentary and structural associations implies that similar sedimentary and tectonic conditions occurred globally during Archean time and that these conditions differed markedly from those characteristic of Phanerozoic time. In particular, the widespread metamorphism and the presence of ultramafic magmatic rocks indicate higher temperatures in the Earth during Archean times. The petrology of the ultramafic magmas implies that they formed by about 50 percent melting of a mantle source at temperatures

---

[5] Archean comes from the Greek *archi*, beginning; Proterozoic comes from the Greek *proteron*, before, and *zoe*, life, an illusion to the original, and erroneous, idea that these rocks were unfossiliferous.

[6] The greenschist facies is characterized by the presence of chlorite and actinolite. The amphibolite facies characteristically includes hornblende plus or minus aluminosilicate minerals and garnet. Pressure and temperature conditions for greenschist facies are approximately 400–500°C and 200–500 MPa and for amphibolite facies are about 500–650°C and 200–500 MPa.

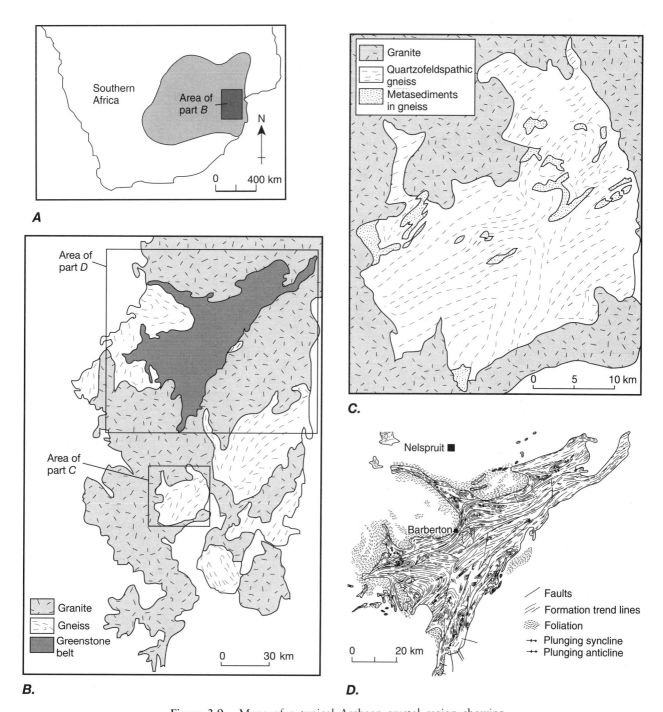

Figure 3.9 Maps of a typical Archean crustal region showing gneissic terranes with associated metasedimentary units, granitic rocks, and a greenstone belt: a portion of the Kalahari craton, southern Africa. *A.* Regional map of Kalahari craton showing areas of granite, gneiss, and the Barberton greenstone belt. *B.* Detailed map of part of the Kalahari craton showing areas of granite, gneiss, and the Barberton greenstone belt. *C.* The Mankayane inlier, showing infolds in gneiss of metasedimentary and metaigneous rocks consisting of a mafic-ultramafic unit structurally overlain by a sedimentary unit of metaquartzite; quartzofeldspathic, pelitic, and calcareous schist; and iron formation. *D.* Detailed map of Barberton greenstone belt showing internal structure. *(A., B., and C. after Jackson, 1984; D. after Anhaeusser, 1984)*

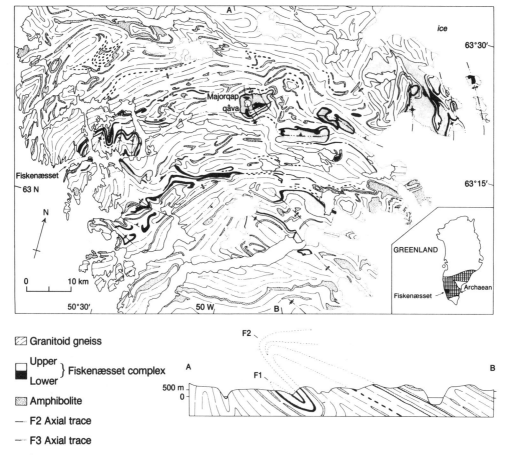

Figure 3.10 Map and cross section of a portion of the Fiskenaesset region, southern Greenland, showing refolded folds. The Fiskenaesset complex is a mafic-ultramafic stratiform sequence; the lower unit is peridotite, the upper unit is gabbroic. *(After Myers, 1984; in Kroner and Greiling, 1984)*

of approximately 1500°C. Theoretical heat-budget calculations suggest that the rate of increase of temperature with depth in the Earth (the geothermal gradient) was approximately two or three times the present one.

Proterozoic Terranes

Proterozoic terranes display both slightly deformed stable regions and highly deformed mobile areas, in contrast to the ubiquitous evidence for mobility displayed by Archean terranes. Regions of the Earth's crust that have achieved tectonic stability, called cratons,[7] first appeared in Proterozoic time. In these areas, vast deposits of weakly deformed, unmeta-

morphosed Proterozoic sediments typically overlie a basement of deeply eroded, deformed, and metamorphosed Archean rocks.

Proterozoic cratonic sediments display evidence of relatively stable tectonic environments; mature sediments such as quartzites and quartz-pebble conglomerates are common in regionally extensive stratigraphic units. Quartzites are often intercalated with abundant iron formations composed of interstratified iron-rich oxides, iron carbonates, and iron silicates. These mostly undeformed cratonic sediments are deposited on preexisting older basement. They host many vast Precambrian placer gold and uranium deposits, as well as most of the world's iron ore deposits. These sedimentary sequences extend over large areas.

Proterozoic deformed belts are of two general types. Some display multiply deformed regions rich

---

[7] From the Greek *kratos*, power.

in volcanic rocks, reminiscent of Archean terranes, as well as many Phanerozoic (post-Precambrian) volcanic-rich orogenic belts. Others exhibit thick sedimentary sequences deposited in linear troughs, presumably along ancient continental margins, and subsequently deformed into linear fold-and-thrust belts similar to those of the Phanerozoic orogenic belts.

Proterozoic igneous rocks also display distinctive differences when compared with older and younger terranes. Many Archean regions are cut by regionally extensive Proterozoic dike swarms of basal-tic composition.

A number of Proterozoic dike systems are associated with extensive mafic-ultramafic stratiform complexes, such as the Muskox and Bird River complexes in Canada, the Stillwater complex in the United States, and the Bushveld complex in South Africa (Fig. 3.11). These complexes are essentially undeformed masses of layered igneous rocks hundreds to tens of thousands of square kilometers in area. They resemble the layered igneous complexes

of the Archean but differ in that they are only weakly deformed. Figure 3.12 shows an example of such a feature, the Bushveld complex. The vertical columnar sections from three widely separated locations in the complex illustrate the amazing continuity of the distinctive layers. The dark layers in the columnar sections shown in Figure 3.12 represent individual layers of chromite within a sequence of gabbroic cumulate rocks that can be traced for tens of kilometers.

Large intrusive massifs of anorthosite[8] form another distinctive igneous-metamorphic rock suite that appeared in late Proterozoic time, 1000–2000 Ma BP (before present) (indicated by the symbol A in Fig. 3.11). In some cases, these rocks are clearly of igneous origin; in other cases, deformation and recrystallization have so modified the primary textures

---

[8] An igneous rock composed almost completely of plagioclase feldspar.

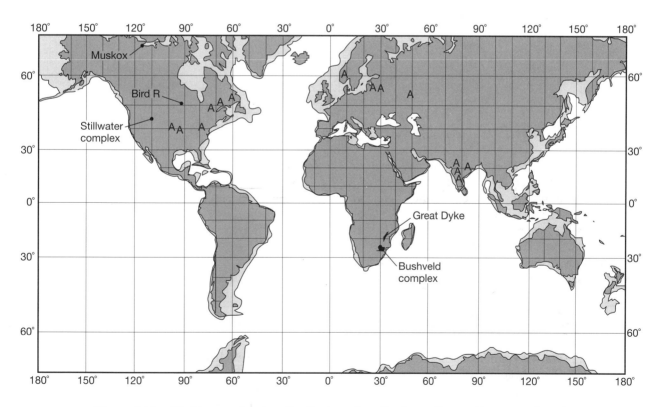

Figure 3.11. World map showing distribution of mafic-ultramafic stratiform complexes (black dots and blobs) and of anorthosite complexes (indicated by A), all of Proterozoic age. *(After Stanton, 1972)*

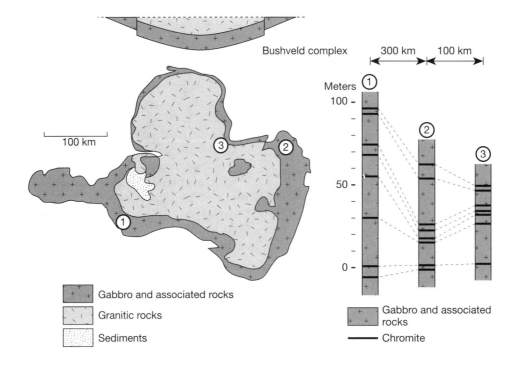

Figure 3.12. Simplified map and cross section of Bushveld complex, South Africa, with selected stratigraphic sections. Dark lines are chromite layers. Note the similarity in the sections over great distances. *(After Stanton, 1972)*

and structures of the rock that their origin is difficult to decipher.

Many Proterozoic regions exhibit a series of linear sediment-filled grabens called **aulacogens**[9] that generally strike at high angles to the trend of adjacent orogenic belts. The aulacogens that have been mapped on the North American continent are shown in Figure 3.13A. Careful mapping of several aulacogens shows that their sediments correlate with the thick sediments in the adjacent orogenic belt, as well as with neighboring thinner undeformed platform sediments (Fig. 3.13B). The sediments are typically undeformed or only slightly folded, with fold axes trending parallel to the axis of the trough.

Thus, tectonic conditions during the Proterozoic apparently differed from those in the Archean. Widespread undeformed platform sequences indicate the

existence in Proterozoic times of large, stable continental regions. Regional dike swarms and linear sediment troughs, such as aulacogens, indicate that these regions could undergo brittle extension. In this respect, Proterozoic tectonics more closely resembled Phanerozoic tectonics than Archean tectonics. Some workers have even suggested that the Archean-Proterozoic transition is the single most important tectonic event in all of Earth's history. What caused this transition, and what does it indicate? Why was the Archean so different from later times? How did the Proterozoic differ from later times? What were the characteristics of global tectonics during Proterozoic and Archean times? Did the plate tectonics of either period have any resemblance to Phanerozoic plate tectonics? We return to these questions briefly in Section 3.5 and again in Chapter 12, after we have discussed in more detail the Phanerozoic tectonic processes for which the evidence is so much clearer.

---

[9] The term is derived from the Greek *aulax*, furrow.

Figure 3.13 Aulacogens. *A.* Map of North America showing inferred and documented Proterozoic aulacogens. *B.* Stratigraphic-structural cross section of the Amargosa aulacogen, Death Valley, California. The rocks probably are approximately 0.6 to 1.2 billion years old. Lithologies: Crystal Spring formation, predominantly carbonate with interlayered basalts. Kingston Peak formation, pebbly mudstone (diamictite), sandstone, and subordinate iron formation. Note thickening of units toward axis of trough. Also note normal faults truncated by (and older than) overlying Kingston Peak and Noonday rocks. *(A. after Burke, 1981; B. after Wright and Troxel, 1971)*

A.

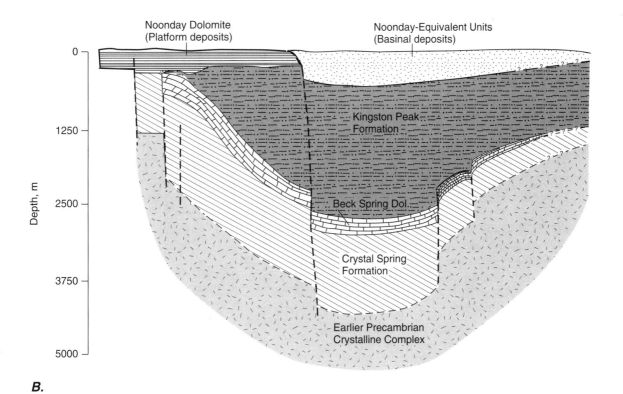

B.

3.5 Phanerozoic Regions

As we consider Phanerozoic (Cambrian and younger) features of the Earth, the available evidence increases greatly, and we can obtain a much more detailed picture of the structural characteristics of the younger parts of continents than we can of the Precambrian areas. In this section we briefly describe features of continental platforms, orogenic belts, continental rifts, and continental margins.

Continental Platforms

All continents contain regions of interior lowlands and cratonic platforms where relatively thin sequences of sedimentary rocks overlie Precambrian rocks that are subsurface continuations of the shields. With minor exceptions, these sedimentary rocks are flat-lying and are composed of lithologic units that are continuous over vast areas larger than the Precambrian shields themselves. They are mostly plains that stand a few hundred meters above sea level. To a structural geologist, these regions are relatively monotonous. Yet their economic wealth in coal, petroleum, mineral deposits, and agricultual resources is such that a large body of knowledge about them has accumulated, giving rise to what the late American tectonicist P. B. King called "the science of gently dipping strata."

Most interior platform sedimentary sequences begin with middle Cambrian or younger deposits; lower Cambrian or older Phanerozoic rocks are generally found only at the edges of the platforms. Throughout much of the world, and especially in North America, the contact with the underlying Precambrian shield rocks is a profound unconformity, commonly called a great unconformity, that marks a worldwide transgression of the sea over older continental interiors. In most places, this unconformity represents a time gap of tens to hundreds of millions of years.

Most platform sediments are marine and represent deposition in epeiric seas.[10] A major exception is the platform sequence of much of Gondwana,[11]

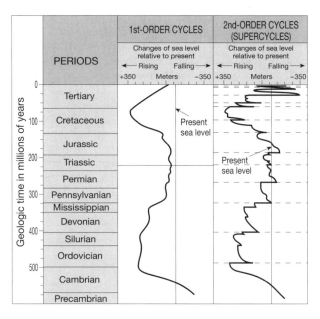

Figure 3.14 Transgression-regression curves for the world's continents. The curves show major long-term changes in sea level (first-order cycles) and more detailed changes (second-order cycles). *(After Cloetingh, 1976; Vail et al., 1977)*

which is mostly nonmarine in origin. The marine sediments record major periods of transgression and regression throughout the Phanerozoic, which in turn reflect major fluctuations in the level of the oceans relative to that of the continents (Fig. 3.14). The main structural characteristic of these platforms is a group of cratonic basins separated by intervening domes or arches (Fig. 3.15). Many of these features exhibit evidence of vertical movement of the crust lasting intermittently over tens to hundreds of millions of years. Arches served as sources of sediment during some stratigraphic intervals and were covered in others, but with thinner stratigraphic sequences than the surrounding platforms. Basins may contain thicker sequences deposited in deeper water. In times of general regression, these basins show evidence of restricted circulation and even dessication.

North America provides numerous examples of these features (see Fig. 3.15). The Transcontinental Arch stood high above the surrounding area throughout most of Paleozoic time. The sedimentary facies of some stratigraphic intervals show that at times the arch was emergent in an otherwise flooded continental region. Conversely, the Michigan and Illinois basins were relatively depressed features

---

[10] From the Greek *epiros*, continent.

[11] The supercontinent formed by India, Africa, Australia, and Antarctica. The name was first used for these regions by the Austrian geologist Eduard Suess, who took the name from an area of India, meaning "land of the Gonds."

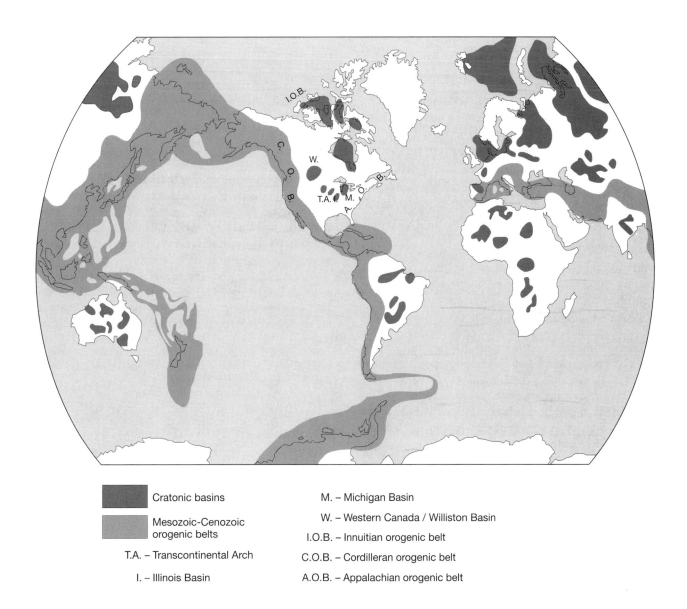

| | Cratonic basins |
| | Mesozoic-Cenozoic orogenic belts |

T.A. – Transcontinental Arch

I. – Illinois Basin

M. – Michigan Basin

W. – Western Canada / Williston Basin

I.O.B. – Innuitian orogenic belt

C.O.B. – Cordilleran orogenic belt

A.O.B. – Appalachian orogenic belt

Figure 3.15 World map showing the distribution of cratonic basins. Basins and arches of the interior platform of North America are identified. *(After Bally et al., 1979)*

throughout most of the Paleozoic. During times of high sea level, sediments in these basins were of deeper water origin and thicker than sediments on the surrounding platforms. During times of low sea level, basin sediments record evidence of restricted circulation. During some regressive periods, evaporite deposits developed.

The existence of domes and basins on the continental platform and the reflection of fluctuations of sea level in platform sediments have been known for decades. In the light of our present understanding of plate tectonics and its operation during part or all or Phanerozoic time, two questions come to mind: What tectonic processes have caused these domes and basins to form? In response to what plate tectonic processes?

Orogenic Belts

Orogenic[12] belts are some of the most prominent tectonic features of continents, and they have been the primary focus of work in structural geology for the past century. These belts are characteristically formed of thick sequences of shallow-water sandstones, limestones, and shales deposited on continental crust and oceanic deposits characterized by deepwater turbidites and pelagic sediments, commonly with volcaniclastic sediments and volcanic rocks. Typically, orogenic belts have been deformed and metamorphosed to varying degrees and intruded by plutonic rocks, chiefly of granitic affinity.

Structurally, most orogenic belts display a crude bilateral symmetry that is manifest by a linear central area of thick deformed and metamorphosed sedimentary and/or volcanic accumulations bordered on either side by undeformed regions, either oceanic or continental. In the past, much significance was attached to the symmetry of orogenic belts. Recent work, however, has demonstrated that the symmetry is more apparent than real, since in many cases the structures are of different ages on either side of the center.

The application of the plate tectonic model to the study of orogenic belts has revolutionized our ideas about how these belts form. We now believe that orogenic belts form at convergent margins either as a result of long periods of subduction beneath the margin or as a result of the collision of two continents, of a continent with an island arc, or of a continent with other thick crust of oceanic origin. Different types of mountain belts form depending on the character of the colliding blocks and on which side overrides the other.

Thus, the tectonic history of an orogenic belt may record some aspects of the history of plate tectonic activity. By studying the tectonic history of young orogenic belts, we can discover the relationship between orogenic structures and associated plate tectonic activity. Similar structures in inactive or older orogenic belts can then be used to infer the existence of similar plate tectonic activity in the geologic past.

Continental Rifts

Active continental rifts are marked by abundant normal faulting, shallow earthquakes, and mountainous topography. The North American Basin and Range province and the East African Rift are examples that we discuss in more detail in Chapter 5. In such regions, the continental crust is undergoing extension; according to the geologic record, such extension has often preceded the breakup of continents and the formation of new ocean basins. Ancient rifts are marked by comparable structures, but only rare earthquakes.

Modern Continental Margins

The margins of the present continents are apparently marked by a relatively sharp transition from continental to oceanic crust that is poorly exposed and difficult to resolve with common geophysical techniques. Seismic refraction, which uses layered models, cannot be applied where the layers are discontinuous, as at the margins of continents. Only recently has it become possible to penetrate the thick marginal sedimentary sequences with seismic reflection techniques and produce images of the continental-oceanic crust transition. Consequently, the structure of continental margins is still not well known.

Four types of continental margins are recognizable, however, based upon their tectonic environment (Fig. 3.16): passive, or Atlantic-style; convergent, or Andean-style; transform, or California-style; and back-arc, or Japan Sea–style. The geographic name sometimes used to refer to each style is taken from a region where it is characteristically developed.

Passive margins (**rifted margins**) or **Atlantic-style margins** are present on both sides of the Atlantic, as well as around the Indian and Arctic oceans and around Antarctica. They develop as continents rift apart to form new ocean basins. They initiate at a divergent plate boundary, but as spreading proceeds and the ocean basin widens, they end up in a midplate position (Fig. 3.17).

Passive margins include a coastal plain and a submarine topographic shelf of variable width, generally underlain by a thick (10–15 km) sequence of shallow-water mature clastic or biogenic sediments. Along some margins, an outer ridge is present in the

---

[12] From the Greek *oros,* mountain, and *genesis* origin, birth. We use the term to refer to areas that are major belts of pervasive deformation. The term *mobile belt* means approximately the same thing, being a region that has been tectonically mobile. *Mountain belt* is a geomorphic term referring to areas of high and rugged topography. Most mountain belts are also orogenic belts, and thus the two terms are often used interchangeably. Not all orogenic belts, however, are mountainous.

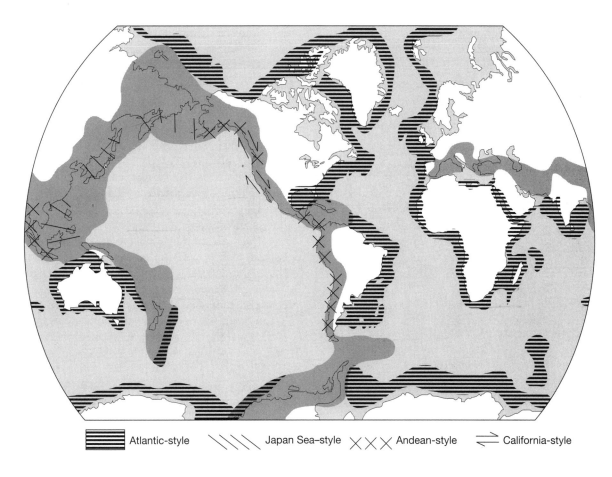

| ☰ Atlantic-style | \\\\ Japan Sea–style | ✕✕✕ Andean-style | ⇌ California-style |

Figure 3.16. World map showing present Atlantic-style, Andean-style, Japan Sea–style, and California-style continental margins. *(After Bally et al., 1979)*

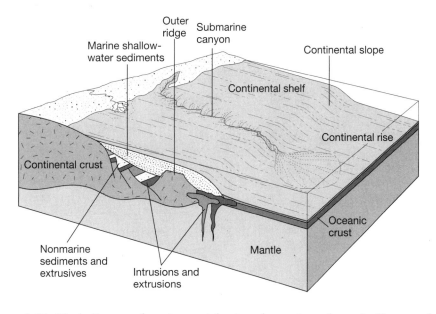

Figure 3.17. Block diagram of passive, or Atlantic-style, continental margin. Not to scale.

thick sedimentary sequence, generally at the point where the shelf passes into a steeper topographic slope toward the ocean basin. A relatively thick (roughly 10 km) sequence of sediments is generally present along the continental rise and slope (see Fig. 3.17). Normal faults, including growth faults, are the most characteristic structural features found in the sediments along these margins.

Convergent margins or **Andean-style margins** are present where consuming plate boundaries are located along a continental margin. They exhibit an abrupt topographic change from a deep-sea trench offshore to a high belt of mountains within 100 to 200 km of the coast. Continental shelves tend to be narrow or absent. The mountains along these margins are characterized by a chain of active volcanoes of principally andesitic composition (Fig. 3.18). Active deformation results in thrust complexes near the trench, high-angle normal faults near the volcanic axis, and either normal or thrust faults between the volcanic axis and the continent.

Transform margins or **California-style margins** are also characterized by sharp topographic differences between ocean and continent. They are marked by active strike-slip faulting, sharp local topographic relief, a poorly developed shelf, irregular ridge-and-basin topography, and many deep sedimentary basins. Figure 3.19 shows schematically the development of such topography by strike-slip displacement on two faults along an irregular continental margin. As the faults move, they progressively displace portions of the continent from each other, thereby producing in some places an alternation of narrow ocean basins and continental fragments.

Back-arc margins or **Japan Sea–style margins** consist of a passive Atlantic-style margin separated by a narrow oceanic region from an active island arc. The Japan Sea is a narrow ocean between the passive east coast of Asia and the active volcanic arc of Japan. Both the passive and the active margins of the composite margin have the features of the individual margins described above (Fig. 3.20).

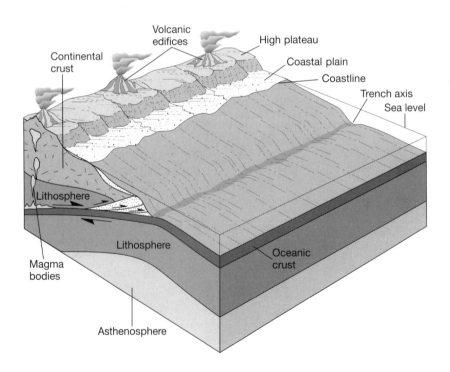

Figure 3.18. Block diagram of convergent, or Andean-style, continental margin. Not to scale.

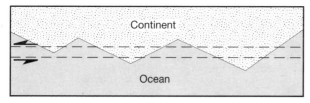

A.

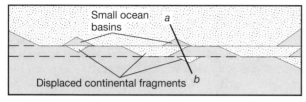

B.

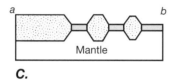

C.

Figure 3.19. Maps and cross section illustrating development of a transform, or California-style, continental margin. Not to scale. *A.* Irregular continental margin and a two-fault strike-slip system. *B.* After motion on both faults of the system, portions of the continent are displaced to new positions, *C.* Cross section *ab* showing ridge-and-basin structure.

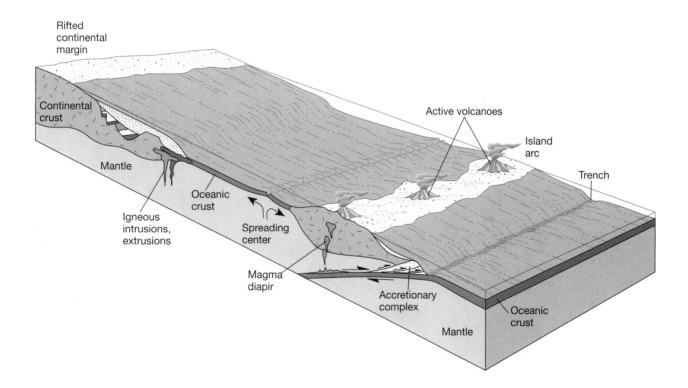

Figure 3.20. Block diagram of Japan Sea–style margin. Not to scale.

Additional Readings

Burchfiel, B. C. 1983. The continental crust. In *The Dynamic Earth*. R. Siever, ed. Special issue of *Scientific American*. September.

Francheteau, J. 1983. The oceanic crust. In *The Dynamic Earth*. R. Siever, ed. Special issue of *Scientific American*. September.

Hoffman, P. 1988, United Plates of America. *Ann. Rev. Earth and Plan. Sci.* 16:543–603.

King, P. B. 1977. *The Evolution of North America*, rev. ed. Princeton: Princeton University Press.

Kröner, A., and R. Greiling, eds. *Precambrian Tectonics Illustrated*. Stuttgart: Schweizerbartsche.

National Academy of Sciences-National Research Council. 1980. *Continental Tectonics*. Washington, D.C.

Nisbet, E. G. 1987. *The Young Earth: An Introduction to Archaean Geology*. Boston: Allen and Unwin.

Pakiser, L. C., and W. D. Mooney, eds. 1989. *Geophysical Framework of the Continental United States*. Geological Society of America Memoir 172. Boulder: Geological Society of America.

Sclater, J. G., C. Jaupart, and D. Galson. 1980. The heat flow through oceanic and continental crust and the heat loss of the Earth. *Rev. Geophys. and Space Phys.* 18:269–311.

Uyeda, S. 1978. *The New View of the Earth*. San Francisco: W. H. Freeman and Company.

Windley, B. F. 1995. *The Evolving Continents* 3d ed. New York: Wiley.

Zonenshain, Lev. P., M. I. Kuzmin, and L. M. Natapov. 1990. Geology of the USSR: A plate-tectonic synthesis. American Geophysical Union Geodynamic Series, vol. 21, B. M. Page, ed.

CHAPTER

4   Plate Tectonics

4.1   Introduction

This chapter describes the geometry of plate tectonics in more detail. In particular, we discuss the relative motions of two or more plates on a sphere, examine the possibility of determining absolute plate motions, and outline the pattern of plate tectonic motions through geologic history.

Figure 4.1 shows the seven major and five minor plates of the Earth and their present-day relative motions. Much of the evidence supporting the model of plate motion and interaction is geophysical. Magnetic anomaly patterns are symmetrical across mid-ocean ridges and thus demonstrate that the oscillations in the Earth's magnetic field are recorded by symmetrical spreading at the divergent margins. Earthquake foci concentrate at the plate boundaries where adjacent plates slide past each other. First-motion studies of plate-boundary earthquakes define the orientation of these faults and the sense of relative motion across them. High heat flow near divergent boundaries and low heat flow near consuming margins reflect the upwelling and downwelling of mantle material (see Chapter 2, Sec. 2.7).

If the motion of each plate is a rigid-body motion—that is, if there is no deformation within any individual plate—then all the consequences of plate interaction are confined to the plate boundaries. This model turns out to be a good first approximation of most large-scale tectonic motions. But areas adjacent to plate boundaries are clearly not rigid, as demonstrated by the widespread faulting and folding that records the plate interaction.

4.2   Relative Motion of Two Plates on a Sphere

Assuming that plates (or shells) on a sphere move as rigid bodies, it is possible to apply a theorem of spherical geometry[1] to describe their relative motions. The theorem states, in effect, that any displacement of a spherical plate over a spherical surface from one position to another can be accomplished by a rotation, denoted by ROT[E, Ω], of that plate by Ω degrees about a specific axis E passing through the center of the sphere[2] (Fig. 4.2).

---

[1] Attributed to the Swiss mathematician Euler (pronounced "oiler") in 1776 (see Le Pichon et al., 1973, p. 28).

[2] In this notation, we follow the convention of Cox and Hart, 1986.

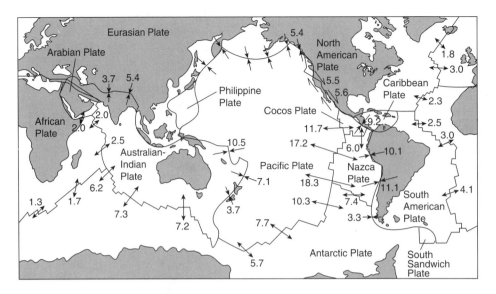

Figure 4.1 World map (Mercator projection) showing the seven major and five minor plates of the Earth, the types of boundaries, and generalized relative motion of the plates. Boundaries with two arrowheads pointing toward each other are convergent margins; with two arrowheads pointing away from each other are divergent margins; and with two parallel but opposing arrows are transform fault boundaries. *(After McKenzie and Richter, 1976)*

For example, Figure 4.2*A* shows the initial position and orientation S of a plate and its final position and orientation S_1. The displacement of the plate from S to S_1 is expressed by a rotation of the plate by Ω_1 degrees about an axis E_1 through the center of the Earth. In contrast, a different rotation by Ω_2 about a different axis E_2 would carry the plate from the same initial state S to a different final state S_2. Similarly, any other initial and final positions and orientations can be connected by a single rotation about a single axis.

The axis of rotation (E_1 or E_2 in Fig. 4.2) intersects the Earth's surface at the **Euler pole**, which is defined by the latitude and longitude (λ, ϕ) of the intersection point. The axis itself is taken to be of unit length, so in fact it is completely described by the Euler pole. Thus, the rotation of any plate A is defined by the latitude and longitude of the Euler pole and by the amount of the rotation

$$\text{ROT}_A = (\lambda_A, \phi_A, \Omega_A) \tag{4.1}$$

The relative rotation of one plate with respect to another is denoted $_A\text{ROT}_B$, where in this and similar notations, the left subscript indicates the plate "held fixed" (that is, on which the reference coordinates are fixed), and the right subscript indicates the plate whose rotation or motion is described from those coordinates. *In other words, we read $_A\text{ROT}_B$ as the rotation of B with respect to A.*

We can also use this geometry to describe the *instantaneous* motion of plate B relative to plate A in terms of the relative angular velocity vector $_A\omega_B$. The angular velocity vector is parallel to the Euler pole $_AE_B = (_A\lambda_B, _A\phi_B)$, and the rate of rotation about the Euler pole is $_A\omega_B$ measured in units of angle per unit time, commonly degrees or radians per year or million years (Fig. 4.3*A*). Thus, the relative angular velocity vector is defined by[3]

$$_A\boldsymbol{\omega}_B = (_A\lambda_B, {}_A\phi_B, {}_A\omega_B) \tag{4.2}$$

where $_A\omega_B$ is the magnitude of the vector $_A\boldsymbol{\omega}_B$. The vector $_A\boldsymbol{\omega}_B$ must be equal in magnitude but opposite in sign to $_B\boldsymbol{\omega}_A$,

$$_A\boldsymbol{\omega}_B = -_B\boldsymbol{\omega}_A \tag{4.3}$$

The relative *angular* velocity between two plates is the same along their entire common boundary. The relative *linear* velocity between the plates, however, is measured as distance per unit time, and it varies along their common boundary according to the relationship

$$_A V_B = R \sin \theta \; _A\omega_B \tag{4.4}$$

---

[3] Although the right sides of Equations (4.1) and (4.2) are similar, ROT_A is not a vector like $_A\boldsymbol{\omega}_B$. See the brief discussion in Box 4.1.

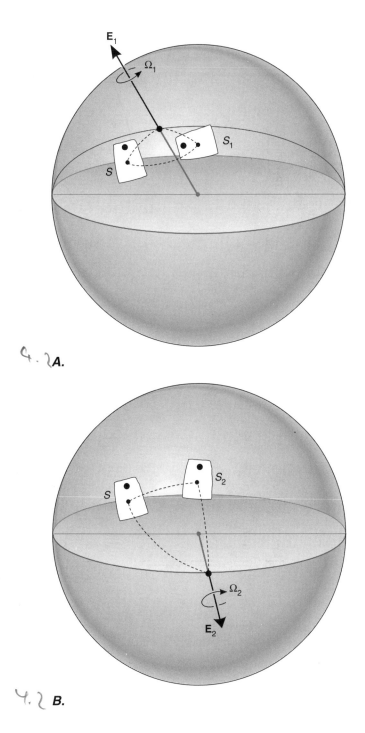

4.2 **A.**

4.2 **B.**

Figure 4.2 Motion of a small plate on a sphere. *A.* A plate moves from position and orientation *S* to position and orientation S_1 by a rotation of Ω_1 degrees about the Euler pole $\mathbf{E}_1$. The rotation path of any individual point is a small circle about the pole. *B.* The plate moves from the same position and orientation *S* to a position that is similar to that in *(A)* but a different orientation S_2. The rotation is Ω_2 degrees about a different Euler pole $\mathbf{E}_2$. Any given point rotates along a small circle about $\mathbf{E}_2$.

$$V \ [\text{mm/year}] = V \ [\text{km/Ma}]$$
$$= 6380 \ [\text{km/radian}] \sin\theta \ \omega \ [\text{radians/Ma}]$$
$$= 111 \ [\text{km/degree}] \sin\theta \ \omega \ [\text{degrees/Ma}]$$
$$(4.5)$$

The orientation of plate boundaries with respect to both the relative velocity and the location of the Euler pole depends in part on the type of boundary. Transform faults are approximately segments of small circles concentric about the Euler pole and are therefore parallel to the relative linear velocity vectors. Divergent plate boundaries (ridges) usually lie along segments of great circles that intersect at the Euler pole and are perpendicular to transform faults and to the relative linear velocity. Consuming margins have no particular orientation with respect to either the relative velocity of the plates or the location of the Euler pole, except that they cannot be strictly parallel to the relative plate motion. A Mercator projection about the Euler pole illustrates these properties well (Fig. 4.3*B*). In such a projection, the horizontal lines are small circles about the Euler pole, and the vertical lines are great circles that all intersect at the Euler pole (see the appendix for a description of a Mercator projection). Transform margins and relative velocity vectors are parallel to the small circles; divergent margin segments are parallel to the great circles; and convergent margin segments have an arbitrary orientation. Because of the distortion of the Mercator projection, the linear displacement vectors for a given time increment appear constant along the entire length of a divergent or convergent boundary.

The Euler pole for the relative velocity between two plates is a geometric construct that need not be located on either plate. It is, however, essentially fixed with respect to both plates for any small increment of motion. We can determine its approximate location by constructing great circles, called Euler great circles, perpendicular to the transform faults separating the two plates; the circles should intersect at the Euler pole. Figure 4.4*A* shows the Euler pole constructed in this manner for the relative rotation between the

where $_A V_B$ is the relative linear velocity at a given point along the boundary, *R* is the radius of the sphere, θ is the spherical angle between the Euler pole and the point on the boundary, and $_A \omega_B$ is the relative angular velocity of the two plates (see Fig. 4.3*A*). Because of the dependence on the sine function, the relative linear velocity is a maximum at $\theta = 90°$ from the Euler pole and vanishes at the pole itself where $\theta = 0°$. For the Earth, this equation reduces to

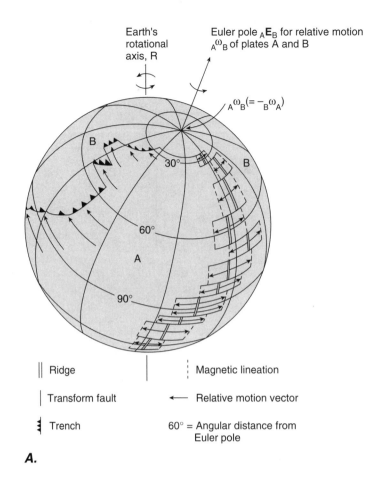

Earth's rotational axis, R

Euler pole $_A\mathbf{E}_B$ for relative motion $_A\omega_B$ of plates A and B

$_A\omega_B(=-_B\omega_A)$

B

30°

B

60°

A

90°

‖ Ridge

Magnetic lineation

| Transform fault

← Relative motion vector

Trench

60° = Angular distance from Euler pole

A.

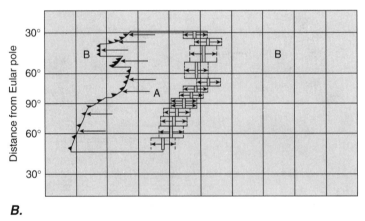

Distance from Eular pole

30°

B

60°

A

90°

60°

30°

B

B.

Figure 4.3 The motion of two plates A and B with respect to each other on a sphere. *A.* The motion of plates on sphere is described by a relative rotation about an Euler pole. The boundary at right is a spreading ridge with transform faults. The boundary at left is a subduction zone with transform faults. Linear velocities along divergent and convergent boundaries vary as the sine of the angular distance from the Euler pole. Linear velocities at the divergent boundary are spreading velocities shown relative to the boundary. Linear velocities at the convergent boundary are of the downgoing plate relative to the overriding plate. Also shown is the pole of relative rotation $_A\mathbf{E}_B$ between the two plates. Note that the pole of relative rotation of the plates does *not* correspond to the pole of rotation of the Earth. *B.* Mercator projection of the globe shown in *(A).* Projection is made on the pole $_A\mathbf{E}_B$, so that transform faults and relative motion vectors are parallel to latitude lines, and ridges and magnetic lineations are parallel to longitude lines. Note that the Mercator projection distorts the dimensions so that the relative displacement arrows now are all the same length. *(After Uyeda, 1978)*

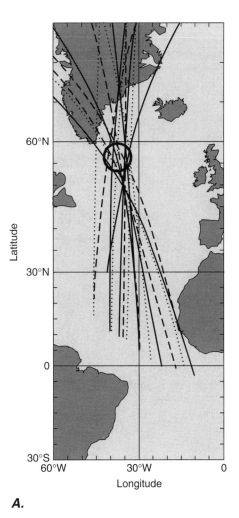

Figure 4.4 Determination of the relative angular velocity between the African and South American plates. *A.* Determination of the position of the Euler pole of relative plate rotation by construction of normals to the transform faults in the central Atlantic. With one exception, these lines pass within the circle centered at 58°N, 36°W. *B.* Comparison of the measured values of spreading rates for the Atlantic Ocean with the predicted values for a pole of relative rotation at 62°N. Points are rates determined from magnetic anomalies. The dashed line shows the rate measured parallel to the direction of spreading, and the solid line shows the rate perpendicular to the trend of the ridge, for a pole at 62°N, 36°W. *(After Morgan, 1968)*

A.

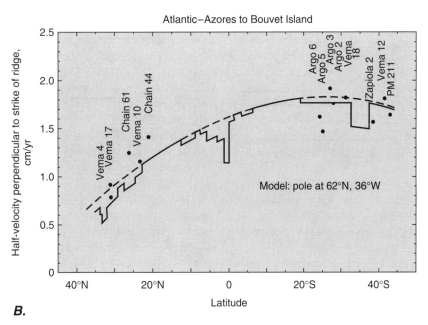

B.

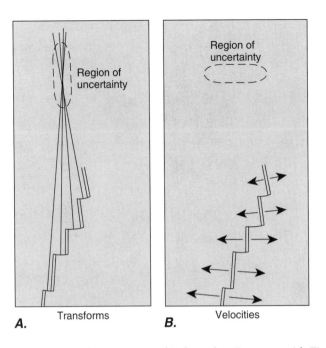

A. Transforms   **B.** Velocities

Figure 4.5 Errors in determination of Euler poles. Compare with Figure 4.4. *A.* Region of uncertainty in the determination of Euler poles from great circles constructed normal to transform faults (Euler great circles). *B.* Region of uncertainty for determining the Euler pole from the distribution of spreading velocities along a divergent margin. *(After Cox and Hart, 1986)*

African and South American plates (see geometry of mid-Atlantic ridge in Fig. 4.1).

We can check the location of the Euler pole by comparing the relative velocities between the plates along their mutual boundary with the velocities predicted for a particular Euler pole by Equation (4.4) or (4.5) (Fig. 4.4B). Along a divergent boundary (a ridge), we can obtain the half-spreading rate at a given point by measuring the distance from the plate boundary to a point of known age and then dividing that distance by the age. Ages of 10^6 years or less give better approximations than ages of 10^7 years or more. We obtain approximate ages for a particular location on the ocean floor by identifying critical fossils in the sediment, by radiometric dating of the igneous rocks, or by identifying the local magnetic anomaly stripe and consulting the magnetic reversal time scale.

Figure 4.4B shows a plot of spreading velocities versus angular distance from the assumed Euler pole for the Mid-Atlantic Ridge. The comparison between the measured velocities and those calculated from Equations (4.4) or (4.5) provides a check on the lo-

cation of the Euler pole determined in Figure 4.4A. The solid and dashed lines show the theoretical predictions for the spreading velocity perpendicular to the trend of the ridge and parallel to the direction of spreading, respectively. The dots show the spreading velocities computed from magnetic anomaly profiles. The measurements agree very well with the predictions, considering the errors inherent in the velocity determinations. The difference between the dashed and solid lines is a measure of the departure from 90° of the angle between the spreading direction and the ridge trend.

In constructing great circles normal to transform faults, the angle between any pair of great circles tends to be rather small, so that small errors in the orientation of the great circles produce large variations in the point of intersection. The result is that the region of uncertainty for the location of the Euler pole is elongate along the general orientation of the great circles (Fig. 4.5A). Fitting the observed spreading velocities to the required function of angular distance from the Euler pole (Eq. 4.5), however, gives better constraint on the angular distance from any point to the Euler

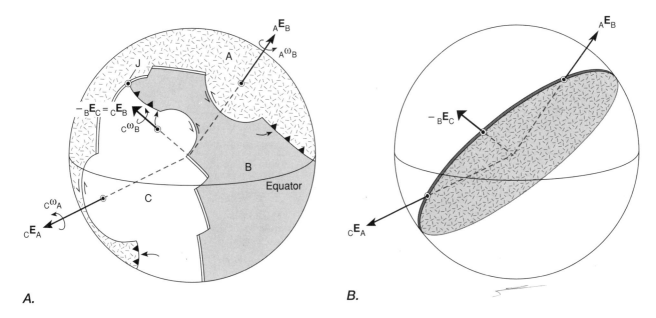

Figure 4.6 Globe with three plates A, B, and C. *A.* This diagram shows plates with boundaries and corresponding poles of rotation $_A\mathbf{E}_B$, $-_B\mathbf{E}_C$, and $_C\mathbf{E}_A$ with appropriate angular velocities $_A\boldsymbol{\omega}_B$, $_B\boldsymbol{\omega}_C$, and $_C\boldsymbol{\omega}_A$. We plot the negative of the axis $_B\mathbf{E}_C$ and $_B\boldsymbol{\omega}_C$ for the sake of clarity in the diagram. The vectors $_B\mathbf{E}_C$ and $_B\boldsymbol{\omega}_C$ in fact point in the opposite direction from $-_B\mathbf{E}_C = _C\mathbf{E}_B$ and $_C\boldsymbol{\omega}_B$. *B.* This diagram illustrates the fact that the three poles of rotation are coplanar. Equal angle projections, upper hemisphere.

pole, but poorer constraint in a direction roughly normal to the Euler great circles (Fig. 4.5*B*). Thus, whenever possible, both constraints are used together to improve the precision of location of the Euler pole.

4.3 Triple Junctions and the Relative Motions of Plates on a Sphere

Kinematics of Triple Junctions

Because the Earth's lithosphere comprises a mosaic of interlocking plates, there are several places, known as triple junctions, where three plates come together. The diagrams in Figure 4.6 shows a three-plate geometry where plates A, B, and C are in motion relative to one another. The relative motion of each two-plate pair is characterized by its own Euler pole, labeled $_A\mathbf{E}_B$, $-_B\mathbf{E}_C = _C\mathbf{E}_B$, and $_C\mathbf{E}_A$. Each Euler pole has its corresponding relative angular velocity $_A\boldsymbol{\omega}_B$, $_B\boldsymbol{\omega}_C$, and $_C\boldsymbol{\omega}_A$. Any circuit around a triple junction intersects three plate boundaries. The sum of the relative veloc-

ities at those three boundaries must be zero,[4] because in essence this sum expresses the relative velocity of one plate with respect to itself. Thus, for the relative angular velocities,

$$_A\boldsymbol{\omega}_B + _B\boldsymbol{\omega}_C + _C\boldsymbol{\omega}_A = 0 \qquad (4.6)$$

and for the relative linear velocities,

$$_A\mathbf{V}_B + _B\mathbf{V}_C + _C\mathbf{V}_A = 0 \qquad (4.7)$$

To satisfy these relations, the three angular velocity vectors, and thus the three Euler poles, must be coplanar (Fig. 4.6*B*). Thus, in any three-plate geometry, if two of the relative velocities are known, it is possible to find the third by simple vector addition.

The type of triple junction depends upon the various combinations of the three basic types of plate

---

[4] We express the velocities by the direction of the circuit around the junction. Where the circuit crosses a boundary from the previous plate (P) to the next plate (N), the relative velocity is expressed as the velocity of plate N relative to plate P and is designated $_P\boldsymbol{\omega}_N$.

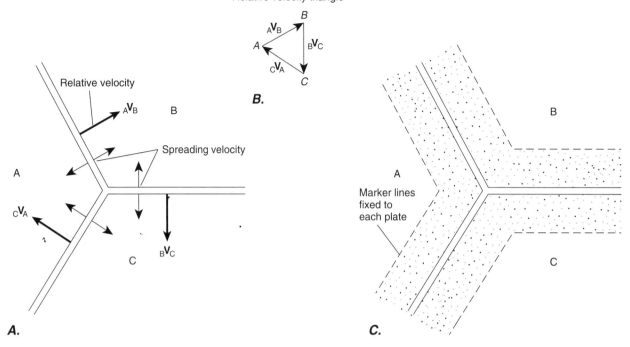

Figure 4.7 Evolution of a triple junction at which three divergent boundaries meet, known as an RRR (ridge-ridge-ridge) triple junction. *A.* Diagram in physical space showing three plates and relative orientation of plate boundaries and plate velocities. Both spreading velocities (thick arrows) and relative velocities $_A\mathbf{V}_B$, $_B\mathbf{V}_C$, and $_C\mathbf{V}_A$ (thin arrows) are shown. *B.* Velocity triangle showing that the addition of velocity vectors must form a closed triangle, thereby summing to 0. *C.* The triple junction after an interval of time. Shaded bands indicate the new crust created during this time. The configuration of the triple junction, however, is unchanged.

boundary: divergent, convergent, and strike-slip. A convergent boundary can have either of two polarities, depending on which is the overriding plate, and a strike-slip boundary can have either of two shear senses. Thus, there are a total of five different boundary geometries. At a triple junction, three plate boundaries come together, so there are 125 combinations of five boundary geometries taken three at a time ($5 \times 5 \times 5$) that we can imagine. Of these, only 16 are kinematically possible or distinctly different, and only 14 can actually exist for any geologically significant length of time. These 14 types of junctions are called **stable triple junctions.** (If we include triple junctions for which the boundaries must be oriented at specific angles relative to one another, then there are 19 types of junction that can be stable.)

The criteria for stability of a triple junction involve subtle relationships between the relative velo-

cities of the plates and the orientations of the plate boundaries (Box 4.2). To illustrate this point, consider Figures 4.7 and 4.8, which show the evolution of different configurations of triple junction. For each case, the relative velocity vectors $_A\mathbf{V}_B$, $_B\mathbf{V}_C$, and $_C\mathbf{V}_A$ are shown at the boundaries. If the relative velocity vectors are constructed head to tail, they must form a closed triangle to satisfy Equation (4.7), and this relative velocity triangle is shown for each triple junction in Figures 4.7 and 4.8.

The triple junction illustrated in Figure 4.7*A* is the junction of three divergent margins, referred to as a ridge-ridge-ridge or RRR triple junction. The velocity triangle is shown in Figure 4.7*B*. After an increment of time, new surface area indicated by the shading in Figure 4.7*C* is created around the triple junction, but the *configuration* of the junction remains unchanged, and the triple junction is stable.

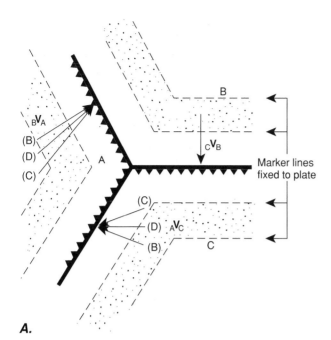

A.

Figure 4.8 The evolution and stability of a triple junction where three convergent boundaries meet. *A.* Generalized initial configuration for a trench-trench-trench (TTT) triple junction. The dashed lines are marker lines fixed to the plates. The relative velocity vectors at the three boundaries are shown. The consequences of the three different velocities shown for $_BV_A$ and $_AV_C$ are illustrated in Parts *(B)*, *(C)*, and *(D)*. *B.* If $_AV_C$ has an upward component relative to plate A, the triple junction is unstable, and it migrates up to a stable configuration along the boundary of plate A. *C.* If $_AV_C$ has a downward component relative to plate A, the triple junction is unstable, and it migrates down to a stable configuration along the boundary of plate A. *D.* If $_AV_C$ is parallel to the *BC* boundary, the triple junction is stable.

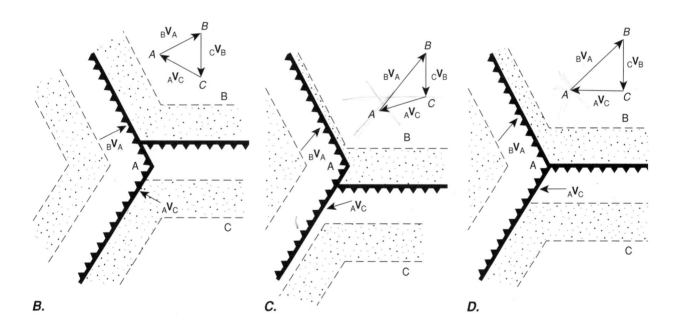

B.　　　　　　　　　　**C.**　　　　　　　　　　**D.**

By contrast, consider the evolution of the junction shown in Figure 4.8*A*. It is the junction of three convergent margins, known as a trench-trench-trench or TTT triple junction. In this configuration, because plate C overrides plate B, and plate A overrides them both, it is the velocity of plate C relative to A, $_AV_C$, that determines the geometric evolution of the triple junction. Three possibilities exist depending on the orientation of the relative velocity $_AV_C$ with respect to the *BC* boundary, as illustrated in Figures 4.8*B*, 4.8*C*, and 4.8*D*.

If $_AV_C$ has an upward component on the diagram relative to plate A (Fig. 4.8*B*), then the *BC* boundary, which is a fixed edge of plate C, must migrate

Box 4.1 The Difference between Finite Rotation and Angular Velocity

A finite rotation and an angular velocity are similar in that both can be described as a rotation about an axis, the angular velocity being simply the amount of rotation per unit time, or the rate of rotation. An angular velocity takes on unique properties, however, because it is defined in terms of an infinitesimal rotation in an infinitesimal increment of time, and the very small magnitude of the rotation permits a significant simplification in the description compared to a finite rotation. We describe this difference in technical terms below.

A finite rotation can be defined by the orientation of the axis of rotation and the magnitude of the rotation. The orientation of the axis of rotation is specified by a unit vector $\mathbf{E}$ that passes through the center of the unit sphere and pierces its surface at the Euler pole. In polar coordinates, the Euler pole is located at (λ, ϕ), and the components of the unit vector parallel to the axis of rotation are $[\lambda, \phi, 1]$. Here λ and ϕ are comparable to longitude and latitude on the unit sphere, and the third component 1 is the vector length. Because the length of the vector is 1 by definition, we can specify its orientation simply by specifying the location of the Euler pole, and the term *Euler pole* is often used synonymously with *axis of rotation*. The magnitude of the rotation is indicated by Ω. We designate a finite rotation by $\mathrm{ROT}(\mathbf{E}, \Omega)$, but the unit vector $\mathbf{E}$ and the finite rotation Ω *do not* define a vector of length Ω parallel to $\mathbf{E}$, as we show below.

The *rate* of rotation is also defined by the orientation of the axis of rotation that parallels the unit vector $\mathbf{E} = [\lambda, \phi, 1]$ through the Euler pole at (λ, ϕ) and the magnitude ω of the angular velocity. A unit vector $\mathbf{E}$ and an angular velocity of magnitude ω *do* define a vector of length ω parallel to $\mathbf{E}$. This difference from a finite rotation exists because of the difference between a finite and an infinitesimal rotation.

The difference is easiest to explain using Cartesian coordinates. A rotation is described by the relationship between a reference set of three mutually orthogonal axes and a second set that has been rotated relative to the reference axes. The rotation is defined by the cosines of the angles between each rotated axis and each of the three reference axes. There are nine such cosines, three for each rotated axis, which are arranged in a 3×3 rotation matrix denoted by R_{Kj} (for K and $j = 1, 2,$ and 3):

$$R_{Kj} = \cos \theta_{Kj}$$

where θ_{Kj} is the angle between the Kth coordinate of the reference frame and the jth coordinate of the rotated frame. In general, the nine components are all different. All nine, however, may be determined from the Cartesian components of the Euler pole $\mathbf{E} = [E_1, E_2, E_3]$ and the rotation magnitude Ω by

$$R_{Kj} = \begin{bmatrix} R_{11} & R_{12} & R_{13} \\ R_{21} & R_{22} & R_{23} \\ R_{31} & R_{32} & R_{33} \end{bmatrix}$$

where the three columns in the matrix are defined by

$$\begin{bmatrix} R_{11} \\ R_{21} \\ R_{31} \end{bmatrix} = \begin{bmatrix} E_1E_1(1 - \cos \Omega) + \cos \Omega \\ E_2E_1(1 - \cos \Omega) - E_3 \sin \Omega \\ E_3E_1(1 - \cos \Omega) + E_2 \sin \Omega \end{bmatrix}$$

$$\begin{bmatrix} R_{12} \\ R_{21} \\ R_{32} \end{bmatrix} = \begin{bmatrix} E_1E_2(1 - \cos \Omega) + E_3 \sin \Omega \\ E_2E_2(1 - \cos \Omega) + \cos \Omega \\ E_3E_2(1 - \cos \Omega) - E_1 \sin \Omega \end{bmatrix}$$

$$\begin{bmatrix} R_{13} \\ R_{23} \\ R_{33} \end{bmatrix} = \begin{bmatrix} E_1E_3(1 - \cos \Omega) - E_2 \sin \Omega \\ E_2E_3(1 - \cos \Omega) + E_1 \sin \Omega \\ E_3E_3(1 - \cos \Omega) + \cos \Omega \end{bmatrix}$$

(4.1.1)

upward relative to plate A. The configuration of the triple junction changes, showing the original configuration to be unstable. Similarly, if $_A\mathbf{V}_C$ has a downward component on the diagram relative to plate A (Fig. 4.8C), the BC boundary must migrate downward relative to Plate A, and again the configuration of the triple junction changes. Only if the velocity vector $_A\mathbf{V}_C$ is exactly parallel to the BC boundary (Fig. 4.8D) can the original configuration of the triple junction be stable. Such precise align-

ment is rarely found in nature. Thus, in general, the triple junction of Figure 4.8A would be unstable. Box 4.2 discusses the general method for determining the stability of a triple junction.

Stability of a triple junction does *not* imply that the location of the junction is fixed on the Earth's surface or on a plate boundary. For example, in Figures 4.8B and 4.8C, the final triple junction configuration is stable, but the junction migrates along the boundary of plate A. Box 4.2 also discusses how to deter-

Because all nine components are different and necessary to the description of the rotation, there is no way to reduce the description to a simple vector quantity with only three components.

The angular velocity, or rate of rotation, is the infinitesimal rotation that occurs in an infinitesimal increment of time. An infinitesimal rotation Ω' can be described in a manner similar to Equation (4.1.1). In the infinitesimal limit, however,

$$\cos \Omega' \approx 1 \quad \text{and} \quad \sin \Omega' \approx \Omega' \quad (4.1.2)$$

where the second relation holds if Ω' is expressed in radians. With these simplifications, Equations (4.1.1) reduce to

$$R_{Kj} = \begin{bmatrix} R_{11} & R_{12} & R_{13} \\ R_{21} & R_{22} & R_{23} \\ R_{31} & R_{32} & R_{33} \end{bmatrix} =$$

$$\begin{bmatrix} 1 & +E_3\Omega' & -E_2\Omega' \\ -E_3\Omega' & 1 & +E_1\Omega' \\ +E_2\Omega' & -E_1\Omega' & 1 \end{bmatrix} =$$

$$\begin{bmatrix} 1 & 0 & 0 \\ 0 & 1 & 0 \\ 0 & 0 & 1 \end{bmatrix} + \begin{bmatrix} 0 & +E_3\omega & -E_2\omega \\ -E_3\omega & 0 & +E_1\omega \\ +E_2\omega & -E_1\omega & 0 \end{bmatrix} \delta t$$

$$(4.1.3)$$

We obtained the right-hand side by expressing the infinitesimal rotation Ω' as the product of an angular velocity and an infinitesimal time increment ($\Omega' = \omega \, \delta t$) and by separating the simplified rotation matrix into the sum of the identity matrix and an antisymmetric matrix. In the identity matrix, the components on the principal diagonal equal one and all others are zero; in an antisymmetric matrix, the components on the principal diagonal equal zero, and each off-diagonal component is the negative of the component located symmetrically across the principal diagonal. Thus, there are only three independent components in Equation (4.1.3) rather than the nine in Equation (4.1.1),

and this simplification is possible only because the rotation Ω' is an infinitesimal rotation. Using these three independent components, the rotation rate can be defined by an angular velocity vector $\boldsymbol{\omega} = [w_1, w_2, w_3] = [E_1\omega, E_2\omega, E_3\omega]$. When this vector is expressed in polar coordinates, its components are simply $\boldsymbol{\omega} = [\lambda, \phi, \omega]$.

Thus, a finite rotation can be expressed either as an Euler pole $\mathbf{E}$ and a scalar rotation Ω, or by nine Cartesian components R_{Kj}. An angular velocity, on the other hand, can be expressed in three different but related ways: (1) as an Euler pole $\mathbf{E}$ and a scalar rotation rate ω; (2) as the sum of an identity matrix and an antisymmetric matrix; or (3) as a vector that can be defined by three Cartesian components $\boldsymbol{\omega} = [w_1, w_2, w_3]$ or three polar components $\boldsymbol{\omega} = [\lambda, \phi, \omega]$. From the above discussion, we see that the angular velocity can be expressed as a vector only because in the infinitesimal limit, the rotation matrix reduces to an antisymmetric matrix plus the identity matrix.

The difference is important, as illustrated by the commutative property of vector addition. For two angular velocity vectors $\boldsymbol{\omega}_A$ and $\boldsymbol{\omega}_B$, we can write

$$\boldsymbol{\omega}_A + \boldsymbol{\omega}_B = \boldsymbol{\omega}_B + \boldsymbol{\omega}_A$$

The addition of finite rotations, however, is not commutative, which means that the order in which the rotations are performed makes a difference. For example.

$$\text{ROT}[\mathbf{E}_1, 90°] + \text{ROT}[\mathbf{E}_2, 90°] \neq$$
$$\text{ROT}[\mathbf{E}_2, 90°] + \text{ROT}[\mathbf{E}_1, 90°]$$

This fact is illustrated in Figure 4.1.1, where a book is subjected to two successive 90° rotations about axes $\mathbf{E}_1$ and $\mathbf{E}_2$. In Figure 4.1.1A, the first rotation is about a horizontal axis $\mathbf{E}_1$ and the second is about a vertical axis $\mathbf{E}_2$. In Figure 4.1.1B, the first rotation is about the vertical axis $\mathbf{E}_2$ and the second is about the horizontal axis $\mathbf{E}_1$. In each case the rotations are the same, although doing

mine the velocity of the triple junction relative to the different plates.

This point is important, because along the margins of plates such as plate A in Figure 4.9, we expect the orientation of structures to reflect the orientation of the relative motion at the boundary. As the triple junction migrates past some point P on plate A, for example, the direction of convergence changes. First plate B converges toward P with a small right lateral component (Fig. 4.9A), and then, after passage of

the triple junction, plate C converges toward P with a large left lateral component (Fig. 4.9B). Thus, we might expect overprinting at P of one set of structures by a later set, each set having formed under a different sense of relative motion. These structures would not imply a fundamental change in the tectonic evolution of the area or a global change in plate motions, but simply a change associated with the continuous motion of the plates and the migration of a triple junction.

them in different order results in a different final orientation.

This result explains why a finite rotation ROT($\mathbf{E}$, Ω) cannot be a vector. If it could, then we would be able to express the vector in terms of its components parallel to three mutually perpendicular coordinates (Cartesian coordinates), and this would imply that the finite rotation could be represented as the sum of three rotations, one about each of the coordinate axes. But this is not possible, because a vector sum is commutative but the sum of finite rotations is not.

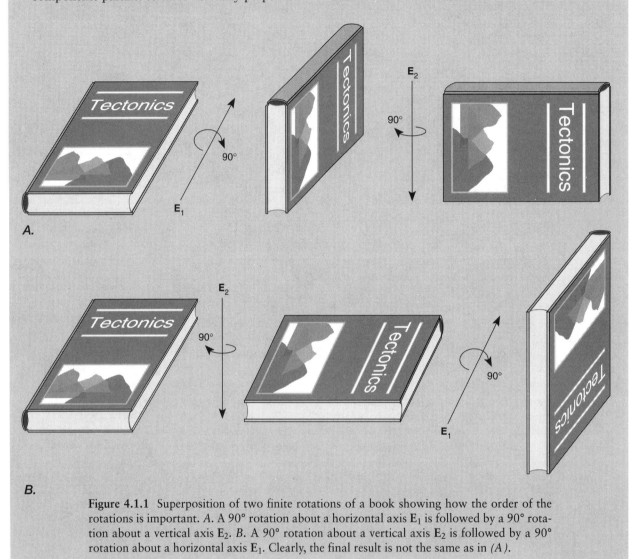

Figure 4.1.1 Superposition of two finite rotations of a book showing how the order of the rotations is important. *A.* A 90° rotation about a horizontal axis $\mathbf{E}_1$ is followed by a 90° rotation about a vertical axis $\mathbf{E}_2$. *B.* A 90° rotation about a vertical axis $\mathbf{E}_2$ is followed by a 90° rotation about a horizontal axis $\mathbf{E}_1$. Clearly, the final result is not the same as in *(A)*.

Global Plate Motions

We can extend the principle that the sum of the motions of plates about a triple junction must equal zero (Eq. 4.6) to state that the sum of the relative angular velocities at all plate boundaries crossed by a continuous closed circuit on the Earth's surface must be zero, or

$$_A\boldsymbol{\omega}_B + {}_B\boldsymbol{\omega}_C + \ldots + {}_X\boldsymbol{\omega}_A = 0 \qquad (4.8)$$

where the X indicates an arbitrary number (≥ 3) of plates along the circuit. This expression must be true, because any such sum around a closed circuit ultimately expresses the angular velocity of plate A relative to itself.

If sufficient information is available about the relative plate velocities, it is possible to use the principle expressed in Equation (4.8) to help calculate a com-

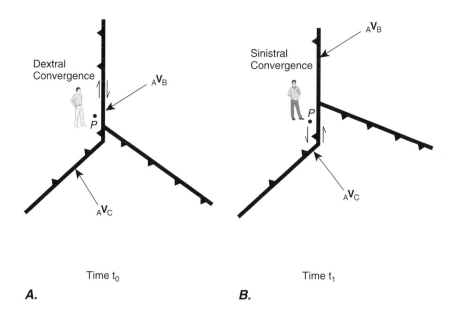

Dextral
Convergence

$_A\mathbf{V}_B$

P

$_A\mathbf{V}_C$

Time t_0

A.

Sinistral
Convergence

$_A\mathbf{V}_B$

P

$_A\mathbf{V}_C$

Time t_1

B.

Figure 4.9 The change of relative motion at a point P along the boundary of plate A, as a result of migration of a triple junction. *A.* Convergence at point P has a small dextral component. *B.* After the triple junction has passed, convergence at P has a large sinistral component.

plete global system of relative plate motions. The principal sources of information about the magnitude and orientation of relative plate velocities include (1) modern or recent spreading rates calculated from magnetic anomaly data and the magnetic reversal time scale; (2) the orientation of transform fault boundaries; and (3) first-motion studies of earthquakes on all three types of boundaries. Table 4.1 shows one estimate of the relative angular velocities for a ten plate model that provides the basis for the relative velocities shown in Figure 4.1.

4.4 Finite Plate Motions

So far, we have discussed the kinematics of plate tectonics in terms of the relative plate velocities. In effect, these motions are *instantaneous* or *infinitesimal* motions; in practice, we determine these motions by measuring the small displacements that occur over relatively short intervals of geologic time (on the order of 10^7 years or less). We are also interested, however, in reconstructing the configurations of the continents at various times in the geologic past. Such reconstructions require large rotations that accumulate over long intervals of geologic time (greater than 10^7 years). We show next that, in general, an Euler pole $_A\mathbf{E}_B$ for the angular velocity of plate B relative to

plate A can migrate relative to both plates, and so it cannot be used to describe the finite rotation that accumulates over a geologically long period of time.

To illustrate the problem, consider the kinematics of an RRR triple junction over a long period of time. Figure 4.10A shows the spreading of three continental masses attached to the three plates A, B, and C. For this model, we assume plate A is fixed, and the motions are plotted relative to that plate. Each of the angular velocity Euler poles $_A\mathbf{E}_B$, $_B\mathbf{E}_C$, and $_C\mathbf{E}_A$ must be *instantaneously* fixed on the two plates whose relative motion they describe. Assume now that for a *finite* amount of time, the pole $_A\mathbf{E}_B$ remains fixed with respect to plate A. During that time, plate B rotates progressively about $_A\mathbf{E}_B$ (from position B1 to B5, Fig. 4.10A), so that $_A\mathbf{E}_B$ is also fixed with respect to plate B. Assume furthermore that $_C\mathbf{E}_A$ also remains fixed with respect to plate A. Plate C rotates progressively about $_C\mathbf{E}_A$ (from position C1 to C5, Fig. 4.10A), and therefore $_C\mathbf{E}_A$ also is fixed with respect to plate C.

Consider now the pole $_B\mathbf{E}_C$. Equation (4.4) requires that this pole be fixed with respect to the other two poles. Therefore, it is fixed with respect to plate A, because the other two poles are both fixed with respect to plate A. But $_A\mathbf{E}_B$ is the only Euler pole that can be fixed with respect to both A and B, and $_C\mathbf{E}_A$ is the only Euler pole that can be fixed with respect to both A and C. Thus $_B\mathbf{E}_C$, which is fixed with respect to plate A, cannot remain fixed through time with respect to either plate B or plate C, and the pole of

Table 4.1 **Relative Angular Velocities for a 10-Plate Earth**

| Plate boundary | Pole Location | | Rate (°/Ma) |
	Latitude °N	Longitude °E	
NA-PA	50.9	−66.3	0.75
CO-PA	41.3	−108.1	2.02
CO-NA	31.8	−123.3	1.42
CO-NZ	−2.9	−135.1	0.77
NZ-PA	56.6	−85.6	1.64
AN-NZ	−37.6	90.2	0.67
SA-AN	−77.7	78.2	0.38
PA-AN	−68.7	100.4	1.03
NZ-SA	51.9	−91.4	0.99
PA-IN	−59.8	178.0	1.26
IN-AN	10.7	31.6	0.67
IN-AF	18.7	44.8	0.58
IN-EA	23.0	33.9	0.65
AF-AN	−19.0	−13.3	0.32
AF-NA	80.1	23.5	0.32
AF-EA	29.6	−25.7	0.14
AF-SA	57.4	−37.5	0.37
NA-SA	−3.0	−53.3	0.18
EA-NA	69.3	128.0	0.27
EA-PA	65.3	−69.8	0.91
IN-AR	7.4	65.5	0.39
AR-AF	30.5	8.1	0.27

Plate symbols: AF, African; AN, Antarctic; AR, Arabian; CO, Cocos; EA, Eurasian; IN, Indian; NA, North American; NZ, Nazca; PA, Pacific; SA; South American

The first-named plate moves counterclockwise relative to the second one, about the pole indicated.

After Minster et al., 1974.

instantaneous rotation $_B E_C$ must migrate relative to plates B and C. This is evident from Figure 4.10A, in which all three Euler poles and plate A remain fixed through time, while plates B and C move. Clearly, the location of $_B E_C$ relative to plate B and to plate C must change.

Figure 4.10B shows the geometry of the three plates at the time of B5 and C5. As plate B rotates progressively from position B1 to B5 in Figure 4.10A, an observer stationary on plate B would observe the position of the pole $_C E_B$ (= $-_B E_C$) to migrate along a small circle from P_{B1} to P_{B5} in Figure 4.10B.

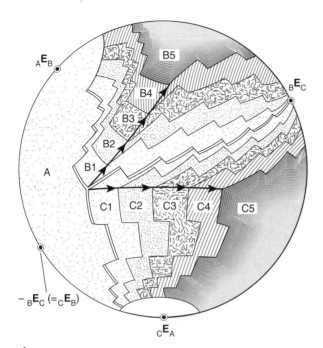

A.

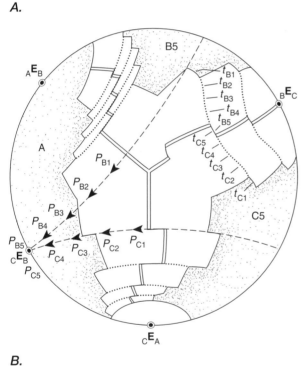

B.

Figure 4.10 Upper hemisphere equal-area projection showing migration of plates and an angular velocity Euler pole in a three-plate system. *A.* Diagram showing the initial position of the ridge-transform system along which the plates break apart and the positions of plates B and C with respect to A for successive times 1 to 5. $_A E_B$ and $_A E_C$ are fixed with respect to plate A. $_B E_C$ is the current Euler pole, which is fixed with respect to the other two Euler poles and thus must migrate with respect to plates B and C. *B.* Position of plates at time 5. The positions of $_{-B} E_C$ ($= _C E_B$) with respect to plate B are at P_{B1}, P_{B2}, P_{B3}, P_{B4}, and P_{B5} for times from the past to the present; with respect to plate C, the positions of the Euler pole $_C E_B$ are at P_{C1}, P_{C2}, P_{C3}, P_{C4}, and P_{C5} for corresponding times. These paths are small circles (dashed lines) about $_A E_B$ and $_C E_A$, respectively. Inactive transform fault zones are shown as dotted lines. Labels t_{B1} through t_{B5} and t_{C1} through t_{C5} indicate the parts of the transform that were active at the time when the angular velocity Euler pole was located at the point with the same index. *(After Dewey, 1975)*

Similarly, as plate C rotates progressively from C1 to C5 in Figure 4.10*A*, an observer stationary on plate C would observe the position of the pole $_C E_B$ to migrate along a small circle from P_{C1} to P_{C5} in Figure 4.10*B*. The locations of P_{B5} and P_{C5} coincide at the current time. Thus, the past positions of the angular velocity Euler pole determined from plate B are different from those determined from plate C, and the pole positions for a given time in the past, as viewed from each plate, coincide only when the plates are rotated back to their correct relative positions for that time.

We have described a very special case in which the three Euler poles are fixed with respect to one of the plates. In general, however, *for a system of three or more plates, we must expect the angular velocity Euler poles to migrate with respect to the plates whose motion they describe, and therefore the relative plate motion must change through time.* We discuss further the consequences of this conclusion below.

In the example just given, the motion of plate C relative to plate B is described by infinitesimal rotation increments about an Euler pole $_C E_B$ that migrates continuously with respect to both plates. For any given large interval of time, however, the final position of plate C relative to plate B can be described by a single rotation about a single finite rotation Euler pole. This finite rotation is the net result of the entire sequence of incremental rotations that occurred in the time interval. In general, the finite rotation pole does not coincide with any of the angular velocity poles, and the process of rotation about the finite rotation pole does not duplicate the actual history of the relative motions of plates C and B.

We approximate the history of relative rotations by defining a series of **stage poles** or **finite-difference poles**. Each stage pole defines the axis of finite relative rotation connecting the initial and final positions and orientations of a plate over a stage, which is a

Box 4.2 Stability of Triple Junctions

How can we determine, in general, whether a particular configuration of triple junction is stable? As the example in Figure 4.9 illustrates, we must consider both the relative velocity vectors and the type and orientation of the boundaries. Consider first a system of two plates, A and B, separated by a linear divergent boundary (Fig. 4.2.1A). If we choose a reference frame (x,y) such that the origin is always on the boundary, and the angles between the boundary and the axes (x,y) are constant, the configuration of the boundary in that frame always looks the same. Thus, if the reference frame is fixed with respect to the boundary, or if it has any velocity parallel to the boundary, the configuration of the boundary always remains the same.

It is useful to consider this condition for a constant boundary configuration by plotting the velocity vectors relative to one another (Fig. 4.2.1B). Taking A to be the reference plate, the velocity of B with respect to A is shown as $_A\mathbf{V_B}$. The velocity of the ridge with respect to plate A is shown by the vector $_A\overline{\mathbf{V}}_{AB}$, where AB indicates the boundary between plates A and B. $_A\overline{\mathbf{V}}_{AB}$, is exactly half of $_A\mathbf{V_B}$ for symmetrical spreading. The line ab is the locus of all velocity vectors $_A\mathbf{V}_{AB}$ having one component equal to the ridge minimum velocity $_A\overline{\mathbf{V}}_{AB}$, and having an arbitrary component parallel to the boundary AB. The configuration of the divergent boundary looks the same through time only if the velocity of the reference frame is one of the vectors $_A\mathbf{V}_{AB}$, because the origin of the reference frame can remain on the ridge only if it moves with one of these velocities. Several of these reference frame velocity vectors are shown in Figure 4.2.1B.

We can apply the same concept to determine the stability of a triple junction. The junction maintains a stable configuration only if there is at least one reference frame in which the configura-

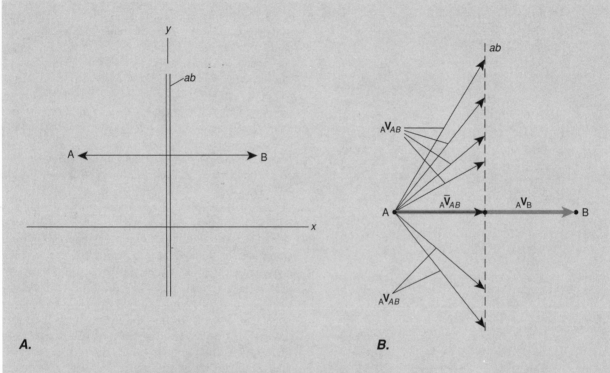

A.

B.

Figure 4.2.1 The stable configuration of a single boundary. *A.* Diagram of two plates A and B separated by a divergent boundary indicated by the double line. A coordinate system x-y has its origin on the boundary. As long as the origin stays on the boundary and the axes do not rotate with respect to the boundary, the axes can travel with any velocity, and the boundary looks the same in that coordinate system. *B.* Diagram of velocities relative to plate A. $_A\mathbf{V_B}$ is the velocity of plate B relative to plate A; $_A\overline{\mathbf{V}}_{AB}$ is the minimum velocity of the AB boundary with respect to plate A; ab is the locus of all velocity vectors $_A\mathbf{V}_{AB}$ for which the configuration of the boundary looks the same.

Figure 4.2.2 Stability of an RRR triple junction. *A.* Physical space showing plate boundaries and relative velocity vectors. *B.* Velocity diagram showing the relative velocities $_A V_B$, $_B V_C$, and $_C V_A$ of plates A, B, and C. The lines *ab*, *bc*, and *ca* are the loci of velocity vectors of the reference frames in which the *AB*, *BC*, and *CA* boundaries, respectively, maintain a stable configuration. These three lines intersect at a point *J*, which means that there is one reference frame in which all three boundaries maintain a stable configuration. The velocities $_A V_J$, $_B V_J$, and $_C V_J$ define the rate at which the triple junction moves with respect to plates A, B, and C, respectively.

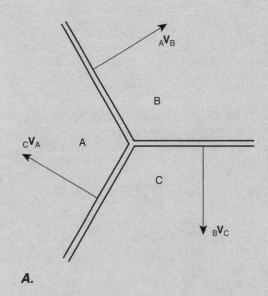

A.

tion of all three boundaries is stable. We illustrate the construction using a ridge-ridge-ridge (RRR) triple junction (Fig. 4.2.2*A*). Figure 4.2.2*B* shows the velocity triangle for the three plates. The line *ab* shows the range of acceptable reference frame velocities $_A \mathbf{V}_{AB}$ for which the configuration of the ridge between plates A and B is unchanged. Similarly, the lines *bc* and *ca* define the reference frame velocities $_B \mathbf{V}_{BC}$ and $_C \mathbf{V}_{CA}$ for which the configurations of boundaries *BC* and *CA*, respectively, remain stable. The three loci of acceptable reference frame velocities—*ab*, *bc*, and *ca*—intersect at the point *J*, which means there is one reference frame in which the configuration of all three boundaries, and therefore of the triple junction, remains unchanged. Therefore, the triple junction configuration must be a stable configuration.

The vectors $_A \mathbf{V}_J$, $_B \mathbf{V}_J$, and $_C \mathbf{V}_J$ define the velocity of the reference frame with respect to each of the three plates A, B, and C, respectively. Because the triple junction configuration is stable in this reference frame, these velocities define the motion of the triple point with respect to each plate.

We can apply the same technique to the trench-trench-trench (TTT) triple junctions portrayed in Figures 4.8*A*, 4.8*B*, and 4.8*C*. First we examine the initial configuration of the triple junction and use the velocity vectors from Figure 4.8*B* (Fig. 4.2.3*A*). We construct the loci of acceptable reference frame velocities on the velocity triangle. Plate C overrides plate B, and plate A overrides the other two. Thus, the boundaries *AB* and *CA* are fixed on plate A, and the boundary *BC* is fixed on plate C. For the *AB* boundary, one reference frame in which the configuration is stable is fixed to plate A. On the velocity diagram, therefore, the locus *ab* of acceptable reference frame velocities

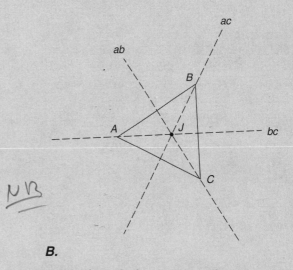

B.

must pass through the point *A*, indicating a zero velocity relative to plate A, and *ab* also must be parallel to the boundary *AB*. A similar argument shows that the locus *ca* must also pass through point *A* and be parallel to the *AC* boundary and *bc* must pass through point *C* and be parallel to the *BC* boundary. The loci of acceptable reference frame velocities do not meet in a point. Thus, there is no single reference frame in which all three boundary configurations are stable, and therefore the triple junction is not stable.

The condition for stability of this triple junction is easy to determine, because for stability, the three loci of acceptable reference frame velocities *ab*, *bc*, and *ca* must meet at a point. If we assume that the orientations of the boundaries are fixed as given, then the orientations of the loci are also fixed. The only way these lines can meet at a point is if the relative velocity $_A\mathbf{V}_C$ is parallel to the BC boundary, as shown in Figure 4.2.3B. In this case, the three loci meet at the point $J = A$, the triple junction is stable, and the triple point does not move with respect to plate A.

Using the relative velocities shown in Figure 4.2.3A, the triple junction evolves to the configuration shown in Figure 4.2.3C. The velocity triangle for this example is identical to that in Figure 4.2.3A, but loci ab and ac are now identical because, adjacent to the triple junction, the boundaries AB and AC are parallel. The three loci ab, bc, and ca thus meet at a point J, and the triple junction is stable. The velocity of the junction with respect to plate A is given by the vector $_A\mathbf{V}_J$.

These examples illustrate the procedure to follow to determine the stability of a given triple junction:

1. In physical space, construct an accurate diagram of the triple junction, showing the boundary orientations, the type of boundary, and the relative velocities of the pairs of plates around the junction.
2. Construct the velocity triangle.
3. For each pair of plate boundaries, construct the locus of reference frame velocities in which the configuration of the boundary is stable. If the three loci meet in a point, the triple junction is stable.

Not all of the 16 possible triple junction configurations are present in the modern world. Figure 4.2.4 displays all the possible junction types, including those already discussed, as well as the conditions of their stability and examples from the present plate tectonic regime of the Earth.

Figure 4.2.3 Stability of a TTT triple junction. *A.* A TTT triple junction with the geometry of Figure 4.9A and the relative velocities of Figure 4.9B. On the velocity triangle, the three loci of acceptable reference frame velocities for configurational stability of the three boundaries do not intersect at a point, indicating that the configuration of the triple junction is unstable. *B.* The condition for stability of the TTT triple junction in (*A*) is that the velocity of plate C with respect to plate A, $_A\mathbf{V}_C$, should be parallel to the BC boundary. In that case, the three loci for acceptable reference frame velocities meet at a point $J = A$. *C.* The TTT triple junction configuration into which (*A*) would evolve. The junction in this case is stable because the AB and AC boundaries are parallel, which means that the loci of acceptable reference frame velocities ab and ca are coincident. The three loci ab, bc, and ca meet at J, and the velocity $_A\mathbf{V}_J$ gives the rate at which the triple junction migrates along the boundary of plate A.

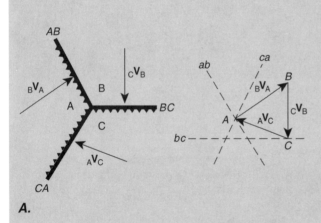

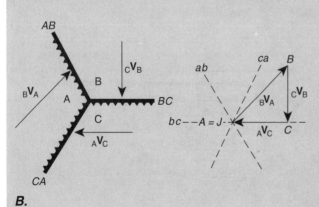

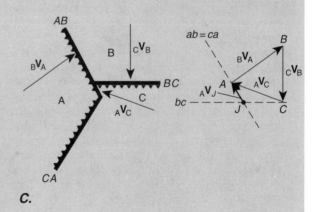

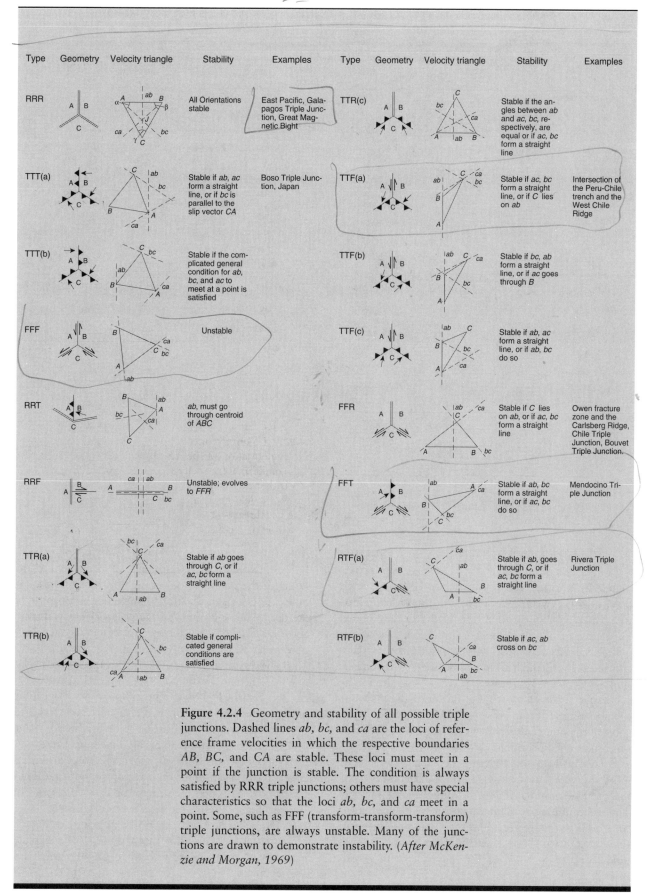

Figure 4.2.4 Geometry and stability of all possible triple junctions. Dashed lines *ab*, *bc*, and *ca* are the loci of reference frame velocities in which the respective boundaries *AB*, *BC*, and *CA* are stable. These loci must meet in a point if the junction is stable. The condition is always satisfied by RRR triple junctions; others must have special characteristics so that the loci *ab*, *bc*, and *ca* meet in a point. Some, such as FFF (transform-transform-transform) triple junctions, are always unstable. Many of the junctions are drawn to demonstrate instability. (*After McKenzie and Morgan, 1969*)

Table 4.2 Backward-Motion Stage Poles for Motion of North America versus (A) Eurasia and (B) Africa

Time interval (Ma)	North America		Eurasia		Rotation rate (10⁻⁷deg/Ma)
A.	Latitude	Longitude	Latitude	Longitude	
0–9	68.0N	137.0E	68.0N	137.0E	2.78
9–38	63.5N	131.1E	63.5N	131.6E	1.76
38–53	27.6N	155.7E	29.0N	161.1E	1.84
53–63	67.3N	152.0W	72.3N	150.1W	4.55
63–72	76.0N	49.8W	76.8N	11.1W	8.56
72–81	76.0N	49.8W	70.4N	15.1E	8.56
81–					
	North America		Africa		
B.	Latitude	Longitude	Latitude	Longitude	
0–9	69.7N	33.4W	69.7N	33.7W	4.00
9–38	77.1N	53.5E	78.4N	56.7E	2.18
38–53	72.2N	16.0E	72.4N	19.8E	2.78
53–63	66.2N	11.8W	66.2N	18.2E	3.15
63–72	57.8N	42.0W	54.4N	29.7W	8.41
72–81	57.8N	42.0W	54.4N	29.7W	8.41
81–155	59.4N	22.8W	59.6N	13.5W	4.99
155–180	56.8N	73.3E	72.7N	165.0W	3.57

The table column "Coordinates of pole fixed to" spans North America and Eurasia (and North America and Africa).

After Pitman and Talwani, 1972.

span of time of finite length but of geologically short duration. The motion over the next stage is approximated by another rotation about a different stage pole. In effect, we approximate the smooth plate motion and the smooth migration of the instantaneous Euler pole with a series of small finite rotations about a series of different poles whose positions jump discontinuously through time. These **stage rotations** or **finite-difference rotations** provide "snapshots" at successive times of the configuration of the plates at specific times during smooth plate evolution. The shorter the stage, the closer the stage pole will approximate the instantaneous pole position.

To illustrate the function of such poles, Table 4.2*A* shows a series of stage poles and rotations for the relative motion of North America and Europe. The rotation for one stage carries the plate configuration from the more recent to the older time listed in the time interval. The top part of the table shows stage poles for the rotation of Europe relative to North America, as well as for North America relative to Europe. Note that the amounts of rotation are the same for the same stages, but the positions of the stage poles are different, illustrating the effect of opening of the ocean and migration of the pole as observed from coordinate frames on the two plates involved (compare poles P_{B1}–P_{B5} and P_{C1}–P_{C5} in Fig. 4.10*B*). Table 4.2*B* shows similar rotations for North America relative to Africa and Africa relative to North America.

Figure 4.11 shows a north polar stereographic view of the location, on a reference frame fixed to North America, of the Europe–North America stage poles for various time intervals listed in Table 4.2. The successive positions of these stage poles provide a crude approximation of the path of the instantaneous pole of rotation, in much the same way as a series of straight lines can approximate a smooth curve.

The migration of the poles of rotation during plate motion means that the geometry of the plate boundaries must be constantly changing, and this

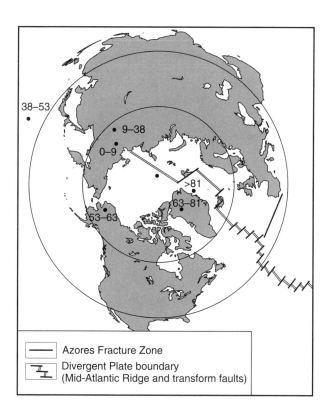

—— Azores Fracture Zone

⊥⊤ Divergent Plate boundary
(Mid-Atlantic Ridge and transform faults)

Figure 4.11 Location of Euler poles for the relative motion of North America and Eurasia, in a reference frame fixed with respect to North America. Data from Table 4.2. *(After Cox and Hart, 1986)*

change affects the structure of the boundaries. In Figure 4.10*B*, the ridge segments on the *BC* boundary lie on great circles through the current angular velocity Euler pole $_C E_B$. Thus, if spreading is orthogonal to the ridge, any magnetic anomaly should lie along a great circle through the location of the angular velocity pole that was active at the time the anomaly formed. Because the pole migrates, however (see P_{B1}–P_{B5} and P_{C1}–P_{C5} in Fig. 4.10*B*), different anomalies will lie along the great circles through different points, and thus the anomalies of different ages must have different orientations.

Similarly, any active transform fault on the *BC* boundary should lie on a small-circle arc about the current Euler pole $_C E_B$. Inactive segments of the transform fault on either plate, however, should lie along small circles about the corresponding older Euler pole positions, which for plate B are poles P_{B1}, P_{B2}, P_{B3}, P_{B4}, P_{B5}, and so forth, and for plate C are poles P_{C1}, P_{C2}, P_{C3}, P_{C4}, P_{C5}, and so forth. Therefore, it is apparent that if motion on the active part of a transform fault is to remain purely strike-slip through time, the transform fault must continually adjust its orientation. Thus, the inactive trace of a transform fault cannot lie exactly along any single small circle. The transform fault zones in Figure 4.10*B* are indicated by dotted lines. On one of the fault zones crossing

from plate B to plate C, the approximate age when the segment of the transform was active is labeled along the fault zone with the times t_{B1} through t_{B5} on plate B and t_{C1} through t_{C5} on plate C. These times correspond to the locations of the angular velocity pole $_C E_B$ having the same index number. Thus, the shape of the transform is determined by the fact that at any given time, the transform must have been parallel to a small circle about the corresponding angular velocity pole. Because those poles migrate in time, the transform fault zone must be a complex curve, not a simple small circle about the most recent angular velocity pole. Moreover, because the poles have different locations on the different plates, the transform fault zone generally will have a different shape on opposite sides of the current ridge; its shape is not symmetrical on the plates B and C, although the equal angular velocities about $_A E_B$ and $_C E_A$ assumed for the construction of Figure 4.10*B* do not illustrate this fact. If the orientation of a transform fault cannot adjust easily to the changing position of the angular velocity pole, then long-term motion on the active part of the transform fault cannot be purely strike-slip.

The relative motion of plates B and C in Figure 4.10 clearly cannot be described by a simple rotation about a single axis. Is there any way to describe this motion? Let us return to the simple model in Figure 4.10 for which plate A is assumed to be stationary in some absolute reference frame and in which the motions of plates B and C are described by a constant angular velocity about the respective poles $_A\text{E}_B$ and $_A\text{E}_C$, which are also stationary in the absolute reference frame. Points on plate B trace out a small-circle path about $_A\text{E}_B$ and points on plate C trace out a small-circle path about $_A\text{E}_C$ (see the two sets of multiple arrows in Fig. 4.10A). What path, then, do points on plate B trace out when viewed from plate C?

All points fixed in the absolute reference frame, such as those on plate A in this example, must appear to trace out small-circle paths about $_A\text{E}_C$ when viewed from plate C, except for the point at $_A\text{E}_C$ itself, which appears stationary. Thus, when viewed from plate C, the pole $_A\text{E}_B$, which is fixed relative to plate A, must also appear to trace out a small-circle path about $_A\text{E}_C$. But points on plate B trace a small-circle path about $_A\text{E}_B$, while $_A\text{E}_B$ itself appears to trace out a small-circle path about $_A\text{E}_C$. In other words, plate B rotates about a pole $_A\text{E}_B$ that itself appears from plate C to be rotating about $_A\text{E}_C$. Thus, points on plate B appear to trace a spherical cycloidal path when viewed from plate C.

A planar analogy to spherical cycloidal motion is the motion seen by a stationary observer watching a point on a bicycle tire as the tire rolls down the street. In the case of plate tectonics, however, the motion occurs on the surface of a sphere. The exact shape of the trajectory depends upon the location of the point relative to the absolute poles and the relative rates of motions of the two plates about their respective absolute poles. This motion is considerably more complex than the instantaneous relative motion between the two plates for which points on both plates move along a small-circle path about the relative pole of instantaneous rotation $_B\text{E}_C$. Those instantaneous small-circle paths are tangents to the long-term cycloidal paths.

4.5 Absolute Plate Motions

Throughout this discussion, we have focused mostly on *relative* plate motions, which are the motions of one plate relative to a coordinate frame fixed with respect to another plate. It would be useful to be able to describe the angular velocity of each individual plate in absolute terms; that is, relative to an absolute reference frame R whose orientation is independent of the motions of all the plates. The angular velocity

of a given plate such as A, for example, could then be expressed relative to the reference frame R as $_R\omega_A$. The relative angular velocity of plate B with respect to plate A could be obtained by vector addition. Applying Equation (4.6),

$$_A\omega_B + {}_B\omega_R + {}_R\omega_A = 0$$

$$_A\omega_B = {}_R\omega_B + {}_R\omega_A \qquad (4.9)$$

where we used the relation $_R\omega_B = -{}_B\omega_R$. Using such a reference frame would circumvent some of the complications associated with relative motions and would simplify the evaluation of models of the driving force of plate tectonics (see Sec. 4.6). The proposition is difficult because first an absolute coordinate system must be defined, and then there must be some way of determining the orientation of that coordinate frame with respect to a given plate.

The Earth's axis of rotation might appear to be one possibility for an absolute reference axis. The assumption that the magnetic pole parallels the Earth's rotational axis would allow the orientation of the axis to be determined through time using paleomagnetism. There are two problems with this possibility, however. First, this technique can determine only the change in plate orientation and the change in latitude relative to the magnetic pole; the longitudinal component of any motion is not defined by the paleomagnetic vectors. Thus, the location of the pole of absolute plate rotation and the magnitude of the absolute velocity vector cannot be defined uniquely from paleomagnetic data. Motion relative to a second absolute reference axis is needed.

The second problem with using the Earth's rotation axis as an absolute reference axis is that true polar wander of the Earth's rotation axis can occur and probably has occurred in the past. Polar wander happens when the entire Earth rolls relative to the rotation axis to keep the axis of maximum moment of inertia parallel to the rotation axis. The axis of maximum moment of inertia might change because the distribution of mass within the Earth could change with changes in the pattern of convection cells in the mantle and the core or with shifts in position of lithospheric plates. Such true polar wander would result in a wholesale shift of the magnetic pole relative to the Earth's deep interior. In such circumstances, the paleomagnetic data would include a component of plate motion unique for each plate and a component resulting from true polar wander, which would be the same for all plates. We are interested primarily in the former component, and separation of the two components is not easy.

Some workers have proposed that the absolute motions of the deep mantle may be small compared to those of the plates, and therefore a surficial feature whose origin is tied directly to the deep mantle could provide an approximate location for an absolute coordinate axis. W. Jason Morgan proposed in 1972 that convection plumes rising from the deep mantle (Fig. 4.12) cause **hot spots** on the Earth's surface (Fig. 4.13) that give rise to localized regions of volcanic activity. Iceland and Hawaii are two of the best known of such regions. If plumes are indeed fixed with respect to the deep mantle, the associated hot spots should not move with respect to one another, and they could be used to define an absolute reference frame. As a lithospheric plate moves over the mantle plume, igneous rocks accumulate on the surface, producing an aseismic topographic ridge (see Fig. 4.12). The aseismic ridge would then mark the trajectory of the plate relative to the absolute reference frame. If the motion of a plate is a rotation about a fixed absolute-angular-velocity pole, then an aseismic ridge on the plate should be an arc of a small circle about that pole, and dating the volcanic rocks along the ridge would permit determination of the rate of rotation of the plate.

Testing whether hot spots move relative to one another is easiest when two or more hot-spot traces are present on a single plate. The test is further simplified if it is assumed that the pole of absolute angular velocity is fixed for finite periods of geologic time. The distance between points of the same age on aseismic ridges produced by different hot spots on the same plate should remain the same distance apart through time.

Figure 4.14 shows the result of applying this test to the Pacific, South Atlantic, and Indian oceans. For the Pacific, two absolute rotation poles are chosen to fit the well-defined Hawaiian Islands–Emperor Seamount aseismic ridge. The change in pole of rotation corresponds with the "elbow" in the Hawaiian-Emperor ridge and indicates a change in motion of the Pacific plate relative to the Hawaii hot spot in Eocene time (approximately 45 Ma). The solid lines in Figure 4.14A show predicted locations of aseismic ridges that would develop from three proposed hot spots in the Pacific. The Tuamotu–Line Islands trace from the Easter Island hot spot seems to fit the prediction well, but the Marshall–Gilbert–Austral Islands ridge from the Macdonald hot spot does not. The misfit can be accounted for by a motion of the Macdonald hot spot relative to the other two hot spots by approximately 2 cm/yr. Thus, hot spots apparently cannot provide a completely reliable absolute reference frame.

A second test of the fixed-hot-spot hypothesis comes from comparison of paleomagnetic data with the motion of a plate determined from hot-spot traces. Figure 4.14B shows a test of the fixed-hot-spot hypothesis with respect to the African Plate. The solid lines are predicted hot-spot traces using a rotation of the African plate of 27° about a pole located at 25°N, 55°W. The fit to the topographic ridges seems reasonably close. The predicted latitudinal component of the motion has been compared with that indicated by paleomagnetic data from the African continent, and the results are compatible.

A third test of the fixed-hot-spot hypothesis is to compare relative positions of the hot spots at times in the past, after reconstructing the plate positions from oceanic magnetic data. For the past 10 Ma, this test reveals no discernible motion of the hot spots shown in Figure 4.13 relative to one another. Thus, for the geologically "instantaneous" motions of the past 10 million years, hot spots seem to provide a reasonably good approximation of an absolute reference frame. Table 4.3 shows one version of calculated angular velocities of the plates relative to a fixed-hot-spot frame. These poles give rise to the map of absolute velocity vectors presented in Figure 4.15. The map shows several suprising relationships. The Pacific and Indian plates are converging rapidly toward Southeast Asia. Convergence rates between Eurasia and plates to the south increase from the Mediterranean eastward. The slowest moving plates are all predominantly continental. Among the fastest moving plates, the Pacific, Nazca, Cocos, and Philippine plates bear no continental masses at all, and the Australian-Indian Plate is less than 50 percent continental.

A comparison of Figures 4.1 and 4.15 also demonstrates that spreading centers and convergent zones reflect only the relative motions across a given plate boundary. These boundaries may themselves move in the absolute reference frame.

Estimates of the motion of hot spots for times prior to 10 Ma are controversial. Some workers have argued that motion of as much as 2 cm/yr can take place between individual hot spots, whereas others believe that the hot spots provide a good average reference frame. Nevertheless, the hot-spot reference frame has been used in numerous analyses of finite plate motions. Despite the provisional nature of such analyses, they emphasize the fact that plate configurations, as well as continental positions, have been radically different in the past.

Many attempts have been made to estimate the past positions of the continents, based upon the reconstruction of successive configurations of the

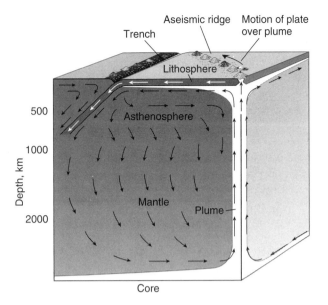

Figure 4.12 Block diagram illustrating the structure of a plume that originates in the lower mantle and rises to the asthenosphere, where it spreads like a thunderhead and produces a hot spot on the Earth's surface, with its associated aseismic ridge. The possible convection return flow is indicated schematically by the arrows. It occurs throughout the mantle as well as at the subduction zone. The line of volcanoes in the aseismic ridge is parallel to the motion of the plate over the plume. Approximately to scale, except surface features vertically exaggerated.

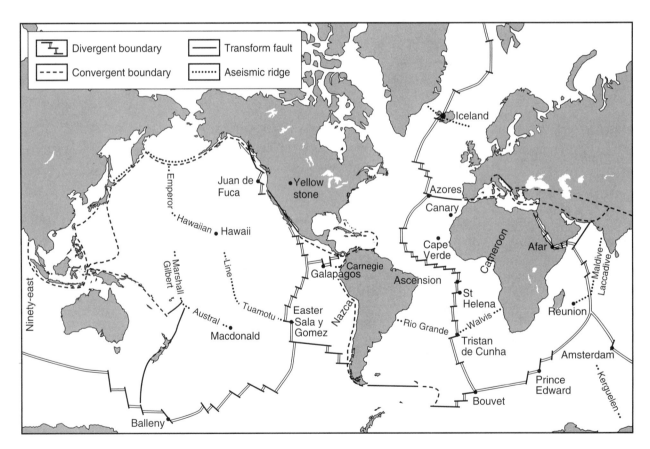

Figure 4.13 Plate tectonic map showing the locations of the principal proposed hot spots and the aseismic ridges thought to result from the motion of plates over them. *(After Morgan, 1972)*

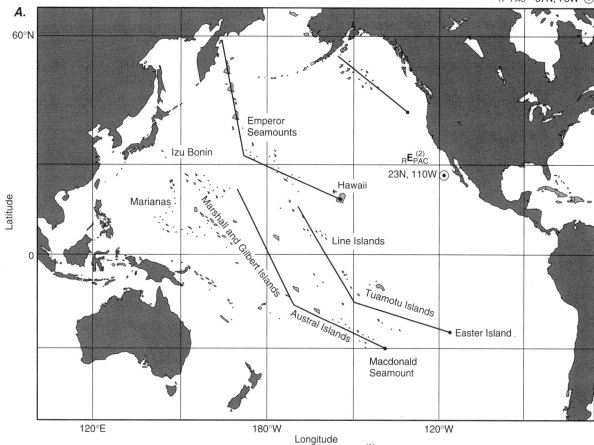

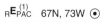

$_R\mathbf{E}_{PAC}^{(1)}$ 67N, 73W ⊙

$_R\mathbf{E}_{PAC}^{(2)}$
23N, 110W ⊙

$_R\mathbf{E}_{PAC}^{(1)}$ = Pole of rotation of Pacific Plate relative to hot
spot reference frame until approximately 45 Ma

$_R\mathbf{E}_{PAC}^{(2)}$ = Pole of rotation of Pacific Plate relative to hot
spot reference after 45 Ma

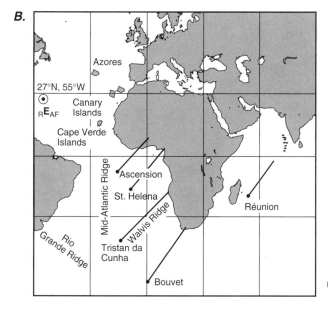

Figure 4.14 Hot-spot trajectories compared with aseismic ridges. *A.* Map of Pacific Ocean showing trajectories constructed by rotation of the Pacific Place 34° about a pole at 67°N, 73°W, then 45° about a pole at 23°N, 110°W. *B.* Trajectories in the Atlantic and Indian oceans constructed by rotation of the African Plate 27° about a pole at 25°N, 55°W. *(After Morgan, 1972)*

$_R\mathbf{E}_{AF}$ = Pole of rotation of African Plate
relative to hot spot reference frame.

Table 4.3 Angular Velocities of Plates Relative to Hot-Spot Framework

Plate	Euler vector		
	Latitude	Longitude	Rate (10^{-7} deg/yr)
African	32N	61 W	2.0
Antarctic	58N	145 W	1.5
Arabian	40N	7 W	4.9
Cocos	23N	117 W	14.7
Eurasian	19N	109 W	1.1
Indian	30N	32 E	6.6
North American	37S	71 W	2.5
Nazca	36N	94 W	8.7
Pacific	65S	107 E	8.8
South American	71S	131 W	2.3
Somalian	44N	54 W	2.3
Philippine	36S	46 W	8.9

After Chase, 1978.

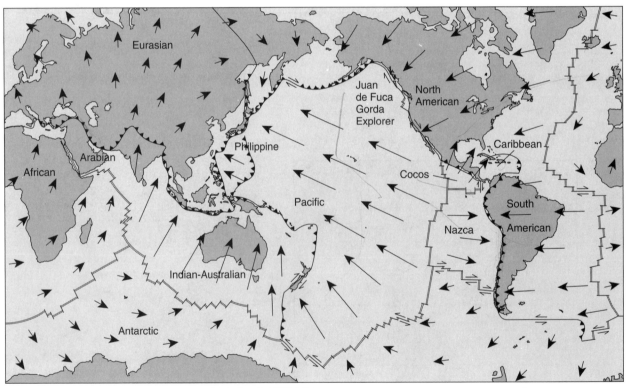

═══ Divergent boundary	⇄	Transform fault (relative movement)	
▲▲▲ Convergent boundary	←	50 mm/yr	

Figure 4.15 Velocity of the Earth's plates relative to the hot-spot frame of reference from data in Table 4.3. The length of the arrows is proportional to the linear velocity. *(After Cox and Hart, 1986; Chase, 1978)*

Atlantic and Indian oceans. As an example, in Figure 4.16 we combine estimates of the positions of the continents in the past 200 Ma with estimates of the positions of plate boundaries in the proto-Pacific basin, called Panthalassa,[5] at the same time. The estimates for the Pacific Plate configuration are clearly approximate, as are the estimates of plate boundaries in the pre-Alpine-Himalayan, or Tethyan, ocean. Also, they account only for major plates, not for smaller ones that may have existed between the major ones.

A comparison of the width of magnetic anomalies with the magnetic reversal time scale indicates that there have been substantial changes in the worldwide rate of plate motion through geologic time. Especially in the late Cretaceous, there was a period of global increase in relative plate motions. This period also was a time during which the magnetic field did not undergo changes in polarity, suggesting a relationship between upper mantle processes (spreading rates) and core processes (magnetic field behavior).

4.6 Properties of the Mantle

In the previous sections, we have discussed the motions of plates, or the kinematics of plate tectonics. Ultimately, we want to gain some insight into the forces that drive the plates, or the mechanics of plate tectonics. To understand the driving forces, we need to understand the mechanism of plate tectonics, and to understand the mechanism, we must understand the properties of the mantle.

The Composition of the Mantle

Our knowledge of the deep mantle depends largely upon remote sensing using seismic waves, experimental investigations into the behavior of minerals at very high pressures, and considerations from heat flow measurements. The measurement of the variation of seismic velocity with depth shows that there are two distinct worldwide seismic velocity anomalies at depths of approximately 400 km and 700 km where the increase in seismic wave velocity with depth is anomalously large (Fig. 4.17). Most geophysicists believe that these discontinuities are caused by phase changes in mantle minerals which are changes in crystal structure that cause increases in density and in seismic velocity.

We have evidence of the mantle composition above 400 km from orogenic belts such as the Alpine-Himalayan system where mantle rocks have been tectonically uplifted and exposed at the surface by erosion. Samples of the mantle also have been carried to the surface during volcanic eruptions as solid fragments suspended in the lava. Above 400 km depth, the chief constituent mineral is olivine $(Mg,Fe)_2SiO_4$. The 400-km discontinuity is consistent with a phase change of olivine to a structure called the spinel structure in which the oxygen atoms remain in a tetrahedral arrangement around the silicon atoms but are more closely packed together (see Fig. 4.17). The structure takes its name from its similarity to the structure of the mineral spinel.

The 700-km discontinuity may represent another transition of olivine from the spinel structure to the still more dense perovskite structure. This phase transformation involves an increase in the number of oxygen atoms arranged about each silicon atom from four to six, a change from a tetrahedral to an octahedral structure. Some scientists have argued, however, that the 700-km discontinuity is the result of changes in the composition of the mantle, in which case the mantle may be compositionally stratified, with the more dense material lying stably below less dense material.

Convection in the Mantle

Our studies of plate kinematics imply that material moves down into the mantle at subduction zones and up from the mantle at spreading centers. These motions must be part of some sort of convection system involving the mantle (Fig. 4.18), presumably driven by density differences. How deep into the mantle does this convection extend? Is it limited to some upper portion of the mantle (Fig. 4.18*A*), or does it involve the entire mantle (Fig. 4.18*B*)?

Arguments have raged over whether convection in the mantle can penetrate the 700-km discontinuity. A phase change might or might not impede convection, depending on the kinetics of the change and on whether the phase change is exothermic or endothermic. A density stratification associated with a change in composition could be a barrier to through-going convection, requiring one set of convection cells in the upper mantle and another set in the lower mantle. We know that earthquakes are present to depths of 700 km in some areas where lithosphere is being subducted. Other seismic evidence indicates that a seismically higher velocity slab continues down to depths of at least 1200 km, suggesting that in some areas, convection penetrates deeper than the 700-km discontinuity.

[5] From the Greek *pan*, all, everything; and *thalassa*, sea.

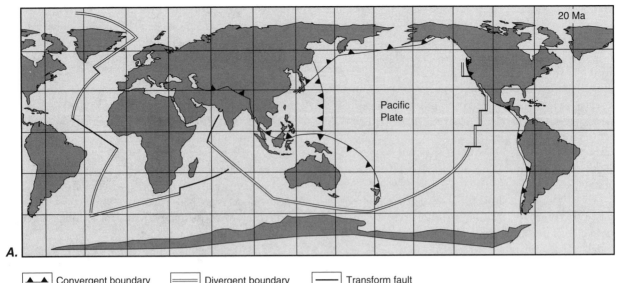

20 Ma

Pacific Plate

A.

▲▲ Convergent boundary   ═══ Divergent boundary   ── Transform fault

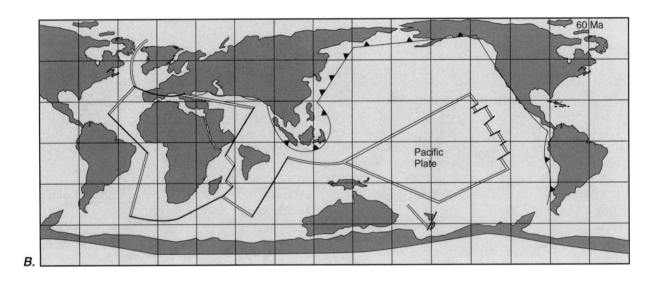

60 Ma

Pacific Plate

B.

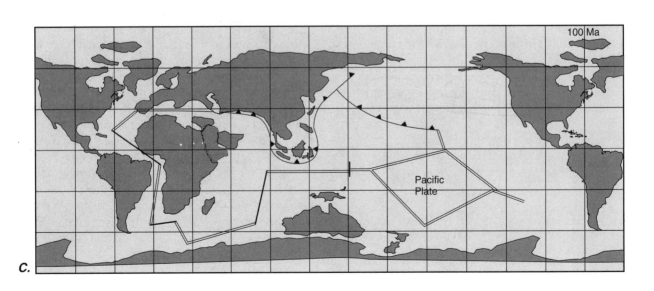

100 Ma

Pacific Plate

C.

Figure 4.16 World maps showing the possible configuration of continents at 20, 60, 100, 140, and 180 Ma, together with generalized plate margins in oceans. Double lines, divergent margins; single lines, transform faults; barbed lines, convergent margins. Note the changes in plate margins as pieces of Gondwana separate from one another and from Antarctica. Note also the growth of the Pacific Plate from a small center within an RRR triple junction to its present size. *(Data from Smith et al., 1981; Morgan, 1981; Hilde et al., 1977; Engebretson et al., 1981)*

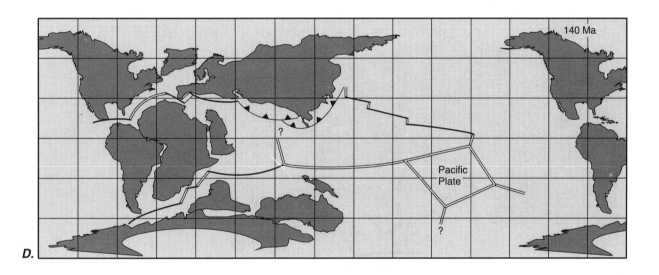

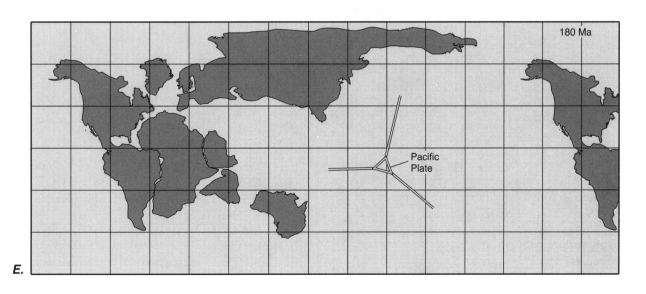

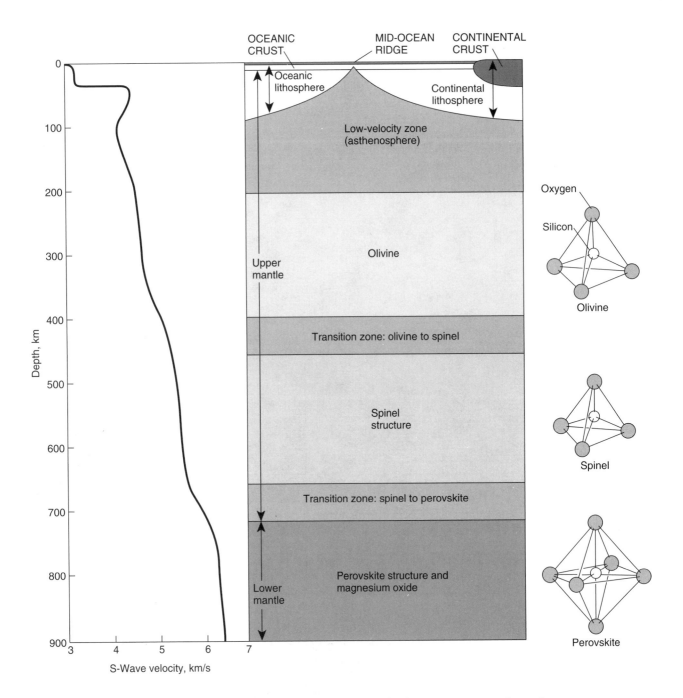

Figure 4.17 Graph of S-wave velocity (P_S) in the Earth, showing increase at base of crust; low-velocity zone; and transition zones at 400 and 700 km, thought to correspond to transitions of olivine from its low-pressure structure to the spinel structure and then to the perovskite structure, as illustrated at right. *(After McKenzie, 1983)*

Recent investigations using sophisticated seismic tomographic techniques allow the average seismic velocity within a specific restricted volume of the mantle to be determined. Mapping the mantle velocities with these techniques suggests that there are large-scale horizontal variations in seismic velocity within the mantle, some as deep as 2300 km. The

lower-mantle seismic velocity seems to be abnormally high beneath most of the Pacific rim, in the down-dip projection of many subducted slabs. This evidence suggests that the slabs penetrate aseismically into the lower mantle and that at least in places individual convection cells extend from the top to the bottom of the mantle. The existence of whole-mantle convection

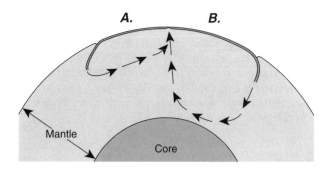

Figure 4.18 Cross section of Earth's mantle showing possible convection geometries. *A.* Shallow layer convection limited to the upper 700 km of the mantle. Such a structure would presumably require independent convection cells in the deep mantle as well. *B.* Deep convection extending all the way to the core-mantle boundary.

would support the idea that the 400-km and 700-km seismic discontinuities are probably the result of phase changes. This debate, however, is not yet resolved.

Heat Transfer and Tectonics

Convection is an efficient mechanism for transporting heat out of the interior of the Earth. The heat budget of the Earth, therefore, is the primary factor in the existence of plate tectonics and the long history of mobility of the planet's surface. Thus, the study of the heat budget can provide significant insight into tectonic activity.

The mean heat flow for all continental crust is about 56.5 mW/m² (1.35 hfu). The mean for oceanic crust is about 78.2 mW/m² (1.87 hfu). Roughly half of the heat flow from continental crust can be attributed to the heat production of radioactive minerals within the crust. In the oceanic crust, however, only about 0.5 percent of the heat flow can be attributed to radioactive decay in the basaltic crust. These numbers are important, because they imply that the average heat flux from the mantle into the continental crust is only slightly more than a third of that into the oceanic crust. Thus, the mantle beneath the continents must be substantially cooler than that beneath the ocean basins.

Within the continental crust, heat flow varies with tectonic environment. High heat flows are characteristic of active volcanic arcs and of actively extending terranes such as the Basin and Range province of the western United States. Relatively low heat flows are characteristic of old, stable continental shields. The magnitude of the heat flow also tends to decrease with increasing age of the rocks and to be proportional to the radioactive heat production of the surface rocks. Both these results are consistent with a decrease in the radioactive heat production with depth in the crust, although the radioactive heat production of continental rocks is highly variable, and such an inference is undoubtedly an oversimplification.

If the temperature at depth is calculated from a straight-line extrapolation of the crustal gradients of 20°C–30°C per km, it exceeds the melting point of mantle rocks at all depths below roughly 100 km. The fact that mantle rocks transmit seismic shear waves, however, shows that they are solid, not liquid. Thus the temperature must not exceed the melting point, and the temperature gradient must therefore decrease with depth. Figure 4.19 shows a first-order model of a temperature profile within the Earth. The curve results from assuming a conductive temperature gradient within the lithosphere and an **adiabatic temperature gradient** within the underlying mantle. An adiabatic gradient is the increase in temperature with increasing depth that a thermally insulated body would undergo simply because of an increase in pressure (one can experience an adiabatic temperature increase with a bicycle tire pump. When the air in the pump is rapidly compressed, it heats up and makes the pump warm). An adiabatic temperature profile is a reasonable approximation for a body that is convecting heat much more rapidly than it can conduct heat because under those circumstances any volume of material moving vertically, and therefore changing pressure, behaves as if it were almost perfectly insulated. Of course lateral variations in this geothermal gradient may exist, and these in fact are suggested by the variations in seismic velocity.

Radioactive elements decay into stable elements. Because there is no mechanism on Earth for producing radioactive elements from stable ones, the total amount of radioactive elements in the Earth must decrease over time. Thus, there must have been greater production of heat in the past than there is now. In fact, heat production 3 billion years ago was about twice what it is today, and early in Earth's history, radiogenic heat could have caused a period during which the Earth actually heated up and possibly melted throughout.

The decreasing heat production over time has interesting implications for Earth's tectonic evolution. If heat production has been decreasing, then the internal temperatures must also have been

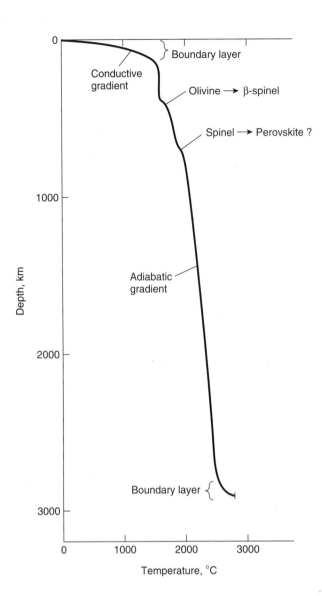

Figure 4.19 Model temperature-depth curve for the Earth's mantle. Boundary layers occur at the crust-mantle and mantle-core boundaries. The curve shows rapid increase of temperature with depth, declining to an adiabatic gradient for depths greater than about 100 km. Bumps in the curve at 400 and 700 km are for phase transitions to denser assemblages thought to be present in these regions. *(After Jeanloz and Richter, 1979)*

decreasing. Because the effective viscosity of solid-state flow is highly temperature-sensitive, however, lower temperatures mean higher viscosities and thus less vigorous convection. This creates a feedback loop, because less vigorous convection means a lower rate of heat loss, which in turn leads to higher internal temperatures and a lower viscosity. Thus, the coupling of temperature and viscosity provides a buffer to changes in both quantities. The viscosity must be sufficiently low (and thus the temperature sufficiently high) to let the excess heat be convected away, but not so low that the resulting heat loss lowers the temperature and raises the viscosity to the point that convection cannot remove the excess heat.

This feedback mechanism leads to the expectation that the temperature profile of the mantle should be self-regulated to maintain a nearly constant viscosity from top to bottom. Although we do not know enough about the deformation mechanisms through-out the mantle to use this expectation as a basis for inferring a temperature profile, it is unlikely to result in an adiabatic profile. Thus, there may be a trade-off in the actual temperature profile of the mantle between that required by the mechanics of convection and that required by the thermodynamics. Recent work indicates that the mantle viscosity probably increases slightly from top to bottom.

Despite this buffering of the mantle temperature and viscosity, mantle convection must have been more rapid and vigorous in the past because heat production, and therefore heat loss, must have been larger. This implies that the conductive temperature gradient in the thermal boundary layer (the lithosphere) must have been steeper, and thus at any given depth in the lithosphere, the temperature was higher than it is today. Higher temperatures in the lithosphere, in turn, imply that the thermal boundary layer was thinner.

What can be expected from tectonic processes that involved a hotter and thus a more ductile crust, a thinner and hotter lithosphere, and more rapid internal convection? How might such processes differ from those we observe today? These questions are difficult to answer from theoretical considerations alone, but we must be prepared to find differences in tectonic style in the ancient (for example, the Archean) geologic record and to interpret the data with models that are not applicable to the present-day Earth.

4.7 Driving Forces of Plate Tectonics

The rates of plate motion through time have not been constant. What forces cause the plates to move, and why does the rate vary through time? With the discussion of the properties of the mantle as a background, we can now consider two principal categories of model for the driving forces of plate tectonics: the mantle plume model, and models based upon force balance calculations.

The Mantle Plume Model

The mantle plume model, as originally proposed by W. J. Morgan, holds that an important source of convective heat transfer from the lower to the upper mantle is plumes, which are rising columns of hot material a few hundred kilometers in diameter that spread out into the asthenosphere like a thunderhead beneath the lithospheric plates (see Fig. 4.12). The complementary return flow would involve a uniform sinking of the entire mantle below the asthenosphere, in addition to the more localized downflow associated with subduction zones, as shown schematically in Figure 4.12.

The lateral spreading of material in the asthenosphere away from a plume produces a radial shear stress on the bottom of the overlying lithosphere. If a number of plumes are aligned, then the dominant asthenospheric flow would be laterally away from the line of plumes, and the shear stress would act to pull the lithosphere apart, creating a spreading center along the line of plumes. Indeed, as illustrated in Figure 4.13, many plumes are located on or near spreading centers, with a few notable exceptions, such as Hawaii, Yellowstone, and Macdonald.

Force-Balance Models

Although the plume model is intriguing and may account for part of the force that drives the plates, it is incomplete in that it does not consider all the possible forces that can act on a lithospheric plate. Moreover, it is possible that mantle convection can occur in patterns reflected only indirectly in plate motion. To develop a more complete understanding of the driving forces of plate tectonics, it is necessary to account for all the possible forces on a plate, as shown schematically in Figure 4.20. Models incorporating these factors are called **force-balance models**. We discuss each of the possible forces in turn.

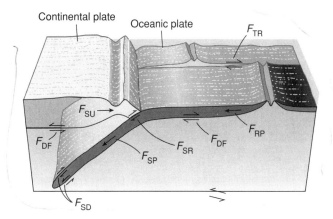

Figure 4.20 Diagram of plate boundaries with forces used in force-balance models labeled. *(After Forsyth and Uyeda, 1975)*

The **ridge-push force** (F_{RP}) represents a push from a divergent plate margin. It originates from the topographic slope of the ocean bottom created by isostatic uplift at a spreading center. At a given depth in the mantle above the level of isostatic compensation, the pressure P_1 at a depth below point 1 close to the spreading center is higher than the pressure P_2 at a point below point 2 farther away, because the weight of the overlying column of rock and water is larger (Fig. 4.21). The resulting horizontal pressure gradient provides a force that tends to drive the lithospheric plates apart at the spreading center. This force therefore depends on the average topographic slope of the ridge flank.

The **mantle-drag force** (F_{DF}) is the shear force exerted on the base of the plate by the relative motion of the underlying mantle. It reflects the viscous coupling between the asthenosphere and the lithosphere. This force could either drive the plate forward or resist its movement, depending on the direction of the mantle's relative velocity with respect to the lithosphere. In the plume model above, the mantle-drag force is

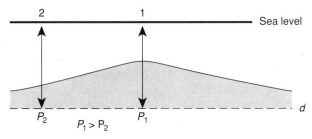

P_1 = Pressure at depth d below locality 1
P_2 = Pressure at depth d below locality 2

Figure 4.21 Cross section of a ridge illustrating the origin of the ridge-push force. Pressure at point P_1 at depth d underneath the ridge is greater than pressure at point P_2 at the same depth underneath the abyssal plane, because there is more rock and less water above P_1 than above P_2, and therefore the weight of the rock-water column is greater above P_1 than above P_2.

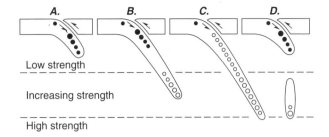

Figure 4.22 Cross sections of slabs illustrating dominant first-motion solutions for earthquakes in subducted slabs. Open circles indicate down-dip contraction. Closed circles indicate down-dip extension. Opposing arrows indicate thrust fault solutions. *A.* Slabs extending to shallow depths show down-dip extension. *B.* Slabs extending to intermediate depths show down-dip extension in the shallower part of the slab, and down-dip contraction in the deeper part. *C.* Slabs extending to great depths are dominated by down-dip contraction. *D.* Segmented slabs show down-dip extension in the shallow part and down-dip contraction in the deep part. *(After Isacks and Molnar, 1972)*

a major driving force. Many scientists accept the hypothesis that the lithosphere is thicker under continents than it is under oceans and that the asthenosphere under continents is thinner or nonexistent. If so, continents may exert a restraining drag on plate motion. The low absolute velocities exhibited by plates that are predominantly continental (see Fig. 4.15) are consistent with this hypothesis. The magnitude of the force depends on the viscosity of the asthenosphere and on the velocity of the plate relative to the mantle below the asthenosphere.

The **slab-pull force** (F_{SP}) is caused by the tendency of the colder and denser lithosphere to sink into the underlying mantle. This force should be greater for older lithosphere than for younger, because older lithosphere is colder and therefore more dense. As the slab sinks, it tends to pull the surficial part of the plate behind it. Of course, the stress in the slab is not an absolute tensile stress, but the minimum principal stress is less than the lithostatic pressure and is oriented parallel to the slab. The result is that the minimum principal *deviatoric stress,*[6] which is tensile, is parallel

to the slab dip. First-motion studies of earthquakes that occur in slabs extending to depths of less than about 200 to 300 km show predominantly down-dip extension, indicating that slab pull can be an important force (Fig. 4.22*A,D*). The magnitude of this force depends on the density difference between the slab and adjacent mantle, and on the length of slab that has been subducted.

The **slab-drag force** (F_{SD}) arises from the resistance of the mantle to the slab as it sinks. This force should be less in the asthenosphere, where the mantle viscosity is relatively low, than below the asthenosphere, where the viscosity increases. First motions showing down-dip contraction are characteristic of many deep earthquakes and of earthquakes throughout slabs that are continuous to depths of 600 to 700 km (Fig. 4.22*B,C,D*). These data suggest that the resistance to sinking is important at greater depths and that, in slabs that are continuous to these depths, it can even dominate the slab-pull stress (Fig. 4.2.2*C*). The magnitude of this force depends on the viscosity of the adjacent mantle and on the velocity of the plate being subducted.

The **transform-resistance force** (F_{TR}) is a resistance to strike-slip motion along a transform fault. The presence of seismic activity along transform faults is evidence of friction on the fault that resists the fault motion. At deeper levels, the resistance is associated with ductile shearing. Thus, the magnitude of this force depends on the effective coefficient of friction and the normal stress across the fault at shallow depths, and on the effective viscosity and the rate of shearing at greater depths.

The **subduction-resistance force** (F_{SR}) results from the shearing between an overriding plate and a downgoing plate. This force gives rise to the many shallow earthquakes that occur along subducting plate margins. The magnitude of the force presumably depends on the effective coefficient of friction between the plates.

---

[6] The deviatoric stress $\Delta\sigma_{ij}$ is the applied stress σ_{ij} minus the mean normal stress. Using the geologic sign convention (compression positive)

$$\begin{bmatrix} \Delta\sigma_{11} & \Delta\sigma_{12} & \Delta\sigma_{13} \\ \Delta\sigma_{21} & \Delta\sigma_{22} & \Delta\sigma_{23} \\ \Delta\sigma_{31} & \Delta\sigma_{32} & \Delta\sigma_{33} \end{bmatrix} =$$

$$\begin{bmatrix} \sigma_{11} & \sigma_{12} & \sigma_{13} \\ \sigma_{21} & \sigma_{22} & \sigma_{23} \\ \sigma_{31} & \sigma_{32} & \sigma_{33} \end{bmatrix} - \begin{bmatrix} \sigma_n & 0 & 0 \\ 0 & \sigma_n & 0 \\ 0 & 0 & \sigma_n \end{bmatrix}$$

where $\sigma_n = (1/3)(\sigma_{11} + \sigma_{22} + \sigma_{33})$. Thus, each normal stress component on the principal diagonal is reduced by the mean of the normal stresses, and the shear stress components remain unchanged. In two-dimensional stress, the Mohr circle for deviatoric stress is always centered on the origin. At least one of the principal deviatoric stresses must be positive and one negative.

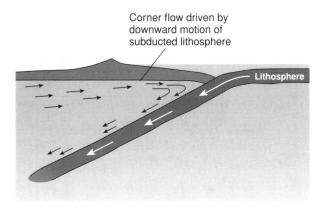

A.

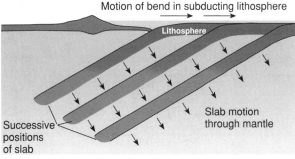

B.

Figure 4.23 Possible origins of the trench-suction force. *A.* The downgoing slab drives a corner flow in the overlying mantle, which in turn drives the overriding lithosphere toward the trench. *B.* A slab sinks through the mantle, producing a seaward motion of the bend in the plate. This sinking causes a flow of the mantle and lithosphere above the slab toward the trench, the so-called trench-suction force. *(A. From Richardson et al., 1979; B. from Zoback and Zoback, 1980)*

The **trench-suction force** (F_{SU}) tends to draw the overriding plate toward the trench. Its importance may be indicated by the observation that all circum-Pacific trenches are moving toward the ocean basin, and the Pacific basin is getting smaller. The implication of this observation is that, *above* the subducting slab, there is a flow of the mantle and lithosphere toward the trench. Such a flow could be set up as a result of the entrainment of the mantle by the downward motion of the subducted slab, which would cause a corner flow in the overlying mantle wedge (Fig. 4.23A). Less probably, this flow might compensate for the sinking of the downgoing slab in a direction perpendicular to its surface (Fig. 4.23B). This broadside sinking would produce a low pressure above the slab that would tend to "suck" the overriding plate toward the trench. It is very difficult to estimate the magnitude of this force, although presumably it is governed by the viscosity of the mantle and the rate of the flow.

These forces apparently encompass all the forces that could affect the motion of a plate. The magnitudes of the forces depend on a variety of factors, which are not the same for all forces or all plates. The factors include the slope of the plate surface, the velocity of the plate relative to the lower mantle, the plate area, the length and density of the subducted

slab, the viscosity of the mantle, and the frictional resistance to sliding. Estimates of their relative importance are based upon models of the way in which these factors govern the various forces.

Any successful model of the forces acting on the plates must account for the absolute velocities determined for the various plates. Thus, the method for determining these velocities is very important. Some calculations have used the hot-spot frame of reference described above. Others have used a reference frame in which the relative motion of the plate boundaries with respect to the lower mantle is a minimum. The rationale behind this approach is that at these boundaries, particularly the divergent and convergent ones, the coupling between the lithosphere and the lower mantle is the greatest. In an efficient plate kinematic system, the velocity of the lithosphere relative to the lower mantle should be a minimum at these boundaries. A third approach adopts a reference frame in which the net torque on the lithosphere is zero, which is equivalent to assuming that the net angular momentum of the lithosphere is constant and that, therefore, that there is no net angular acceleration of the plates.

It is not possible to calculate directly the magnitudes of all the forces acting on the plates. For example, there are no data on which to base a calculation

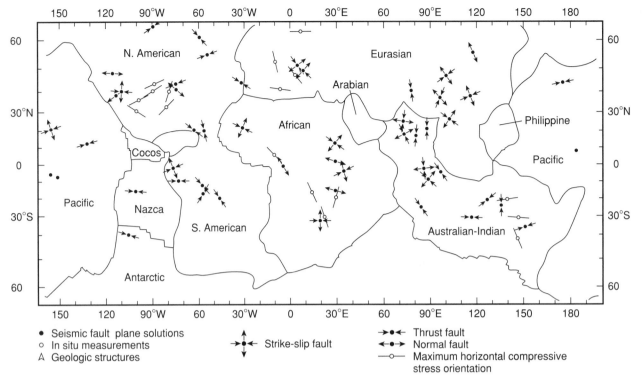

A.

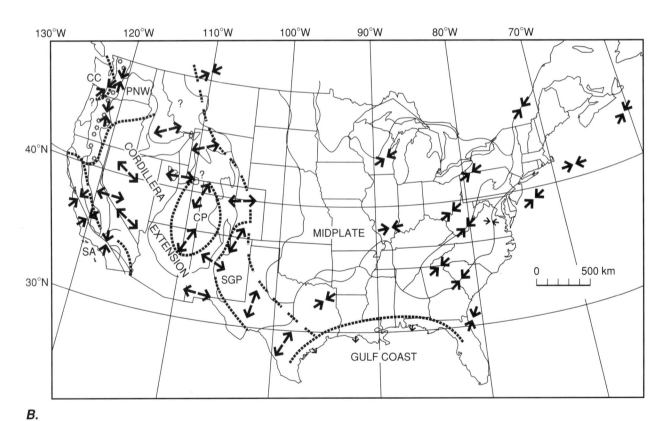

B.

Figure 4.24 In situ stress measurements. *(From Richardson et al., 1979; reprinted in Twiss and Moores, 1992)*

of the trench-suction force. Therefore, the unknown forces remain in the model as undetermined variables that can be adjusted to provide the best fit to the absolute velocities. These force-balance models also vary depending on the relative importance assigned to such factors as mantle drag, particularly on continental lithosphere, and on the values of the material constants used to describe the viscous and frictional effects. Thus, the solutions are not unique.

Most force-balance models indicate that the slab-pull and ridge-push forces are the most important ones driving the plates and that slab-pull dominates. Mantle drag may be important locally, especially as a resistance to motion under continents, but the driving force caused by mantle plumes is relatively minor. Other forces, such as subduction resistance, are important locally, but not globally.

The stresses acting on the boundaries and the bottoms of plates must determine the distribution of stress within the plates. Thus, the measurement of intraplate stresses provides another constraint on the force-balance models. It is not obvious in Figure 4.24 how these intraplate stresses may relate to plate boundaries, although there is a slight tendency for maximum compression axes to be oriented perpendicular to ridges and trenches, consistent with the ideas of ridge-push and subduction-resistance forces.

The subjects of absolute plate motions and driving mechanisms are controversial and no doubt will be the subject of much future research. It is entirely possible that the real situation is quite different from our current ideas.

Additional Readings

Chase, C. G. 1978. Plate kinematics: The Americas, east Africa, and the rest of the world. *Earth. Plan. Sci. Lett.* 37:355–368.

Cox, A. 1972. *Plate Tectonics and Geomagnetic Reversals.* San Francisco: W. H. Freeman and Company.

Cox, A., and R. B. Hart. 1986. *Plate Tectonics: How It Works.* Palo Alto: Blackwell's.

Cronin, V. S. 1987. Cycloid kinematics of relative plate motion. *Geology* 15:1006–1009.

Cronin, V. S. 1991. The cycloid relative-motion model and the kinematics of transform faulting. *Tectonophysics* 187:215–249.

Cronin, V. S. 1992. Types of kinematic stability of triple junctions. *Tectonophysics* 207:287–301.

DeMets, C. 1993. Earthquake slip vectors and estimates of present-day plate motions. *J. Geophys. Research* 98: 6703–6714.

Demets, C., R. G. Gordon, D. F. Argus, and S. Stein. 1990. Current plate motions. *Geophys. J. International* 101: 425–478.

Dewey, J. F. 1975. Finite plate implications: Some implications for the evolution of rock masses at plate margins. *Am. J. Sci.* 275A:260–284.

Dziewonski, A., et al. 1987. A window into the lower mantle. *Science* 236:41–45.

Engebretson, D. C., A. Cox, and R. G. Gordon. 1985. Relative motions between oceanic and continental plates in the Pacific basin. GSA Special Paper 206.

Engebretson, D. C., K. P. Kelley, H. J. Cashman, and M. A. Richards. 1992. 180 Million Years of Subduction. *GSA Today* 2(6):93–95, 100.

Hilde, T. W., S. Uyeda, and L. Kroenke. 1977. Evolution of the western Pacific and its margin. *Tectonophysics* 38:145–165.

Isacks, B., and P. Molnar. 1972. Mantle earthquake mechanisms and the sinking of the lithosphere. From *Nature* 223:1121–1124, reprinted in A. Cox, Ed., *Plate Tectonics and Geomagnetic Reversals.* San Francisco: W. H. Freeman and Company, p. 401–406.

Jeanloz, R., and F. Richter. 1979. Convection, composition and the thermal state of the lower mantle. *J. Geophys. Res.* 81:5497–5504.

Larson, R. L., and W. C. Pitman III. 1972. World-wide correlation of Mesozoic magnetic anomalies and its implications. *GSA Bull.* 83:3645–3662.

Le Pichon, X., and J. Francheteau. 1976. *Plate Tectonics.* New York: Elsevier.

McKenzie, D. P. 1983. The Earth's mantle. In The dynamic Earth. R. Siever, ed. *Scientific American.* September.

Minster, J. B., and T. H. Jordan. 1978. Present-day plate motions. *J. Geophys. Res.* 83:5331–5354.

Minster, J. B., T. H. Jordan, P. Molnar, and E. Haines. 1974. Numerical modeling of instantaneous plate tectonics. *Geophys. J. Roy. Astron. Soc.* 36:541–576.

Morgan, W. J. 1968. Rises, trenches, great faults, and crustal blocks. *J. Geophys. Res.* 73:1959–1982.

Morgan, W. J. 1972. Plate motion and deep mantle convection. *GSA Mem.* 132:7–22.

Patriat, P., and V. Courtillot. 1984. On the stability of triple junctions and its relationship to episodicity in spreading. *Tectonics* 3:317–332.

Pitman, W. C., III, and M. Talwani. 1972. Sea-floor spreading in the North Atlantic. *GSA Bull.* 83: 619–646.

Press, F., and R. Siever. 1986. *Earth.* 4th ed. New York: W. H. Freeman and Company.

Uyeda, S. 1978. *The New View of the Earth.* San Francisco: W. H. Freeman and Company.

5 Divergent Margins and Rifting

5.1 Introduction

Divergent margins include the mid-oceanic rift system and those continental regions of extension that appear to be continuations of the oceanic rift system. In this chapter, we examine the structural characteristics of divergent margins and the evidence pertaining to their development.

The mid-oceanic rift system is one of the largest structural features on Earth, with a total length of some 40,000 km (Fig. 5.1). To become familiar with its geography, we trace the system over the Earth's surface starting at one end of the main rift and proceeding to the other. We begin in the northeast Pacific Ocean just west of Canada, where a small ridge segment, the Explorer Ridge (E in Fig. 5.1) abuts the southeast end of the Queen Charlotte Islands transform fault (QC). The Explorer Ridge extends southwest and is transformed successively into the Juan de Fuca (JF) and Gorda (G) ridges. The latter ends against the Mendocino transform fault (M), which joins the San Andreas fault (SA) at the Mendocino transform-transform-trench triple junction (MTJ). Together, these two transform faults offset the ridge system approximately 3000 km southeast to the mouth of the Gulf of California, where the Rivera ridge-trench-transform triple junction (RTJ) is located.

Southwest of this triple junction, the East Pacific Rise extends for about two thousand kilometers, offset by a number of transform faults, to the Galapagos ridge-ridge-ridge triple junction (GTJ). From this junction, the Galapagos Rift (Ga) extends eastward until it connects with a transform fault that joins it with the Middle America Trench. South of the Galapagos triple junction, the East Pacific Rise continues to the Chile ridge-ridge-transform triple junction (CTJ) connecting with the Chile Rise. Southwest of this junction, the Pacific-Antarctic Ridge (PAC-ANT R) separates the Antarctic plate to the south from New Zealand and the Indian-Australia plate to the north. It becomes the Southeast Indian Ridge (SE IND R) south of Australia, which leads to another ridge-ridge-ridge triple junction, the Indian triple junction (ITJ), in the central Indian Ocean. From this junction, a branch of the rift system, the Carlsberg Ridge (C), extends northwest into the Gulf of Aden to the Afar ridge-ridge-ridge triple junction (ATJ), from which extend the Red Sea (RS) and East African (EA) rifts.

From the Indian triple junction (ITJ), the Southwest Indian Ridge (SW IND R) extends in a series of short ridge and long transform fault segments around the southern tip of Africa into the South Atlantic Ocean to the Bouvet ridge-transform-

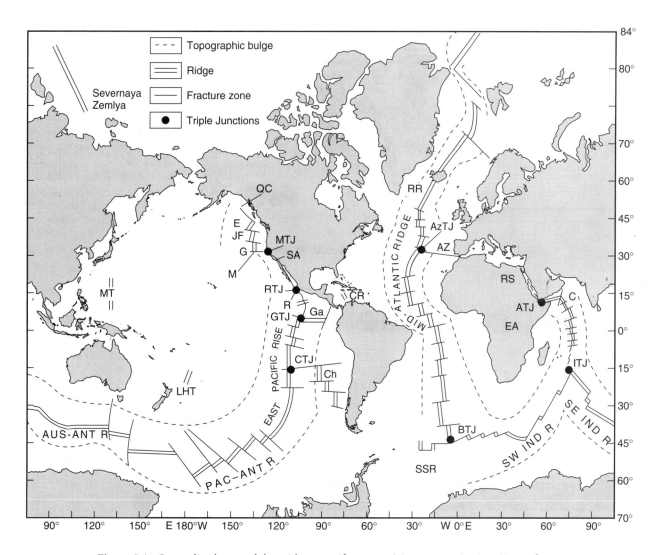

Figure 5.1 Generalized map of the mid-ocean rift system. Mercator projection. Key to letter symbols for ridges and transform faults: AZ, Azores; C, Carlsberg; Ch, S Chile Ridge; CR, Caribbean Plate; CT, Cayman Trough; E, Explorer; EA, East African Rift; G, Gorda; Ga, Galapagos; J, Juan de Fuca; LHT, Lau-Howe Trough; M, Mendocino; MT, Mariana Trough; QC, Queen Charlotte Islands; RR, Reykjanes Ridge; RS, Red Sea; SA, San Andreas; SSR, South Sandwich Rift; PAC-ANT R, Pacific-Antarctic Ridge; SE IND R, SE Indian Ridge; SW IND R, SW Indian Ridge; AUS-ANT R, Australia-Antarctic Ridge. Triple junctions: ATJ, Afar; AzTJ, Azores; BTJ, Bouvet; CTJ, Chile; GTJ, Galapagos; ITJ, Indian Ocean; MTJ, Mendocino; RTJ, Rivera. *(After Menard and Chase, 1970)*

transform triple junction (BTJ). A ridge system extends into the consuming margin of the South Scotia arc. North of this junction, the Mid-Atlantic Ridge extends northward through a series of transform fault offsets near the equator to the Azores triple junction (AzTJ). From this junction, the Azores fracture zone (AZ) extends through the Straits of Gibraltar into the Mediterranean. The main ridge system extends northward along the Reykjanes Ridge

(RR) south of Iceland, through the northern Atlantic into the Arctic Ocean, where it finally ends against a complex fault system near Severnaya Zemlya.

In oceans bordered by passive (Atlantic-style) continental margins, the ridge is located in the middle of the ocean basin and roughly parallels the margins of the continents on either side of it. Thus, throughout most of its length, the ridge system lies roughly along the median line of the local ocean basin. The

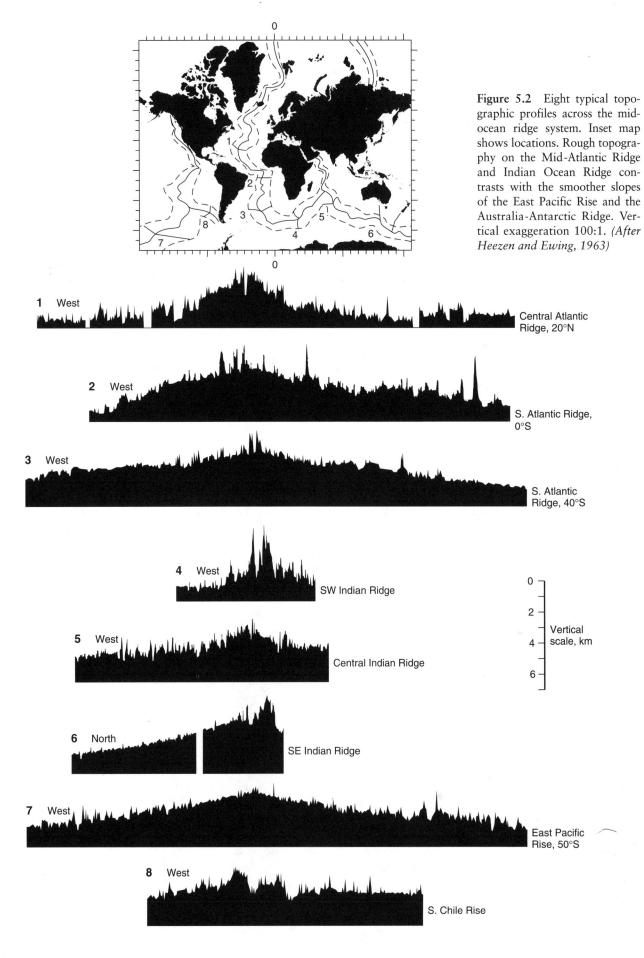

Figure 5.2 Eight typical topographic profiles across the mid-ocean ridge system. Inset map shows locations. Rough topography on the Mid-Atlantic Ridge and Indian Ocean Ridge contrasts with the smoother slopes of the East Pacific Rise and the Australia-Antarctic Ridge. Vertical exaggeration 100:1. *(After Heezen and Ewing, 1963)*

1 West — Central Atlantic Ridge, 20°N

2 West — S. Atlantic Ridge, 0°S

3 West — S. Atlantic Ridge, 40°S

4 West — SW Indian Ridge

5 West — Central Indian Ridge

Vertical scale, km

6 North — SE Indian Ridge

7 West — East Pacific Rise, 50°S

8 West — S. Chile Rise

exceptions are portions of the Pacific and Arctic oceans. In the Pacific, involvement of oceanic lithosphere with consuming margins along the eastern Pacific margin has resulted in the migration of the ridge into the eastern part of the ocean basin. In the Arctic, the ridge currently is located in the eastern part of the ocean basin.

Topographically, the ridge stands approximately 2 to 2.5 km above the flanking abyssal plains (Fig. 5.2). The height of the ridge at any particular location depends upon the age of the lithosphere. Thus, the width of the topographic bulge varies depending upon the spreading rate. The topography is highest near the crest, and the elevation decreases with the square root of age ($t^{1/2}$), reaching abyssal depths at ages of about 80 Ma (Fig. 5.3). The age-depth relationships shown in Figure 5.3 can be explained by conductive cooling and consequent thermal contraction of the lithospheric plate, which we discuss in more detail in Section 5.7.

As mentioned above, the main rift system has several subsidiary branches that are connected to the main system at triple junctions. These branches—which include the Galapagos Rise, the South Chile Rise (Fig. 5.2, diagram 8), the Carlsberg Ridge, the Red Sea Rift, and the Azores Rift—display topography similar to the main ridges, but the simple relationships shown in Figure 5.3 are obscured by abundant transform faults or by nearness to continental margins. In two regions, the rift system extends into continents: the Basin and Range province of North America and the East African Rift system.

In addition, there are several small ridge segments that are not directly connected to the main ridge system. Chief among these are back-arc spreading ridges associated with consuming margins—such as the Mariana Trough (MT), the Lau-Havre Trough (LHT), and the South Scotia Ridge (SSR). In addition, the Cayman Ridge (CR) is a short ridge segment associated with a transform fault boundary between the Caribbean and North American plates.

We begin our discussion of rift structure by considering these areas of continental rifting, because it has been possible to investigate them in more detail and because they are useful analogues for the structure of less accessible oceanic rifts. Our interest in using continental geology to understand pre-Mesozoic plate tectonics also makes modern continental rifts an important source of information. We then discuss other parts of the rift system, proceeding from the young, narrow ocean basins, exemplified by the Red Sea, to the more mature basins, looking first at the rifted continental margins and then the ocean floor itself. In this way, we can trace the evolution of

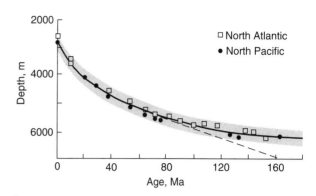

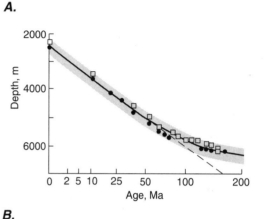

A.

B.

Figure 5.3 Relationship between elevation of the ridges and age for the North Pacific and the North Atlantic. A. Plot of depth versus age showing mean depth and age for the two oceans. The shaded area shows an estimate of the scatter in the original points used to produce the curve. The solid line shows the theoretical topography, assuming that the lithosphere reaches a limiting thickness at about 80 Ma. The dashed line shows the theoretical topography for an ever-thickening lithosphere. B. Plot of depth versus square root of age, illustrating the dependence of elevation on the square root of age ($t^{1/2}$) up to ages of about $t = 80$ Ma. The dashed line shows the extrapolation of the $t^{1/2}$ relationship, which would apply if the plate cooled and thickened indefinitely. *(After Sclater et al., 1980)*

rift features from intracontinental rifting through the initial stages of ocean basin development to mature ocean basins bordered by rifted continental margins. The sequence outlines a coherent model of the rifting process based on presently active examples of all stages of rift development. With such a model, we can interpret the data pertaining to two ancient rifted margins that are found in the geologic record and that have been modified by subsequent tectonic events, one on the east coast of North America and the other on the west coast.

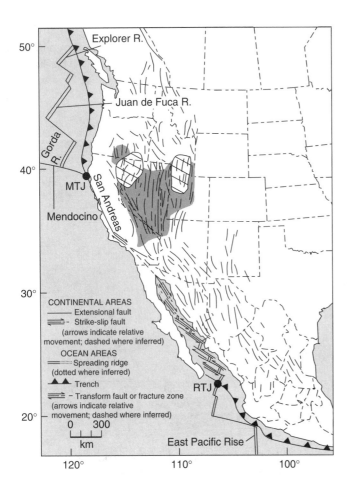

Figure 5.4 Map of western North America schematically showing major normal faults of the Basin and Range province. Cross-ruled areas are relatively low, less than 1500 m in elevation. The shaded area is the Great Basin, a region of internal drainage. *(After Stewart, 1980, 1978; Mayer, 1986)*

5.2 Continental Rift Zones

The Basin and Range Province of North America

The Basin and Range province (Fig. 5.4) extends from northern Mexico through the United States and into Canada, a distance of approximately 3000 km. Its width reaches a maximum of approximately 1000 km in the western United States, but it is narrower at both north and south ends. This major structural province consists of a series of alternating ranges and basins, elongate in a northwest or a north-northeast direction. The province is part of a broad, topographically high region, whose average elevation is 1 to 2 km above sea level. A large portion of the province within the United States is a region of internal drainage, the so-called Great Basin, that includes most of Nevada and adjacent areas of Utah, California, and Oregon (shaded area in Fig. 5.4). Within the Great Basin, topographic depressions on the west and east (diago-

nal ruling in Fig. 5.4) flank a region of relatively high elevation in the center.

Figure 5.4 also illustrates the current plate tectonic setting of the Basin and Range province. The southern part of the province lies east of the north end of the East Pacific Rise, where it enters the Gulf of California and is offset by the San Andreas transform fault system. The northern part of the province lies east of a subduction zone located off the northwest continental margin of the United States and is associated with the volcanoes of the Cascade Range.

The dominant structure of the Basin and Range province consists of a series of normal fault blocks, most of which are tilted. The valleys are mostly grabens or half-grabens, and the ranges are horsts or the high ends of tilted fault blocks. Most ranges are bounded by normal faults, many of which dip steeply at the surface but have a listric geometry overall. Several faults extend deep into the continen-

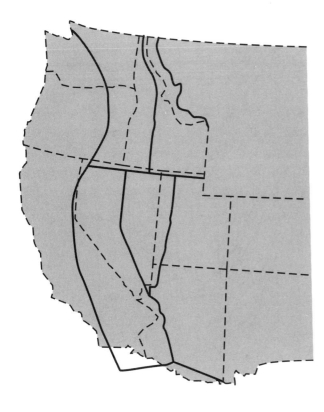

Figure 5.5 Diagram of western United States illustrating extension of the Basin and Range province. Solid lines indicate possible approximate locations of state boundaries in Eocene time, prior to Basin and Range extension. Dashed lines indicate modern state boundaries. *(After Hamilton, 1978)*

tal crust; others are more surficial. The displacement on individual range-bounding faults is as much as 8 to 10 km, and displacements as large as 15 to 20 km have been proposed for some listric faults that involve more than one range.

Estimates of the total amount of displacement on the faults across the entire Basin and Range province range from 60 km to more than 300 km. Estimates have tended to increase as the true extent of listric faulting becomes better documented. Thus, the structural evidence indicates that the Basin and Range province is an area where the continental crust is extending, and it may have extended to as much as twice its original width. Figure 5.5 shows an estimate of the outlines of the boundaries of the western states before this extension occurred.

Figure 5.6 Focal mechanism (first-motion) solutions of intraplate earthquakes and regional distribution of heat flow values for the western United States. *(After G. P. Eaton, 1980)*

Geophysical evidence supports the structural conclusion that the Basin and Range province is a region undergoing extension. Heat flow in the region is high, as much as three times normal for continental areas (Fig. 5.6). Most seismic activity in historical time has been concentrated in two north-trending zones along the western and eastern margins of the province. Earthquake first-motion studies yield predominantly normal fault solutions, except in the southwestern marginal area, where strike-slip solutions are also present. Throughout the province, extension axes from first-motion solutions tend to be oriented east-west. Seismic refraction results reveal a thin crust overlying a low-velocity mantle and show little or no evidence of a high-velocity lithosphere immediately below the crust (Fig. 5.7A). The thinnest parts of the crust (Fig. 5.7B and 5.7C) lie along the zones of active seismicity (see Fig. 5.6) and correspond with the two topographic lows in the Great Basin (see Fig. 5.4).

Thus, the geophysical, structural, and topographic evidence all indicate that the Basin and Range province is currently undergoing extension. Earliest extension in a roughly northeast-southwest direction began at about 20 Ma. The current phase of east-west extension began at approximately 6 to 10 Ma

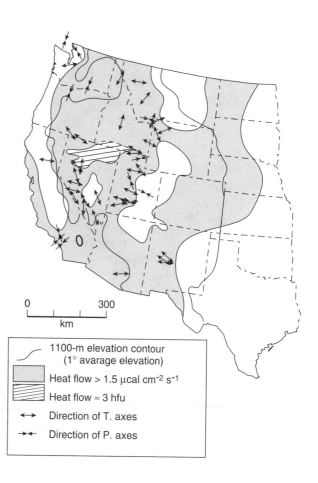

0   300

km

    1100-m elevation contour
    (1° avarage elevation)

    Heat flow > 1.5 μcal cm⁻² s⁻¹

    Heat flow ≈ 3 hfu

↔ Direction of T. axes

→← Direction of P. axes

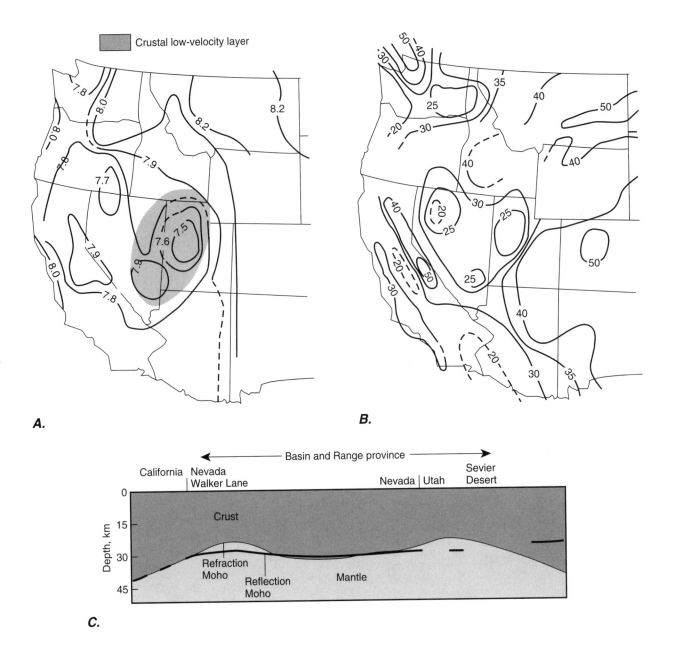

Figure 5.7 Crust and mantle characteristics of the Basin and Range province. *A.* Map of sub-Moho mantle P-wave velocity V$_P$. The shaded area shows where there is a low-seismic-velocity layer within the crust, and the velocities below the Moho beneath this area are also abnormally low. *B.* Map of crustal thicknesses in the western United States. Contours are drawn on the Moho. *C.* Profile of the crustal structure across the Basin and Range province. The cause of the difference between the seismic reflection Moho and seismic refraction Moho is unknown. Vertical exaggeration ≈ 4.5:1. (*A. and B. after Smith, 1978; C. after Allmendinger et al., 1987*)

in the Great Basin, but possibly earlier to the south. The data suggest an overall extension rate of 1–5 cm/yr for the Great Basin region, compared with a rate on mid-ocean ridges of 2–20 cm/yr.

Rocks forming today in the Basin and Range province reflect the tectonic setting. The sharp local relief and dry climate combine to produce coarse alluvial fan and landslide deposits bordering the

Figure 5.8 Complete cross sections accounting for the extension in provinces of normal faulting, such as the Basin and Range province. *A.* The shallower crust extends by brittle normal faulting. The deeper crust extends and thins by ductile deformation. The extension is accommodated in the mantle by the convective inflow of material. The province of shallow normal faulting is symmetrical. *B.* Extension occurs by displacement along a normal detachment fault that extends completely through the crust. The brittle shallow crust extends by brittle imbricate listric normal faulting. Faulting on the detachment at depth is by ductile shear. The extension is accommodated in the mantle by the convective inflow of material. The province of shallow normal faulting is asymmetrical. *(After Lister and Davis, 1989)*

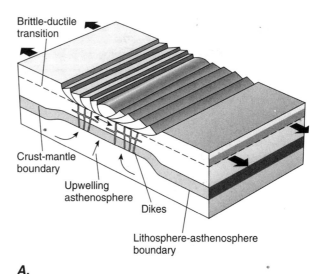

A.

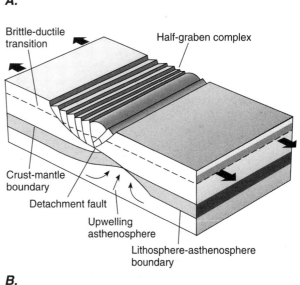

B.

ranges. Away from the ranges, these deposits become finer grained and interfinger with lake sediments. The lowest part of nearly every valley in the Great Basin is occupied by a dry lake or **playa,** which contains water only during wet seasons. Saline deposits that form in these environments include abundant halite and gypsum and less abundant salts of strontium or boron. The latter are economically important.

Most recent volcanic rocks in the Basin and Range province display a distinctly bimodal distribution in silica content, consisting of dominant basalts and subordinate rhyolite. These recent volcanic rocks contrast strongly with older Tertiary deposits that are intermediate in composition and that may reflect subduction processes along the Pacific coast (see Chapter 7).

Taken together, the geology and structure of the Basin and Range province suggest one of the two crustal models illustrated in Figure 5.8. These models have several features in common. They indicate a thinned continental crust including Tertiary sediments and volcanic rocks overlying a sequence of deformed Paleozoic-Mesozoic sedimentary and volcanic rocks and, in the eastern and southeastern part of the province, Precambrian crystalline rocks. Upper crustal rocks are brittlely deformed by listric normal faults that flatten downward and converge in a gently dipping major mylonitic zone. The lower crust is made up of strongly metamorphosed Paleozoic-Mesozoic and older rocks. Whether lower crustal rocks have extended by uniform stretching and by intrusion of dikes (Fig. 5.8A), or by shear on a low-angle fault that cuts through the lithosphere (Fig. 5.8B) is currently much debated. The lithosphere may include a thin layer of mantle material, or the lithospheric mantle may be completely missing and the asthenosphere may be directly below the crust

(Fig. 5.8A). In this upper mantle material, extension is accommodated by intrusion and ductile stretching. The amount of stretching is greatest in the regions where the crust is thinnest.

Africa and the East African Rift

Africa and its margins exhibit a number of features that illustrate various stages in the rifting of continents. The entire continent itself stands high above sea level. Its average elevation of 1000 m is greater than that of any other continent except Asia. Much of the African crust is characterized by a series of structural domes and basins, which are tens to hundreds of kilometers in dimension (Fig. 5.9). Two basins, the Chad and Congo, are as large as the Illinois or Michigan basins (see Chapter 3). Arrayed around these basins are a series of uplifts of Jurassic

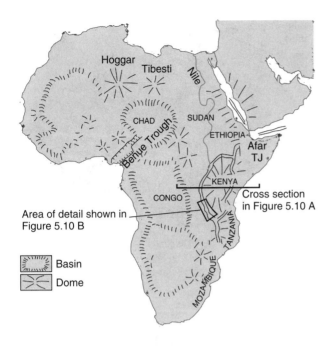

Figure 5.9 Map of Africa showing domes and basins, the East African Rift, the Benue Trough, and other adjacent tectonic features. The location of the cross section and detail that are shown in Figures 5.10A and 5.10B are also labeled. Afar TJ = Afar triple junction. *(After Clifford and Gass, 1970; Burke and Whiteman, 1972, 1973)*

gion of the continent that apparently is underlain by thin crust and thin lithosphere, as shown in the model in Figure 5.10A. Faulting seems to be concentrated along the two rift valleys.

The area bounded by the two rifts is approximately equivalent to that of the Basin and Range province. Although faulting in the Basin and Range is apparently more complex, the two areas are similar in that seismic activity in the Basin and Range is concentrated along the east and west margins of the region (see Fig. 5.6), and the thin lithosphere and crust are very similar to that in Africa (compare Fig. 5.7C with Fig. 5.10A).

Geologic information suggests that the rift began as early as the Eocene or even the Jurassic but that the current phase of extension began in the mid-Miocene, similar in time to the inception of the Basin and Range faulting. The tectonic activity included doming, development of the rifts, and volcanism early in the process. Stratigraphic evidence suggests that the rifting and doming occurred concurrently and that the floors of the rift valleys have been subsiding since late Miocene time.

Rocks deposited in the East African Rift are reminiscent of those associated with the Basin and Range province. Marginal fanglomerates give way in the center of the valleys to saline lake deposits and fluviatile sediments. Active volcanoes also characterize the region, principally the eastern rift; they are dominantly alkaline or peralkaline in character, which means they are of low silica and high alkali (K, Na) compositions.

Recent work in Lake Tanganyika in the western rift (see Fig. 5.9) has shown that its structure may be characterized by a series of alternating and in part overlapping half-grabens. This structure, shown in the diagrams in Figures 5.10B and 5.10C, suggests that the faults associated with the rift are arcuate in plan view and listric in three dimensions. The area between overlapping fault segments is a topographic ridge that appears on a cross section as a horst (section A–A′, Fig. 5.10B), or a dome within a graben (section B–B′, Fig. 5.10B), depending upon whether the faults in the overlap zone dip away from or toward each other.

Geophysical and structural data indicate that this region is actively undergoing extension. The total amount of extension is not great, less than 50 km, indicating a rate of extension of 0.5 cm/yr or less. The absolute angular velocity vectors for Africa and Somalia are not very different (see Table 4.3), so that the difference between them, which gives the relative angular velocity (see Eq. 4.9), is small.

At present, it is not clear whether the Basin and Range province and the East African Rift are strictly

to Neogene age that form domelike swells in the erosional surface cut mainly into Precambrian rocks. These domes in the continental surface are approximately 1 km high, 50 to 200 km wide, and 100 to 500 km long. Recent alkalic (high K, low Si) volcanism is associated with the domes. Several domes exhibit a rift structure with two or three grabens that intersect at the crest of the dome. The most prominent domes are the Tibesti and Hoggar massifs in the Sahara Desert and a series of domes aligned along the eastern part of the continent, from the Red Sea south to Tanzania.

The East African Rift extends along the axis connecting these East African domes, from the Afar triple junction south for approximately 3000 km through the eastern part of Africa. This area is the second large region, in addition to the North American Basin and Range province, where the world rift system extends into a continent, and it has been called an incipient plate boundary. Southwest of the Afar triple junction, the rift divides into two branches. The eastern branch is more seismically active than the western branch, although both show evidence of recent faulting. Both rifts are part of an uplifted re-

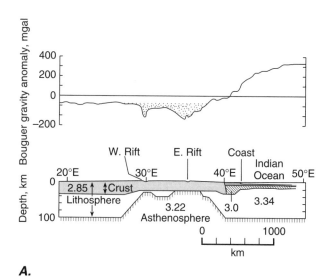

A.

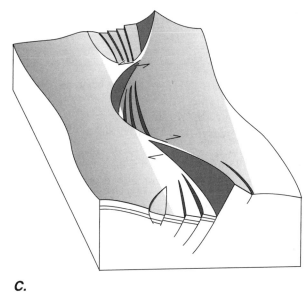

C.

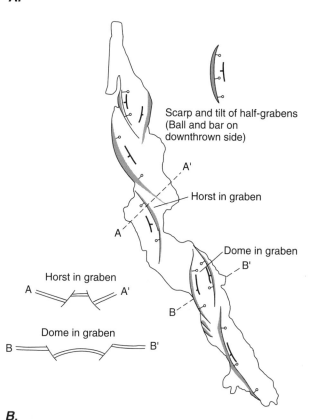

B.

Figure 5.10 Details of the East African Rift. *A.* Cross section of eastern Africa showing the Bouguer gravity anomaly and a crustal and upper mantle model based upon gravity and seismic information. Vertical exaggeration 5:1. *B.* Map and selected cross sections of Lake Tanganyika (See Fig. 5.9) showing alternating and overlapping half-grabens. Cross section A–A′ shows a central horst bounded by two normal faults. Cross section B–B′ shows a dome developed in a graben. Not to scale. *C.* Three-dimensional sketch of the geometry of faulting shown in part *B.* (*A. after Fairhead, 1986; B. after Rosendahl, 1986; C. after Rosendahl, 1986*)

comparable. The Basin and Range province displays a more uniformly complex structure; its extension is oblique, whereas that in East Africa is approximately perpendicular to the trend of the province. The Basin and Range province may be simply a later stage in the development of an area such as that between the east and west branches of the East African Rift, or its history may be unique, possibly influenced by subduction along the west coast of North America and the oblique extension.

5.3 A Young Ocean Basin: The Red Sea

The Red Sea is a narrow ocean basin separating Africa from Arabia. It is approximately 3000 km long and about 100 to 300 km wide (Figs. 5.9 and 5.11A). The margins of the Red Sea are steep fault scarps, as much as 3 km high, that rise sharply from the coast. Elevation generally is higher on the eastern than the western side. Most of the seafloor is shal-

low; only the south-central portion of the sea approaches abyssal depths (Fig. 5.11A). The shallow parts of the Red Sea appear to be floored by continental crust thinner than that of the flanking continental areas. The thinning apparently has occurred on a series of listric normal faults (Fig. 5.11B). This thinned continental crust is overlain by a layer of salt, chiefly halite, as much as 1 km thick.

Stratigraphic evidence indicates that the opening of the Red Sea began in early Tertiary time with regional uplift accompanied by subsidence of the crust

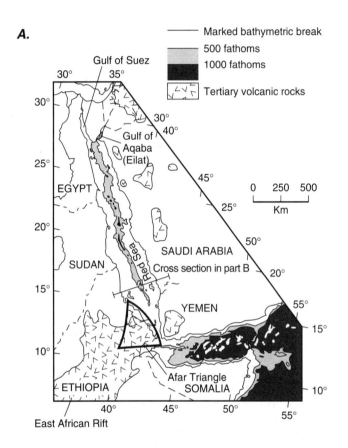

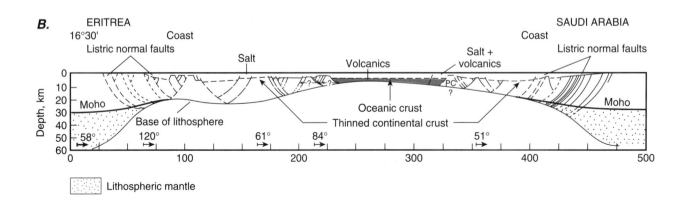

Figure 5.11 Structure of the Red Sea. *A.* Map of the Red Sea and neighboring region. *B.* Cross section of the southern Red Sea showing listric normal faults, the base of the crust, and the base of the lithosphere. *(A. after Lowell and Genik, 1974; Dixon et al., 1989; B. after Lowell and Genik, 1974)*

along an axial rift zone. The zone of maximum uplift and volcanic activity generally is east of the Red Sea, in the Arabian platform. Igneous activity began early in the uplift and rift regions. Rocks include basaltic dikes and flows, the latest of which are associated with fissures approximately parallel to the present rift. Thick basaltic sills are found on land along the margins of the rift, where they intrude Precambrian crustal rocks and in some places comprise a high percentage of the rift-margin rocks. The sills can be traced toward the sea where they extend underneath the salt deposits. Compositionally, these basalts are similar to those being erupted along the active central axis of the rift.

In the central part of the southern Red Sea, the bottom reaches truly oceanic depths (5 km. or so; see Fig. 5.11A). Seismic evidence suggests that the crust here is of oceanic thickness. Magnetic surveys in this central region reveal the presence of symmetrical magnetic anomalies, which indicates that typical ocean spreading has been taking place for the past 5 million years. Thus, the Red Sea provides an example of an ocean basin that has newly formed and has only recently, in geologic terms, evolved past the initial stage of continental rifting.

The structure of the transition from continental to oceanic crust observed in the Red Sea (see Fig. 5.11B) provides a model for understanding the evolution of other continental margins, which we apply in our discussion in Section 5.6.

5.4 Mature Rifted Continental Margins: Passive, or Atlantic-Style Margins

So far, we have discussed the rifting of continents from the intracontinental rift stage (the Basin and Range province and the East African Rift) through the new ocean stage (the Red Sea). With the addition of new material to the plate at the spreading center, the rifted continental margin becomes part of the plate interior. The underlying lithosphere cools, contracts, and subsides isostatically. As the continental margin subsides with it, a thick sequence of shallow-water marine sediments builds up on top of the rifted margin.

Examples of this type of margin include the present-day Atlantic and Gulf Coast margins of North America and the Atlantic margins of Europe and Africa. Figure 5.12 shows these margins essentially at the Red Sea stage of their development. The present geography, bathymetry, and structural cross sec-

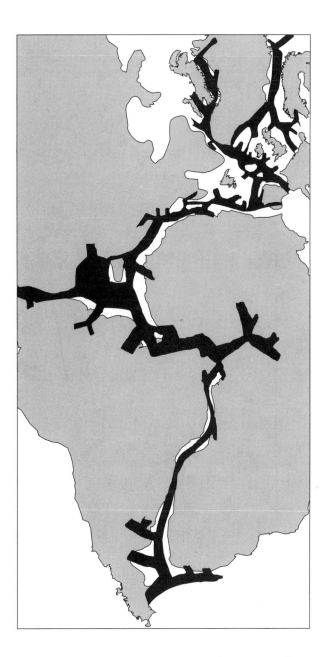

Figure 5.12 Prerifting reconstruction of continents about the Atlantic, showing development of subsidiary rift valleys. *(After Burke, 1980)*

tions of the northeastern continental margin on the North American Plate and the northwestern continental margin on the African Plate are shown in Figure 5.13.

Several features typify the Atlantic style of continental margin. The sequence of shallow-water marine sediments typically accumulates in areas underlain by older, commonly crystalline continental

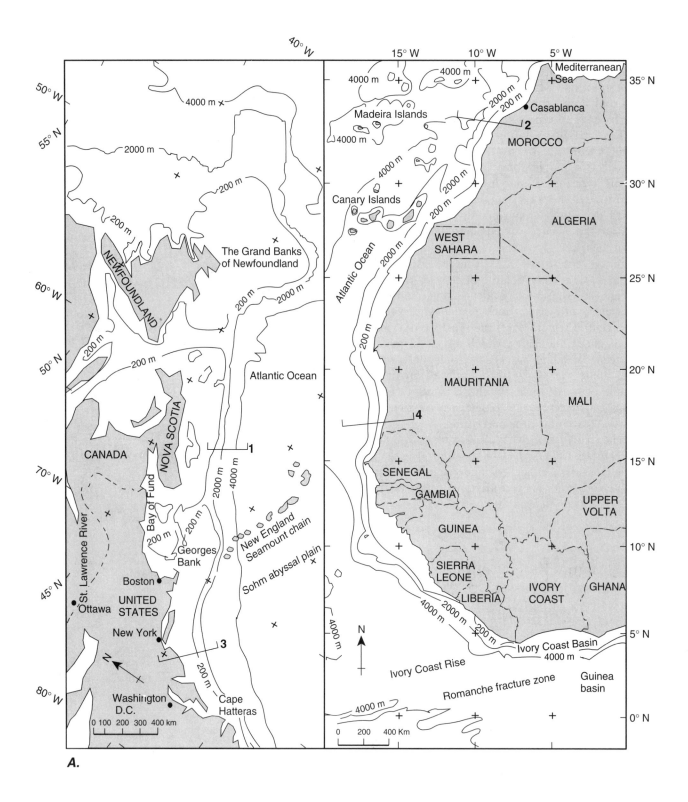

Figure 5.13 The North American and African continental margins. *A.* Geography and bathymetry of the formerly adjacent sections of the North American and African continental margins (compare Fig. 51.2) Numbers indicate the location of cross section in part *B.* *B.* Cross sections of the North American continental margin (1, 3) and the West African continental margin (2, 4), all at the same scale. Vertical exaggeration 5:1. *(After Schlee, 1980)*

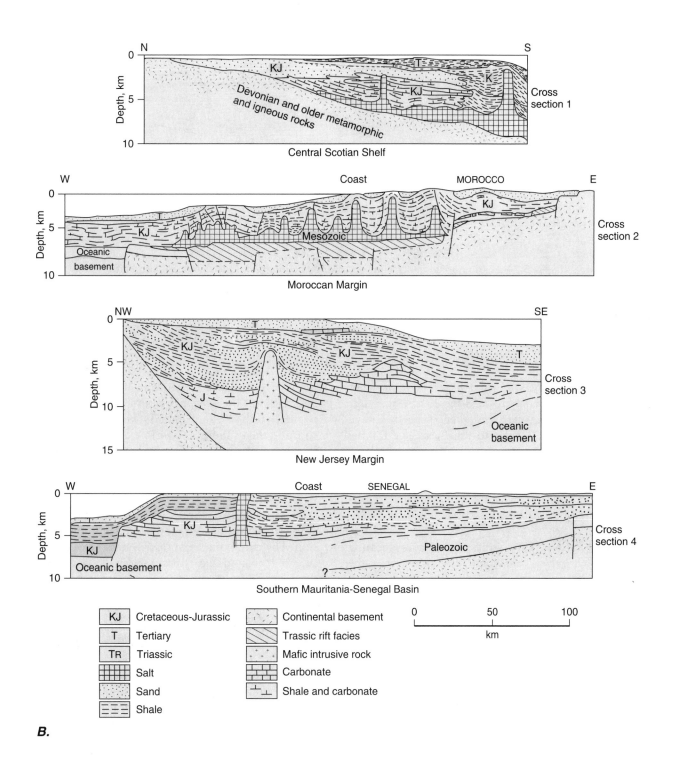

KJ Cretaceous-Jurassic

T Tertiary

TR Triassic

Salt

Sand

Shale

Continental basement

Trassic rift facies

Mafic intrusive rock

Carbonate

Shale and carbonate

0   50   100

km

B.

basement (Fig. 5.13B), and the sediments are generally deposited directly on the basement, forming a profound unconformity. This continental basement, originally of normal thickness, is thinned toward the margin as a result of extensive listric normal faulting, as exemplified by seismically controlled sections in the Bay of Biscay (Fig. 5.14). In a number of regions, fault-bounded basins of continental deposits are present atop the crystalline basement. Along the eastern North American continental margin, these basins contain Triassic-Jurassic sediments. Basaltic rocks interfinger with the upper stratigraphic levels of these rift valley sediments. All these rocks in turn were tilted and eroded prior to deposition of the first shallow-water marine deposits, indicating continued faulting on listric faults.

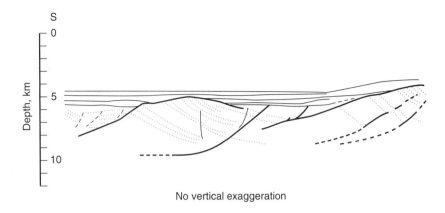

S

Depth, km

0

5

10

No vertical exaggeration

Figure 5.14 Listric normal faults along a modern rifted margin. Line-drawing interpretation of the seismic reflection record in the Bay of Biscay off the southwest Atlantic coast of France. *(After Montadert et al., 1979)*

Shallow marine sediments deposited upon older rocks thicken markedly toward the continental edge (Fig. 5.13B). Many marine sequences begin with a thick layer of salt, similar to that described for the Red Sea. This salt layer subsequently produces belts of abundant salt domes, such as those found on the Gulf Coast of the United States and along the western margin of Africa, which are of economic importance because of the oil traps they form (Fig. 5.13B, sections 1, 2).

Many sedimentary sequences thin significantly over buried ridges that commonly occur near the outer edge of the continent. These outer rises could indicate buried horsts or tilted fault blocks (Fig. 5.14; compare with Fig. 5.13), some sort of intrusion such as a salt diapir or volcanic rocks (Fig. 5.13B, sections 1 and 2), or buried carbonate reefs (Fig. 5.13B, section 3). At the continental slope itself, sediments once again thicken seaward and consist of a sequence of deep-sea fan deposits shed off the edge of the continent onto the ocean floor. Growth faults are common in most sequences.

As in the Red Sea, the early stages of rifting are characterized by the emplacement of basaltic igneous rocks in two distinct modes: as dikes in the rifting continental crust and, in places, as voluminous outpourings of magma that are present along some continental margins as the seaward-dipping reflectors present in many seismic reflection profiles.

Along the margins of the central Atlantic Ocean, dikes within the basement rocks form a roughly radial pattern in the reconstructed plate geometry, as shown in Figure 5.15A. The mechanics of dike injection is comparable to that of hydrofracting, in

that the pressure of magma in a magma chamber increases until a tension fracture forms in the walls of the chamber. The magma intrudes along the fracture and then solidifies to form a dike or sill. Because the fracture is a tension fracture, it is oriented normal to the minimum compressive stress $\hat{\sigma}_3$ and parallel to the maximum and intermediate compressive stresses $\hat{\sigma}_1$ and $\hat{\sigma}_2$.

Figure 5.15B shows an interpretation of the stress trajectories based on this model of dike formation. The maximum compressive stress $\hat{\sigma}_1$ is radially oriented around the tripartite join between North America, South America, and Africa, and the minimum compressive stress $\hat{\sigma}_3$ has a circumferential orientation. This distribution of stress is what we would expect for a large-diameter (2500 km) domal uplift of the crust at the triple junction, and it is reminiscent of the domes observed along the East African Rift (compare Figs. 5.9 and 5.11). The volcanism that accompanies the doming, as suggested above, and the volcanic deposits that appear only in the upper part of the section of rift sediments along the northeast coast of the United States suggest different times for the onset of volcanism in the rifting process, and thus differences in the governing process (see Secs. 5.6 and 5.7).

A volcanic-rich continental margin with seaward-dipping reflectors is a recently discovered phenomenon that is now recognized along most passive margins. Figure 5.16A is a world map displaying the regions where these features have been found. Figure 5.16B shows a schematic cross section of a volcanic-rich margin with the seaward-dipping reflectors separating a region of faulted continental margin and

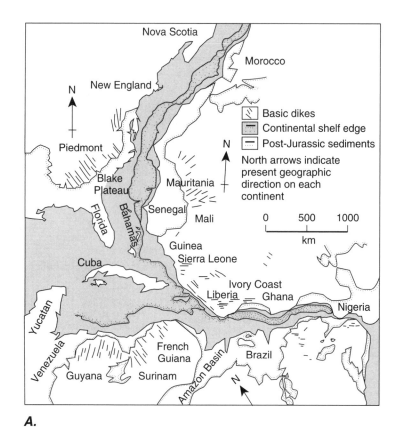

A.

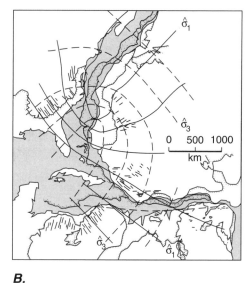

B.

Figure 5.15 Prerifting map of the central Atlantic region, showing orientations of mafic dikes. *A.* Mafic dike orientations in North America, South America, and Africa appear to be radially arranged about the triple junction before rifting separated the three continents. *B.* Assuming that dikes form as tension fractures, the stress trajectories for maximum and minimum principal stresses ($\hat{\sigma}_1$ and $\hat{\sigma}_3$, respectively) can be inferred. The maximum principal stress is radially oriented, and the minimum principal stress is circumferentially oriented. *(After May, 1971)*

"normal" oceanic crust. The reflectors are thought to represent volcanic (probably basaltic, but possibly also silicic) flows that thin toward the continent, thus giving rise to the dipping pattern. The transition between these volcanic-rich margins and their volcanic-poor counterparts is not yet understood.

The rifted continental margins discussed so far are ones for which spreading has been more or less at a high angle to a linear continental margin. Other continental margins display sharp bends, as exemplified by the formerly adjacent margins of South America and Africa (see Fig. 5.12). In the reconstructed fit, the E-W and N-S parts of the African margins form two parts of what was originally a three-part rift system. The third part is a NE-trending rift structure within Africa, the Benue Trough, that contains a thick marine sedimentary sequence (see Fig. 5.9). Early in the breakup of Africa from South America, these three rift structures formed a ridge-ridge-ridge triple junction. Two of the arms of this triple junction spread until they became parts of the Atlantic Ocean. The third, the Benue Trough, failed to develop and even closed up again, as in-

dicated by fold axes parallel to the trough and by some subduction-related volcanic deposits. Topographically, it is now only a linear depression, trending at a high angle to the sharp bend in the continental margin, which contains the mouth and delta of the large Niger River.

Sharp bends in continental margins, such as the Africa–South America bend, typically display a trough extending away from the apex of the bend in the concave continental margin that may represent the failed arm of a ridge-ridge-ridge triple junction. Figure 5.12 shows a number of these features around the margins of the Atlantic Ocean. Many workers consider these failed arms to be the modern counterparts of the aulacogens of continental platforms that we discussed in Chapter 3 (Sec. 3.5; compare Fig. 3.13 with Fig. 5.12).

Oceanic and continental rocks differ substantially in structure and composition, as the discussion in Chapter 3 makes evident. The examples of rifted margins in various states of development that we describe above provide some indication of the character of the transition from one type of crust to the

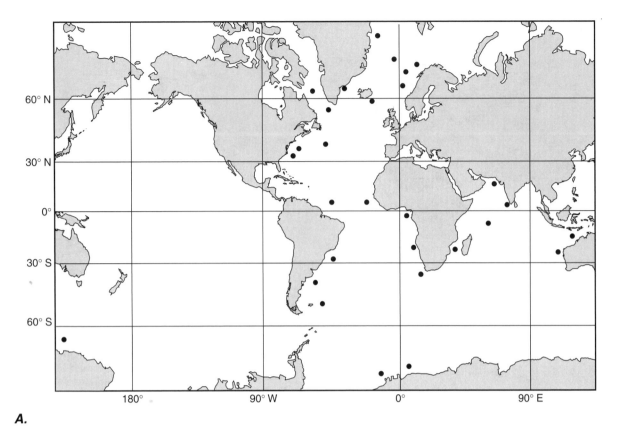

A.

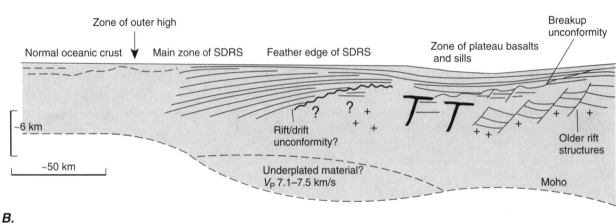

B.

Figure 5.16 Volcanic-rich passive margins. *A.* World map showing regions (black dots) where seaward-dipping reflectors (SDRS) have been reocgnized; these are thought to represent volcanic flows. *B.* Cross section of volcanic-rich rifted margin showing the presence of landward thinning and seaward-dipping seismic reflectors separating a zone of rifted continental basement from normal oceanic crust. *(After Joides, 1991)*

other at continental margins, but the complete answer still eludes us. Published seismic reflection profiles generally yield little information about the deeply buried continent-ocean transition (note the lack of information about the transition in Figs. 5.13*B* and 5.14). Seismic refraction techniques, which employ models with horizontally continuous

layers, are poorly suited to deciphering the complexities of this region, where the character of the rocks must change drastically over short distances.

Geophysical surveys of continental edges around the world show that they have characteristic gravity and magnetic anomaly profiles (Fig. 5.17). The gravity field of a typical continental margin displays

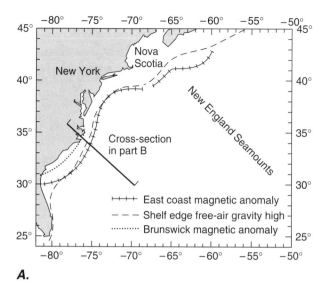

Figure 5.17 Geophysical features of the continental margin of North America. *A.* Map of the east coast of North America showing the location of the gravity anomaly and magnetic anomaly that mark the edge of the continent. *B.* Gravity and magnetic profiles across the continental margins showing the characteristic gravity high and the complex magnetic profile across the continental margin. (*A. after Rabinowitz, 1974; B. after Sheridan et al., 1988*)

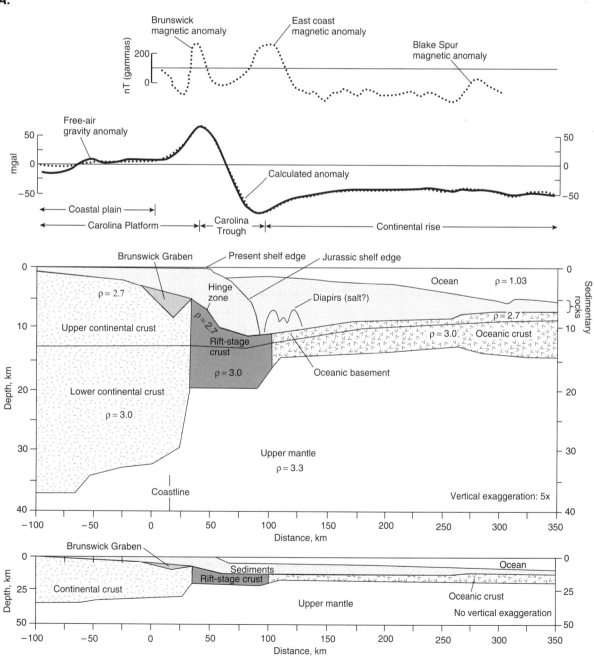

a distinctive high superimposed upon an otherwise more or less smooth gradation from high oceanic to low continental values.

Magnetic anomaly profiles are more complex. Some show more than one magnetic high between oceanic and continental crust, whereas others show a smooth, featureless transition between continental and oceanic patterns.

These gravity and magnetic profiles are useful, however, because they are like ones found in mountain belts within continents that may reflect ancient continental margins like the modern rifted Atlantic-style ones.

5.5 Oceanic Crust and Ocean Spreading Centers

Beyond the continental margins, the deep ocean floor consists of rocks formed chiefly at divergent plate margins. Figure 5.18 shows the world oceanic rift system in generalized form differentiated according to spreading rate. For the purpose of discussion, it is useful to separate the divergent margins into slow- (1–5 cm/yr), intermediate- (5–9 cm/yr), and fast- (9–18 cm/yr) spreading margins, as determined by magnetic anomaly data. These three types of spreading center differ in topography, distribution of rocks, and structure (Fig. 5.19). They also exhibit distinctive magnetic and gravity anomaly patterns, seismic structures, and heat flow distributions, as described briefly below.

Our understanding of the composition and structure of the ocean floor comes from numerous dredge-haul and drill-core samples, and it is enhanced by the study of ocean-floor sequences thrust up and exposed on land. These sequences, known as ophiolites, are the basis for detailed models with which we can interpret some of the geophysical data, and they allow us to infer some of the processes that operate during spreading.

Topography

In two dimensions, slow-spreading ridges (Fig. 5.19A) exhibit a deep axial graben that drops 1.5 to 3 km below the adjacent ocean floor and a rough faulted topography that is approximately symmetrical about the ridge axis (compare Fig. 5.2, diagrams 1–4). Within the axial graben itself, elongate shield volcanoes are intermittently present, as shown in the Figure 5.19A diagram; they tend to occur in clusters. This topography is superimposed on a broad axial topographic high that defines the crest of the ridge itself.

Divergent margins having intermediate spreading rates display a more subdued topography than that of slow-spreading margins (Fig. 5.19B; compare Fig. 5.2, diagrams 5, 6, 8). A central graben is present, but it is generally not more than 100 to 200 m deep. Central volcanoes are nearly continuous, commonly with extension fractures arranged en echelon along their crests (Fig. 5.19B).

Fast-spreading ridges typically display no axial valley (Fig. 5.19C; compare Fig. 5.2, diagram 7). The volcanic edifice at the center of the rift is more continuous, reminiscent of Hawaiian volcanoes, with a small summit ridge or graben. The topography is smooth, unlike that of slow-spreading rifts.

Although the mid-ocean ridge system is more or less continuous around the Earth, some workers argue that on a regional scale, all ridges display consistent patterns of discontinuities in topography along the strike that have been classified into first-, second-, third-, and fourth-order discontinuities (Fig. 5.20). "First-order" discontinuities are transform faults, spaced an average of 300–500 km apart along the ridge, that offset the axial valley or ridge by at least 20 km and generally more than 50 km. Second-order discontinuities are present at a spacing of 50 to 300 km. On slow-spreading ridges, these discontinuities are seen as a bend or jog in the rift valley. On fast-spreading ridges, second-order discontinuities occur as the overlapping tips of spreading axes (Fig. 5.20A). Typically, the amount of overlap is a few kilometers to tens of kilometers, and the spacing between the overlapping segments is generally about one-third the overlap distance. The area between the overlapping segments is a basin a few hundred meters deep. Third-order discontinuities on fast-spreading ridges are small overlaps and on slow-spreading ridges are gaps

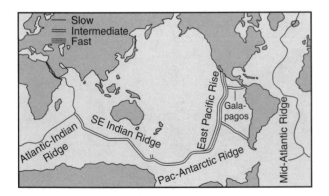

Figure 5.18 Map of world's ridges, classified according to spreading rate. *(After Macdonald, 1982)*

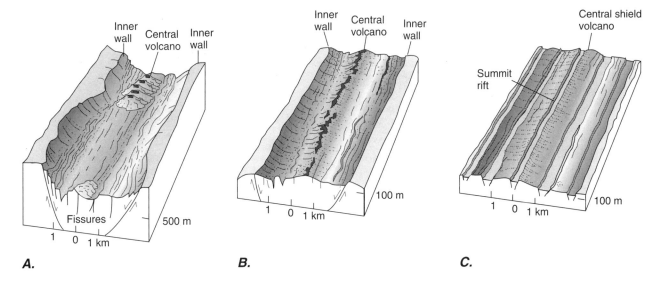

Figure 5.19 Diagrams showing the topography and structure of the axial zone on ridges of varying spreading rate. Note that the block diagrams have a vertical exaggeration of 2:1. *A.* Slow-spreading ridge (1–5 cm/yr). *B.* Intermediate-spreading ridge (5–9 cm/yr). *C.* Fast-spreading ridge (9–18 cm/yr). *(After Macdonald, 1982)*

between chains of volcanoes. Fourth-order discontinuities are seen as minor bends in the trend of the axis (also known as devals) on fast- to intermediate-spreading ridges and as spacing within the volcano clusters on slow-spreading ridges. The consistency of these discontinuities on the ridges has given rise to a model for the development of mid-ocean ridges that relates spreading to magma supply, as discussed below. Ridges are also variable in elevation along the strike, being relatively higher between the discontinuities than near them.

Magnetic Anomaly Patterns

The ocean floor on either side of ocean spreading centers is characterized by a pattern of linear magnetic anomalies that is symmetrical across the ridge axis. The width of individual anomaly bands is proportional to the rate of spreading (Fig. 5.21). The details of the pattern are better resolved, and the boundaries between positive and negative anomaly bands are sharper, at fast-spreading ridges such as the East Pacific Rise (Fig. 5.21C) than at slow-spreading ridges such as the Mid-Atlantic Ridge (Fig. 5.21A).

The main zone of magnetization that produces the anomaly pattern appears to be restricted approximately to the upper 2 km of the oceanic crust. A few drill holes in the floor of the North Atlantic Ocean have revealed the presence of interlayered normally and reversely magnetized basalts. These data suggest that in some instances magnetizations for the different polarity periods are not well separated horizontally, thereby accounting in part for the poor resolution of the anomaly pattern on slow-spreading ridges. Direct observations and paleomagnetic measurements on slow-spreading ridges also demonstrate the presence of rotated fault blocks, presumably resulting from movement on listric normal faults. Such rotations would also affect the orientation of the paleomagnetic vectors preserved in the rock and would contribute to the loss of resolution in the magnetic anomaly pattern.

A distinctive feature of the magnetic anomaly patterns on all ridges is the decrease in amplitude from the axial anomaly to flanking anomalies (see Fig. 5.21). Several explanations for this puzzling feature have been proposed, such as metamorphism or oxidation of previously magnetized crust; intrusion of magmas at depth away from the spreading axis; interlayering of normally and reversely magnetized material away from the spreading axis, as mentioned above for the Mid-Atlantic Ridge; and rotation of the magnetic vector during listric faulting.

In a number of places, offsets occur in marine magnetic anomalies that are not parallel to transform faults and bear no clear relationship to plate

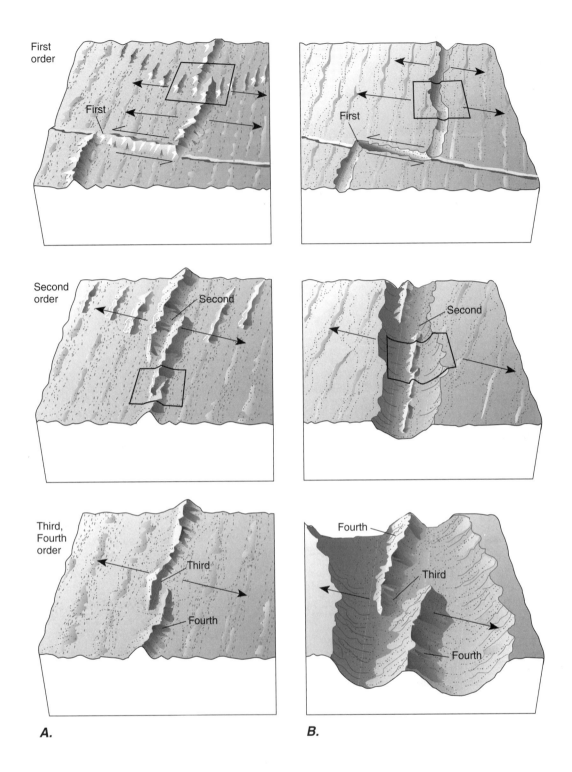

First
order

First

Second
order

Second

Third,
Fourth
order

Third

Fourth

First

First

Second

Fourth

Third

Fourth

A.

B.

Figure 5.20 Block diagrams illustrating discontinuities in the mid-ocean ridges. First-order discontinuities are transform faults that continue beyond the ridge intersections as fracture zones. Second-order discontinuities on fast-spreading centers are overlapping rifts that are preserved as V-shaped trains of ridge tips. Third- and fourth-order discontinuities are lesser overlaps, offsets in volcanoes, or slight deviations in strike. *A.* Fast-spreading center. *B.* Slow-spreading center. *(After Macdonald and Fox, 1990)*

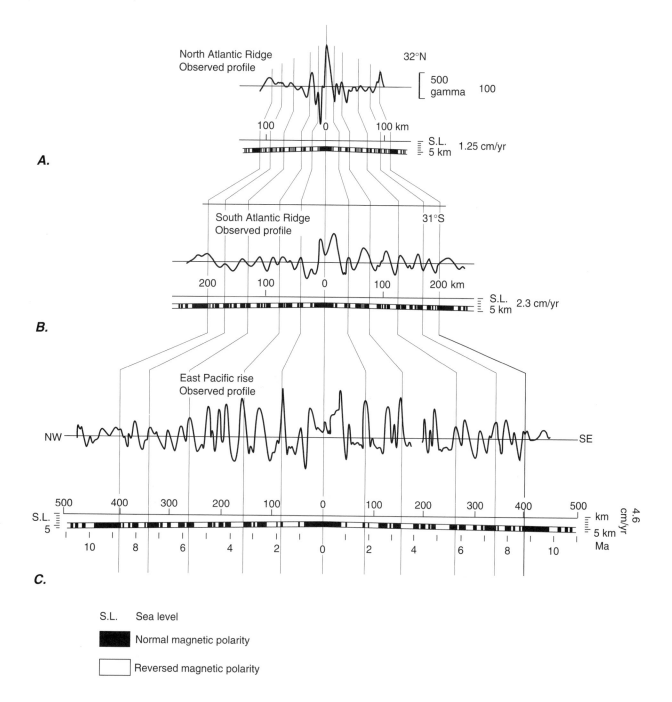

Figure 5.21 Sample magnetic profiles from three oceanic areas of different spreading rates. The bar pattern shows the magnetic anomaly reversal history for the past 11 Ma, repeated symmetrically about the ridge axis at a scale appropriate for the local average spreading rate. Note the difference in detail and intricacy of patterns. *A.* Magnetic anomaly pattern at 32°N in the North Atlantic, a slow-spreading ridge. *B.* Magnetic anomaly pattern at 21°S in the South Atlantic, an intermediate-spreading ridge. *C.* Magnetic anomaly pattern at the East Pacific Rise, a fast-spreading ridge. *(After Vine, 1967)*

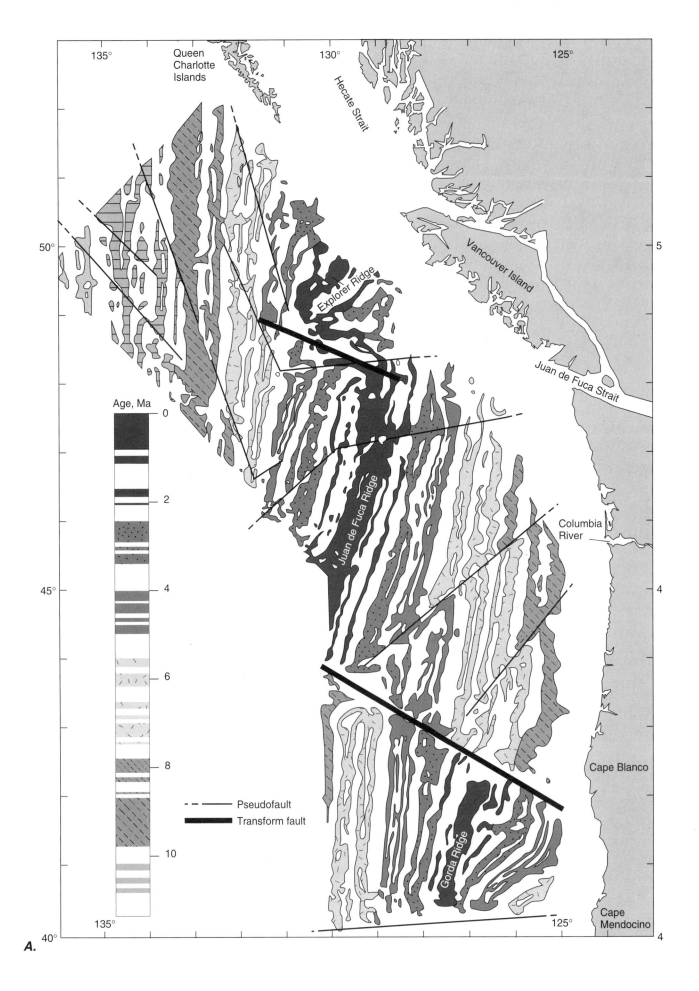

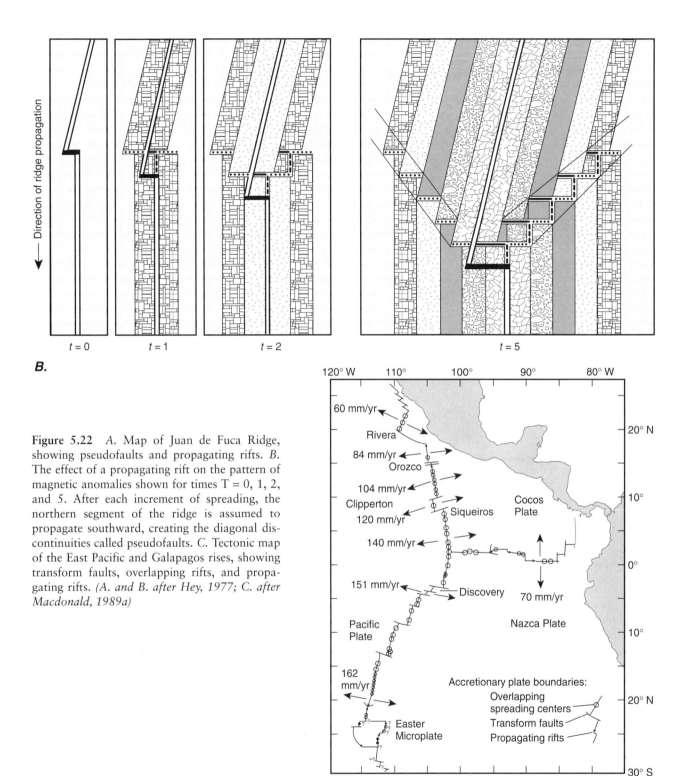

t = 0   t = 1   t = 2   t = 5

B.

Figure 5.22 *A*. Map of Juan de Fuca Ridge, showing pseudofaults and propagating rifts. *B*. The effect of a propagating rift on the pattern of magnetic anomalies shown for times T = 0, 1, 2, and 5. After each increment of spreading, the northern segment of the ridge is assumed to propagate southward, creating the diagonal discontinuities called pseudofaults. *C*. Tectonic map of the East Pacific and Galapagos rises, showing transform faults, overlapping rifts, and propagating rifts. *(A. and B. after Hey, 1977; C. after Macdonald, 1989a)*

120° W   110°   100°   90°   80° W

60 mm/yr
Rivera
84 mm/yr
Orozco
104 mm/yr
Clipperton
120 mm/yr   Siqueiros   Cocos Plate
140 mm/yr

151 mm/yr   Discovery   70 mm/yr

Pacific Plate   Nazca Plate

162 mm/yr

Easter Microplate

20° N
10°
0°
10°
20° N
30° S

Accretionary plate boundaries:
Overlapping spreading centers
Transform faults
Propagating rifts

C.

motions (Fig. 5.22). Each of these apparent offsets is produced by a **propagating rift;** that is, a segment of a ridge that grows longer while an adjacent segment, connected by a transform fault, grows shorter. Figure 5.22B shows schematically the evolution of such a

rift with the configuration of the ridge-transform system at t = 0 and at three subsequent increments of time (t = 1, 2, 5). Between t = 0 and t = 1, the spreading ridge produces the lithosphere marked with mosaic pattern lines. At t = 1, the southern tip of the

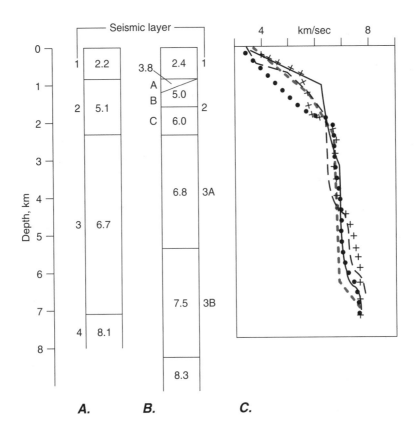

Figure 5.23 Seismic structural models of oceanic crust. *A.* Standard model with three layers in the crust above the Moho (compare Fig. 3.3A). *B.* More detailed model based upon sonobuoy surveys in which crustal layers 2 and 3 are subdivided. *C.* Gradient models of oceanic crustal structure in which velocity increases continuously, but with discontinuities in the slope of the velocity-depth curves. *(After Christiansen and Salisbury, 1975; Rosendahl, 1976; McClain, 1981; and Atallah, 1986)*

northern ridge segment propagates to its new position. There it spreads until $t = 2$, producing the lithosphere marked with hatched pattern. Alternation of spreading and lengthening of the northern ridge segment produces the pattern of crust indicated for $t = 5$. The result is a V-shaped pattern of offset magnetic anomalies called **pseudofaults** that are thought to be a complex mixture of spreading and intrusion. Although these features are not completely understood, they are related to the discontinuities in topography just discussed. As exemplified by the East Pacific Rise (Fig. 5.22C), overlapping and propagating rifts are widespread.

Seismology

Seismic data from the ocean ridges include the geographic distribution of earthquakes (see Fig. 4.2), the results of first-motion studies, and the results of seismic refraction and reflection studies.

Seismic refraction studies originally resulted in the three-layered model of the oceanic crust (Fig. 5.23A; compare with Fig. 3.3A). Subsequent refinement of this model with techniques that provide more detailed information, such as the use of anchored seismic receivers (sonobuoys), has given rise to a more complex layered model with a two- or three-part subdivision of layer 2 (designated layers 2A, 2B, and 2C) and a two-part subdivision of layer 3 (designated layers 3A and 3B) (Fig. 5.23B). This refined seismic model suggests an oceanic crust considerably more complex than previously supposed. A further refinement of the interpretation of the seismic data involves modeling the structure as a continuous variation of seismic velocities with depth, with the different layers bounded by discontinuities in the seismic velocity gradient, rather than by a sharp jump in velocity between layers (Fig. 5.23C).

These seismic models clearly have significance for our interpretation of the structure and petrology

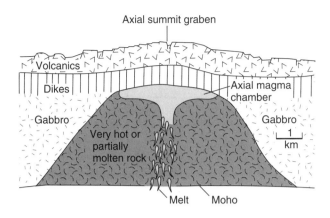

Figure 5.24 Schematic cross section of the East Pacific Rise, showing the shape of an axial magma chamber flanked by very hot or partially molten rock. *(After Detrick et al., 1987; Macdonald, 1989b)*

of the oceanic crust. Layer 2A, with a low velocity that rapidly increases with depth, may represent highly fractured basaltic extrusives with cracks that close up as pressure increases with depth. Layer 2B, with a constant seismic velocity, may represent the main bulk of the extrusive section. Layer 2C, with a sharp increase in seismic velocity with depth, may represent a change in metamorphic assemblage, a change in igneous lithology, or a change from faulted rocks above the discontinuity to unfaulted rocks below.

The presence and size of a magma chamber in the crust at mid-ocean spreading centers can be inferred from a crustal low-velocity zone or a region of pronounced attenuation of S waves. Seismic evidence suggests that under the axes of modern intermediate- and fast-spreading ridges, there are magma chambers approximately 4 km wide and a few hundred meters thick (Fig. 5.24). Under slow-spreading ridges, on the other hand, magma chambers are presumably intermittent—no magma chamber has been found so far beneath any slow-spreading ridge.

Oceanic crustal structure can also vary along a ridge axis. In particular, on a slow-spreading ridge within a distance of about 30 km of a transform fault, the thickness of the oceanic crust near the ridge decreases from the normal 5 to 6 km to almost 3 km adjacent to the fault. On intermediate- to fast-spreading ridges, the change of crustal thickness near transform faults is less, but the ridges do tend to lose their nonrifted character and develop an axial valley reminiscent of slow-spreading ridges. We discuss related features more fully in Section 6.2.

Seismic evidence from the mantle directly below ocean ridges indicates that the lithosphere there is very thin or nonexistent. It thickens away from the ridge axes themselves, reaching a thickness of approximately 100 km at a lithospheric age of approximately 80 Ma (Fig. 5.25; compare Figs. 3.22 and 5.3).

Gravity and Heat Flow

Ship-borne measurements of gravity at ocean ridges indicate the presence of a pronounced symmetrical negative Bouguer anomaly and a small positive free-air anomaly (Fig. 5.26), which also is observed in satellite measurements. The negative Bouguer anomaly indicates a mass deficiency at depth, relative to the adjacent crust and mantle, and the small positive-free air anomaly suggests that the ridge is nearly, but not completely, in isostatic equilibrium and that there is a small excess in the topographic elevation of the ridge.

Heat flow measurements show a marked increase in the thermal flux near ridge crests, where values as high as 10 hfu are found. In contrast, the average ocean values are in the range of 1.0 to 1.5 hfu (Fig. 5.26). These data imply a large heat source in the region below the ridge axis.

Direct observations of the ocean floor from manned submersibles reveal the presence of numerous hot springs near active spreading centers. Extensive hydrothermal alteration is also present in fossil rift zones. These observations indicate that a significant quantity of heat is carried out of the crust by the pervasive convection of pore water. This heat

Figure 5.25 Thickening of lithosphere away from a ridge by growth at the expense of asthenosphere. The vertical exaggeration varies depending on the spreading rate for the ridge represented, but it is approximately 10× for a spreading rate of 5 cm/yr. *(After Sclater et al., 1981)*

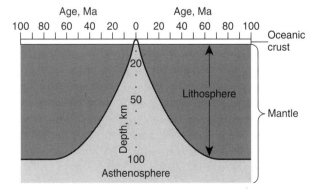

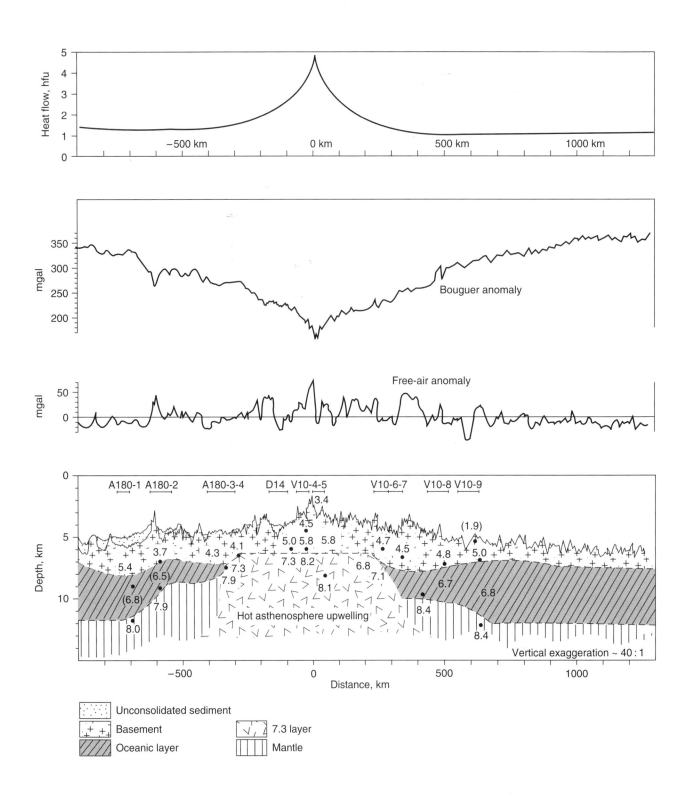

Figure 5.26 Seismic, gravity anomaly, and heat flow profiles across a typical mid-ocean ridge. The Bouguer anomaly was calculated using a density of 2.60 g/cm³ for the basement rocks in the crust. The numbers in the seismic velocity structure profile show the seismic velocities in km/s; assumed values are in parentheses. *(After Talwani et al., 1963)*

loss is not included in heat flow measurements, which only detect conductive heat loss. Thus, measured values of heat flow on the ocean floor are minimum values. The amount of heat lost by hydrothermal convection is very difficult to determine because the amount and extent of hydrothermal circulation is difficult to measure.

Composition of Oceanic Crust and the Ophiolite Model

Since the explosive growth of marine research after World War II, thousands of samples of oceanic crust have been obtained by dredging and core sampling carried out by the Deep Sea Drilling Project (DSDP) and the Ocean Drilling Program (ODP). These rocks are predominantly abyssal sediments at the ocean bottom, overlying a series of igneous rocks. Tholeiitic basalt, low in K, is the dominant rock type found in the oceans so far. Important subordinate rocks include gabbro, peridotite, diabase, and their metamorphosed equivalents.

Despite the abundance of samples, it is important to remember that our sampling of the oceanic crust and mantle is still poor. Drilling has been limited mostly to seismic layers 1 and 2 of the oceanic crust (see Fig. 5.23); only a few holes have penetrated into rocks thought to represent layer 3 or the upper mantle. The data, scanty as they are, confirm the presence of dikes and other intrusive rocks at depth. Dredge sampling of deeper crustal levels is possible along fault scarps, principally along transform faults, and along the ridge-parallel faults of axial grabens, but these sites may not represent the oceanic crust as a whole. In general, we have only fragmentary knowledge of the rocks that make up the lower crustal levels at ridges that do not have axial grabens.

Fortunately, there exist on land a number of examples of distinctive rock sequences that may be fragments of oceanic crust and mantle formed at ocean spreading centers. These sequences, called **ophiolites**[1], display rock types and overall structures similar to those inferred to exist in the oceanic crust (Fig. 5.27). The ideal ophiolite sequence includes the following units from bottom to top:

A. At the base of the sequence is tectonized peridotite, which is typically composed of harzburgite (olivine + orthopyroxene) but also lherzolite

[1] From the Greek *ophis*, snake, an allusion to the mottled green snakeskinlike appearance of most serpentinites, and *lithos*, rock.

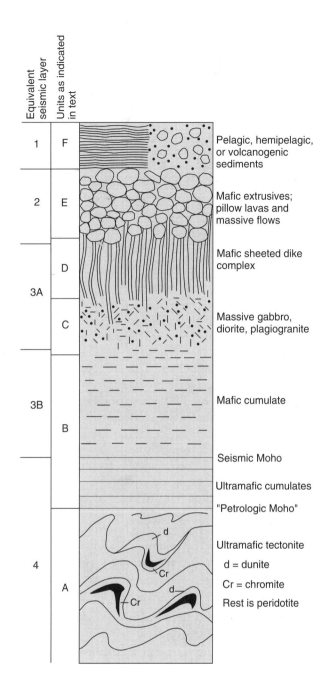

Figure 5.27 Idealized ophiolite sequence. The ophiolite layers are numbered and named on the left and correspond to the discussion in the text. The correlation with the seismically determined layers is shown on the right. *(After Moores, 1982)*

(olivine + orthopyroxene + clinopyroxene), with subordinate dunite (>95 percent olivine) and chromitite (concentrations of chrome spinel). Compositionally, these rocks are fairly uniform, and the Mg-Fe ratio in olivines and pyroxenes is approximately 9:1. These rocks display tectonite fabrics characterized by textures that reveal a history of recrystallization during ductile deformation, by a pronounced foliation defined by preferred orientation of elongate mineral grains, and by laminar concentrations of subordinate pyroxene and spinel grains. Observation under a microscope reveals that the fabric is characterized by the strong crystallographic preferred orientation of olivine. Layers of pyroxene, chromite, or dunite are typically folded into isoclinal (class 2) folds, and in places such folds are refolded. These rocks are sometimes called Alpine peridotites from their characteristic occurrences in the Alps. We infer that they represent the upper mantle lithosphere.

B. A mafic-ultramafic stratiform plutonic complex overlies the tectonite or Alpine peridotite. This plutonic complex is usually rich in olivine and pyroxene at the base and grades upward into units rich in plagioclase and, near the top, even quartz or hornblende. These rocks are compositionally layered, but their layers are more regular than those of the underlying tectonite. The mineral grain textures are characteristic of crystallization from a magma, rather than of recrystallization during metamorphism or ductile flow. Peridotite rock and mineral compositions are more iron-rich than the underlying Alpine peridotites. Layering and compositions suggest that many of these rocks formed by accumulation of crystals at the bottom of a magma chamber by gravity settling. Thus, this contact marks the true crust-mantle boundary because it is the contact between the solid, ductilely deformed mantle and the oceanic crust formed from magmatic rocks.

C. The top of the plutonic complex is generally composed of coarse- to fine-grained varitextured gabbro, diorite, and leucocratic quartz diorite (often called plagiogranite). These rocks are probably the result of in situ crystallization rather than the gravity settling that characterizes the stratiform part of the plutonic complex. We interpret these rocks to represent the top of a plutonic body.

D. A finer grained mafic dike complex overlies the plutonic complex in many instances. This complex consists of dikes intruded into plutonic rocks at the base; areas of 100 percent dikes in the middle called a **sheeted dike complex,** where dikes have intruded earlier dikes; and, at the top, dikes intruded into extrusive rocks, commonly pillow basalts (Fig. 5.28A). In areas of multiple dike injection, dikes intrude either along the margin or into the center of preexisting dikes (Fig. 5.28B). Where a later dike splits an earlier dike in half, presumably it is because the older dike is still hot and therefore weak in the central region but cooler and stronger along its margins. Thus, a new fracture develops where the rock is weakest. This process is repeated many times resulting in the development of dikelike bodies that mostly are half-dikes, bounded on each side by chilled margins having a polarity from the margin into the chilled magma pointing in the same direction (Fig. 5.28C).

E. Overlying the dike complex with a gradational contact is a sequence of extrusive volcanic rocks, generally mafic in composition. These rocks consist of massive or pillowed flows, with a few sills or dikes, and scattered breccias. We infer that these sequences originate by submarine extrusion of lavas.

F. Pelagic sediments, here and there interbedded with metal-rich chemical sediments, typically overlie the extrusive sections. These sediments include thinly bedded cherts in some places; elsewhere, limestones rest directly on the volcanic basement and themselves are overlain by cherts.

This idealized ophiolite section agrees well with the structure of oceanic crust deduced from seismic data and from direct samples. The thicknesses of the units within ophiolites inferred from field work correlate fairly accurately with the seismically identified layers of the oceanic crust. As shown in Figure 5.27, we identify the sediments (F) with seismic layer 1; the volcanic extrusive rocks (E) with layer 2; the dike complex (D) and upper plutonic complex (C) with layers 3A and 3B, respectively; and the olivine-rich plutonic rocks (B) and mantle tectonite (A) with layer 4. Implicit in this correlation is the important fact that the igneous-tectonite contact between ophiolite layers A and B, sometimes called the petrologic Moho, does not correspond to the seismic Moho, which is located between seismic layers 3 and 4 and which lies within the plutonic complex (ophiolite layer B), between the olivine-rich cumulate rocks below and the olivine-poor cumulate rocks above (see Fig. 5.27). The sharpness of this petrologic transition from rich to poor olivine compositions is a measure of the sharpness of the seismic Moho. In many ophiolites, this transition ranges in thickness from 50 m to a few hundred meters.

If ophiolites are indeed sections of oceanic crust thrust onto continents and preserved, then studying them should tell us a great deal about the processes by which new plates are formed. In particular, their

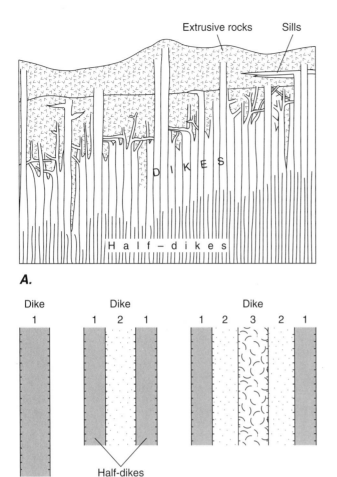

A.

B.

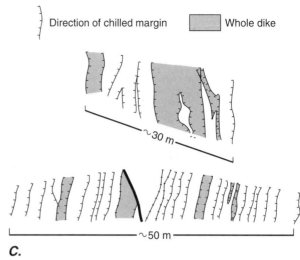

C.

Extrusive rocks   Sills

D I K E S

H a l f – d i k e s

Direction of chilled margin   Whole dike

~30 m

~50 m

Dike
1

Dike
1   2   1

Dike
1   2   3   2   1

Half-dikes

Figure 5.28 Sheeted dike complexes. *A.* Troodos sheeted dike complex and its relationship to the overlying extrusive rocks. *B.* Three steps showing the development of dike-within-dike relationships by successive intrusion of new dikes up the center of an earlier dike. The final result is areas of dike intrusions characterized by chilled margins all of the same polarity. *C.* Representative cross sections through the Troodos sheeted dike complex showing the polarity of the chilled margins. *(A. after Wilson, 1959; B. and C. after Moores and Vine, 1971)*

internal structure should reveal how magmatic and structural processes interact. Several ophiolite complexes display tilted fault blocks at shallow to intermediate levels in the rock column. These faults do not appear to be present in the overlying sediments; thus, they may be faults that were active during formation of the oceanic crust. These faults display listric geometry and in places flatten into ductile shear zones near the interface between the plutonic complex and the dike complex (ophiolite layers C and D). These ductile shear zones may be characterized by mylonitic rocks. Such structures are reminiscent of the listric normal faults in detachment terranes discussed above that characterize regions such as the U.S. Basin and Range province. Figure 5.29 shows schematically the relationships developed in the Troodos ophiolite complex, Cyprus, which seem to typify many ophiolites. The figure shows that magmatic accretion fol-

lowed by structural thinning and further magmatic accretion can produce the structure found in the Troodos complex.

Detailed mapping of a number of ophiolite complexes suggests that individual magma chambers 1 to 2 km wide can be identified. The complex intrusive observed in some ophiolites suggest formation by multiple injection of discrete magma bodies, whereas a few others display no such intrusive complexity and may have formed by a steady-state, continuously fed magma chamber, such as those postulated for fast-spreading ridges.

Ophiolites are fairly controversial bodies of rock, in part because differences do exist between their composition and the compositions thought to be representative of modern spreading systems. They nevertheless provide a structural model by which observations of ocean spreading centers can be interpreted.

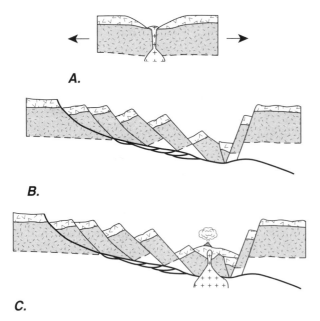

A.

B.

C.

Figure 5.29 Schematic cross section of dike-extrusive complex of part of the Troodos complex, showing tilted fault blocks deformed on listric normal faults bottoming out in a horizontal shear zone representing the brittle-ductile transition in oceanic rocks. *A.* Extension by magmatic accretion. *B.* Extension by structural thinning. *C.* Extension by renewed magmatic accretion. *(After Varga and Moores, 1990)*

With the ophiolite occurrences as a guide to the processes at spreading plate margins, we can suggest that divergent plate boundaries are regions where deformation such as listric normal faulting occurs simultaneously with magmatic events, giving rise to a crust of complex structure. It is perhaps surprising in view of this situation that the seismic structure of oceanic crust is so uniform around the world and that the linear magnetic anomalies are as consistent and coherent as they are. Perhaps the size of the tectonically rotated blocks is small compared with the width of the magnetic anomalies, so that the complexity is only a second-order effect. As we discussed in Chapter 3, however, our principal source of information about the structure of the oceanic crust comes from seismic refraction, which tends to smooth out local complexity and average the seismic velocities over a wide region.

Ophiolites are unique in one important respect: in contrast to much of the other oceanic crust produced at spreading centers through time, they have been preserved rather than recycled back into the mantle at a subduction zone. The problem of ophiolite emplacement and preservation is an important one to which we will return in Chapter 9.

5.6 Model of the Evolution of Rifted Continental Margins

The Basin and Range province, eastern Africa, the Red Sea, and the Atlantic margins provide examples of rifted continental margins in various stages of development ranging from *incipient* to *fully mature.* These examples allow us to make some general inferences about the sequence of events that characterizes the formation of rifted continental margins.

The Initiation of Rifting

The first event may be either the formation of a topographic and structural dome in a continental platform, or subsidence along the line of incipient rifting. The doming is characteristic of the domes of northern Africa; the East African Rift; the Afar triple junction at the southern end of the Red Sea; and possibly the triple junction from which North America, South America, and Africa separated. Where rifting is initiated by doming, igneous activity tends to occur early in the sequence, and the rocks tend to be high in alkalis and low in silica, indicating a deep-

Figure 5.30 Model for development of volcanism and topography around the Red Sea. *A.* Mantle upwelling occurs under the continent somewhat away from the zone of weakness that will become the future break. *B.* Magma is emplaced at the base of the crust (cross-hatched area) and causes doming, crustal thickening, and some surface volcanism. Rifting begins in the zone of weakness. *C.* The continent migrates over the asthenosphere bulge as volcanism wanes and spreading continues in rift region. *D.* Oceanic crust forms over the asthenosphere bulge. *(After Dixon et al., 1989)*

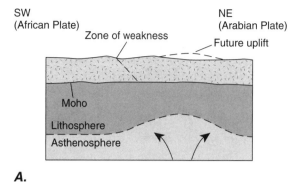

A.

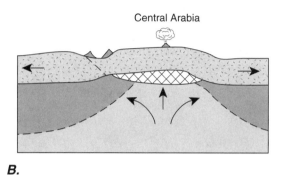

B.

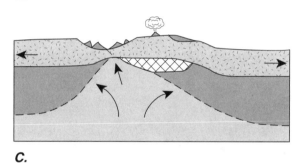

C.

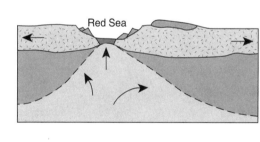

D.

seated source. Where rifting begins with subsidence, igneous rocks tend to occur later in the rifting history and are basaltic in composition, indicating a shallow source.

Rifting

As extension proceeds, fault structures may be either symmetric or asymmetric across the rift, depending on the geometry of the fault system. If a horizontal detachment at the brittle-ductile transition separates a lower zone that extends by ductile deformation from an upper zone that extends by listric normal faulting, then a structurally symmetric rift would develop (see Fig. 5.8*A*). If a major low-angle detachment cuts through the entire crust and possibly into the lower lithosphere, an asymmetric margin would develop (Fig. 5.30; compare Fig. 5.8*B*).

For example, the Red Sea Rift provides some features of both the symmetric and the asymmetric models. Figure 5.30 shows a model of the progressive evolution of the Red Sea Rift. The onset of rifting (Fig. 5.30*A*) was marked by the formation of a topographic bulge in Oligocene time, presumably over an upwelling in the asthenosphere. The bulge corresponded to the future rift in the south, but toward the north, the bulge was east of the Red Sea Rift, in the Arabian Peninsula. Possibly, underplating of magmatic rocks derived from the mantle resulted in thickening of the crust beneath the bulge, as indicated in Figure 5.30*B*. Rifting of the Red Sea itself took place westward of the bulge, possibly along a preexisting zone of weakness. As rifting occurred, the location of the asthenospheric bulge shifted, so that at present it is beneath the axis of the Red Sea. Normal faulting was accompanied by the intrusion of basalts, producing crust transitional between continental and oceanic (Figs. 5.30*B* and 5.30*C*). Finally, true oceanic crust began to form at the rift axis (Fig. 5.30*D*), and continental margins approximately

symmetrical in structure across the ocean basin drifted apart passively.

The Red Sea is interesting from another point of view. Rifting is older and more complete in the south than in the north. Indeed, rifting has only just started in the Gulf of Suez, at the northwest end of the Red Sea, and the continental crust is continuous across the rift region (see Fig. 5.11A). Some workers have proposed that the sea opened by propagation of the rift along strike from south to north.

Figure 5.31A shows a schematic cross section of a rifted ocean basin formed by the asymmetric extension of the continent on a dominant low-angle detachment fault (compare Fig. 5.8B). In this model, the wedge of continental crust in the lower plate margin (footwall block; left side, Fig. 5.31A) below the lithospheric detachment is extended and thinned by large amounts. The upper plate margin (hanging-wall block; right side, Fig. 5.31A) hardly deforms at all. Thinning in the lower plate margin results in isostatic adjustment and rise of the Moho and the lithosphere-asthenosphere boundary. The continental crust separates where it is thinnest, and oceanic

crust forms. One might expect to see dike and sill intrusion into both margins, but it should be most abundant in the footwall block, where thinning of the crust is most extreme and uplift of the asthenosphere is the greatest.

The resulting continental margins on opposite sides of the ocean basin are of markedly different structure. One side is composed of the footwall block with overlying fragments of the hanging wall block separated by bowed-up parts of the detachment faults. The other side is composed of the remainder of the hanging-wall block cut by steep normal faults of relatively minor displacement. The continental drainage divide on the upper plate margin would be near the coast. An example of such upper plate margins might be the southern Brazil

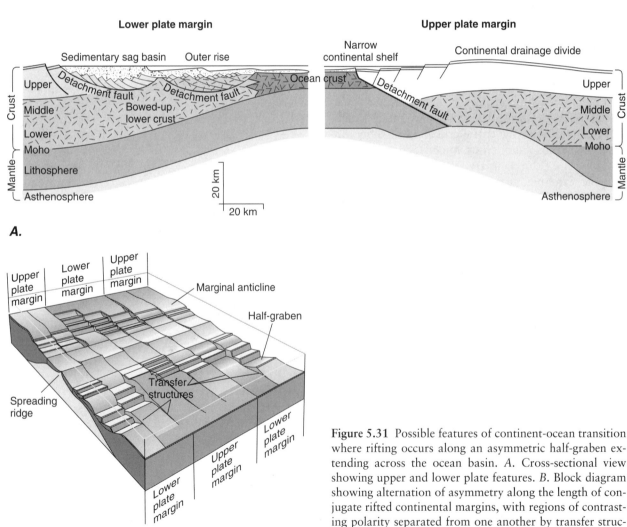

Figure 5.31 Possible features of continent-ocean transition where rifting occurs along an asymmetric half-graben extending across the ocean basin. *A.* Cross-sectional view showing upper and lower plate features. *B.* Block diagram showing alternation of asymmetry along the length of conjugate rifted continental margins, with regions of contrasting polarity separated from one another by transfer structures. *(After Lister et al., 1987)*

margin, which has high elevations and a continental drainage divide near the coast.

Along strike, the half-graben formed by the asymmetric rifting might be expected to terminate at a transfer structure and change to an asymmetric rift of the opposite polarity, as illustrated in Figure 5.31B, in a manner reminiscent of the pattern observed in fault systems in the East African Rift (Sec. 5.2). Thus, a given continental margin might display alternating upper-plate and lower-plate characteristics along its length.

In either case, grabens or half-grabens form in association with listric normal faulting (see Figs. 5.8 and 5.31), and they generally trap continental sediments that range from coarse clastics along their margins to fine fluviatile sediments and saline lacustrine deposits in the centers.

Extension and thinning of the continental crust eventually cause the floor of the rift valley to drop below sea level, as has occurred in Death Valley, California, and the Afar region of Ethiopia. When the ocean gains access to the topographic depression, it forms shallow seas having restricted circulation, and marine sedimentation begins. If the climate at the rift is hot and dry, the first marine deposits may be salt formed by the evaporation of seawater, as in the initial phases of formation of the Red Sea (see Figs. 5.30B and 5.30C) and the Gulf of Mexico. Salt thicknesses of 1 km or more may accumulate before the ocean basin is large enough to permit open circulation, which prevents the precipitation of salt. In cooler, wetter climates, salt should be absent, but continental and marine sediments may be interbedded at first.

Finally, as the rift widens, open marine conditions prevail. The thinned and intruded continental crust ultimately drifts away from the spreading center as new oceanic crust is formed, as we see happening today in the Red Sea. As the continental margin moves away from the plate boundary and assumes its role as a passive margin, its underlying lithosphere cools and contracts, and the margin subsides. Subsidence is accompanied by the accumulation of an increasingly thick sequence of shelf, slope, and rise sediments, ultimately resulting in the mature continental margins presently found on either side of the Atlantic.

If a rift begins by the formation of a three-armed graben on a dome, as occurred at the junction of the Red Sea, Gulf of Aden, and East African Rift, then either of two histories is possible. If all three arms of the rift continue to spread, the result is three separate continents and a mid-oceanic ridge-ridge-ridge triple junction (Fig. 5.32A). If, however, one branch ceases to spread, forming a failed arm such as the Benue Trough (see Fig. 5.9), then one rift spreads more or less perpendicular to the continental margins, and the other must spread obliquely (Fig. 5.32B). This model accounts for the present ridge-transform geometry in the equatorial Atlantic.

5.7 Models of the Seafloor Spreading Mechanism

The data have given us a reasonable picture of the evolution of rifts and continental margins, but what do they imply about the driving mechanisms of the process? We can suggest some models that are consistent with much of the data, but there are still aspects that are not understood.

The Initiation of Rifting

The difference between doming and subsidence at the initiation of rifting, and the difference in the timing and composition of the first volcanism, suggests that different mechanisms can operate, at least locally, along a rift.

Initial doming could be accounted for by a convective plume or diapir in the mantle and asthenosphere impinging on the base of the lithosphere, perhaps caused by a deep mantle plume. The convection transports deeper, hotter rocks to shallower levels, causing them to melt. Because convection is the driving mechanism, it is active before the doming and rifting occur, and the onset of igneous activity therefore occurs early in the rifting process. The deep-seated convection and the large initial thickness of the lithosphere account for the melting at relatively deep levels, producing the alkali-rich melts. Convection could provide a local driving force for the subsequent rifting.

Subsidence of the crust at the onset of rifting, on the other hand, could be explained if rifting begins with a stretching and thinning of the crust with the consequent subsidence of the surface. This model implies that the driving force is applied to distant boundaries of the lithospheric plates, and the local effect is a passive separation. The horizontal separation of the plates initiates an upwelling flow in the mantle below the rift, with the shallower mantle being affected more and earlier than the deeper mantle. Because the upwelling is driven by the horizontal motion of the plates, it begins after the initiation of rifting. Thus, magmatic activity starts some time after rifting, and because the initial mantle flow is shallow and the lithosphere is already partially

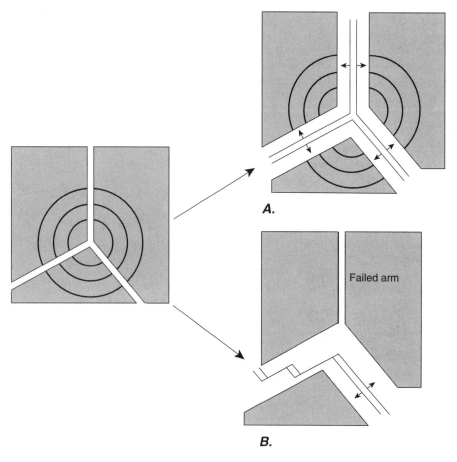

Figure 5.32 Evolution of a rift starting as a three-armed graben on a domal uplift. *A.* Breakup of a continent into three masses with the development of a ridge-ridge-ridge triple junction. *B.* Breakup of a continent into two masses with a predominantly ridge boundary, a predominantly transform boundary, and a failed arm of a ridge forming a trough that extends into one continent at the concave angle in the continental margin.

thinned, the melting occurs at relatively shallow depths, producing basaltic magmas.

The evidence suggests an alternation of these mechanisms along a rift between regions underlain by plumes, which produce abundant early alkalic volcanism, to regions of passive extension, which produce only late basaltic volcanism. The persistence of such differences after the opening of an ocean basin could account for some variability along mid-ocean ridges.

Magnetic Anomalies

The oceanic crust forms by intrusion and extrusion of magma that originates by decompression melting during diapiric upwelling of the mantle. The crust becomes magnetized parallel to the ambient magnet-

ic field of the Earth at the time the crystallized magma cools below the Curie temperature, and it moves away from the spreading center, together with the mantle below, as new crust is created. Data from oceanic rocks and from ophiolites suggest that most of the magnetization is in lavas and in fresh gabbros.

Heat Flow and Topography

The hottest part of the subcrustal mantle is where upwelling brings hot material from the interior closest to the surface. Thus, heat flow would be expected to be a maximum at the ridge crest. During subsequent spreading, the crust and upper mantle cool as heat is transported out by conduction and by the convective circulation of ocean water, which permeates the crust possibly as deep as the upper mantle.

Lower internal temperatures mean lower thermal gradients and a consequent drop in the heat flux, as observed.

The lithosphere-asthenosphere boundary is a rheological boundary dependent on temperature. At the ridge axis, the base of the lithosphere is almost at the surface. As the lithosphere and underlying asthenosphere cool, the boundary moves downward, making the lithosphere thicker.

Cooling also results in thermal contraction, which lowers the equilibrium level of the ocean bottom. Thus, the topographic high at mid-ocean ridges, and the decay of topography as the square root of the age of the ocean floor, can be explained by the conductive cooling of the lithosphere. The large-scale topography of the ocean floor therefore reflects the underlying thermal state of the lithosphere, which correlates directly with the age of the lithosphere for ages younger than about 60 to 80 Ma (see Fig. 5.3). For older ages, the lithosphere does not continue to thicken, nor does the depth of the ocean bottom continue to increase, indicating that a thermal steady state in the lithosphere has been reached that permits the steady conductive outflow of heat from the Earth's interior.

Gravity

The gravity data from mid-ocean ridges also can be explained by a similar model of convective upwelling in the solid mantle beneath active spreading centers. The hot rising mantle is less dense than the adjacent mantle, accounting for the negative Bouguer anomaly. The upward force generated by the convectively rising material supports a small part of the ridge topography, accounting for the small positive free-air anomaly. This evidence tends to support the model in which convection under the ridge drives the separation of the plates, rather than the one in which the separation of plates is driven by distant forces and the mantle wells up passively at ridges to fill in the volume created. In the latter case, a small negative free-air anomaly would be expected.

Complications and Problems

The model outlined above generally agrees well with the data available. As with any model, however, it is a simplified view. A number of complications and problems are worthy of mention.

The heat loss determined from the measured values of heat flow at the ridges, although high, is considerably less than predicted theoretically from models involving convective upwelling of hot mantle under a ridge. The difference probably can be accounted for by the heat lost through convective circulation of pore water through the oceanic crust, which is not determined by normal measurements of heat flow (see Sec. 5.5). The extent of heat loss by this mechanism, however, is difficult to determine and is not yet clear.

Old oceanic crust—such as in the central and western Pacific, the margins of the central Atlantic, and the Caribbean—displays topographic highs and lows that are not related to the theoretical age-depth curve, which is nearly horizontal for that age (see Fig. 5.3). Such features must have a different explanation. They may reflect either topography on the lithosphere-asthenosphere surface or the pattern of second-order convection cells in the upper mantle.

These regions of old crust also contain oceanic plateaus (Fig. 5.33), where the crust is markedly thicker than normal oceanic crust. In some cases, this crust appears to be of the same age as the neighboring region and to result from a greater magmatic activity at a spreading center. In other cases, such as the Caribbean and the Nauru Basin–Line Islands regions, anomalously thick crust has resulted from voluminous basaltic intrusion and extrusion in mid-plate regions, which yet we do not fully understand. One theory to explain such large volumes of plateau basalts is that they reflect the arrival at the base of the lithosphere of a very large plume from the deep mantle.

The mechanisms that account for a steady-state axial valley at the crest of slow-spreading ridges, and the absence of such a valley along fast-spreading ridges, are also poorly understood. We know that at a slow-spreading ridge, as a given point drifts away from the ridge axis, the surface of the crust must increase its elevation from that of the axial valley to that of the ridge crest, for only in this way can the axial valley remain as a steady-state feature of the topography. We also know, however, that the axial valley is a graben characterized by listric normal faults that accommodate extension and thinning of the crust. Such faults have been identified by normal first motions on shallow earthquakes at the ridges, by paleomagnetic evidence of the rotation of fault blocks, by detailed examination of the structure of ophiolites, and by direct sampling in at least one Ocean Drilling Project hole.

The problem, therefore, is how to rationalize topographic uplift with listric normal faulting. One possibility is that the topography is affected by competing processes that tend to depress or elevate it. The evidence shown in Figure 5.3 indicates that the thermal expansion of new hot crust accounts for the overall high elevation at the ridge axis, and that thermal contraction of the lithosphere during cooling

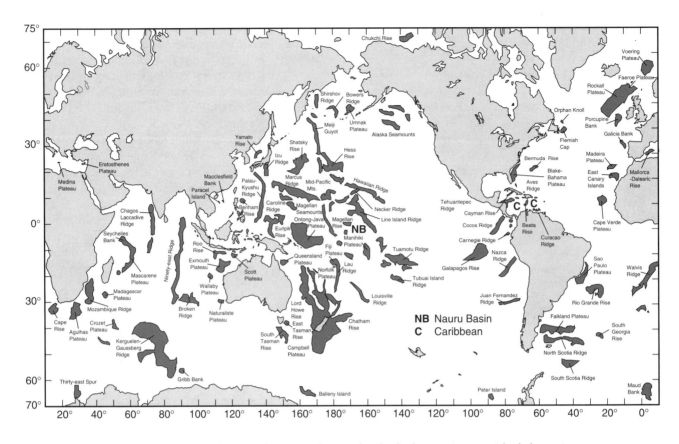

Figure 5.33 The distribution of regions of anomalously thick oceanic crust (shaded areas) throughout the world's ocean basins. The areas include large plateaus and narrower aseismic ridges. *(After Nur and Ben-Avrahem, 1982)*

dominates the large-scale ridge topography. The isostatic uplift of crust during thickening by magmatic extrusion and intrusion would also lead to an increase in elevation close to the ridge axis where the crust is being formed. On the other hand, a lithosphere that is actively being pulled apart as it forms would create a topographic depression above the zone of active extension. The expression of this extension in the upper crust may be a system of listric normal faults.

Thus, if all the extension is concentrated in a narrow zone near the ridge and if the thickening of the crust by magmatic processes occurs across a wider zone (Fig. 5.34A), then the extension dominates locally, creating a topographic low at the ridge axis, and the isostatic uplift of the magmatically thickening crust creates the sides of the axial valley. These competing processes could maintain a steady-state axial valley. If, on the other hand, the extension is spread across a wider zone than the zone of magmatic thickening of the crust (Fig. 5.34B), the topographic depression is small and wide, and no pronounced axial valley would form.

The relationship between axial discontinuities and topography outlined above (Sec. 5.5; Fig. 5.20) has suggested a model that relates these features to the presence of major magma chambers. Regions between second-order discontinuities or between first- and second-order discontinuities are thought to represent individual magma chambers that are thickest at the midpoints between discontinuities and are thinner and deeper near the discontinuities (Fig. 5.35). According to this model, new pulses of magma inflate the chamber and propagate along the strike of a ridge. As two new pulses propagate toward each other, they will interact in one of three

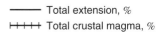

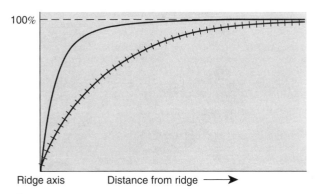

A.

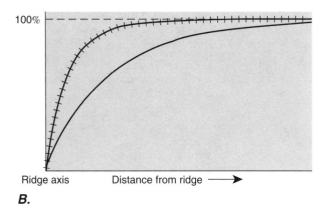

B.

Figure 5.34 The presence or absence of an axial valley at an ocean spreading center may depend on the competing processes of crustal extension by normal faulting and crustal thickening by magmatic intrusion and extrusion. *A.* Crustal extension is concentrated in a narrow zone above the ridge axis, but crustal thickening occurs across a relatively wide zone. Locally the extension dominates, creating a topographic depression, but isostatic uplift due to crustal thickening elevates the sides of the rift valley. *B.* Crustal extension is distributed across a relatively wide zone over the ridge axis, and magmatic crustal thickening occurs in a relatively narrow zone. Under these conditions, an axial valley would not develop.

possible ways (Fig. 5.36). If they meet head-on, they join in a saddle along the ridge (Fig. 5.36*A*). If they are misaligned originally, as they continue to propagate, they may join up (Fig. 5.36*B*) or not (Fig. 5.36*C*). In either case, a ridge tip is cut off or "de-capitated." As the process repeats, V-shaped trains of decapitated ridge tips are preserved in the oceanic crust (t_n in Figs. 5.36*B* and 5.36*C*), which may correspond to the V-shaped pseudofaults of propagating rifts, outlined in Sec. 5.5 (see Fig. 5.22).

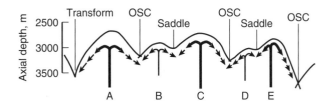

Figure 5.35 Generation of overlapping rifts by propagation outward from magmatic centers. *(After Macdonald et al., 1987)*

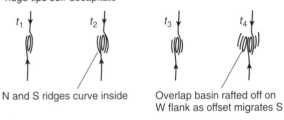

A. Magmatic pulses meet head-on, creating saddle point

t_1 t_2

Low saddle, minimal discordant zone on flanks

B. Magma pulses misalign but eventually link

t_1 t_2 t_3

Decapitated ridge tip

C. Magma pulses misalign and do not link; ridge tips self-decapitate

t_1 t_2 t_3 t_4

N and S ridges curve inside

Overlap basin rafted off on W flank as offset migrates S

Figure 5.36 Model for evolution of ridge-axis discontinuities. *A.* Ridges propagate toward each other, finally meeting in a saddle. *B., C.* Ridges propagate toward each other and overlap, with relict tips developing by decapitation. *(After Macdonald, 1987a)*

5.8 Miogeoclines: Ancient Atlantic-Style Continental Margins?

In Chapter 3, we briefly describe miogeoclines—thick sequences of shallow-water marine sediments. One obvious modern analogue for miogeoclines is an Atlantic-style continental margin. We examine in this section two ancient miogeoclinal sequences that are preserved in the Cordilleran mountain system in western North America and in the Appalachian mountain system in eastern North America. According to models based on observations of modern tectonic environments, these miogeoclines accumulated along rifted continental margins. This discussion provides examples of the value and the difficulties of using models based on the modern record to aid our interpretation of the ancient rocks.

The Cordilleran Miogeocline

The Cordilleran miogeocline (Fig. 5.37) extends continuously from northern Canada to southern California and possibly into northern Mexico. It had its inception in two distinct rifting events, one at about 1450 Ma in the northern United States and western Canada, and another at about 750–850 Ma in the

United States and southern Canada. The timing of each event is difficult to ascertain precisely because of the complex subsequent history of the rocks and the paucity of fossils in Precambrian time. The dating is based primarily on radiometric ages of igneous rocks associated with the sediments.

The oldest sediments, which are roughly dated between 1350 and 850 Ma, are those of the Belt-Purcell sequence; they consist mainly of thinly laminated argillite, limestone, dolomite, and shale, probably of marine origin. They are best exposed in western Canada and Montana in a sinuous but discontinuous belt that presumably was once continuous (dotted pattern, Fig. 5.37). Other occurrences that may be part of this sedimentary sequence are found farther south, in the Uinta Mountains, the Grand Canyon area, and the Apache group in central Arizona, but their correlation with the Belt-Purcell sequence is difficult, because of the lack of continuity and the imprecision of dating in Precambrian rocks.

The main evidence for a rifting event around 1450 Ma is the emplacement of basaltic extrusive and intrusive rocks about this time and the beginning of deposition of the thick sequences of Belt-Purcell rocks. The sediments are deposited unconformably on crystalline basement in some places such as the eastern lobes of Belt-age rocks in Montana, the Uinta

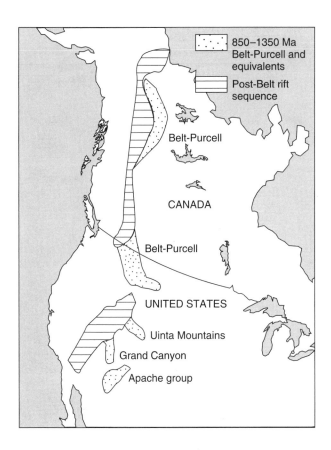

Figure 5.37 Diagram of rifted margin of western North America, showing generalized locations of the two rift sequences—the Belt-Purcell and the late Precambrian–early Paleozoic post-belt. *(After Dickinson, 1977)*

Mountains, the Grand Canyon suite, and in central Arizona. In most other areas, however, the base of the sediments is a thrust fault, and there is no direct evidence of what they were deposited on, although one reconstruction suggests it may have been oceanic crust.

The eastern lobes of sediment in Montana, the Uinta Mountains, and the Grand Canyon area are troughs of sediment oriented at a high angle to the general trend of the sedimentary belt. These troughs are thought to be aulacogens and to be related to the main rift in a manner similar to the relationship between the Benue Trough in central Africa and the Mid-Atlantic Rift (see Fig. 5.9).

In the late Precambrian, another rifting event occurred along the entire length of the Cordillera. This event initiated the deposit of a new sequence of post-Belt miogeoclinal sediments in a nearly continuous belt from northern Canada to the southwestern United States. The age of these sediments spans the time from late Precambrian (about 750–850 Ma) to the early Mesozoic (about 240 Ma) (Fig. 5.38A).

The main evidence for this rifting event includes minor extrusive rocks and more abundant dike rocks of basaltic composition that cut the Belt-Purcell sequence of rocks and are dated at about 850 Ma.

Along their eastern margin, the sediments unconformably overlie crystalline basement as well as the older Belt-Purcell rocks. The base of the sequence is characterized by a widespread pebbly mudstone or **diamictite,** thought to be of glacial origin. Fault-related deposits and saline deposits are rare.

Following these initial deposits, a vast sequence of carbonate and subordinate quartz-rich clastic sediments was laid down in a prism that increases in total thickness from 3 km on the platform to the east to 10 or 15 km farther west. The edge of the platform where the thickness of sediment begins increasing rapidly toward the west is known as the Wasatch Line (Fig. 5.38B). Because these sediments are fossiliferous, their ages and environments of deposition are known much more precisely than those of the earlier Belt-Purcell rocks. For the most part, they are shallow-water deposits that pass westward into deeper water sediments. The actual western boundary of these sediments is a tectonic contact; hence, the original configuration of the environments at or beyond the edge of the continent remains unknown.

Thus, the Cordilleran miogeocline exhibits features both similar to and different from those of the modern rifted margins. Similarities include the prisms of sediment unconformably overlying continental

Upper Precambrian and lower Cambrian rocks absent

Data from outcrop and drill-hole information:
- Complete
+ Incomplete

Stratigraphic section:
- - -   Isopach (contour interval 5000 feet)
D   Outcrop of diamictite unit
V   Outcrop of volcanic rock
S   Diabase and gabbro dikes and sills dated as late
        Precambrian by K-Ar methods

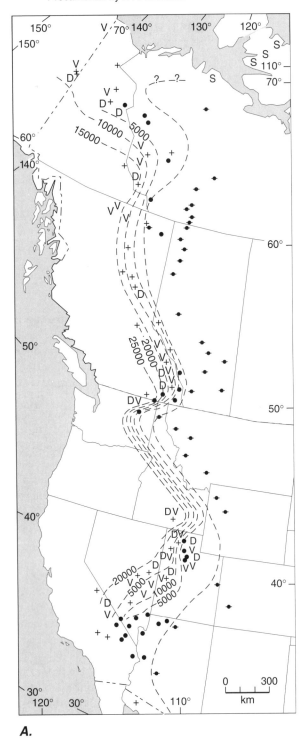

A.

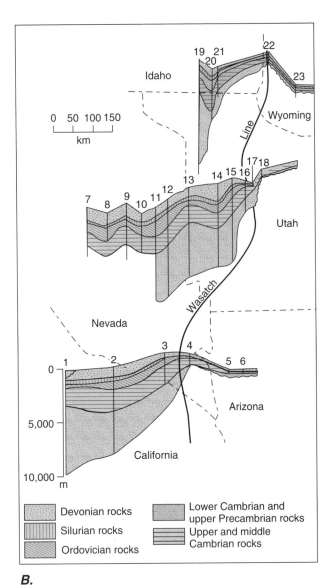

B.

Figure 5.38 North American Cordilleran miogeocline of post-Belt rocks. *A.* Map of the miogeocline, showing isopachs of thickness for the upper Precambrian–lower Cambrian strata, as well as locations of outcrops of diamictite and volcanic rocks. *B.* Representative stratigraphic cross sections of the Nevada-Utah-Wyoming region, showing thickening of sections to the west. State boundaries are deformed by removal of the Basin and Range extension. *(After Stewart, 1972)*

basement, thick sequences of rock types characteristic of passive margin sequences, the presence of aulacogens filled with the apparent equivalents of Belt-Purcell rocks, the early basaltic igneous rocks associated with the younger rifting event, and the thickening of the post-Belt sequence away from the interior of the continent.

Differences include the overall lack of well-developed normal faults and graben fillings below the main miogeoclinal deposits, the lack of saline deposits, the lack of preservation of any kind of oceanic crust beneath these sediments or their deeper water equivalents, the lack of volcanic sequences thick enough to be equivalent to the seaward-dipping reflector sequences in the modern oceans, and the presence of diamictite at the base of the post-Belt sequence. The lack of oceanic crust as basement to any of the sediments may be accounted for by thrust faulting that

separated most of the Belt-Purcell rocks from their basement and buried the western exposures of the post-Belt rocks. The lack of saline deposits is also found along some modern passive margins and is attributable to inappropriate climatic conditions. Other differences, such as the lack of normal faulting in the basement or of very thick and extensive volcanic sequences, however, may result from differences through geologic time in tectonic and/or igneous processes that are still poorly understood.

The Appalachian Miogeocline

The Appalachian miogeocline and its continuation to the southwest, the Ouachita miogeocline, had their inception in latest Precambrian time (Fig. 5.39). The earliest deposits are clastic sedimentary rocks and associated volcanic rocks that lie unconformably on

Figure 5.39 Map of Appalachian region showing late Precambrian–early Paleozoic rifted margin sequence, extending from Newfoundland to the southeast United States, and showing the present western edge of the deformed zone, the eastern edge of miogeoclinal deposits, and an inferred ancient continental margin. *(After Williams and Stevens, 1974)*

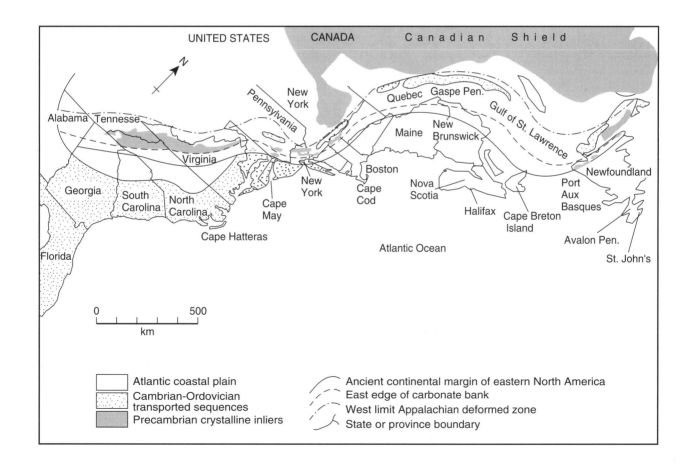

crystalline basement or occur in large thrust sheets. This unconformity, marking the onset of deposition of large thicknesses of sediment, along with the volcanic rocks, provides the main evidence for the rifting event. There is little evidence preserved for normal faulting, fault-bounded continental basins, or initial thick sequences of salt.

The basal clastic sediments are succeeded by a Cambrian-Ordovician carbonate sequence as much as 3000 m thick. This sequence resembles that of the modern Atlantic continental margin in general thickness distribution (Fig. 5.40). Tectonically overlying these miogeoclinal deposits to the east are thrust complexes that contain thick sequences of slightly to strongly metamorphosed deeper water clastic deposits and associated volcanic rocks. These sequences are up to several thousand meters thick and are thought to represent an ancient continental rise or oceanic rocks.

Deformation has obscured the real character of the transition from shallow-water sedimentary rocks of the Appalachian miogeocline to deeper water sediments beyond, and it has introduced large masses of allochthonous metamorphic and ophiolitic rocks that complicate the interpretation of the continental margin. The lower Paleozoic sediments of the Appalachian system, however, are similar to our model of a passive continental margin such as currently exists on the North American Atlantic coast, and we can conclude that this margin was a rifted continental margin.

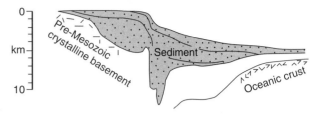

Seismic refraction profile of present Atlantic margin at Cape May

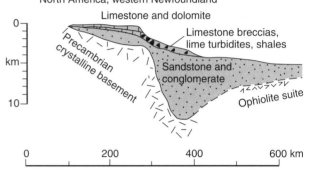

Restored cross section, ancient margin of eastern North America, western Newfoundland

Figure 5.40 Comparison of generalized sedimentary records of modern Atlantic continental margin with that represented by the latest Precambrian–early Paleozoic Appalachian miogeocline. *(After Williams and Stevens, 1974)*

Additional Readings

Allmendinger, R. W., T. A. Hauge, E. C. Hauser, C. J. Potter, S. L. Klemperer, K. D. Nelson, P. Knuepfer, and J. Oliver. 1987. Overview of the COCORP 40°N transect, western United States: The fabric of an orogenic belt. *GSA Bull.* 98:364–372.

Burke, K. 1980, Intracontinental rifts and aulacogens. In *Continental Tectonics.* Washington, D.C.: National Academy of Sciences.

Burke, K., and J. T. Wilson. 1976. Hot Spots on the Earth's Surface. In *Continents Adrift and Continents Aground.* Readings from Scientific America. J. T. Wilson, ed. San Francisco: W. H. Freeman and Company.

Coward, M. P., J. F. Dewey, and P. L. Hancock, 1987. Continental Extension Tectonics. Geol. Society of London, Special Publication 28.

Dixon, T. H., E. R. Ivins, and B. J. Franklin. 1989. Topographic and volcanic asymmetry around the Red Sea: Constraints on rift models. *Tectonics* 8:1193–1216.

Eaton, G. P. 1980. Geophysical and geological characteristics of the crust of the Basin and Range Province. In *Continental Tectonics.* Washington, D.C.: National Academy of Sciences.

Hamilton, W. 1978. Mesozoic tectonics of the western United States. In Soc. Econ. Paleontol. Mineralog. Pacific Coast Paleogeogr. Symp. 2. Los Angeles.

Hey, R. N. 1977. A new class of pseudofaults and their bearing on plate tectonics: A propagating rift model. *Earth Planet Sci. Lett.* 37:321–325.

King, P. B. 1977. *Evolution of North America.* rev. ed. Princeton: Princeton University Press.

Lister, G., M. A. Etheridge, and P. A. Symonds. 1986. Detachment faulting and the evolution of passive continental margins. *Geology* 14:246–250.

Macdonald, K. C. 1982. Mid-ocean ridges: Fine scale tectonic, volcanic and hydrothermal processes within the plate boundary zone. *Ann. Rev. Earth Planet. Sci.* 10:155–190.

Macdonald, K. C. 1989a. Tectonic and magmatic processes on the East Pacific Rise. In *Geology of North America*. vol. N, East Pacific Ocean and Hawaii. E. L. Winterer, D. M. Hussong, and R. W. Decker, eds. Boulder, Colo.: Geological Society of America.

Macdonald, K. C., and P. J. Fox. 1990. The mid-ocean ridge. *Scientific American* 72–79, June.

Mayer, L. 1986. Topographic constraints on models of lithospheric stretching of the Basin and Range province, western United States. GSA Special Paper 208.

Moores, E. M. 1982. Origin and emplacement of ophiolites. *Rev. Geophys. and Space Physics* 20:735–760.

Richardson, R. M., S. C. Solomon, and N. H. Sleep. 1979. Tectonic stress in the plates. *Rev. Geophys. and Space Phys.* 17:981–1019.

Rosendahl, B., D. J. Reynolds, P. M. Lorber, C. F. Burgess, J. McGill, D. Scott, J. J. Lambiase, and S. J. Derksen. 1986. Structural expressions of rifting: Lessons from Lake Tanganyika, Africa. In *Sedimentation in the African Rifts*. L. E. Frostick et al., eds. Geological Society of America, Special Publication 25.

Sclater, J. G., C. Jaupart, and D. Galson. 1980. The heat flow through oceanic and continental crust and the heat loss of the Earth. *Rev. Geophys. and Space Phys.* 18:269–311.

Sheridan, R. E., J. A. Grow, and K. D. Klitgord. 1988. Geophysical data. In *Geology of North America*. vol. I-2. The Atlantic continental margin: U.S. R. E. Sheridan and J. A. Grow, eds. Geological Society of America. Boulder, Colo.: 177–196.

Stewart, J. H. 1978. Basin-range structure in western North America: A review. *GSA Memoir* 152:1–31.

Varga, R. J., and E. M. Moores. 1990. Intermittent magmatic spreading and tectonic extension in the Troodos ophiolite: Implications for exploration for black smoker-type ore deposits. In J. Malpas, E. Moores, A. Panayiotou, and C. Xenophontos, eds. *Ophiolites: Oceanic Crustal Analogues*. Nicosia: Geological Survey of Cyprus. 53–64.

Williams, H., and R. K. Stevens. 1974. The ancient continental margin of eastern North America. In *The Geology of Continental Margins*. C. A. Burk and C. L. Drake, eds. New York: Springer.

6 Transform Faults, Strike-Slip Faults, and Related Fracture Zones

6.1 Introduction

Transform faults are plate margins characterized by strike-slip faulting, where adjacent plates move horizontally past each other and lithosphere is neither created nor destroyed. In principle, the active faults are everywhere parallel to small circles about the instantaneous pole of relative rotation between the two plates. Fracture zones are prominent breaks in the oceanic crust that include both transform faults and their inactive extensions into the plate interior beyond the actively shearing segment. The inactive parts of a fracture zone, therefore, preserve the structures created at the transform fault. These fracture zones appear as prominent linear topographic features on the ocean floor that mark discontinuities in the linear magnetic anomaly stripes. Because the Euler pole of relative angular velocity (or instantaneous relative rotation) for two plates tends to migrate with respect to both plates (see Sec. 4.4), the fracture zones created at a particular transform fault between those plates generally do not lie along a single small circle. Instead, a series of different small circles can be fit to segments of the fracture zone, reflecting the formation of the segments about different Euler poles.

Figure 6.1 shows the locations of major transform faults and associated fracture zones. Several different types of transform faults are possible, depending upon the type of plate boundaries that are connected by the fault.

A **ridge-ridge transform fault** (Fig. 6.2) connects two segments of a ridge or divergent plate margin; it is the most common of all transform fault types. Indeed, on average, such faults are present approximately every 100 km along the strike of a ridge axis. If, as is usual, the spreading is symmetrical about a nonpropagating ridge, and the linear rate of spreading increases

Figure 6.1 Generalized world map (Mercator projection) showing major oceanic fracture zones, active transform faults, and active and inactive continental strike-slip faults. *(After Menard and Chase, 1970)*

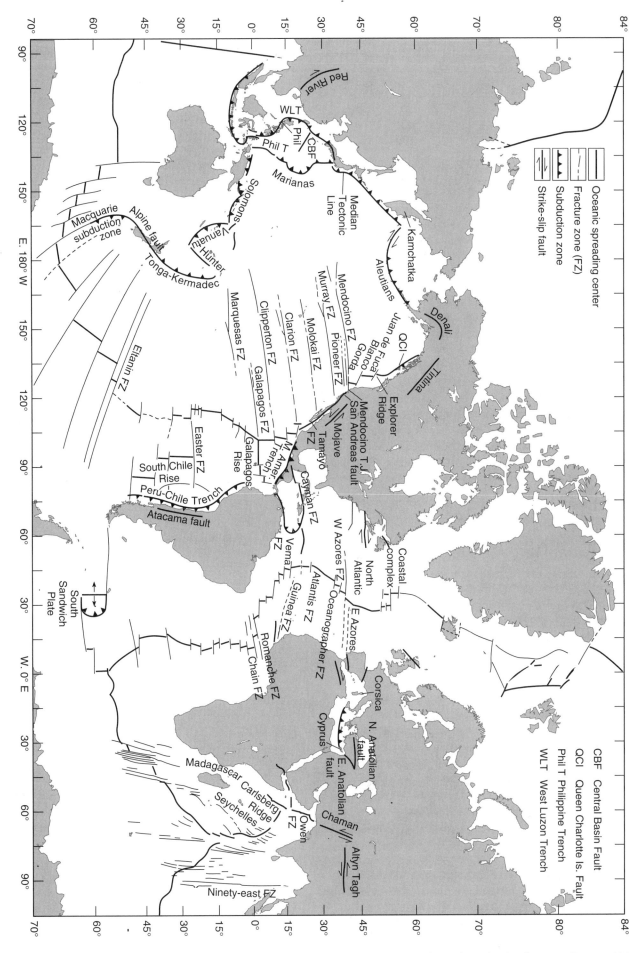

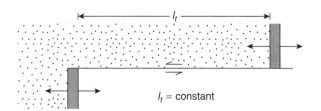

$l_t = \text{constant}$

Figure 6.2 A ridge-ridge transform fault. The transform fault is at a constant angular distance from the Euler pole of relative velocity; it is not active beyond either ridge segment.

as the sine of the angular distance from the Euler pole of relative angular velocity, each transform fault maintains a constant length between ridge segments. Examples of such transform faults are especially prominent in the equatorial Atlantic and Indian oceans, but they are present on other ridges as well (see Fig. 6.1).

Ridge-trench transform faults are of two types, depending on the polarity of the trench (Fig. 6.3). The first type (Fig. 6.3A) connects a ridge with the overriding side of a convergent boundary. The fault lengthens at a rate equal to the half-spreading rate of the ridge. An example is found southeast of the tip of South America, where transform faults on the north and south margins of the tiny South Sandwich Plate connect the trench to the short spreading axis just behind the trench (see Fig. 6.1).

The second type of ridge-trench transform fault (Fig. 6.3B) connects a ridge with the downgoing side of a trench. In this case, the length of the fault changes at a rate equal to the difference between the half-spreading rate and the subduction rate. Thus, if the half-spreading rate exceeds the subduction rate, the ridge migrates away from the subduction zone, and the length of the transform fault increases. If the subduction rate exceeds the half-spreading rate, the ridge migrates toward the trench and the length of the transform fault decreases. Examples of this type of transform fault include the Queen Charlotte Island fault, which connects the Explorer Ridge to the subduction zone off the northwest coast of the United States; the transform fault connecting the South Chile rise to the South American subduction zone, and the transform fault that connects the Galapagos Rise to the Central American subduction zone (see Fig. 6.1).

Trench-trench transform faults are of three types distinguished by the polarities of the two convergent margins connected by the transform (Fig. 6.4). If the two subduction zones dip toward each other (Fig. 6.4A), the two subduction rates are necessarily equal and the transform fault lengthens at this same rate. Examples (see Fig. 6.1) include the Alpine fault in

New Zealand between the Kermadec Islands and Macquarie Ridge (a nascent subduction zone), and an incipient boundary of this type in Luzon between the West Luzon Trench and the Philippine Trench.

If the two subduction zones dip away from each other (Fig. 6.4B), then the two subduction rates are equal and the transform fault shortens at the same rate. No clear examples of this kind of transform fault are present on current plate boundaries, perhaps because this geometry must eventually evolve to the geometry shown in Figure 6.4A.

Finally, if two subduction zones dip in the same direction (Fig. 6.4C), the transform fault remains constant in length. A modern example is the fault between the western Aleutian Islands and Kamchatka along the North American–Asian plate boundary (see Fig. 6.1).

These examples of transform faults include all possible types that connect two simple boundaries. Other types end at triple junctions. In these cases, it is not possible to predict how the fault will evolve and whether the fault will lengthen or shorten without knowing the geometry of the triple junction and the relative velocities of the three plates involved (see Box 4.2). The San Andreas transform fault in California is one that terminates at each end in a triple junction. We discuss this fault system in more detail in Section 6.3.

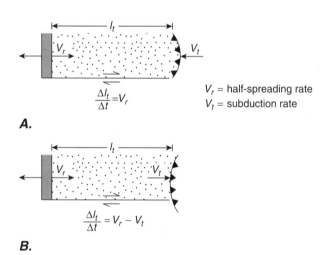

$$\frac{\Delta l_t}{\Delta t} = V_r$$

V_r = half-spreading rate
V_t = subduction rate

A.

$$\frac{\Delta l_t}{\Delta t} = V_r - V_t$$

B.

Figure 6.3 Ridge-trench transform faults. *A.* The transform fault connects a ridge and an overriding trench margin. The length of the fault increases at a rate equal to the half-spreading rate. *B.* The transform fault connects a ridge to the downgoing side of a trench margin and separates two plates that are being subducted at different rates. The rate of change of fault length is the difference between the half-spreading rate and the subduction rate of the stippled plate.

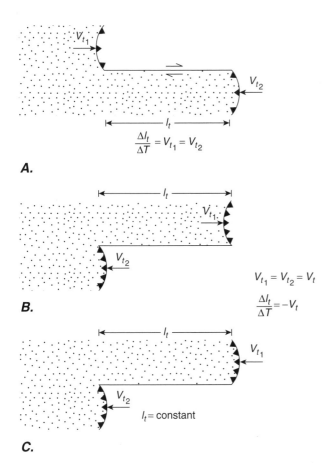

A.

B.

C.

$$\frac{\Delta l_t}{\Delta T} = V_{t_1} = V_{t_2}$$

$$V_{t_1} = V_{t_2} = V_t$$

$$\frac{\Delta l_t}{\Delta T} = -V_t$$

$l_t =$ constant

Figure 6.4 Trench-trench transform faults. *A.* A transform fault connects two subduction zones that dip toward each other. The subduction rates are equal ($V_{t_1} = V_{t_2}$). The fault lengthens at a rate equal to the subduction rate. *B.* A transform fault connects two subduction zones that dip away from each other. The subduction rates are equal, and the transform fault shortens at a rate equal to the subduction rate. This geometry should eventually evolve to that shown in (*A*). *C.* A transform fault connecting two subduction zones of the same polarity. The fault remains constant in length.

6.2 Oceanic Transform Faults and Fracture Zones

Our limited knowledge of oceanic transform fracture zones is derived from bathymetric measurements; geophysical measurements; dredge sampling; and, in a few areas, detailed surveys by submersibles. In some places, however, oceanic crust is exposed on land where a more detailed investigation of the structure of these faults is possible.

Physiography and Structure

Oceanic transform fracture zones are typically regions of sharp topographic discontinuities, where steep ridges or scarps form the sides of narrow, deep basins. Like other strike-slip faults, a transform fracture zone is not a single plane through the Earth's crust but rather can be as much as several kilometers wide.

Fault scarps are especially apparent on ridge-ridge transform fracture zones and are most pro-

nounced on long transform fracture zones where there is a large age difference across the fault. The younger, hotter lithosphere has an equilibrium elevation higher than that of the older, cooler lithosphere (Fig. 6.5), and the difference in temperature (reflecting the difference in age) across the transform fracture zone tends to show up as a topographic scarp. Because of this relationship, the high side of the fault changes midway between the ridge segments.

The Tamayo fracture zone is a ridge-ridge transform fault that illustrates the characteristics of these features. It is located at the mouth of the Gulf of California, where the ridge-ridge Tamayo transform fault joins segments of the East Pacific Rise. Figure 6.6A shows the topography of an area along this fracture zone, and Figure 6.6B shows several topographic profiles across the zone. Its width ranges from 1 km (at the intersection with the southern ridge axis) to 30 km a few tens of kilometers away. The younger side of the

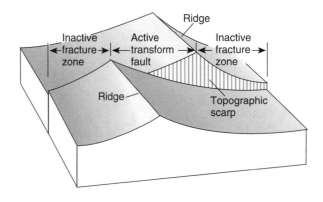

Figure 6.5 A transform scarp is formed because of the contrast in age and temperature of the lithosphere across the fracture zone. Younger, hotter lithosphere stands topographically higher than older, cooler lithosphere. Not to scale.

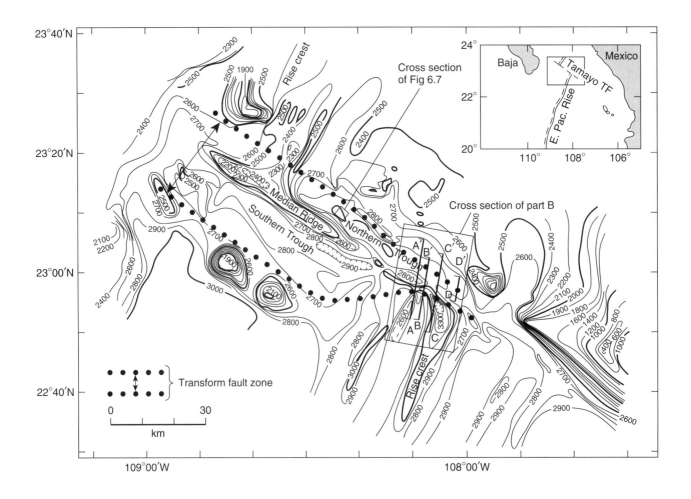

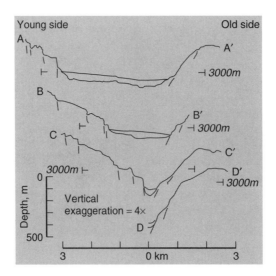

Figure 6.6 The Tamayo transform fracture zone in the mouth of the Gulf of California (shown by inset map). *A.* The topography of the Tamayo transform fracture zone in the vicinity of the two ridge segments. The contours are the depth below sea level, so high contours are topographic lows. *B.* Profiles across the Tamayo fracture zone. *(A. after Robinson et al., 1983; B. after Macdonald et al., 1979)*

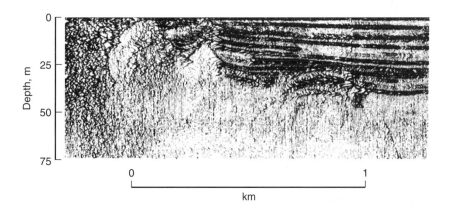

Figure 6.7 Seismic profile across the Tamayo transform fault showing possible serpentinite diapirs. *(After Macdonald et al., 1979)*

Figure 6.8 Characteristics of slow-spreading (less than 5 cm/yr) transform fault zones. *A.* Schematic topographic map showing a transform valley and the presence of a principal transform displacement zone. *B.* Schematic structural cross section of a transform fault zone showing the pronounced contrast in lithospheric thickness across the zone and the characteristic thinning of the oceanic crust under a ridge near the transform fault. The contrast in the lithospheric thickness tends to inhibit both a change in location of the zone and the propagation of a ridge crest across it. *(After Fox and Gallo, 1984)*

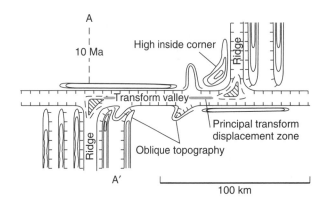

fracture zone is generally higher in elevation than the older side.

Geologically, transform fracture zones appear to be quite complex. Dredge hauls from fracture zones include serpentinites; mylonitized and foliated ultramafic and mafic plutonic rocks; and scattered olivine-rich, low-silica volcanic rocks. Coarse talus breccias and turbidite deposits are also common, derived from the local sharp relief. Seismic reflection profiles within transform fracture zones have revealed diapirlike structures that may be serpentinite diapirs (Fig. 6.7).

The topography of a transform fracture zone and the associated crustal structure depends in part on the spreading rate and the relative age of the lithosphere on either side of the fault. These differences are illustrated schematically in Figures 6.8, 6.9, and 6.10, for slow-, medium-, and fast-spreading ridges, respectively. Transform faults connecting slow-spreading ridges (< 5 cm/yr) have a pronounced valley, within which is a single rather narrow fault zone, generally less than 1 km in width, along which most of the displacement apparently takes place (Fig. 6.8A).

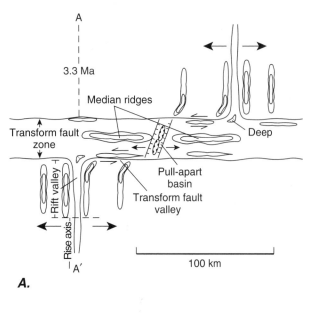

A.

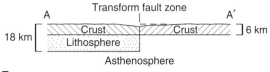

B.

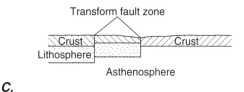

C.

Figure 6.9 Characteristics of transform fault zones slipping at an intermediate rate (5–9 cm/yr). *A.* Schematic map showing a wider transform zone that includes parallel ridges and valleys and one or more pull-apart basins within the transform zone. *B.* Schematic cross section showing the lesser contrast in lithospheric thickness across the transform zone and less crustal thinning under the ridge. *C.* The wide transform fault zone may be reflected in a distribution of shearing on more than one shear in the fault zone. The resulting segments of lithosphere may vary in thickness. *(After Fox and Gallo, 1984)*

sion, respectively. More than one zone of active slip may be present.

The contrast in lithospheric thickness across the fault (Fig. 6.9B) is less pronounced than that across slow-shearing transform faults (see Fig. 6.8B) because age differences across the fault are smaller. For a 100-km transform fault shearing at 6 cm/yr (Fig. 6.9A), the age of the lithosphere at the cross section A–A′ is 3.3 m.y.; the lithosphere is of crustal thickness (6 km) under the ridge and only 18 km thick across the transform fault. Where the fault is segmented, lithospheric thickness may vary in increments across the fault zone (Fig. 6.9C).

Transform faults connecting fast-spreading ridges (9–18 cm/year) are wide zones of complex faulting (Fig. 6.10A). The width of the transform zone varies from tens of kilometers to over 100 km, and the zones include short spreading segments that are not necessarily perpendicular to the spreading direction and may propagate across the zone. There is little contrast in the lithospheric thickness across the fault zone, because the age contrast is small. For example, for a 100-km transform fault moving at 12 cm/yr, the maximum age contrast across the transform is 1.1 m.y. at the section A–A′ (Figs. 6.10A and 6.10B). The lithosphere under the ridge is the usual crustal thickness of about 6 km; across the transform it is only 10 km thick (Figs. 6.10B and 6.10C). Thus, there is little restraint on the exact location of the transform fault. Complications of structure can spread out the transition zone, as illustrated in Figure 6.10D.

In summary, slow-moving transform zones appear to be relatively sharp features that are relatively stable in their location on the plate boundary. As the rate of displacement across the fault increases, the contrast in lithospheric thickness decreases, and the transform fault zones become more complex, with the onset of multiple shear zones, small propagating spreading centers, and volcanism.

There is a pronounced contrast in the thickness of lithosphere across the transform because of the age discontinuity. Older lithosphere is cooler and therefore thicker than younger lithosphere. For example, the cross section A–A′ in Figure 6.8B illustrates the structure of a 100-km transform fault connecting slow-spreading ridge segments (2-cm/yr spreading rate). The young lithosphere under the ridge is no thicker than the crust itself, but across the transform it is 10 m.y. old and 30 km thick. This difference tends to prevent migration of the transform fault and to keep the fault zone narrow.

Transform faults connecting ridge segments that spread at an intermediate rate (5–9 cm/yr) display a somewhat wider fault zone (Fig. 6.9A) than slow-shearing transforms. Within the fault zone, alternating basins and ridges mark areas of intrusion and exten-

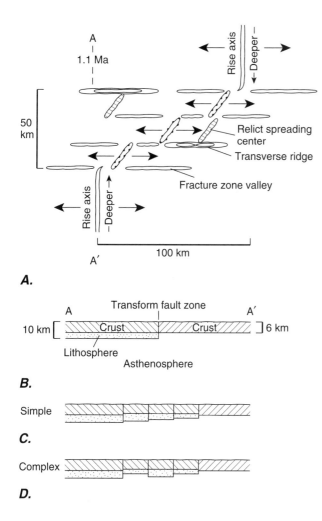

A.

B.

C.

D.

Figure 6.10 Characteristics of fast-spreading (9–18 cm/yr) transform fault zones. *A.* The transform fault zone is a complex zone tens of kilometers wide that includes transverse ridges and valleys with numerous active or inactive pull-apart basins or spreading centers and distributed shearing. *B, C, D.* Schematic cross sections showing little or no contrast in lithospheric thickness across the transform fault zone, segmented lithosphere with variable thickness, and little crustal thinning under the ridge in the vicinity of the transform. *(After Fox and Gallo, 1984)*

In all regions near transform faults, there is a tendency for the structural "grain" of the oceanic crust—defined primarily by the orientation of linear features such as faults and topography, but also defined by the orientation of dikes—to turn toward the adjacent

ridge segment. This characteristic is evident in Figures 6.6A, 6.8A, 6.9A, and 6.10A.

Subaerial Exposures of Fossil Oceanic Transform Fault Zones

Examples of oceanic fracture zones now exposed on land are limited to a few occurrences within ophiolites. Because they provide insight into the geologic details along modern oceanic transform faults, we describe three examples: the Arakapas fault–Limassol forest region of the Troodos ophiolite in Cyprus, the Coastal complex in Newfoundland, and the ophiolites of Italy and Corsica.

The Arakapas fault–Limassol forest region is a zone 10–20 km wide characterized by east-west trending faults, talus breccias intruded by dikes, and a large serpentinite diapir (Fig. 6.11A). The dikes in the so-called sheeted dike complex to the north change their trend progressively from north to northeast as they approach the fault zone.

The intrusion of dikes into fault breccias attests to igneous activity in the fault zone, and the change of dike direction approaching the fault zone from the north mimics the change of the structural grain of faults and dikes near transform faults mentioned above. Apparently, the serpentinite diapir was active during an early stage in the evolution of the complex, because it too is intruded by later dikes.

All rocks of the complex are overlain by pelagic sediments of late Cretaceous age. Because the sediments are at most only slightly deformed, most fault activity must have taken place before deposition of the sediments. Figure 6.11B shows a composite cross section through the Arakapas fault belt that illustrates these relationships. These data strongly suggest that this region is a fracture zone raised above sea level and only minimally eroded.

The Coastal complex in Newfoundland (Fig. 6.12A) lies northwest of the large Bay of Islands ophiolite complex, which is exposed in several massifs. The main complex shows a complete ophiolite stratigraphy, with a dike complex striking northwest. The Coastal complex consists of deformed and metamorphosed gabbro, ultramafic rocks, and metasedimentary rocks intruded by plutonic rocks and overlain by less deformed volcanic rocks. Figure 6.12B is a block diagram of the Coastal complex showing the relationships between major rock units, and in particular an intrusive contact between the main Bay of Islands complex and the deformed rocks within the coastal complex. Figure 6.12C illustrates this relationship schematically. Apparently, the entire fault

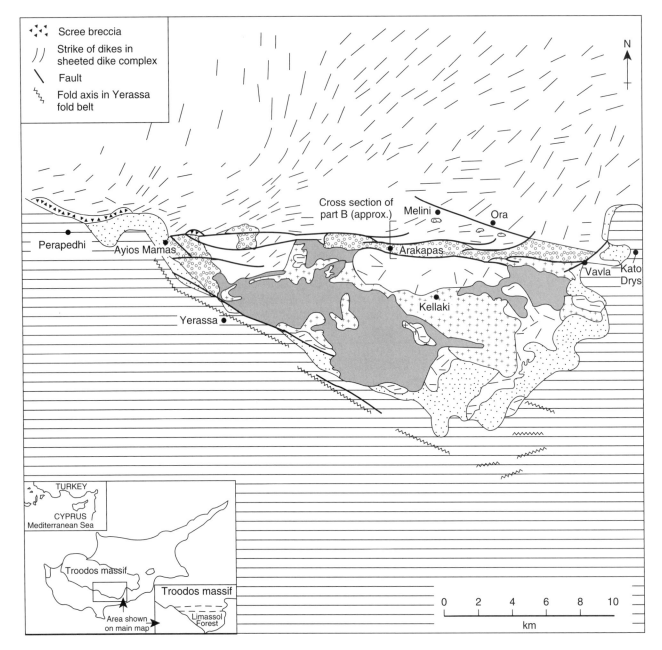

A.

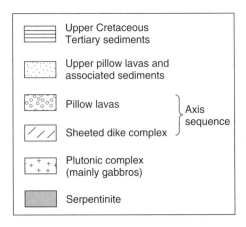

Scree breccia

Strike of dikes in sheeted dike complex

Fault

Fold axis in Yerassa fold belt

TURKEY

CYPRUS
Mediterranean Sea

Troodos massif

Troodos massif

Area shown on main map

Limassol Forest

Perapedhi

Ayios Mamas

Yerassa

Cross section of part B (approx.)

Melini

Ora

Arakapas

Vavla

Kato Drys

Kellaki

| 0 | 2 | 4 | 6 | 8 | 10 |

km

Upper Cretaceous Tertiary sediments

Upper pillow lavas and associated sediments

Pillow lavas

Sheeted dike complex

Plutonic complex (mainly gabbros)

Serpentinite

Axis sequence

Figure 6.11 The Arakapas fault–Limassol forest region, Troodos complex, Cyprus, a possible oceanic transform fault. *A.* Map of the complex. The dikes change orientation from N–S to ENE–WSW with increasing proximity to the fault zone. The serpentinite diapir was active during formation of the complex. B. (opposite page) Schematic N–S cross section through the Arakapus fault zone showing the brecciated basement rocks overlain by talus deposits and interlayered sediments and volcanics. *(After Moores and Vine, 1971; Simonian and Gass, 1978)*

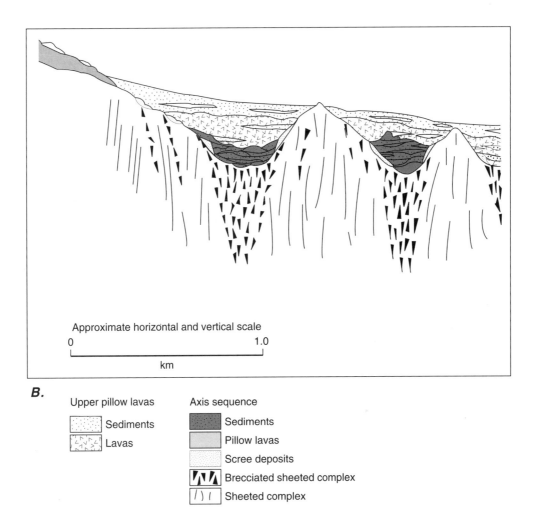

B.

Upper pillow lavas

Sediments

Lavas

Axis sequence

Sediments

Pillow lavas

Scree deposits

Brecciated sheeted complex

Sheeted complex

Approximate horizontal and vertical scale

0   1.0

km

zone was active before and during intrusion of the Bay of Islands rocks. Thus, these rocks may represent a deeper level exposure of a fossil oceanic transform fault that developed as the ophiolite formed during early Ordovician time.

The ophiolite complexes of the Appenines of Italy and the mountains of Corsica are also possible examples of an exposed transform fracture zone. In that region, ophiolite complexes with a peculiar pseudostratigraphy extend discontinuously for several hundred kilometers along the strike of the Italian peninsula (Fig. 6.13A). These bodies typically display serpentinized ultramafic rocks with rare deformed gabbro, overlain predominantly by sedimentary breccias containing coarse angular blocks of serpentinite cemented by calcite. These calcite-cemented breccias, called ophicalcites by the Italian geologists, are in turn overlain or intruded by undeformed volcanic rocks

(Fig. 6.13B). Radiometric age differences of several tens of millions of years are present along the belt. The sequences of rocks observed are characteristic of a transform fracture zone, and the age variations are consistent with those expected from volcanism at a ridge-transform intersection. Thus, these relationships suggest that the ophiolites are the remnants of a large transform fracture zone or zones that were active during Jurassic time.

All three of these examples have some features in common. All have deformed mafic and ultramafic plutonic rocks overlain by breccias, and the entire sequence is then intruded or overlain by mafic dikes and extrusives. Thus, these sections of ophiolitic rocks are quite different from the traditional ophiolite sections outlined in Chapter 5. Where found, such anomalous ophiolite sections may imply the presence of former transform faults.

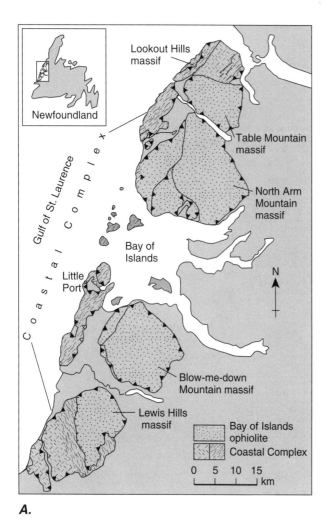

A.

Lookout Hills massif

Newfoundland

Gulf of St. Laurence

C o a s t a l   C o m p l e x

Table Mountain massif

North Arm Mountain massif

Bay of Islands

Little Port

N

Blow-me-down Mountain massif

Lewis Hills massif

Bay of Islands ophiolite

Coastal Complex

0   5   10   15
km

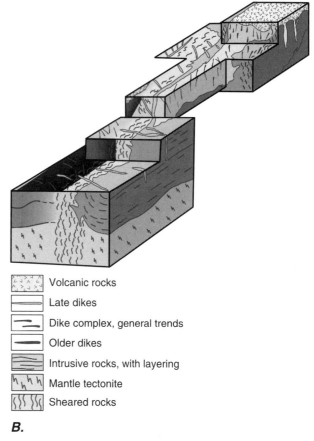

Volcanic rocks

Late dikes

Dike complex, general trends

Older dikes

Intrusive rocks, with layering

Mantle tectonite

Sheared rocks

B.

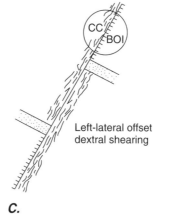

Older, cooler, subsided

Younger, hotter elevated

Coastal complex

Bay of Islands complex

Deformed crust | Undeformed crust

CC   BOI

Left-lateral offset dextral shearing

C.

Figure 6.12 The Bay of Islands complex, Newfoundland, an ophiolite with a possible fossil transform fault zone. *A.* Map of Bay of Islands complex, showing the Coastal complex, a possible oceanic transform fault zone. *B.* Block diagram showing the three-dimensional relationships of the Coastal complex. *C.* Schematic diagram showing intrusive relationship between Coastal and Bay of Islands complexes, indicating a right-offset transform fault. *(After Karson, 1986b)*

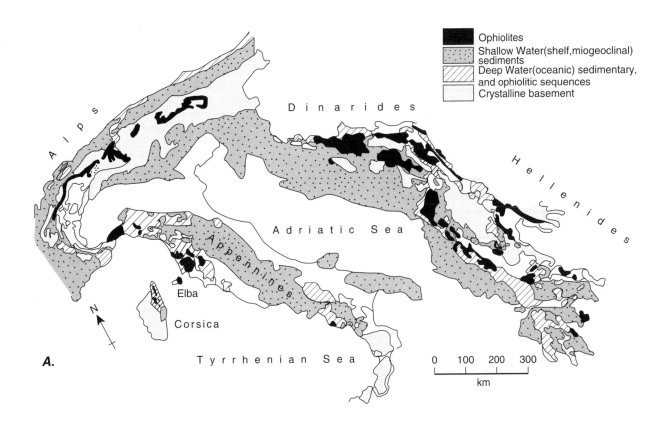

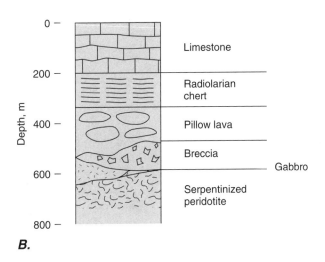

Figure 6.13 The Italian and Corsican ophiolites are possible examples of a transform fault zone. *A*. Map of the Mediterranean region showing location of ophiolite complexes and principal associated rock types. *B*. Schematic columnar section illustrating the characteristic sequence of rocks in the Italian and Corsican ophiolites. Ophiolite complexes in the Dinarides and Hellenides show a more usual sequence, as in Figure 5.27. *(A. after Moores, 1982; B. after Abbate, 1980)*

6.3 Models of Transform Fault Processes

Careful dredging—combined with seismic data, direct obervation from submersibles, and analysis of transform fracture zones in ophiolites—gives a complex picture of a transform fracture zone. A schematic model of some of the main features of such zones is shown in Figure 6.14 as a vertical section parallel to the transform fault (A–B, Fig. 6.14A) and across the fault (N–S, Fig. 6.14B). The normal oceanic crustal sequence, from top to bottom, of pillow basalts and extrusives, dikes and sills, gabbro, and peridotite is partially metamorphosed and cut by zones of deformation, which are typically serpentinized. Serpentinite diapirs intrude the crust in places, and fault slices and horsts of these different rock types occur in the fault zone out of place with respect to the normal sequence. Surficial talus deposits are common along the scarps, and young extrusives may cover the deformed rocks and even the talus deposits. A large area within the fracture zone may be underlain by serpentinite.

The geologic and topographic complexity along transform faults is partly attributable to the fact that the faults tend to be small circles about the Euler pole of instantaneous rotation and the pole migrates with respect to both plates during spreading. Thus, a transform fault must be continually changing its configuration and kinematic character with compressional and extensional deformation developing along the faults during the migration of the Euler pole (see Sec. 4.4; Fig. 4.10).

Another reason to expect complexity along transform fracture zones is the type and timing of the processes that occur along a transform fault, shown schematically in Figure 6.15. Normal oceanic crust is formed by magmatic intrusion and extrusion at each ridge, and it is carried away from the ridge by spreading. Rocks erupted adjacent to a transform fault that drift toward the adjacent ridge segment, however, become deformed within the transform fault zone. Talus breccias accumulate on the low side of the transform fault scarp and cover the older deformed extrusives. Later, the talus deposits themselves may become involved in the deformation. As these rocks move past the other ridge segment, the deformed crust and talus deposits are intruded and/or covered by new magma. Beyond this ridge segment, however, shearing on the transform fracture zone ceases, so that the newest lavas remain undeformed. Thus, transform fracture zones can be expected to display complex intrusive and extrusive contacts between deformed and undeformed igneous and sedimentary rocks.

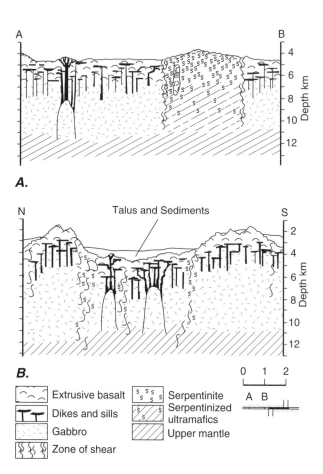

Figure 6.14 The Oceanographer transform fault showing possible geology, based upon dredge samples. A. Longitudinal cross section. B. Transverse cross section. *(After Fox et al., 1976)*

Crustal thinning near transform faults may result from the fact that near such faults, the rising column of asthenosphere is bordered by cooler lithosphere both above and on one side (the side of the transform fault), rather than just above. This situation would enhance the cooling of the rising asthenosphere and result in less partial melting and less efficient separation of the melt from the solid mantle. Less magma would be available to form the magmatic crust, resulting in a thinner crust.

The development of an axial valley on fast-spreading ridges near transform faults may result from a similar process. According to the model discussed in Section 5.7, lower lithospheric temperatures could promote the formation of axial valleys. The presence of a cool side as well as a cool lid to the rising mantle column may make the thermal structure of a fast-spreading ridge near a transform fault resemble

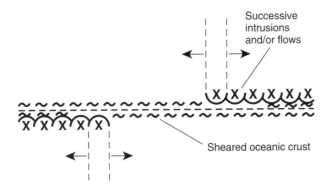

Figure 6.15 Diagram illustrating the complex relationship between intrusion and shearing along a transform fault. The active transform fault between ridge segments shows deformed oceanic crust. The portion of the fracture zone beyond the ridges shows complex igneous intrusion into and flows over previously deformed crust.

that of a slow-spreading ridge, which would account for the formation of the axial valleys (Fig. 6.16).

The change in the strike of topography, faults, and dikes near a transform fault may reflect a change in the orientation of the bulk stress field associated with the ridge system. If we assume an Anderson model for the relationship between faults and principal stresses, then the stress field far from a transform fault should have the orientation characteristic of the normal faults that parallel the ridge axis; that is, we would expect the maximum compressive stress $\hat{\sigma}_1$ to be vertical, the intermediate compressive stress $\hat{\sigma}_2$ to be horizontal and parallel to the ridge, and the minimum compressive stress $\hat{\sigma}_3$ to be horizontal and perpendicular to the ridge. Assuming that dikes are injected normal to the minimum compressive stress $\hat{\sigma}_3$ and parallel to $\hat{\sigma}_1$ and $\hat{\sigma}_2$, this stress field would result in vertical dikes parallel to the ridge axis. At the transform fault, however, the orientation of the principal stresses should be characteristic of strike-slip faults. Thus, we would expect $\hat{\sigma}_2$ to be vertical, $\hat{\sigma}_1$ to be horizontal at an angle of 30° to 45° to the fault, and $\hat{\sigma}_3$ to be horizontal at an angle of 45° to 60° to the fault. This orientation of the principal stresses also may account for the change in the orientation of the dikes in the vicinity of the south Troodos transform fault (see Fig. 6.11A).

Propagating ridges (compare Sec. 5.5, Fig 5.22) may also play a role in the complexity of transform fault zones. As one ridge segment propagates at the expense of the other, material is transferred from one plate to the other. If the ridges are "dueling"—that is, if the segment on one side of the fault propagates first and then recedes as the other side propagates—the result could be a wide transform fault zone with com-

plex age relationships within the zone. Consider, for example, the diagram in Figure 6.17. Two plates, A and B, are separated by a transform fault. At time 1, the dark stripes have just formed. At time 2, the northern ridge has propagated southward and a small bit of plate A (labeled a) has been incorporated into plate B. At time 3, the southern ridge has propagated northward as the northern ridge has receded and piece a and a former piece of plate B (labeled b) have been attached to plate A. As this process develops, inverted L-shaped pieces of lithosphere are formed. At time 6, the northern ridge has propagated back part of the distance toward the south. Continued alternative propagation of the north and south ridges has produced a fault zone between the E–W dotted lines where the age progression is complex and where old blocks of lithosphere (dark shaded rectangles) are present quite near the new lithosphere.

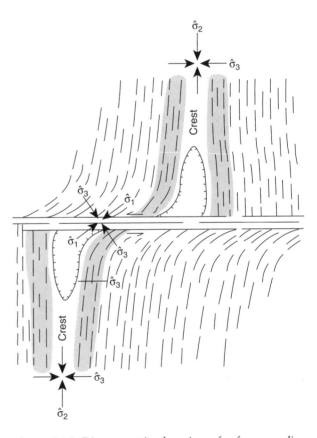

Figure 6.16 Diagrammatic plan view of a fast-spreading ridge showing the development of an axial rift valley near a transform fault. Also shown is the change in the orientation of the structural grain, defined by topography, faults and dikes, and the orientations of the inferred principal stress directions far from and near the transform fault that can account for this reorientation of structures.

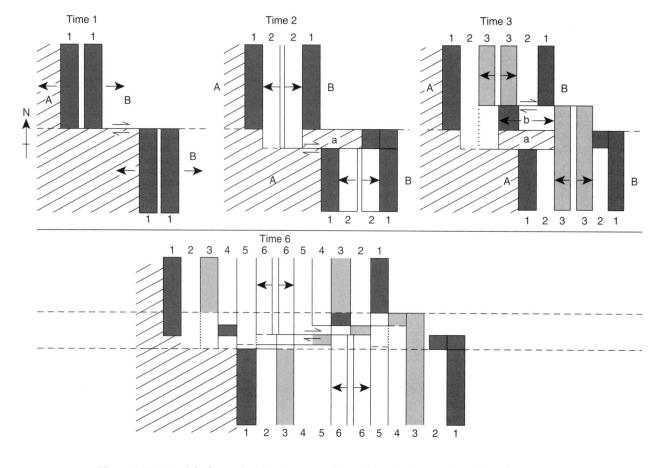

Figure 6.17 Model of complex development of transform fault zones in region of propagating ridges. Time 1, initial configuration; time 2, northern ridge has propagated to south; time 3, southern ridge has propagated to north; time 6, final configuration. Pieces of lithosphere are transferred from one plate to the other and back again during alternating propagation, leading to a complex age distribution in the transform fault zone. *(After Karson, 1986a)*

Such a model can explain the width and complexity of many transform fault zones. In particular, the model can explain why, in places such as in the equatorial Atlantic Ocean, old rocks are found very near mid-ocean ridge crests along some transform fault zones.

6.4 Active Continental Transform Faults

Transform faults that cut continental crust provide another opportunity to study on land the processes associated with transform faulting. They are similar to oceanic transform faults in that the motion is essentially strike-slip and the seismicity is shallow, but the involvement of continental crust affects the character of the faults and adds to their complexity, making it difficult to draw a strong parallel to oceanic transform faults. Nevertheless, the study of continental transform faults is important because of our interest in interpreting ancient structures in continental rocks, not to mention the seismic hazard they present to populated areas.

Transform faults in continents can be expected to follow preexisting zones of weakness, such as older faults that might originally have been a different type. In general, such fault zones are not parallel to a small circle about the current Euler pole of relative rotation. As a result, extensional or contractional strike-slip duplexes should form, depending upon the relative orientations of the fault and plate motion. Figure 6.18 shows a situation in which the fault zone is not

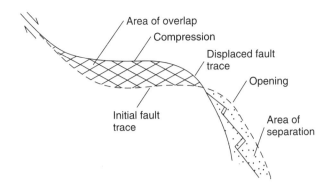

Figure 6.18 Diagram showing the development of compression and opening along an irregular trace of continental transform fault zone. See the text for discussion.

everywhere parallel to the relative velocity vector for the two plates. In principle, the relative plate motion should cause the plates to overlap within the cross-hatched area, thus producing contractional structures such as folds and thrust faults, whereas the motion should cause the plates to separate within the stippled region, thus producing extensional structures such as pull-apart basins and short ridge segments separated by transform faults. Continental transform fault zones also show the structural complexities already discussed for oceanic regions.

To illustrate the complexities that characterize continental transform systems, we describe selected features of the San Andreas fault system in California, the Alpine fault of New Zealand, and the Dead Sea fault zone in the Middle East, which are three well-studied examples of continental transform faults.

The San Andreas–Gulf of California Transform System

The San Andreas–Gulf of California transform fault system is a zone extending approximately 3000 km from the Mendocino transform-transform-trench triple junction off the northwest coast of California, southeast to the Rivera ridge-trench-transform triple junction near the mouth of the Gulf of California (Fig. 6.19). The transform system consists of two main parts: the San Andreas strike-slip fault zone and associated faults to the northwest, and the obliquely rifting Gulf of California to the southeast.

The San Andreas part of the system extends from Cape Mendocino in the northwest to the Imperial Valley in the southeast (Fig. 6.20). This fault zone consists of a number of currently active subparallel, right-lateral strike-slip faults. In addition to the San Andreas fault itself, subsidiary faults include the Hayward, Calaveras, and Green Valley faults in the north

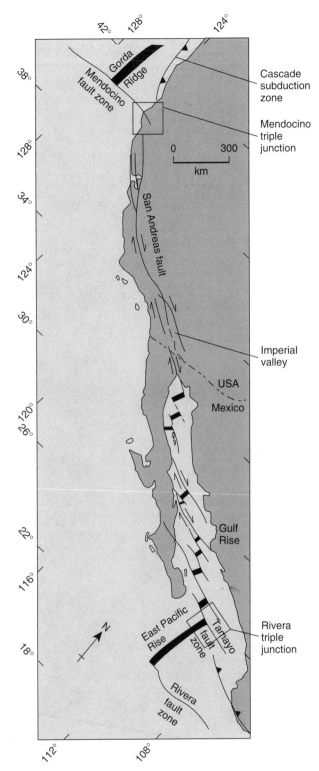

Figure 6.19 San Andreas–Gulf of California transform system. *(After Macdonald et al., 1979)*

and the San Jacinto and San Gabriel faults in the south. Much of the current and most recent movement is taken up on these subsidiary fault zones,

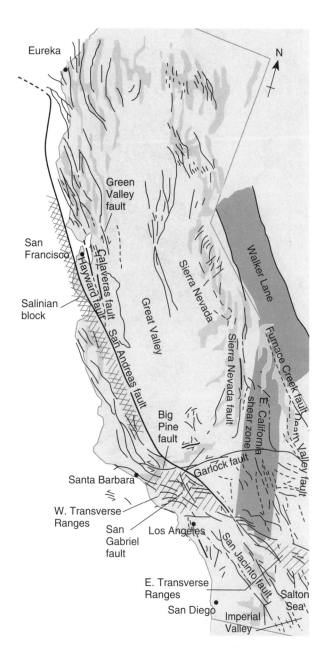

Figure 6.20 Simplified fault map of California showing the San Andreas fault system and principal branches in north and south. *(After Anderson, 1971)*

At its southern end, the San Andreas fault loses its integrity as a single fault and splays out into a series of parallel strike-slip faults in southeastern California (see Fig. 6.20) that continue to the head of the Gulf of California. From there to the mouth of the Gulf, the transform fault system consists of a series of long transform faults and very short ridge segments (see Fig. 6.19).

Most of the floor of the Gulf is fairly shallow, and much of the crust is thinned continental crust. Evidence from deep-sea drilling suggests that the thinning has occurred on listric normal faults. In the fault basins, accumulated sediments have been intruded by sills and overlain by flows of basaltic material. Only at the spreading centers does the bottom of the Gulf approach abyssal depths, and there the crust exhibits oceanic crustal thickness. Magnetic anomalies in the deeps and in the Gulf mouth indicate that it opened approximately 5 million years ago. Since that time, a total of approximately 250 km of oblique opening and concomitant displacement on the transform faults has occurred, suggesting an average displacement rate of 5.0 cm/yr. At present, the displacement rate on the fault system is approximately 5.5 cm/year, based on an analysis of marine magnetic anomalies and offsets of young features.

The origin of the San Andreas system can be inferred from a reconstruction of the plates using the magnetic anomaly stripes and transform fracture zones preserved in the eastern Pacific ocean basin. Figure 6.21A shows that in early Tertiary time the Pacific Plate was completely separated from the North American Plate by the Farallon Plate, which was being subducted beneath the North American Plate. The East Pacific Rise formed the boundary between the Farallon and the Pacific plates. Sometime during the Oligocene epoch (Fig. 6.21B), a ridge-transform intersection on the East Pacific Rise was subducted beneath North America, forming an instantaneous quadruple junction that resolved into two triple junctions (Fig. 6.21C), one a transform-transform-trench triple junction to the northwest (now the Mendocino triple junction) and the other a transform-ridge-trench triple junction to the southeast (now the Rivera triple junction). The transform fault connecting these two triple junctions was the beginning of the San Andreas fault. With continued subduction, the two triple junctions migrated away from each other along the North American Plate margin, increasing the length of the transform fault between them (Figs. 6.21C to 6.21F). Thus, the age of the San Andreas fault decreases in both directions toward the triple junctions, and the maximum displacement would be expected on the oldest segment of the fault.

rather than the San Andreas fault itself, and together they can be considered a single complex fault zone. Two northeast-trending faults, the Garlock and Big Pine faults, are also currently active in southern California and show left-lateral movement. All these faults are part of the same overall tectonic system, which accommodates the motion of the Pacific Plate past the North American Plate, and any valid tectonic model should include all of them.

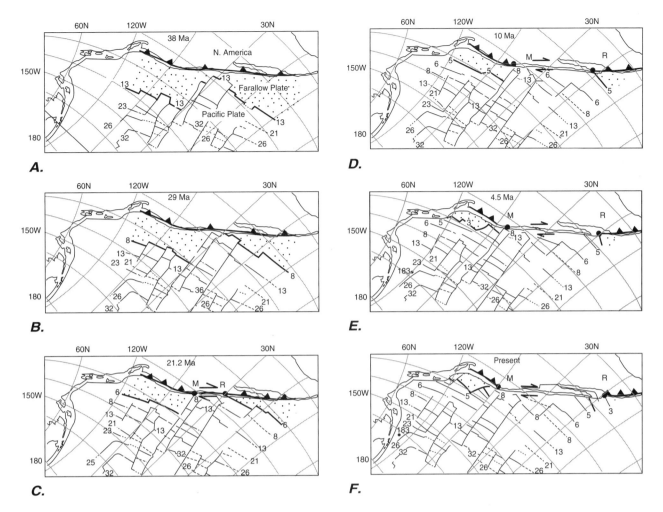

Figure 6.21 A reconstruction of the eastern Pacific magnetic anomalies and their positions at various times, showing the development of the San Andreas system. The eastward subduction of the Pacific–Farallon ridge-transform system under North America resulted, at about 29 Ma, in the separation of the Farallon Plate into two smaller ones, the Juan de Fuca and Cocos plates, connected by the growing San Andreas fault. *(After Atwater and Molnar, 1973)*

The opening of the Gulf of California did not begin until about 5 Ma, rather late in the history of the whole transform fault. Earlier movement along the southern part of the transform system must have been taken up along one or more "proto-San Andreas" faults (Fig. 6.22) that were located west of the present Baja California peninsula and do not now constitute part of the San Andreas fault itself. A number of faults west of the San Andreas fault have been proposed as the southeastern location of the proto-San Andreas fault; for example, the Whittier–Elsinore fault along the line shown in Figure 6.22.

Thus, for reasons we do not yet understand, the transform fault plate boundary propagated or jumped into the continent and began separating off the piece we know as Baja California, which may eventually become a separate microcontinent. The San Andreas–Gulf of California system thus provides a model for the production of small continental fragments surrounded by oceanic crust, which appear to be relatively common in many parts of the ocean, especially in regions where continents have rifted obliquely, such as the western Indian Ocean. There, such islands as Madagascar and the Seychelles archipelago represent continental fragments formerly attached to Africa and India but now separated from those continents by oceans.

Determining the total amount of displacement on a continental transform fault is not as simple as determining it for an oceanic transform, because we do not have the simple magnetic anomaly stripes to provide the answer. Various approaches to the problem with

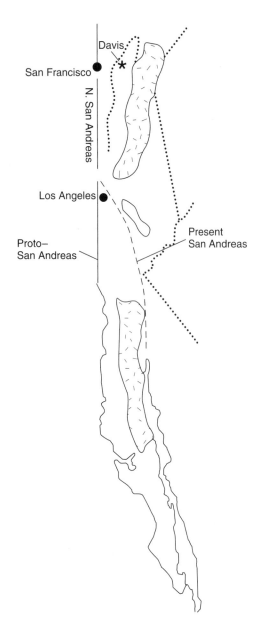

Figure 6.22 Map of proto-San Andreas fault developed before the opening of the Gulf of California. *(After Anderson, 1971)*

sediments along the San Andreas fault alone suggests 300 km of displacement since the Oligocene (about 30 Ma), giving a minimum displacement rate of about 1 cm/yr. The inferred offset of continental basement rocks of the Salinian block (Fig. 6.24) suggests a maximum of 600 km of displacement since Eocene time (50 Ma), giving a minimum of 1.2 cm/yr, but many geologists now think that the Salinian block may not have originated as part of North America at all. If the offset of deep-sea fans on branches of the fault system off the coast of California is included in the total, the estimate of displacement increases to roughly 500 km in the past 15 Ma, or an average of 3.3 cm/yr (see Fig. 6.23).

The total amount of displacment of the Pacific Plate relative to the North American Plate in about the last 30 million years can be calculated by summing the relative plate motions between North America and Africa, Africa and India, India and Antarctica, Antarctica and the Pacific, and the Pacific and North America and setting the sum equal to zero (Eq. 4.8; Sec. 4.4). The results are also plotted in Figure 6.23. The best data available indicate that at the latitude of central California (about 36°N), the total relative

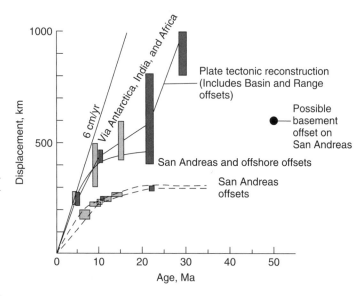

Figure 6.23 Interpretations of total displacement along the San Andreas system. The San Andreas offset is based on land geology. The offshore estimate incorporates the offset of deep-sea fans off the California coast. The plate tectonic reconstruction is derived from the relative motions recorded at the Atlantic, Indian, and Pacific-Antarctic ridges. *(After Atwater and Molnar, 1973)*

the San Andreas fault system have yielded vastly different results and indicate that the transform system is even more complex than is apparent at first.

The total amount of displacement along the fault zone has been estimated from a variety of features (Fig. 6.23). Geologic features cut by the San Andreas fault itself give minimum estimates for the displacement on the fault system. The offset of Tertiary basin

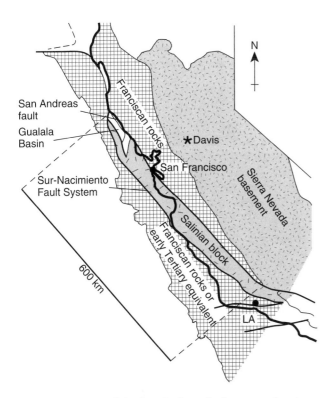

San Andreas
fault

Gualala
Basin

Sur-Nacimiento
Fault System

600 km

Franciscan rocks

★ Davis

San Francisco

Salinian block

Franciscan rocks or
early Tertiary equivalent

Sierra Nevada
basement

LA

N

Figure 6.24 Map of the San Andreas fault system showing major blocks. The offset of 600 km is based upon the correlation of Salinian block metamorphic and granitic rocks with the Sierra Nevada basement. *(After Ernst et al., 1970)*

motion between the North American and Pacific plates is on the order of 48 ± 1 mm/yr, substantially more than can be accounted for by the San Andreas system. The difference is known as the San Andreas discrepancy.

Recently, techniques have been developed that can determine the location of a geographic point on the Earth to within a few millimeters, using either VLBI (Very Long Baseline Interferometry) or the GPS (Global Positioning Satellite) system. In VLBI, a globally dispersed network of radiotelescopes simultaneously records the signals from distant astronomical objects called quasars. These signals arrive at the Earth essentially as planar wave fronts, and by determining the difference in arrival times at the various sites using extremely precise clocks and interferometry, the relative locations of the sites can be measured to within a few millimeters. The GPS system requires the recording of extremely precise timing signals transmitted from a set of orbiting satellites. From a knowledge of the precise orbits of the satellites and

the characteristics of the signals, the location of each recording station can be determined to within a few millimeters. Such precision has provided a revolutionary new tool for the study of ongoing tectonic deformation, because in the span of only a few years, tectonic motions often change the positions of geographic stations by an amount that significantly exceeds the error in determining their locations. Thus, active tectonic deformations can be monitored.

The application of such precise geodetic measurements across the North American Cordillera has demonstrated that the San Andreas fault system accommodates a rate of displacement of about 34 ± 2 mm/yr. Comparing this result with the plate tectonic determination of the Pacific–North American plate motions shows that the San Andreas discrepancy amounts to approximately 14 mm/yr. VLBI measurements have demonstrated that approximately 9 ± 1 mm/yr of this discrepancy is taken up between the Sierra Nevada and the stable North American platform. Although the exact location of this displacement east of the Sierra Nevada has not been proved, it is presumably taken up by strike-slip motion within the eastern California shear zone and the associated Walker Lane belt, and possibly by oblique-slip faulting within the Basin and Range province. The eastern California shear zone splits off from the San Andreas system at the big bend in the fault in southern California (see Fig. 6.20). It trends northward across the Mojave block to the east side of the Sierra Nevada where it joins the Walker Lane belt. The Walker Lane is a belt of major strike-slip displacements just east of the Sierra Nevada in eastern California and western Nevada.

These measurements have reduced the San Andreas discrepancy to approximately 5 mm/yr. Some of this remaining discrepancy could be measurement error, and some could be taken up by faulting offshore west of the San Andreas fault. These measurements have also demonstrated that the rates of motion determined from averages over several million years from plate tectonic data are consistent with those determined from averages over several years from precise geodetic data. This remarkable consistency over many orders of magnitude in time indicates that the process of tectonic displacement is very smooth and continuous.

The partitioning of the Pacific–North American plate motion between the San Andreas fault system and the Basin and Range province implies a very diffuse boundary between the North American and Pacific plates, and this in turn raises the question of whether the situation is unique to this boundary or whether it embodies principles that can be generalized to other continental transform tectonic environments. At this point, the question is not resolved.

The shift in the location of the southern end of the San Andreas–Gulf of California fault system into the continent resulted in a major bend in the fault trace south of the Garlock and Big Pine faults (see Fig. 6.20). In effect, this bend is a major left jog in a right-lateral fault, and the Gulf of California is geometrically equivalent to a very large right jog in a right-lateral fault. Thus, compressional structures develop at the left jog, which are evident as the thrust faults beneath the Transverse Ranges, and extensional structures develop at the right jog, which show up as the rifting of the continental margin and formation of an obliquely opening ocean basin (see Fig. 6.19).

The Alpine Fault of New Zealand

The Alpine fault is another example of a well-recognized continental transform fault. It extends essentially the entire length of both islands of New Zealand and is a trench-trench transform fault system connecting the west-dipping Tonga-Kermadec and the east-dipping Macquarie subduction zones (Fig. 6.25).

Like the San Andreas, the Alpine fault system was discovered many years ago, before its role as a transform plate boundary was recognized. It displays a prominent trace involving many subsidiary faults, much like the San Andreas. A right-lateral separation of 480 km was first suggested on the basis of offset of a belt of serpentinized ultramafic bodies and metamorphic belts. Subsequently, many authors have suggested that the bend in tectonic units illustrated in Figure 6.25 resulted from oroclinal bending during fault movement. Revised estimates of the total amount of movement with this bending removed range from 800 to 1000 km.

The Alpine fault apparently developed as a consequence of relative motions between the Antarctic, Australian-Indian, and Pacific plates, as illustrated in Figure 6.26. New Zealand is part of a series of continental masses east of Australia, including the Lord Howe Rise, the Norfolk Ridge, and the Campbell Plateau (Fig. 6.26A). Seafloor spreading began in this region with separation of the continental mass from the Australia-Antarctica margin of Gondwana by formation of the Tasman Sea and part of the South Pacific from 80 to 38 Ma (late Cretaceous to late Eocene time) (Fig. 6.26B). From 38 to 10 Ma, oblique spreading south of New Zealand translated into transform activity on the Alpine fault. Since 10 Ma, compression across the Alpine fault has accompanied transform motion.

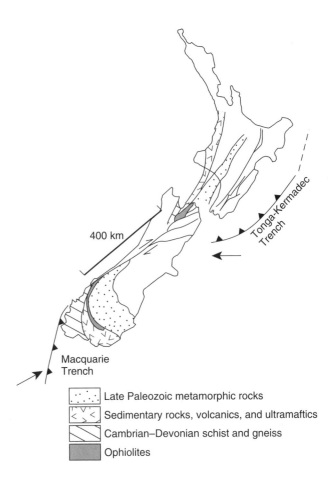

	Late Paleozoic metamorphic rocks
	Sedimentary rocks, volcanics, and ultramaftics
	Cambrian–Devonian schist and gneiss
	Ophiolites

Figure 6.25 Tectonics of the Alpine fault, New Zealand. Map of New Zealand showing west-dipping Tonga-Kermadec and east-dipping Macquarie subduction zones; the Alpine fault, with major strands to the northeast; and the principal lithologic belts whose offset defines the displacement on the fault. *(After Grindley, 1974)*

The Dead Sea Rift

The Dead Sea "Rift"[1] is a 1000-km-long sinistral transform fault system that separates the Gulf of Suez–Red Sea spreading system from the convergent zone in the Taurus-Zagros mountains (Fig. 6.27). Movement on the transform fault began in Tertiary time during the breakup of the African-Araban Shield and formation of the Red Sea (see Sec. 5.3). Two ma-

---

[1] The term rift is commonly used for this fault zone, although it is not a rift in the plate tectonic sense, but a transform fault.

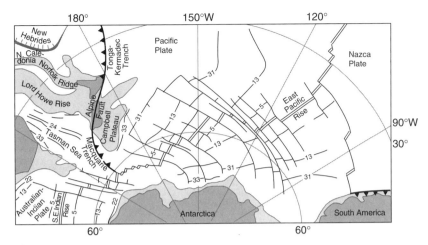

A.

Figure 6.26 Plate tectonic development of the Alpine fault, New Zealand. *A.* Present tectonic setting of New Zealand showing active Australian-Antarctic Ridge, Pacific-Antarctic Ridge, Tonga-Kermadec and Macquarie subduction zones, and Alpine fault. *B.* Tectonic configuration of the New Zealand area at anomaly 25 time (63 Ma; early Paleocene). *C.* Tectonic configuration of the New Zealand area at anomaly 13 time (38 Ma; late Eocene). *D.* Tectonic configuration of the New Zealand area at anomaly 5 time (10 Ma; late Miocene). *(After Carter and Norris, 1976)*

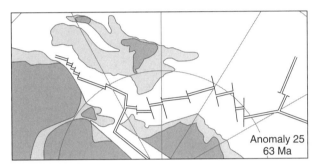

B.

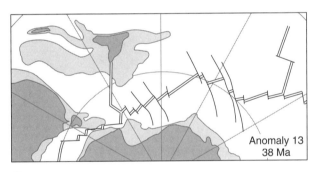

C.

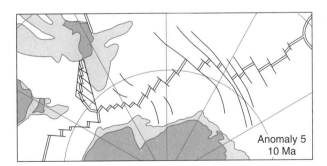

D.

jor pull-apart basins are present along the length of the fault, specifically the Gulf of Elat (Aqaba) and the Dead Sea itself. A major right-step transpression zone in Lebanon corresponds with the Palmyra fold belt.

Offset late Precambrian to late Cretaceous features along the southern part of the fault indicate a total of 105 km of left-lateral movement. Estimates of offsets along the northern part of the fault are less, suggesting that the Palmyra fold belt has taken up some of the movement.

Major Nontransform Continental Strike-Slip Faults

Not all major continental strike-slip faults are necessarily transform faults. Active nontransform strike-slip faults with large displacements occur in a number of regions, chiefly associated either with active subduction systems or with regions of continental collision. Except for the tectonic setting, there seems to be no way to distinguish continental transform faults from these other major strike-slip faults.

Examples of active major strike-slip faults associated with subduction zones are shown in Figure 6.1 and include the Median Tectonic Line in Japan, the Denali fault system of Alaska, and the Atacama fault in Chile. In each of these cases, the strike-slip fault occurs at a subduction zone where convergence is oblique and lies parallel to the subduction zone in the overriding plate. One possible explanation for this phenomenon suggests that where subduction is oblique, the interaction of the downgoing slab with the overriding plate develops a horizontal component of

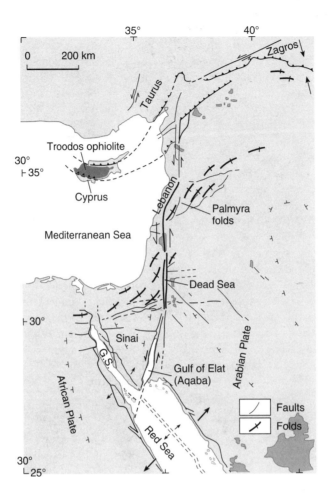

Figure 6.27 Tectonic map of Dead Sea "Rift" showing the fault and related tectonic features. *(After Garfunkel, 1981)*

shear stress that drives the strike-slip faulting behind a small strip at the edge of the overriding plate.

Examples of major active strike-slip faults associated with continental collision are also shown in Figure 6.1 and include the North and East Anatolian fault in Turkey, the Chaman fault in Afghanistan and Pakistan, and the Altyn-Tagh and Red River faults in eastern Asia. Each of these currently active faults characteristically shows hundreds of kilometers of strike-slip displacement. Most workers now agree that they represent tectonic features linked with the complex interactions of continental blocks in the continent-continent collision of Africa and India with Eurasia. As such, they are discussed in more detail in Chapter 9, in the context of these collisions.

6.5 Interpretation of Fossil Transform Faults in the Geologic Record

A number of essentially inactive linear faults or fracture zones are present in both the oceanic crust and the continental crust. Some are clearly fossil transform fault zones; others are more difficult to interpret. Figure 6.1 shows some of the major oceanic and continental fracture zones.

Inactive Oceanic Fracture Zones

Some of the most prominent structures on Earth are the fracture zones on the floor of the eastern Pacific Ocean, such as the Mendocino, Murray, Clarion, and

Clipperton fracture zones, which extend for as much as 10,000 km. These fracture zones clearly offset magnetic anomalies in the ocean floor, indicating that they are fossil ridge-ridge transform faults. The ridge with which they were associated was an extension of the East Pacific Rise and separated the Pacific Plate from the former Farallon Plate. Most of the Farallon Plate and the associated ridge have been subducted under North America, but remnants are preserved as the Juan de Fuca Plate and the Cocos Plate. Comparable fracture zones, such as the Eltanin transform fracture zone and adjacent fracture zones in the southeast Pacific, are still connected with active transform faults on the East Pacific Rise.

Other oceanic fracture zones, such as the Ninety-east Ridge in the Indian Ocean and the Central Basin fault zone in the Philippine Sea, are features whose origins are not so clear. There is some suggestion that they offset magnetic anomalies and thus may also be fossil transform faults. In the case of the Ninety-east Ridge, however, the anomalously thick crust associated with the ridge suggests a more complex origin, perhaps including the action of a hot spot.

Inactive Continental Strike-Slip Fault Zones

Inactive steeply dipping continental fault zones that may have originated as strike-slip faults are abundant.

Their interpretation is complicated, however, by the difficulty in inferring the sense of offset. We do not have the clear magnetic anomaly stripes of the ocean floor to rely on, and there is no seismicity to indicate the sense of shear. Moreover, these zones commonly juxtapose rocks of deep crustal levels that do not reveal clear offsets.

Many of these faults are characterized by mylonite or gouge zones that are 1 km or more wide and that may extend for hundreds of kilometers. In some cases, prominent horizontal stretching lineations are present along the fault zones, and these have been interpreted as evidence of horizontal shearing. With our increasingly sophisticated ability to infer shear sense from many minor structures in sheared rocks, we can anticipate better resolution of some of these problems.

Some prominent fault zones in the Precambrian or Paleozoic regions of several continents separate terranes of different ages and/or metamorphic grades. Table 6.1 lists a number of these faults, which are also labeled in Figure 6.1. All these fault zones display some features of strike-slip faults, and all may have been active as transform faults during their history. Their interpretation, however, is hampered by the lack of clear evidence from which to estimate the sense and amount of movement, and in some cases by a lack of understanding of the exact tectonic configuration in which these faults operated.

Table 6.1 Some major faults in Precambrian and Paleozoic Rocks

Fault Name	Location	Age	Displacement (km)
Tintina	Canada	Cretaceous	300 km
North Atlantic	Europe–North America	Permian-Triassic	Various
Great Glen	Britain	Carboniferous-Permian	400–2000?
Mojave	Southwestern North America	Jurassic	800

Additional Readings

Anderson, D. L. 1971. The San Andreas fault. Scientific American: 52–68 November.

Argus, D. K., and R. G. Gordon. 1991. Current Sierra Nevada–North America motion from very long base-line interferometry: Implications for the kinematics of the western United States. Geology 19:1085–1088.

Atwater, T. 1989. Plate tectonic history of the northeast Pacific and western North America. In The Eastern Pacific Ocean and Hawaii. E. L. Winterer, D.M.

Hussong, and R. W. Decker, eds. The Geology of North America, U.N. Boulder: Geol. Soc. Amer.

Atwater, T., and P. Molnar. 1973. Relative motion of the Pacific and North American plates deduced from sea-floor spreading in the Atlantic, Indian, and South Pacific oceans. In Proceedings of the Conference on Tectonic Problems of the San Andreas Fault System. R. L. Kovach and A. Nur, eds. Stanford: Stanford Univ. Publications.

Casey, J. F., and J. F. Dewey. Initiation of subduction zones along transform and accreting plate boundaries, triple junction evolution, and forearc spreading centres: implications for ophiolite geology and subduction. In Ophiolites and Oceanic Lithosphere. I. G. Gass et al., eds. Geological Society Special Publication 13:269–290. London: Blackwell's.

DeMets, C., R. G. Gordon, D. F. Argus, and S. Stein. 1990. Current plate motions. Geophys. J. International 101:425–478.

Fox, P. J., and D. G. Gallo. 1984. A tectonic model for ridge-transform-ridge plate boundaries: Implications for the structure of oceanic lithosphere. Tectonophysics 104:205–242.

Garfunkel, Z. 1986. Review of oceanic transform activity and development. Jour. Geo. Soc. London 143:775–784.

Menard, H. W., and T. E. Chase. 1970. Fracture zones. In The Sea. Vol. 4. A. E. Maxwell, ed. New York: Wiley.

Simonian, K., and I. G. Gass. 1978. Arakapas fault belt, Cyprus, a fossil transform fault. GSA Bull. 89:1220–1230.

Wallace, R. E., ed. 1990. The San Andreas Fault System, California. U.S.G.S. Prof. Paper. 1515:283.

CHAPTER

7

Convergent Margins

7.1 Introduction

Convergent or consuming plate margins occur where the adjacent plates move toward each other and the motion is accommodated by one plate overriding the other. These margins, also called **subduction zones,** are classified as either **oceanic** or **subcontinental,** depending upon whether the crust of the overriding plate immediately above the convergent margin is oceanic or continental. The crust on the overridden, or subducted, plate is almost always oceanic, because where it is continental a **collision zone** develops and forms an orogenic mountain belt, and subduction generally cannot continue for long. These collision zones are discussed separately in Chapter 9.

Oceanic subduction zones are marked by a **trench**—a linear depression in the ocean floor where the subducted plate bends and starts its descent into the mantle—and by an arcuate chain of volcanoes on the overriding plate, variously called **volcanic arcs, island arcs,** or **oceanic island arcs.** Subcontinental convergent margins are also marked by an oceanic trench and by an arcuate chain of volcanoes that are built on the overriding continental crust. These margins are called **continental arcs,** or **Andean-style continental margins,** reflecting the modern tectonic situation of the Andes along western South America. We

often use the terms *trench, island arc,* and *continental arc* for these features of convergent plate margins. Both oceanic and sub-continental consuming margins have many broadly similar features, reflecting their comparable tectonic settings. There are a number of differences, however, some of which are related to the type of crust on the overriding plate.

7.2 The Geography of Consuming Margins: The Circum-Pacific System

Most consuming plate margins are located around the Pacific Basin (Fig. 7.1). They make up the Ring of Fire around the Pacific Ocean, so called because of the active volcanic arcs above all the consuming margins. Beneath each of these margins is a relatively narrow dipping zone of seismicity that extends in some cases to depths of 700 km (see Fig. 7.4).

We examine the geography of the convergent margins starting in the Gulf of Alaska and tracing the subduction system counterclockwise around the Pacific Basin (Fig. 7.1). Southwest from the Gulf of Alaska along the Aleutian Trench, the Pacific Plate descends

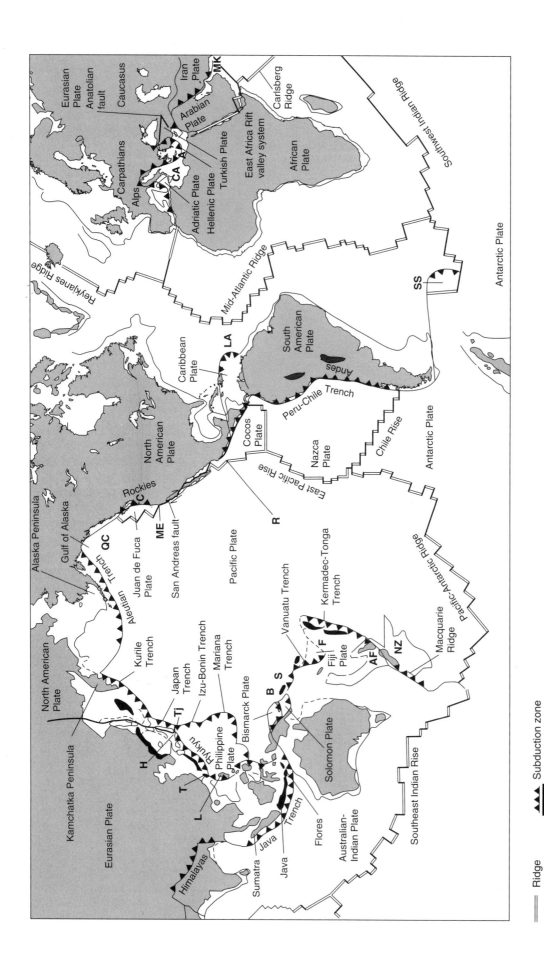

Figure 7.1 Generalized map showing plate boundaries of the world. Note the subduction zones. Key to letter symbols: A, Aegean arc; AF, Alpine fault; B, Bismarck Archipelago; C, Cascades; CA, Calabria; F, Fiji Plateau; H, Honshu; L, Luzon; LA, Lesser Antilles; ME, Mendocino triple junction; MK, Makran; NZ, New Zealand; R, Rivera triple junction; S, Solomon Islands; SS, South Sandwich Plate; T, Taiwan; Ti, Boso triple junction. (*After Dewey, 1976*)

northwestward beneath the North American Plate. Beyond the Western Alaska Peninsula, the margin is an island arc that extends to a cusp, or **syntaxis**, at Kamchatka. The convergent Pacific–North American plate boundary continues to the southwest as the Kurile–Kamchatka island arc system into the northern Japanese Islands to two trench-trench-trench (TTT) triple junctions around central Honshu (H in Fig. 7.1; see also Section 8.6). From there, the Izu-Bonin-Mariana and the Ryukyu island arc systems extend to the south and southwest, respectively, to form the eastern and western borders of the Philippine Plate. At the Izu-Bonin-Mariana convergent margin, the Pacific Plate is subducted beneath the Philippine Plate. At the Ryuku convergent margin, the Philippine Plate is subducted beneath the Eurasian Plate. The Ryuku island arc system ends at the Asian continental margin on the island of Taiwan (T).

South of Taiwan, the geometry of subduction becomes more complex. The subduction zone reverses polarity and becomes an east-dipping zone that extends south from Taiwan past the western side of Luzon (L); this zone is connected by a transform fault to the west-dipping Philippine subduction zone on the east side of the Philippines. This latter zone extends into a very complex region in Indonesia where the boundaries of four major plates—the Philippine, Pacific, Australian-Indian, and Eurasian—interact with one another.

East of this complex region, a northeast-dipping subduction zone extends through the Bismarck Archipelago (B), the Solomons (S), and the Vanuatu Islands to a complex system of small plates and their boundaries that passes through the Fiji Plateau (F) to the north end of the west-dipping Tonga-Kermadec subduction zone. This subduction zone is connected at its southern end to the east-dipping Macquarie Ridge subduction system by the Alpine fault of New Zealand, which is currently a trench-trench transform fault. The Macquarie Ridge ends at a triple junction with the Southeast Indian and Pacific-Antarctic ridges.

On the east side of the Pacific, starting in the Gulf of Alaska and progressing clockwise, the Alaska convergent margin is connected by the Queen Charlotte Islands transform fault (QC) to a short convergent margin beneath the Cascades in the Pacific Northwest of the United States. This convergent margin ends at the Mendocino triple junction (ME). The San Andreas fault extends southeast between this junction and the Rivera triple junction (R) west of Mexico. From there, a subcontinental subduction zone extends along the western margins of Central and South America to southern South America. Two short consuming margins extend into the Atlantic: the Lesser Antilles margin (LA) east of the Caribbean Plate, and the South Sandwich Plate (SS) in the South Atlantic.

Three short segments of convergent plate boundary in the Alpine-Himalayan system involve the subduction of oceanic lithosphere. These segments are the Calabrian (CA) and Aegean (A) boundaries in the Mediterranean Sea and the Makran (MK) boundary along the southwest edge of the Iran Plate.

7.3 Physiography

All convergent margins—regardless of their length, age, or stage of development—display an overall similarity in topography that allows us to define a series of distinctive physiographic zones. Figure 7.2*A* shows a schematic map and cross section of an oceanic convergent margin; Figure 7.2*B* shows a schematic map and cross section of a subcontinental convergent margin. Figure 7.2*C* is a map of the island arcs around the Philippine Plate, and Figure 7.2*D* is a map of the Andes. We describe these zones, proceeding across the margin from the downgoing plate toward the overriding plate.

The **outer swell** is a low topographic bulge in the downgoing plate that develops just outboard from where the plate bends down into the mantle. It generally rises at most a few hundred meters above the neighboring abyssal plain.

From the top of the outer swell, the ocean floor dips downward toward the boundary between the downgoing and overriding plates to form the **outer trench wall**. The trench itself is a deep topographic valley that develops just at the plate boundary. It is typically 10 to 15 km deep (5 to 10 km below the adjacent abyssal ocean bottom) and is continuous for thousands of kilometers, although it may be separated into segments by transverse topographic highs or sills. Trenches tend to be asymmetrical, with a gently sloping outer trench wall (slopes of less than 5°) and a steeper inner trench wall (slopes of greater than 10°).

The **forearc** or **arc-trench gap** includes the entire region between the trench and the volcanic arc itself. It consists of the steep **inner trench wall** or **lower trench slope** that flattens into an area of gentle slope called the **upper trench slope** or **forearc basin**. The two areas are typically separated by a break in topographic slope and in some cases by a small topographic ridge, the **outer ridge**, which is not to be confused with the outer swell on the downgoing plate that was described above.

Beyond the forearc region is the active **volcanic arc** itself, which is built upon a topographically higher

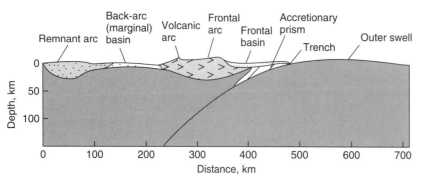

A.

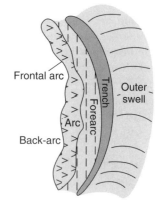

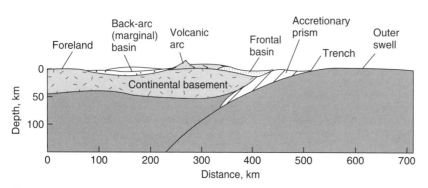

B.

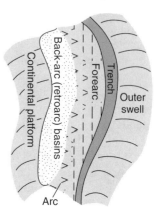

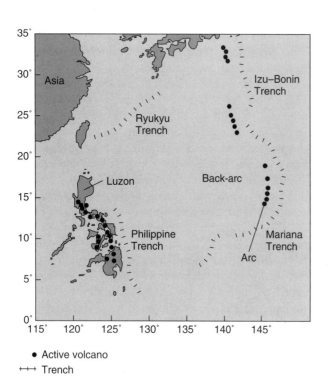

• Active volcano

⊢–⊣ Trench

C.

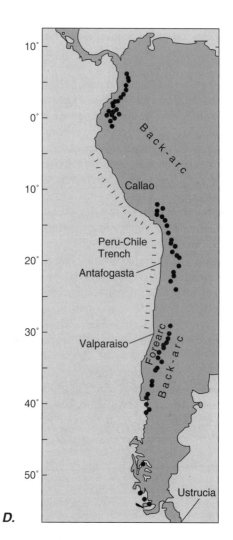

D.

Figure 7.2 Convergent margins. *A.* Schematic map and cross section of an oceanic convergent margin system. *B.* Schematic map and cross section of a continental convergent margin. *C.* Map of the Mariana-Philippine region in the western Pacific. *D.* Map and cross section of the modern Andean continental arc. *(Cross sections in A. and B. after Karig, 1974; C. after Scott and Kroenke, 1980; D. after Hayes, 1974)*

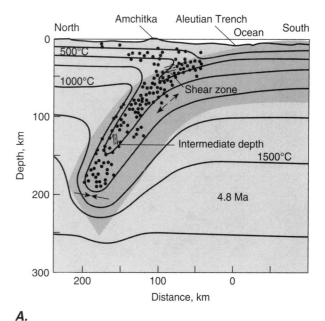

A.

region of older rocks called the **arc basement,** or **frontal arc,** which should not be confused with the forearc region. In island arcs, the arc basement is a shallow marine platform or an emergent region of older rocks. In continental arcs, it is a continental platform of older rocks that stands 1 to 5 km above sea level. Most island arcs consist of volcanoes that have a relatively consistent elevation of about 1 to 2 km above sea level, regardless of the elevation of the arc basement. In contrast, the elevation of volcanoes in continental arcs depends on the elevation of the basement; in this setting, volcanoes consistently rise about 3 km above the basement.

The **back-arc region** is the area behind the volcanic arc. In island arcs, it consists of basins, having oceanic crustal structure and abyssal depths, that may be bordered by **remnant arcs,** which are linear topographic ridges composed of thicker crust comparable to the island arcs. In continental arcs, the back-arc region is the continental platform, which may be either subaerially exposed or inundated to form a shallow marine basin.

7.4 Geophysical Characteristics of Consuming Plate Margins

Geophysical data help us to understand the structure of consuming plate margins and the tectonic processes that operate in these regions. We discuss in this section the seismicity, seismic structure, heat flow, and gravity anomalies that are characteristically associated with these margins.

Seismicity

Nearly all active island arcs exhibit dipping seismic zones called **Benioff zones** or **Wadati-Benioff zones** after the seismologists who discovered them (see the Interlude). The earthquakes in these zones originate from the downgoing lithospheric slab, although the location of seismicity within the slab varies with depth (Fig. 7.3). The slab, being cooler than the surrounding mantle, has a higher seismic velocity than

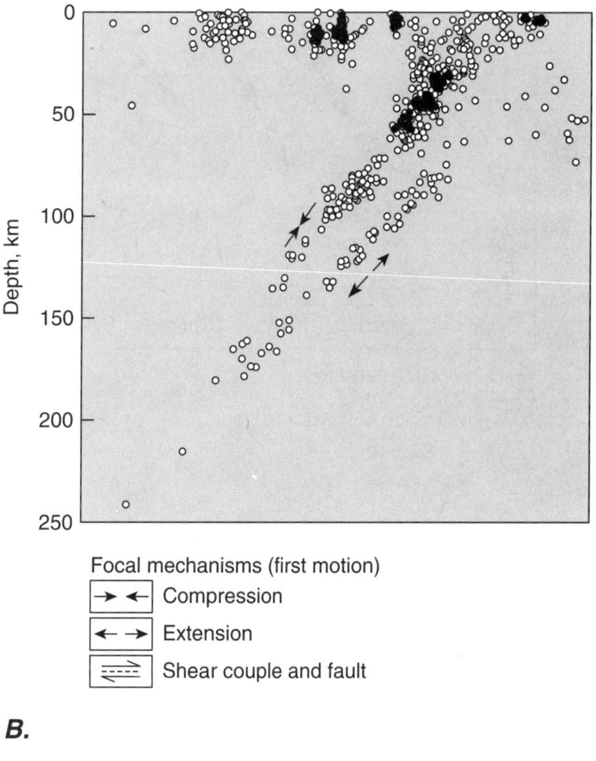

Focal mechanisms (first motion)
→ ← Compression
← → Extension
⇉ Shear couple and fault

B.

Figure 7.3 Distribution of seismicity at subduction zones. *A.* True-scale diagram of the Aleutian arc, showing slab (shaded areas), location of earthquakes within it (dots), focal mechanisms, and possible isotherm distribution. *B.* Cross section of northeastern Japan seismic zone showing the distribution of earthquakes in two planes within the downgoing slab. Arrows indicate focal mechanisms for the lower parts of each plane. *(A. after Toksoz, 1976; B. after Uyeda, 1976)*

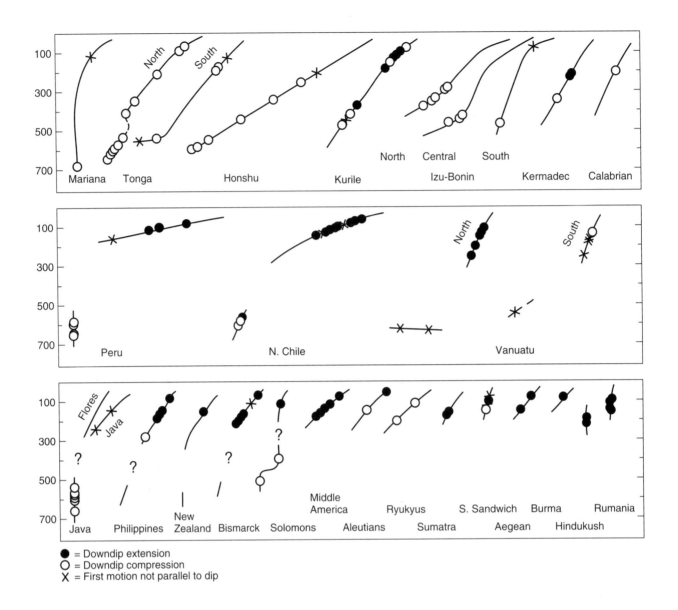

● = Downdip extension
○ = Downdip compression
X = First motion not parallel to dip

Figure 7.4 Cross sections through world's seismic zones. Lines are approximately planes of zones. *(After Isacks and Molnar, 1972)*

its surroundings. Thus, its location at depth can be mapped by tracing the paths of seismic rays that have an anomalously high velocity. Comparing the location of the slab with the location of earthquakes shows that at shallow levels the seismic zone is at the top of the downgoing slab where the overriding and downgoing plates shear against each other (Fig. 7.3*A*). At greater depths, the seismic zone is within the slab. In some cases, such as underneath northeastern Japan, the zone has been resolved into two planar

ones, one at the bottom and one within the plate (Fig. 7.3*B*).

The dip on the seismic zone is generally between 40° and 60°, but considerable variation exists. Figure 7.4 shows a compilation of cross sections through a number of dipping seismic zones around the world with the type of first-motion solution indicated for earthquakes along the zone. Some zones display a shallow dip, whereas others are steep. Some are gently dipping at shallow levels and steepen with depth (for example, the Marianas); others show just the opposite geometry (for example, Tonga) or combinations of the two (for example, Izu-Bonin); still others show distinct interruptions in continuity (for example, Peru, northern Chile, Vanuatu, Java).

The reason for the variation in dip is not exactly clear. Some workers have speculated that it results from horizontal displacement of the top of the slab from the bottom during subduction. Others suggest that interruptions in the slab may be related to interruptions in slab subduction by collision and change, or reversal, of the polarity, as discussed in Chapter 8.

The type of first-motion solutions for earthquakes in subduction zones correlates with their locations. At shallow levels, in the upper part of the downgoing plate or in the adjacent overriding plate, first-motion studies show predominantly thrust fault solutions (see Fig. 7.3A). Within slabs that extend no deeper than about 400 km, earthquakes generally indicate extension parallel to the slab (Fig. 7.4; compare Fig. 4.22). In slabs that extend to greater depths, such as Honshu, however, earthquakes within the slab indicate down-dip extension in the shallow part of the slab but down-dip compression in the deeper parts. The deepest penetrating slabs may show down-dip compression throughout (Fig. 7.4C). In the double seismic zone under northeastern Japan, the upper zone indicates compression parallel to the slab, as in the deep penetrating slabs, while the lower zone indicates extension parallel to the slab.

The various first-motion solutions evident in subduction zones mean that several processes must be active. The origin of the shallow thrust type of first motions is easy to account for, given that thrusting

Figure 7.5 The seismic velocity structure of convergent margins. A. The crustal structure of Japan on a typical cross section. B. The crustal structure across the continental convergent margin of Colombia, northern South America. (A. after Uyeda, 1977; B. after Meissner et al., 1976)

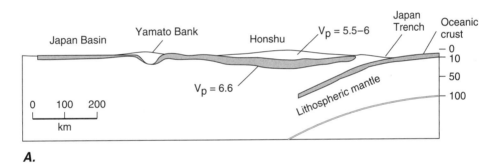

A.

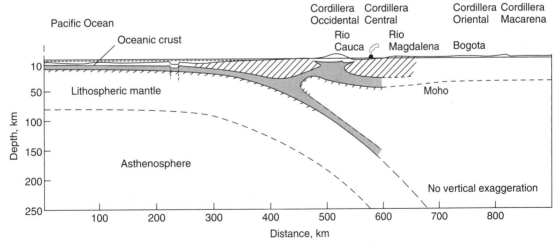

B.

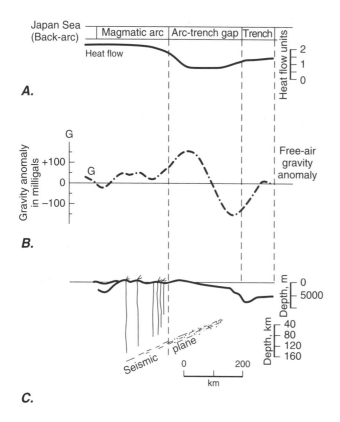

Figure 7.6 Heat flow and gravity anomalies across the Japanese island arc, compared with the location of the seismic plane and the physiographic subdivisions of the subduction zone. *A.* Heat flow anomaly. *B.* Free-air gravity anomaly. *C.* Topography and location of the Wadati-Benioff zone, shown at different vertical scales. *(After Ernst, 1974)*

must occur between the overriding and downgoing lithospheric plates. If the lithosphere is pulled into the mantle by the weight of the downgoing portion of the slab, it should be in down-dip extension. If viscous resistance to its descent increases with depth in the mantle, the deeper parts of the plates should end up being pushed down by the weight of the higher parts of the slab, causing down-dip contraction through the lower part or all of the slab. The first-motion solutions in the double seismic zone have been related to the unbending of the plate as it slides down into the mantle. Why comparable earthquakes do not seem to be associated with the bending at the trench in the first place is not clear.

Crustal Structure

Seismic refraction studies indicate that most mature oceanic island arcs have a crust that is 20 to 30 km thick under the main arc (Fig. 7.5A). In the forearc regions, the crustal thickness is intermediate between that of the main arc and that of standard oceanic crust. In back-arc basins, however, the crust thickness is comparable to standard oceanic crust. Most conti-

nental arcs show an average thickness for continental crust of 25 to 40 km (Fig. 7.5B), including the arc edifice. In some cases, however, the crust can be considerably thicker, such as in the central Andes, several thousand kilometers south of the cross section shown in Figure 7.5B, where the thickness ranges from 50 to 70 km.

Heat Flow and Gravity Anomalies

Heat flow measurements in both oceanic and continental arcs show a negative anomaly in the trench and arc-trench gap regions and a positive anomaly in the active volcanic and back-arc regions (Fig. 7.6A). These measurements are reasonable for the subduction of a cold lithospheric slab at the trench and the intrusion of magmas to form the volcanic arc.

The gravity field over island arcs also shows perturbations related to their unique structure. The trench axis is marked by pronounced local negative value in the free-air or isostatic anomalies (Fig. 7.6A). In arcs marked by a thick accretionary prism, the gravity low is shifted toward the volcanic arc from the trench axis. The anomalies indicate that there is a deficiency of material at depth within the trench and accretionary prism and that the crust is being held down dynamically, out of isostatic equilibrium.

On a regional scale, satellite observations of the Earth's gravity field have recorded a slight positive anomaly over island arc regions. This positive anomaly may result from the mass excess represented by the cold, heavy slab underneath the arc.

7.5 Structural Geology

Each of the physiographic zones described in Section 7.3 displays characteristic structures, which we summarize in this section in order from the outer swell across the trench toward the arc (see Fig. 7.2).

The Trench and Forearc Regions

The outer swell is probably caused by the elastic bending of the plate as it turns to descend into the mantle. It generally exhibits a few normal faults, some of which may be relics from an earlier oceanic history. In principle, old and new structures can be distinguished by their attitude. Faults related to the downbending of the lithosphere are parallel to the trench, whereas pre-existing faults need bear no obvious relationship to the trend of the trench. The few earthquakes that occur along the outer swells typically show extensional axes perpendicular to the axis of bending of the downgoing plate.

Most trenches contain flat-lying turbidite sediments, deposited by currents flowing either down into the trench off the overriding plate or along the axis of the trench.

The forearc region may be underlain either by a thick wedge of mostly deformed sedimentary rocks known as the **accretionary prism** (Fig. 7.7) or by deformed arc basement rocks covered in places by a thin veneer of sediments (Fig. 7.8).

The accretionary prism is apparently the main locus of crustal deformation in a subduction zone. The deformation begins at the foot of the inner trench wall, so this topographic boundary also marks a deformation front. Characteristically, rocks in the accretionary prism are cut by numerous imbricate thrust faults that are synthetic to the subduction zone—that is, they dip in the same direction (see Figure 7.7)—and in most prisms, highly deformed rocks are present at

Figure 7.7 Structure of an accretionary prism: the Sunda arc. The inner trench slope and forearc basin region are underlain by an accretionary prism of imbricate thrusts and trench slope basins. Accreted sediments are thrust over the slope sediments (*insets*). Compared with the inner trench slope area, the thrust faults are steeper under the forearc basin and involve older sediments in the faulting. *(After Moore et al., 1980)*

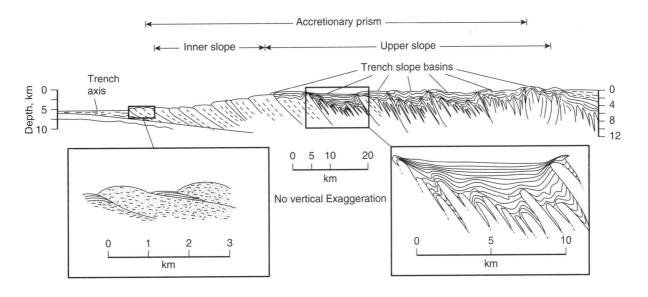

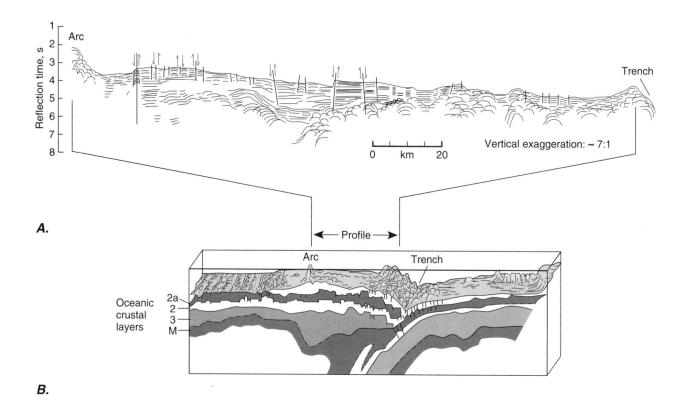

A.

B.

Figure 7.8 Structure of the Mariana arc. The inner trench slope and forearc basin region are underlain by deformed arc basement rocks. Trench sediments are minimal, and little or no accretion has occurred. *A.* Line drawing of seismic reflection profiles of part of the forearc region. *B.* Physiographic diagram and crustal model of the subduction zone and arc. *(After Mrozowski and Hayes, 1980)*

the surface. In a few cases, antithetic faults—for which the dip is in the direction opposite to that of the subduction zone—have been found.

Many of the earthquakes that occur in the accretionary prism display thrust fault focal mechanisms. Most subduction zones are curved, however, so that the relative plate motion is not everywhere perpendicular to the trench. Where the relative plate motion is oblique to the trend of the trench, the faults in the accretionary prism may include a component of strike-slip motion in addition to the thrusting (see Figs. 4.1, 4.15, and 6.1). In many cases, the oblique slip actually seems to be partitioned between two different fault systems, one predominantly strike-slip and the other predominantly thrust.

Accretionary prisms generally display a tapered cross section, the exact shape of which is a result of the mechanical properties of the prism and the forces acting on it. In many respects, the mechanical behavior of accretionary prisms closely resembles that of the

better known fold-and-thrust belts exposed on land (Box 7.1).

The deformed rocks in accretionary prisms are mostly sediments derived from either the overriding or the downgoing plate. In some regions, seamounts from the downgoing plate are being incorporated into the accretionary prisms. Sediments from the arc region are added to the top of the accretionary prism when they are deposited in basins in the forearc area (see Fig. 7.7). They are also carried by turbidity currents into the bottom of the trench. These sediments get added to the accretionary prism as subduction carries them back toward the arc and the basal thrust faults propagate out into the undeformed sediments, a process referred to as **offscraping.** The imbricate thrusts shown in Figure 7.7 formed in this fashion. The deep-sea sediments that accumulated on the downgoing plate may also be incorporated into the accretionary prism by the same process. Offscraping results in progressive widening of the accretionary prism and possibly in decrease in the dip of the subduction zone, as the arc becomes more mature.

At some subduction zones, old rocks of the island arc or continental basement can be traced out to the lower trench slope. Little or no accretionary prism is evident; only thin sediments accumulate in basins on the basement rocks, and very little (if any) sediment occurs in the trench. The arc basement is commonly cut by normal faults downthrown on the trench side (see Fig. 7.8), and although thrust faulting may be present at depth, there is little evidence for it at the surface.

Many trenches show relatively undeformed sediments on the downgoing plate that extend for several kilometers beneath deformed rocks of the trench inner wall. In some cases (for example, south of Java), the sediments on the downgoing plate are added to the accretionary prism when the basal thrust of the prism propagates into the downgoing sediments, forming a duplex structure and adding the sediments to the bottom of the accretionary prism (Fig. 7.9) in a process called **underplating.**

In other cases, little is known about the fate of the downgoing sediments (Fig. 7.10*A*). They may descend into the mantle and become involved in the generation of arc magmas, or they may be carried even deeper. Normal faults in the inner trench walls (see Fig. 7.8) and faults in accretionary prisms where sediments above the décollement show signs of truncation (Fig. 7.10*A*) may indicate the existence of **subduction**

Figure 7.9 Underplating of the accretionary prism off Costa Rica. *A.* Migrated seismic reflection profile off Costa Rica showing complex reflections. Note reflections above downgoing slab. *B.* Line-drawing interpretation of the seismic reflection profile in (*A*) that shows underplating of the accretionary prism by duplex formation. *(After Shipley et al., 1992)*

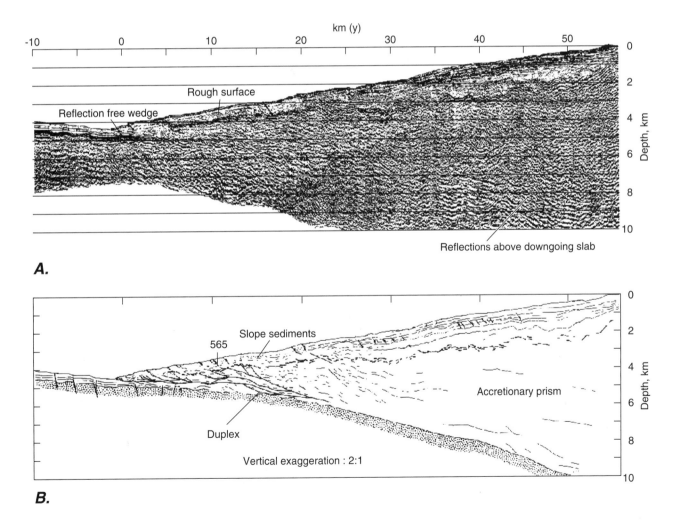

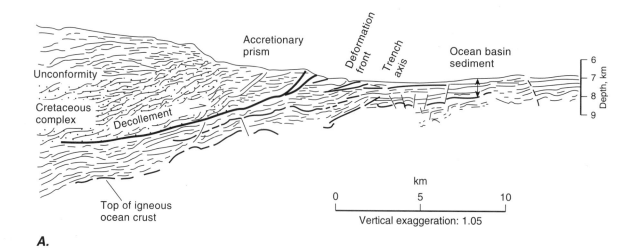

A.

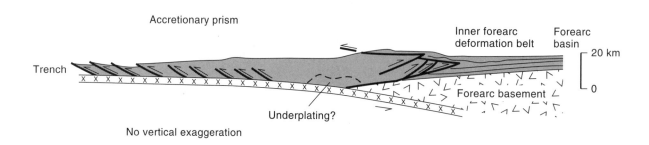

B.

Figure 7.10 Deformation styles in accretionary prisms. *A.* Line drawing of seismic reflection depth section across the Japan Trench off northeast Honshu. Note the undeformed sediments beneath the master décolletement. *B.* Schematic cross section of an accretionary prism, an example of a two-sided accretionary complex. The wedge is largely detached from both the downgoing slab and the forearc crust. *(A. after von Huene and Scholl, 1991; B. after Unruh and Moores, 1991; Torrini and Speed, 1989; Silver and Reed, 1988)*

erosion, a process by which both the sediments from the downgoing plate and the edge of the overriding plate itself are progressively subducted.

In some sediment-rich subduction zones, the accretionary prism is two-sided (Fig. 7.10*B*), and thrusts are both synthetic and antithetic to the subduction zone. In such cases, the accretionary prism is detached from both the downgoing plate and the forearc basement.

Deformational structures in accretionary prisms have been studied by examining samples obtained from drill cores of active accretionary prisms, as well as by examining emergent portions of subduction complexes and inferred fossil accretionary prisms (Fig. 7.11). Such studies reveal an abundance of structures of more than one generation, including slaty cleavage, small-scale folds, and boudinage.

Folding is evident on both large scales (Figs. 7.11*B* and 7.11*C*) and small scales (Fig. 7.11*D*). Some folding is revealed on many seismic profiles, chiefly folds with subhorizontal axes and axial surfaces that trend parallel to the plate boundary and dip moderately to steeply in the direction of subduction, although in places there seems to be little order. Evidence of soft-sediment deformation is abundant. In some cases, the deformation in the accretionary prism is so intense that any preexisting stratigraphic continuity is destroyed. Such chaotic deposits are called *melanges;* they are discussed in more detail in Section 7.6.

The upper part of the arc-trench gap is a gently sloping region commonly underlain by a wide sedimentary basin that is developed above an irregular basement. These sediments are derived mostly from the active arc or from the arc basement rocks. They generally were deposited by turbidity currents that traveled either parallel to the arc along the basin axis or perpendicular to the arc down the regional slope. In addition, isolated sedimentary basins are present on the inner trench slope.

Vertical movement generally dominates in the forearc basin, except in areas of forearc wedging. In some subduction zones, the inner part of the upper slope basin characteristically subsides while the outer edge, near the trench slope break, rises as material accretes beneath the lower slope. In contrast, ocean drilling work on the forearcs of such regions as Japan, Middle America, and the Marianas, which characteristically do not have a large accretionary prism, suggests that subsidence dominates in the entire trench slope region. We discuss models that can account for some of these observations in Section 7.7.

We do not yet know enough about these processes of uplift and subsidence in accretionary prisms to say which should dominate and under what conditions. If imbricate thrusting characterizes a forearc region, the result should be shortening and thickening, and therefore uplift, of the accretionary prism. In areas of extensional forearcs or possible subduction erosion, such as the Marianas, subsidence may be the rule.

Metamorphic rocks from forearc regions or inferred forearc deposits typically contain minerals such as the blue amphibole glaucophane, as well as lawsonite and pumpellyite. These minerals reflect the high-pressure, low-temperature blueschist facies of metamorphism and thus imply a geothermal gradient lower than normal. A low geothermal gradient is consistent with observations of low heat flow in these regions (see Fig. 7.6A) and is what we would expect where a cold slab of lithosphere is being subducted into the warmer mantle. Ancient forearc deposits exhibit an increase in grade of metamorphism going from the outer to the inner parts of the forearc region. Figure 7.12 shows examples of such metamorphic belts from Japan and California. Each of the belts shows lower grade units (zeolite) in the outer zone, grading first into prehnite-pumpellyite and finally into the highest grade rocks (lawsonite in California, blueschist-greenschist in Japan). The direction of inferred subduction is the same as the direction of increasing metamorphic grade, as shown by the arrows in Figure 7.12.

Igneous rocks are rare in forearc regions, except as tectonic units within the accretionary prism. Some forearc areas, however, do contain evidence of magmatic activity. The rocks that are found are generally basaltic in composition. The forearc region of Japan (Fig. 7.13) is one example of the occurrence of basaltic igneous rocks. The origin of such magmas is not well understood. One suggestion is that they rise along conduits formed by preexisting thrust faults from the area of magma generation along the downgoing plate. Another is that forearc igneous rocks result from melting in the forearc region by subduction of a spreading center. Japan, southern Alaska, and the western United States are examples of regions where subduction of a ridge or hot young plate apparently took place in the past few tens of millions of years.

The Arc Basement and the Volcanic Arc

The arc basement, or frontal arc, consists of older, more deformed, and perhaps more metamorphosed rocks on which modern arcs are built. The character of this basement is quite variable. In such arcs as the Tonga-Kermadec or Mariana, scattered exposures of the frontal arcs suggest that they are composed of metamorphosed and deformed oceanic rocks. In Japan, by contrast, a complex older igneous and metamorphic basement includes rocks as old as early Paleozoic or even Precambrian. Continental arcs such as the Andes are built on a typically complex continental basement. Tectonically, the frontal arc seems relatively inactive, except for scattered normal faulting. However, the frontal arc may be the site of igneous and metamorphic processes that are not accompanied by obvious tectonic events. The Kohistan arc in Pakistan may provide a clue as to the internal structure of a typical frontal arc (Box 7.2).

The volcanic arc is the region of active volcanic and plutonic activity marked on the surface by a chain of volcanoes. Most of these volcanoes are large andesitic stratovolcanoes, although basaltic or dacitic volcanic centers are also present. In plan view, the volcanoes are spaced along the arc at a fairly regular interval of approximately 70 km (Fig. 7.14; see Figs. 7.2C and 7.2D). In many arcs, the volcanoes are aligned in groups of 2 to 12 along linear segments that are slightly offset from adjacent segments (Fig. 7.14B). Breaks between linear segments are often associated with discontinuities in the downgoing plate.

The local structural environment of most volcanic arcs is extensional. Mapped faults and minor earthquakes in the regions of volcanic activity are chiefly normal faults, with minor strike-slip motions in some cases. Compressional structures or thrust fault first-motion solutions are almost completely lacking, with the exception of some seismic events in the central Andes. The volcanoes themselves are typically located in grabens, termed **volcanic depressions** by some authors.

In modern volcanic arc regions, either continental or oceanic, we can only surmise whether or not there

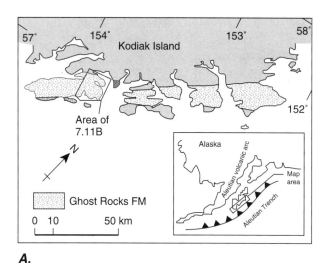

A.

Figure 7.11 Fold styles in forearc of Aleutian arc, Kodiak Island, Alaska. *A.* Map of a portion of the forearc. *B.* More detailed map of an area of the forearc. *C.* Detailed map and cross section of a portion of (*B*). *D.* Fold styles observed in unit A (cross section in (*C*) that formed in partially lithified interlayered sandstone and mudrock. Diagram 4 shows spaced cleavage cutting across the axial surface of earlier folds. Gray units are sandstones. *(After Byrne, 1982)*

Kaiugnak Bay

Kiavak Bay

N

0   3
km

– – – High angle fault
▲▲▲ Thrust fault, teeth
        on upper plate
⟨⟨⟨⟩⟩⟩ Melange

Mafic igneous rocks
H = Hypabyssal
V = Volcanic

B.

is a pluton at depth to correspond with the volcanoes. In inferred ancient continental or island arcs where erosion has exposed deeper levels in the structure, however, the evidence seems clear. Large plutonic bodies of batholithic dimension are rare in ancient oceanic arc complexes. They are abundant, however, in ancient continental arcs. Figure 7.15 shows the large extent of batholithic and associated rocks in the Sierra Nevada, which has been reconstructed by removing the strike-slip motion on the San Andreas.

Metamorphism in arcs and frontal arcs reflects the high heat flow that exists in these regions. The inferred geothermal gradients also suggest that the bottom part of the crust in continental regions may be at or above the minimum melting temperature of granite. This fact alone could explain the abundance of granitic batholiths in continental arc regions.

The presence of metamorphism indicating a high geothermal gradient in arcs and frontal arcs, together with metamorphism reflecting an abnormally low geothermal gradient in forearc regions (see Fig. 7.6), has given rise to the concept of **paired metamorphic belts.** A high-pressure, low-temperature metamorphic belt develops near the trench where the cold downgoing slab depresses the geothermal gradient. Farther back under the volcanic arc, a high-temperature, low-pressure metamorphic belt develops where the intrusion of magmas raises the geothermal gradient. Figure 7.16 shows the pressure-temperature distributions associated with the different metamorphism of these two belts.

In both continental and oceanic arcs, igneous rock compositions may correlate with depth to the seismic zone or to the top of the subducting plate.

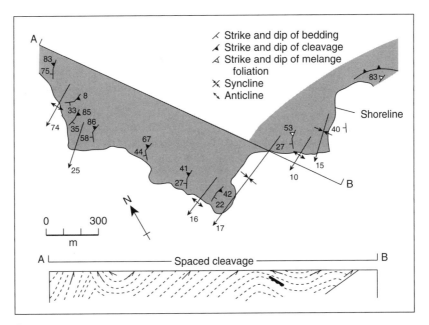

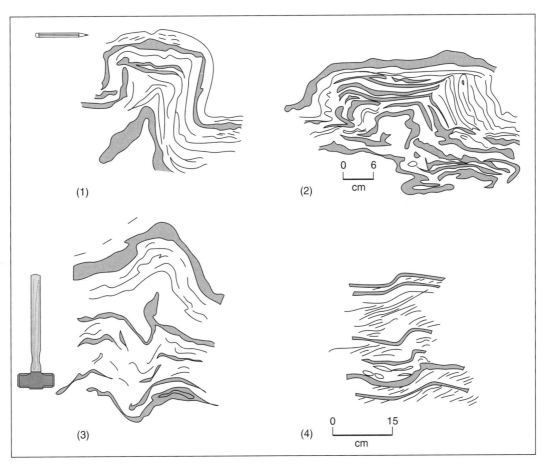

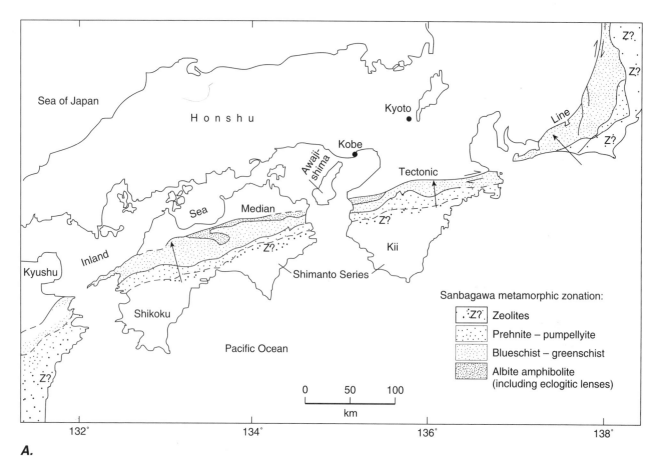

A.

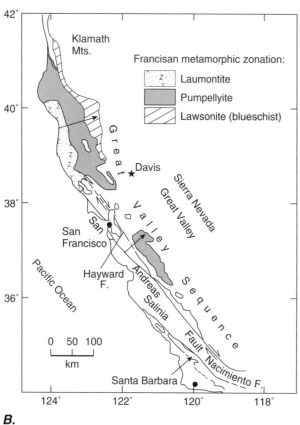

B.

Figure 7.12 Metamorphism in subduction zone regions, showing increase of metamorphism in the inferred direction of subduction (arrows), away from the trench. *A.* Metamorphic zonation in the outer metamorphic belt of southwest Japan. *B.* Metamorphic map of western California. *(After Ernst et al., 1970)*

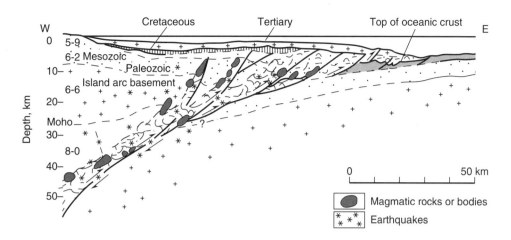

Figure 7.13 Schematic cross section of forearc region of Japan, showing Cretaceous and Tertiary sediments, imbricate thrusts, forearc igneous rocks, and idealized locations of earthquakes. *(After Shiki and Misawa, 1982)*

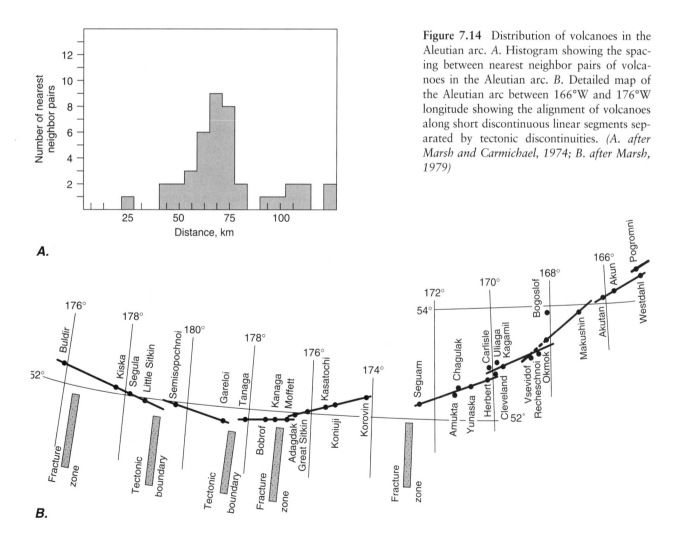

A.

B.

Figure 7.14 Distribution of volcanoes in the Aleutian arc. *A.* Histogram showing the spacing between nearest neighbor pairs of volcanoes in the Aleutian arc. *B.* Detailed map of the Aleutian arc between 166°W and 176°W longitude showing the alignment of volcanoes along short discontinuous linear segments separated by tectonic discontinuities. *(A. after Marsh and Carmichael, 1974; B. after Marsh, 1979)*

Figure 7.15 Ancient continental arc of the western United States. Map of the Mesozoic continental margin of southwest North America prior to displacement on the San Andreas fault, showing accretionary prism (Franciscan), forearc basin (Great Valley, GV), and arc (batholithic belt). *(After Ernst et al., 1970)*

0 100 200 300
km

Approximate base of
continental slope

▲ ▲ ▲
Trench

Franciscan rocks
Granitics ⎤ Sierran
Metamorphics ⎦ type
rocks

GV = Great Valley–type rocks

potentially of great value for inferring the direction of dip of ancient subduction zones.

The success of a plot of K_2O versus depth has led to the widespread application of such compositional indicators to ancient arc deposits. These possible indicators must be used with great care, however, for a number of reasons. First, the same increase in abundance of elements can occur in a single magmatic system as it crystallizes and differentiates. Second, the same chemical differences can be detected in the same place through time as the subduction zone changes its dip or its depth, or as it migrates with respect to the magmatic center. Third, the percentages of many major and minor elements in rocks are highly susceptible to alteration during metamorphism and weathering. These secondary processes severely affect such elements as K, Na, Si, Ca, and (to a lesser extent) Al, precisely those elements that are used to characterize the compositional variations in the primary state. Fourth, some arcs exhibit no variation in composition with depth to the seismic zone. Finally, and perhaps most important, the cause of the variation of chemistry with depth to the seismic zone is not understood, and neither, for that matter, is the exact origin of the magmas produced at consuming margins. In view of this lack of understanding, any uncritical application of such compositional variations to the resolution of tectonic questions is premature. The amount of contamination of magmas from the mantle or crust that occurs as the magma rises to the surface is also unclear. Some workers argue that it is significant, while others suggest that it is minimal.

Sediments associated with island arc complexes are chiefly of two types: clastic and carbonate. Clastic sediments consist of debris from active volcanoes. They range from coarse conglomerates and breccias deposited near the source to fine-grained turbidite deposits that are carried large distances from the source region. Pumice is a common constituent of these sediments. In continental arcs, sediments are predominantly subaerial and subordinately marine. Most island arc sediments are marine, probably reflecting deposition in deep-sea fan complexes that are shed from active volcanic islands.

In tropical regions, active island arcs possess fringing carbonate reefs. These carbonate deposits are flooded intermittently with volcanogenic debris.

Increases in Al, Na, and K for a given rock type have been correlated with Benioff zone depth in a number of arc complexes. Such compositional variations are

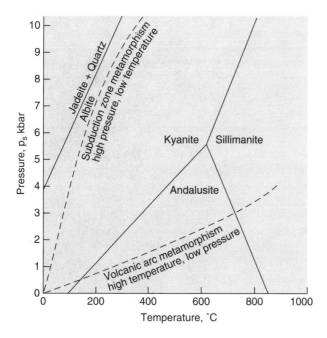

Figure 7.16 Pressure-temperature diagram showing geothermal gradients leading to high-temperature, low-pressure metamorphism characteristic of the arc metamorphic belt and to high-pressure, low-temperature metamorphism characteristic of the forearc, or subduction zone, metamorphic belt. *(After Ernst et al., 1970)*

Thus, carbonate and volcanic deposits interfinger with each other.

The Back-Arc Region

The back-arc regions of most oceanic and continental arcs are marked by extensional tectonics and subsidence relative to the arc itself. In oceanic island arcs, sediments derived from the arc are deposited in an ocean basin behind the island arc, called the back-arc basin. In continental regions, these sediments are deposited in basins that form on top of the platform toward the continental interior. These epicontinental basins are called **foreland basins** or **retro-arc basins.**

Although extension and/or subsidence seem to characterize many continental arcs, there are a few significant exceptions, the central Andes being the

most notable. There, active compressive structures in the back-arc region involve folds and thrusts of the arc toward the interior of the continent.

Back-arc basins characteristic of oceanic island arcs are of two principal types. Some are composed of entrapped old oceanic crust; others are composed of young lithosphere. The latter characteristically show seismic activity, high heat flow, and seafloor spreading (Fig. 7.17). Many currently active back-arc basins, however, do not display the classic symmetrical magnetic anomalies we expect at spreading centers. In many cases, in fact, it is difficult to identify a specific spreading center.

The composition of volcanic rocks in modern back-arc basins is variable. Some rocks, which occur at considerable distances away from an arc, are indistinguishable from those found at mid-ocean ridges. Others, which are found in areas adjacent to the arc that are just beginning to spread, are reminiscent of arc rocks.

Some back-arc basins appear to contain more active faults and more complex spreading than oceanic

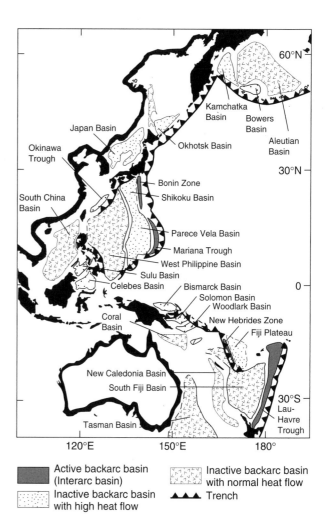

Figure 7.17 Marginal basins of the western Pacific showing basins that are active, inactive with high heat flow, and inactive with normal heat flow. Basins are separated by remnant arcs. Inactive basins with normal heat flow possibly represent entrapped oceanic crust. Others may have formed by spreading. Compare with Figure 7.22. *(After Karig, 1974)*

Box 7.1 **Mechanics of Thrust Sheets**

Critical Coulomb Wedge Theory

Active thrust sheets, such as found in western Taiwan and in the Himalayas, and active submarine accretionary prisms over subduction zones, are tapered in cross section, with the thickness decreasing in the direction of thrusting (Fig. 7.1.1). This shape can be explained as the result of the mechanics of a deformable thrust sheet sliding on a basal décollement.

To understand the various factors that control the taper of a thrust sheet, we examine part of the sheet as a free-body diagram (Fig. 7.1.2). In such a diagram, we represent the mechanical effects of all the material surrounding the isolated portion of the thrust sheet by tractions (force per unit area)

distributed across the boundaries of the body. We assume that the basal décollement is a flat plane that slopes upward toward the foreland at an angle β, and we find from the analysis that the surface of the sheet must slope downward toward the foreland at an angle α. Assuming that the acceleration is small enough to be negligible, the equation of equilibrium requires that the driving forces that push the thrust wedge toward the foreland must be balanced by the resisting forces that tend to prevent motion. The problem then is to account for all the forces that can act on the thrust wedge and to explain how the wedge-shaped geometry is a natural consequence of the mechanics.

Let us first consider the driving forces. The main driving force is contributed by the horizontal normal traction σ_{11} that acts on the rear vertical face of the wedge. To estimate its magnitude, we assume that the stress in the thrust sheet is everywhere as large as possible and thus is at the critical fracture strength of the rock as given by the Coulomb fracture criterion. If we choose reasonable material properties for the rock, the Coulomb fracture criterion allows us to predict how the strength of the rock will increase with pressure and thus how the normal traction σ_{11} will increase with depth. The force on the rear vertical face is simply $\bar{\sigma}_{11} A_r$, the average normal component of the stress $\bar{\sigma}_{11}$ (force per unit area) multiplied by the area of the rear vertical face A_r on which that stress acts. The strength of the rock is strongly affected by the pore fluid pressure internal to the thrust wedge (p_i), however, because an increase in pore fluid pressure tends to counteract the effect of overburden pressure on the rock strength and thus tends to weaken the rock. So we must account for this effect as well.

The surface slope also creates a driving force, because at any given level in the thrust wedge (for example, at $x_3 = 0$), the overburden stress ($\rho g t$) at a

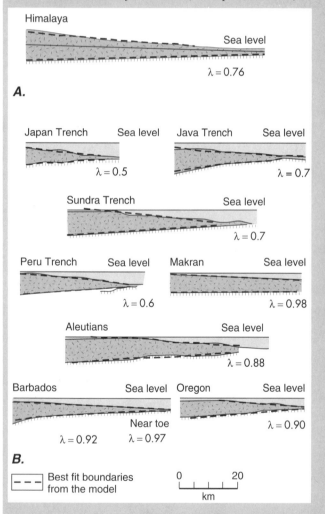

A.

B.

- - - - Best fit boundaries from the model

0   20
|__|__|__|__|
km

Figure 7.1.1 Cross sections of thrust wedges. *A.* Subaerial thrust wedge: the Himalayas. *B.* Several examples of active submarine thrust wedges that occur as accretionary prisms above subduction zones. *(After Davis et al., 1983)*

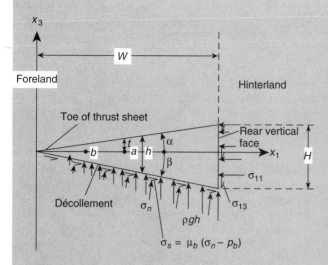

Figure 7.1.2 Free-body diagram of a thrust wedge.

given point at depth t below the surface (for example, point a) is greater than the overburden stress at a point farther toward the foreland (for example, point b) because along any horizontal line, the depth of the sheet t decreases toward the foreland. Because part of the horizontal normal component of the stress σ_{11} is proportional to the overburden, there is a horizontal gradient in σ_{11} that is equivalent to a horizontal force per unit volume of material. This is the same force that makes balls of silicone putty sag into puddles if they are left standing unsupported on a table and that drives the flow of ice sheets and the gravitational collapse of topographic highs. Because the topographic slope is small, however, this component of driving force is not very large.

The resistance to thrusting comes from the décollement. It derives in part from the frictional resistance to sliding on the fault, which is given by the average frictional shear stress on the basal décollement $\overline{\sigma}_s$ multiplied by the area A_b of the décollement. This quantity can be expressed as $\overline{\sigma}_s\, A_b = \langle \mu_b\, [\sigma_n - p_b] \rangle A_b$, where the term between the angled brackets $\langle\ \rangle$ is the product of the coefficient of friction on the décollement μ_b and the effective normal stress $[\sigma_n - p_b]$ on the décollement, and the angled brackets mean the average over the area of décollement. Here again the pore fluid pressure on the décollement p_b is important because it strongly influences the effective normal stress on the thrust and thus the frictional resistance to sliding.

Another part of the resistance to motion derives from the component of the vertical overburden stress that is parallel to the slip direction on the fault ($\rho g h \sin \beta$), which must be overcome to move the thrust sheet. The magnitude of this component of resistance increases with increasing slope of the décollement.

Setting up an equality between the driving forces and the forces of resistance leads to an equation that accounts for the basic wedge shape of a thrust sheet. We can understand the result qualitatively in terms of the dominant driving force on the rear vertical face and the dominant frictional resistance on the fault. If we were to increase the length of the thrust sheet by moving the rear vertical face an increment in the positive x_1 direction, the frictional resistance would increase because the area of the base of the thrust sheet increases. To counteract this resistance and allow the thrust sheet to move, the driving force created by σ_{11} must increase. But σ_{11} itself cannot increase because it is already at the maximum that the rock can withstand. Thus, the only way to increase the magnitude of the driving force is to increase the area over which σ_{11} acts, which means making the thrust sheet thicker. This simple argument shows that the thickness h of the thrust sheet must increase with increasing distance x_1 from the toe. This relationship thus defines the tapered shape of the thrust sheet, although the details add considerable complexity to the analysis.

A more careful mechanical analysis of this problem leads to the following approximate equation for the surface slope angle α (in radians) for subaerial wedges:

$$\alpha = \frac{(1-\lambda_b)\,\mu_b - (1-\lambda_i)\,k\beta}{(1-\lambda_i)\,k + 1} \qquad (7.1.1)$$

where β is the slope of the décollement, λ_b and λ_i are the ratios of pore fluid pressure to overburden pressure along the décollement ($\lambda_b \equiv p_b/\rho g H$) and internal to the thrust sheet ($\lambda_i \equiv p_i/\rho g t$), respectively; μ_b is the coefficient of friction along the décollement, and k is predominantly a measure of the Coulomb fracture strength of the rock in the thrust sheet. A slight modification of the equation is necessary for subaqueous wedges to account for the pressure of the water column. We examine below the factors that affect the slope of the wedge.

Friction and Rock Strength

Let us first assume that the décollement is horizontal ($\beta = 0$) and that pore fluid pressure is everywhere zero. The surface slope is then defined from Equation (7.1.1) by

$$\alpha = \frac{\mu_b}{(k+1)} \qquad (7.1.2)$$

which in essence is a ratio between coefficients that describe the frictional resistance and the Coulomb strength of the rock. Higher coefficients of friction on the décollement μ_b result in higher slope angles α reflecting the need for a larger force on the rear vertical face to overcome the friction and drive the thrust sheet. The effect of friction, however, is counteracted by the strength of the rock in the thrust sheet. The higher the fracture strength, indicated by k, the lower the slope angle required, because the rock can support a higher stress on the rear vertical face and therefore the area need not be as large to obtain the required driving force.

Pore Fluid Pressure

As the pore fluid pressure along the décollement increases—that is, as λ_b increases from 0 to 1—the numerator of Equation (7.1.1) becomes smaller and the surface slope decreases. Physically, increasing the pore fluid pressure on the décollement decreases the frictional resistance to thrusting and thereby decreases the driving force necessary to move any given width of the thrust sheet. Thus, the change in height of the rear vertical face with length of the thrust sheet is smaller, and the surface slope is lower.

Increasing pore fluid pressure within the thrust sheet, however, tends to counteract an increase of pore fluid pressure on the base, because as λ_i increases from 0 to 1, the denominator becomes smaller and the surface slope increases. Physically, increasing pore fluid pressure within the thrust sheet decreases the effective strength of the rock and thus decreases the possible magnitude of the horizontal normal stress σ_{11} that can be applied to the rear vertical face of the wedge. The same driving force on a weaker rock requires a larger area on the rear vertical face than for a stronger rock, because the weaker rock cannot support as large a stress. Thus, for larger values of λ_i the area of the rear vertical face must be larger, and this implies that the surface slope angle of the thrust sheet must be larger.

Dip of the Décollement

If the décollement has a nonzero slope ($\beta \neq 0$), then the surface slope is decreased by a large fraction of β as shown by Equation (7.1.1). The coefficient of β in Equation (7.1.1) reduces to $k/k+1$. This ratio can be close to 1 but is always less than 1. The fact that α decreases as β increases reflects the fact that the rate at which the area of the rear vertical face changes with length W of the wedge depends on β as well as on α. The slope of the décollement also affects the magnitude of the gravitational component of shear stress on the décollement that contributes to the resistance, and it affects the extent to which the driving force on the rear vertical face contributes to the normal stress on the décollement and thus to the frictional resistance. The net effect is that the driving force required to move the thrust sheet increases with increasing β, so that the decrease in surface slope cannot be as large as the increase in β.

The wedge described by this equation is the thinnest body of Coulomb material that can be moved over the décollement. If material is added to the front, or toe, of the wedge, the whole wedge deforms to maintain the critical taper. The deformation takes the form of thrust faults, folds, and fault ramp folds internal to the wedge, all of which provide a net shortening and thickening of the wedge (Figure 7.1.3). If the slope angle of the wedge becomes too large, then the thrust fault will propagate out in front of the wedge to lengthen the thrust sheet and decrease the slope angle, with internal faulting and folding providing the adjustments to the taper throughout the rest of the sheet.

The mechanics of tapered thrust sheets have been compared to the mechanics of dirt or snow wedges that form in front of a bulldozer or snowplow blade. The analogy can be misleading, however, because for it to be applicable, the blade should be a flat vertical blade. In actual practice, the blades are vertically curved, a design that forces the snow or dirt to slide up the blade and fall forward to form a pile whose surface slope is the

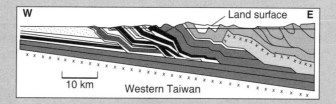

Figure 7.1.3 Cross section of the thrust wedge of western Taiwan showing the thrusts and folds characteristic of the deformation within a critical Coulomb thrust wedge. *(After Davis et al., 1983)*

angle of repose of the material rather than the slope of the critical taper that we are discussing. The angle of repose is the steepest slope angle that loose material can support and is generally about 30°, whereas the steepest slopes predicted by the critical taper model are about 10°.

The results of the critical taper model are indicated in Figure 7.1.4, which shows the dip of the décollement β plotted against the surface slope angle α. The lines are the theoretically predicted relationship for a variety of values for the pore fluid pressure ratio, assuming that the ratio on the décollement equals that within the thrust wedge, $\lambda_b = \lambda_i = \lambda$. The theoretical equation for subaerial wedges is given in Equation (7.1.1). The boxes in Figure 7.1.4 show the geometries of active wedges as labeled. It is clear from this figure that most thrust wedges require a value of λ considerably above the hydrostatic value of about 0.4, implying significant overpressure of the pore fluid. Such values of λ are consistent with measurements made in wells that penetrate into some of these wedges, which lends credence to the theory.

Non-Coulomb Theories for Thrust Wedges

The theory discussed above is based on the assumption that the material in the thrust sheet behaves as a Coulomb material and that the resistance to thrust motion comes predominantly from frictional sliding on the décollement. Other assumptions are possible, of course, and they lead to the prediction of different shapes for a thrust wedge. For a plastic material, for example, the strength of the material is defined by the yield strength, which is independent of confining pressure. Thus, for a plastic thrust wedge, the strength of the material does not increase with depth in the

thrust wedge, and the surface slope would have to be steeper than for a Coulomb thrust sheet, other factors being equal. On the other hand, if sliding on the décollement occurred by plastic deformation, which is independent of pressure, the resistance to thrust motion would be the same under thicker parts of the thrust sheet as it is under thinner parts. Thus, plastic resistance to thrust motion would not increase as rapidly with distance from the toe as it does for friction, and the surface slope angle would be lower. For a plastic material, the pore fluid pressure also would not have a large effect on the mechanical properties. The presence of salt along the décollement, for example, could result in surface slopes as low as 1° and thus could also account for some of the very low slopes observed (see Fig. 7.1.4).

Both temperature and pressure increase with increasing depth in the Earth. The plastic yield strength is relatively insensitive to changes in pressure but decreases dramatically with increasing temperature. The Coulomb strength and frictional resistance, on the other hand, increase significantly with increasing pressure but are relatively insensitive to changes in temperature. Thus, at some depth, probably near 10 to 15 km, we can expect plastic behavior to replace both Coulomb behavior and frictional resistance. If that depth is reached within a single thrust sheet, it should be apparent from a change in the surface slope of the thrust wedge.

The Propagation of Slip Domains

The foregoing analysis of the mechanics of a thrust sheet assumes that the entire mass is at the critical stress for fracture as predicted by the Coulomb fracture criterion and that motion occurs continu-

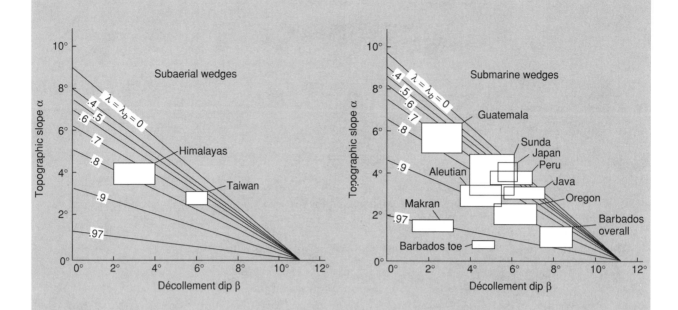

Figure 7.1.4 Plot of the theoretical relationship between the surface slope of a thrust wedge α and the slope angle β of the décollement as a function of the pore fluid pressure ratio λ, assuming that the ratio on the décollement and within the thrust wedge are the same ($\lambda_b = \lambda_i = \lambda$). The angles measured for a number of active thrust wedges are plotted as boxes and suggest the importance of the pore fluid pressure. *(After Davis et al., 1983; Twiss and Moores, 1992)*

ously on the entire fault plane. These assumptions provide a simplified model, however, because the entire thrust sheet neither moves as a rigid block nor undergoes pervasive deformation at one time. Rather, the deformation is accommodated by the discontinuous slip over finite areas of faults within and at the base of the sheet. Such slip events, which are localized in time and space, often cause earthquakes, which we can observe. Only by averaging these events over a long period of time, on the order of perhaps tens of thousands to hundreds of thousands of years, would we see the pattern of pervasive deformation and the slip of the entire thrust sheet on the décollement that we assume for the model.

Physical models of sliding caused by localized events have been constructed by assuming that the frictional resistance on finite areas of a fault varies in magnitude in a random pattern. As the applied stress on the thrust sheet builds up, the weakest areas are the first to slip, and the slipping is limited to the area of weakness. The release of stress on the weak areas requires a concentration of stress in other areas, and these areas can then deform or slip at externally applied stresses that are lower than would be predicted from the strength of these areas. The net result is that the deformation can occur at applied stresses that are lower than would be predicted from an average value of the mechanical properties. The effect is similar to the effect of dislocations in a ductile crystalline material, which allow ductile deformation at stresses well below the theoretical strength predicted on the basis of atomic bonds.

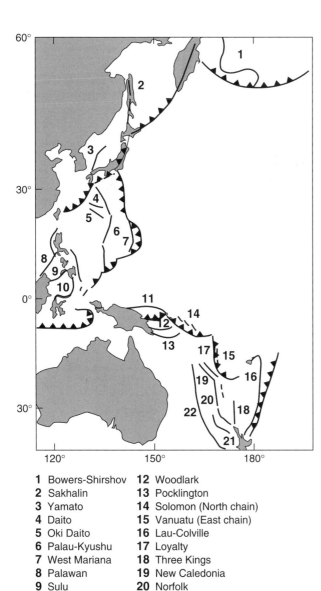

Figure 7.18 Location of remnant arcs in the western Pacific. Individual remnant arcs separate basins of different character, as indicated in Figure 7.17 *(After Karig, 1972)*

1	Bowers-Shirshov	**12**	Woodlark
2	Sakhalin	**13**	Pocklington
3	Yamato	**14**	Solomon (North chain)
4	Daito	**15**	Vanuatu (East chain)
5	Oki Daito	**16**	Lau-Colville
6	Palau-Kyushu	**17**	Loyalty
7	West Mariana	**18**	Three Kings
8	Palawan	**19**	New Caledonia
9	Sulu	**20**	Norfolk
10	Sangihe	**21**	West Norfolk
11	West Melanesia	**22**	Lord Howe

the associated active island arc and the marginal basin. These ridges seem to be the remnants of a pre-existing arc that split to form the sides of a developing marginal basin. Therefore, the origin of remnant arcs is closely related to the origin of back-arc basins themselves, as discussed in Section 7.7.

7.6 Chaotic Deposits

Many modern and inferred ancient forearc regions contain chaotic deposits of predominantly sedimentary rocks in which the normal stratigraphic relationships, such as regular bedding and consistent sequences, have been disrupted or destroyed. Chaotic deposits in orogenic belts have been identified for many years, and they are described by several different terms, each with its own subtle difference in meaning or tectonic implication. In general, chaotic deposits are the product either of very large strains, such as are found in parts of the accretionary complexes of subduction zones, or of submarine landslides.

Chaotic deposits of sedimentary origin may be found within an otherwise normal stratigraphic sequence, typically of shale and/or sandstone. Such chaotic deposits, called **olistostromes,**[1] characteristically contain a disordered arrangement of blocks of one lithology embedded in a matrix of another. For example, Figure 7.19 shows a cross section of a region in the northern Apennines of Italy where olistostromes of ophiolitic lithologies are present in a sequence of shale and calcareous turbidites. Many olistostromes contain mappable blocks of an individual lithology, or **olistoliths.** These olistostromes probably formed by a landslide from a submarine topographic high of oceanic crust, possibly a ridge along a fracture zone, as implied by the interpretation of Apennine ophiolites mentioned in Section 6.2. Other topographic highs that might serve as the source of such slides are scarps along passive continental margins, oversteepened trench walls, and the flanks of midplate volcanic islands such as the Hawaiian Islands.

spreading centers. Several back-arc basins clearly display propagating rifts. These tectonic variations are reflected in major differences in sedimentary deposits. Lavas and pelagic sediments thus interfinger with coarse to fine clastic sediments derived either from the arc or from fault-bounded topographic escarpments within the basin itself.

Back-arc basins behind the same island arc are separated from one another by remnant arcs. There are many such features in the back-arc basins of the Pacific region, as shown in Figure 7.18 (compare Fig. 7.2C). The remnant arcs display the same igneous and sedimentary rocks that characterize active volcanic arcs, with one big difference: they are older than both

---

[1] After the two Greek words, *olistos* = sliding, and *stroma* = bed.

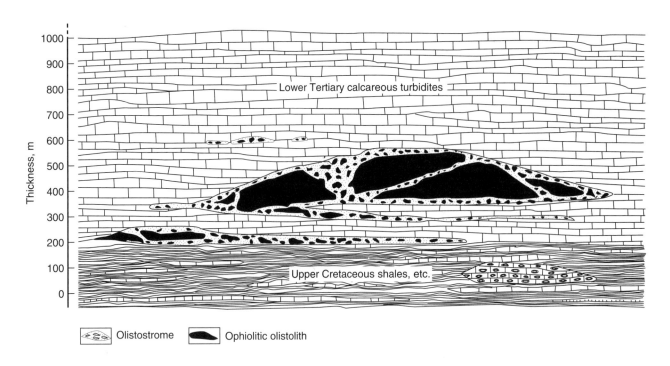

Figure 7.19 Schematic cross section illustrating olistostrome occurrences in the northern Apennines of Italy. Chaotic masses of ophiolitic debris in a pelitic matrix are interbedded with upper Cretaceous–lower Tertiary shales and calcareous turbidites. The larger olisto-stromes contain several olistoliths of ophiolitic lithologies. *(After Abbate et al., 1970)*

Many chaotic deposits are clearly of tectonic origin or are exposed in such a fashion that the origin of the deposit is difficult to decipher from field evidence. For these deposits, the general term **melange**[2] has been adopted. Most melanges are chaotic mixtures of diverse rock types in an irregularly foliated matrix (Fig. 7.20). The blocks range in size from a few millimeters to several kilometers. Blocks of ophiolitic lithologies—such as peridotite, volcanic rock, gabbro, and pelagic chert or limestone—are found in many melanges. Metamorphic rocks and shallow- or deep-water terrigenous sediments are also common. The matrix is generally deformed shale or serpentinite that has an anastomosing disjunctive foliation along which it breaks to form small scalelike fragments. The deformed shale is known as scaly clay, scaly argillite, or the Italian equivalent **argile scagliose.** It is unique to melanges.

Melanges exist in mountain belts as old as late Precambrian (for example, Anglesey, North Wales; Damaran belt, Namibia). In some regions—such as the Coast Ranges of California, Turkey, and Iran—they form vast terranes thousands of square kilo-

meters in area. Elsewhere they are more limited in extent. Melanges also have been identified in a number of modern accretionary prisms. Cores obtained from ocean drilling in areas such as the Aleutian and the Middle America accretionary prisms have contained scaly clay, suggesting at least one origin for the material.

The chaotic character of the rocks in melanges and the general lack of stratigraphic continuity make them difficult to map and interpret. Fossil or radiometric age information obtained from blocks in the melange at best provides a maximum possible age for the incorporation of the block into the melange, since the block must have been formed before it became part of the melange. Melange formation, however, could have started before the rock was formed or long after.

Figure 7.21 shows a number of sites in a subduction zone setting where melanges could form. Sedimentary melanges could form along the margins of forearc or slope basins or along the inner wall of the trench as a result of gravity sliding and disruption. Tectonic melanges could form by disruption associated with dewatering. Dewatering of the downgoing rocks could also result in diapiric rise of the released fluid and entrained lithologies which could bring these

---

[2] After the French word meaning mixture.

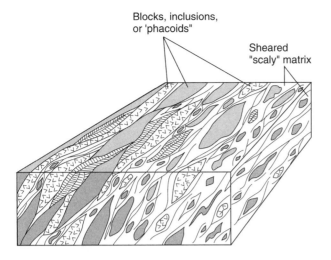

Blocks, inclusions,
or 'phacoids'

Sheared
"scaly" matrix

Figure 7.20 Block diagram of typical melange showing diverse elongate blocks and irregularly foliated (scaly) matrix. The scale could be anything from centimeters to kilometers. The matrix could be sheared clay, a sand-clay mixture, or serpentinite. The blocks could be of sedimentary, metamorphic, or ophiolitic lithologies. *(After Cowan, 1985)*

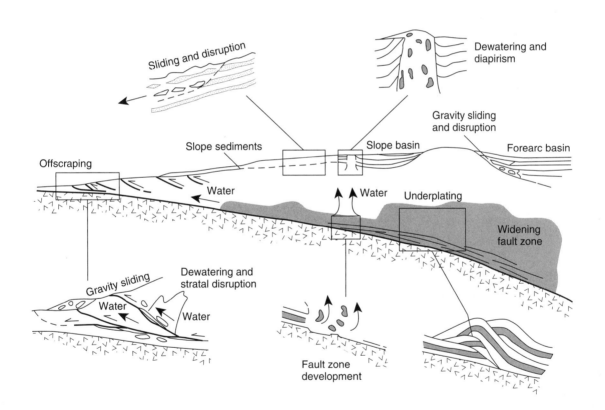

Figure 7.21 Generalized cross section showing possible origins of chaotic deposits. Gravity sliding could occur along trench slopes or in the margins of basins. Dewatering of sediments in offscraping or underplating regions could disrupt rock units, which then could be transported updip with escaping water. Melanges could also form by progressive widening of the fault zone. *(After Cowan, 1985)*

Box 7.2 Case Study: A Cross Section Through an Island Arc—The Kohistan Arc, Himalaya

In our discussion of the crustal structure of island arcs (Sec. 7.5), we mention that the thickness of island arc crust seems to increase with maturity of the arc. In existing island arcs, the mechanism of that thickening is unknown because the deeper regions of the arc crust are unexposed. The Kohistan arc, northwestern Himalaya, provides an insight into the lower crustal levels of an island arc and thus into the mechanism by which arc crust thickens (Fig. 7.2.1).

The Kohistan arc is a generally north-dipping sequence of rocks that is 30–40 km thick. It is exposed between two thrust complexes, labeled the Northern suture and the Indus suture on Figure 7.2.1. The arc comprises two parts. The older sequence, from top to bottom, consists of a series of Cretaceous sediments (the Yasin sediments); a unit of pillow lavas, island arc basalt, andesite, and rhyolite (the Chalt volcanics); a sequence of deformed island arc or ocean floor volcanic and plutonic rocks now metamorphosed to amphibolite facies (the Kamila amphibolites); and a series of metamorphosed mafic and ultramafic rocks (the Jijal-Patan complex). Into this sequence was emplaced a 10-km-thick mafic plutonic intrusive complex, now metamorphosed to granulite facies (the Chilas complex). Older deformed components of the Kohistan batholith apparently are contemporaneous with this older part of the arc.

The younger sequence consists of undeformed parts of the Kohistan batholith that intruded the older part, and the volcanic equivalents to the batholithic rocks, the Dir Utror volcanics. These volcanics include subaerial andesites, dacites, and rhyolites, all interlayered with Eocene sediments.

The relationships exposed in the Kohistan arc suggest that it formed as an intraoceanic island arc in Cretaceous time. Subsequently, but before the arrival of India, it collided against the Eurasian continental margin. Along with the surrounding area, it became the locus of emplacement of batholithic and volcanic rocks as an Andean-style continental margin.

Taken together, the Kohistan rocks provide a model for the construction and thickening of an island arc crust. The model suggests that the crust of an island arc is built up through thickening of the volcanic pile by extrusion and by intrusion of mafic to silicic plutons at various crustal levels. The deepest portion of the Kohistan arc (Jijal-Patan complex) contains significant amounts of ultramafic rocks and perhaps represents rocks from at or near the Moho. Hence, some writers refer to the Indus suture in this region as the Main Mantle Thrust.

Figure 7.2.1 (*opposite page*) The Kohistan arc, northwest Himalaya. *A.* Map of south central Asia showing location of Kohistan arc. *B.* Map of Kohistan arc. *C.* Schematic columnar section showing relationship between various units. *(After Coward et al., 1986; Petterson and Windley, 1985; Bard, 1983)*

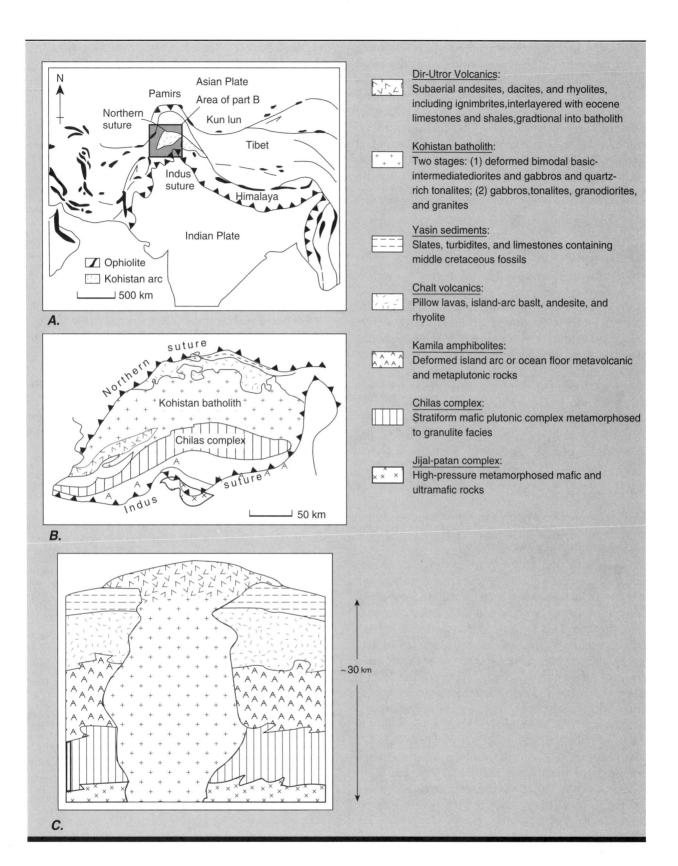

Dir-Utror Volcanics:
Subaerial andesites, dacites, and rhyolites, including ignimbrites, interlayered with eocene limestones and shales, gradtional into batholith

Kohistan batholith:
Two stages: (1) deformed bimodal basic-intermediatediorites and gabbros and quartz-rich tonalites; (2) gabbros, tonalites, granodiorites, and granites

Yasin sediments:
Slates, turbidites, and limestones containing middle cretaceous fossils

Chalt volcanics:
Pillow lavas, island-arc baslt, andesite, and rhyolite

Kamila amphibolites:
Deformed island arc or ocean floor metavolcanic and metaplutonic rocks

Chilas complex:
Stratiform mafic plutonic complex metamorphosed to granulite facies

Jijal-patan complex:
High-pressure metamorphosed mafic and ultramafic rocks

chaotically disrupted rocks to the surface. Widening fault zones in an area of underplating could also result in melange formation. Once formed, a melange can redeform easily because of the mechanical weakness of such an incoherent deposit. Thus, areas of extensive melange exposure, such as the Coast Ranges of California or the Apennines of Italy, also display abundant landsliding. The relative roles of original and subsequent disruption in individual exposures are often difficult or impossible to decipher.

7.7 Models of Subduction Zone Processes

Many workers have proposed models to account for various aspects of convergent margins. Generally, these models deal with only part of the system, the rationale being that reducing a large complex problem to a set of smaller isolated ones can provide insight into an otherwise overwhelmingly complex problem. The models that we discuss below deal with the shapes of plates and volcanic arcs, processes in the accretionary prism, processes of magma formation, and development of back-arc basins.

Shapes of Plates and Volcanic Arcs

Models in this category involve map views of volcanic arcs, the corresponding shapes of slabs at depth, and the origin of the outer swell.

Modeling the lithosphere as an inextensible spherical shell can account for the arc-and-cusp geometry of subduction zones. Such a shell, if forced down into the interior of the sphere, undergoes no change of area; the downbuckled part of the shell must have a spherical shape with a curvature equal but opposite to its original curvature. The geometry is analogous to a dent in a ping-pong ball (Figs. 7.22A through 7.22C), where the dented portion lies along an imaginary intersecting sphere of the same radius as the original ball (Figs. 7.22D through 7.22F). The hinge line of the bend is the arc of a small circle on the sphere. The central angle d of the small circle is equal to the steepest dip δ of the dented portion of the shell (Fig. 7.22G), and the radius r of the small circle is equal to the radius R_e of the Earth times the sine of half the central angle δ, or

$$r = \frac{R_e \sin \delta}{2} \qquad (7.1)$$

The larger the small circle, the steeper the dip of the dented portion of the shell. Along a hinge line of large radius, however, a shell can bend into the sphere at a relatively shallow dip if the hinge line breaks down into a series of arcs and cusps, and the indented portion then assumes the shape defined by a series of smaller dents (Fig. 7.22H).

The dynamics of subduction presumably imposes the angle of dip on the downgoing plate and hence determines the radius of curvature of the arcuate hinge line. If the plate is very large, such as the Pacific Plate, it should conform to a series of spherical surfaces such that the central angles of the various hinge line arcs equal the imposed angle of subduction (Fig. 7.22H). The result is the familiar series of arcs and cusps that characterize, for example, the subduction zones of the western Pacific.

Where the subducting plate does not conform to the required geometry, the plate must deform by extension and tearing if the dip is too large (Fig. 7.23A) or by contraction and buckling if the dip is too small (Fig. 7.23B).

The major Wadati-Benioff zones of the Earth closely fit the inextensible spherical shell model. Observed areas of buckling of the downgoing plate, as well as seismic gaps that correspond to tears in the plate, correspond well with the predictions of the model.

The Outer Swell

Some of the characteristics of the outer swell can be modeled effectively by the bending of an elastic plate. Figure 7.24A shows the downgoing plate with a load **F** and a bending moment **M**. **F** represents the downward force applied by the overriding island arc. **M** represents the action of the deeper part of the subducted plate (not shown in the diagram) on the shallower part of the plate. The moment results in a vertical uplift of the plate beyond the trench by an amount w, reproducing the geometry of the outer swell. The Mariana Trench fits the model very well, as illustrated in Figure 7.24B.

The solution using elastic plate theory seems unlikely, however, because it predicts stresses as large as 900 MPa within the plate, which seem too high by a factor of 5 to 10. Other models using a viscoelastic or an elastic-plastic mechanical model of the lithosphere predict similar geometry but with smaller stresses.

Accretionary Prism Processes

Processes in an accretionary prism that have been modeled include the deformation and subduction of

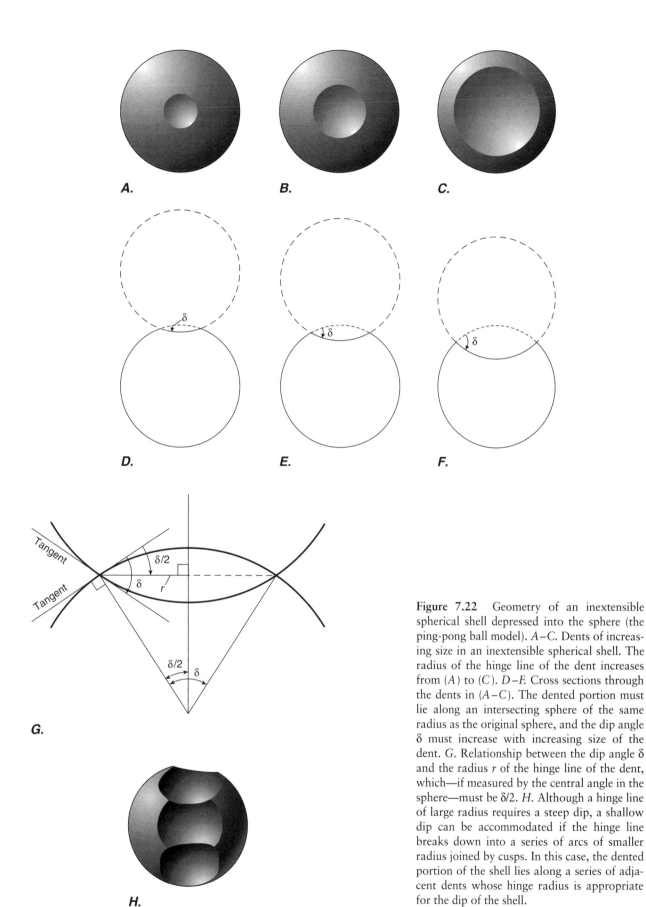

Figure 7.22 Geometry of an inextensible spherical shell depressed into the sphere (the ping-pong ball model). *A–C.* Dents of increasing size in an inextensible spherical shell. The radius of the hinge line of the dent increases from (*A*) to (*C*). *D–F.* Cross sections through the dents in (*A–C*). The dented portion must lie along an intersecting sphere of the same radius as the original sphere, and the dip angle δ must increase with increasing size of the dent. *G.* Relationship between the dip angle δ and the radius *r* of the hinge line of the dent, which—if measured by the central angle in the sphere—must be δ/2. *H.* Although a hinge line of large radius requires a steep dip, a shallow dip can be accommodated if the hinge line breaks down into a series of arcs of smaller radius joined by cusps. In this case, the dented portion of the shell lies along a series of adjacent dents whose hinge radius is appropriate for the dip of the shell.

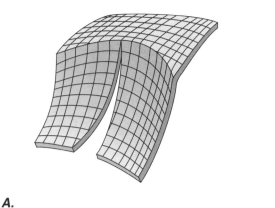

A.

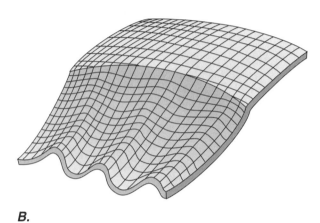

B.

sediments that are carried into the subduction zone, the characteristics of the resulting structures in the prism, and the manner in which high-pressure metamorphic rocks are carried back to the surface and become exposed. We discuss two models for these processes.

The deformation of sediments caught in a subduction zone is analogous to the flow of lubricating grease under a block sliding on a guide rail. Subduction differs from the lubrication problem, however, in three main ways: first, that sediment may be plated onto the bottom of the subduction zone hang-

Figure 7.24 Model of the outer swell in terms of the bending of an elastic plate. *A*. Where a lithospheric plate is bent into the subduction zone, the overriding island arc applies a force **F** and the deeper subducted portion of the plate applies a bending moment **M** across the cross section of the plate. Solution of the equations for the bending of an elastic plate under these conditions show that a bulge, corresponding to the outer swell, must develop. *B*. Solution for the elastic bending of a plate (dashed line) fitted to the topography of the Mariana subduction zone.

Figure 7.23 Deformation of an inextensible spherical shell whose dip is not compatible with the radius of the hinge. *A*. A shell that dips too steeply for the hinge radius must stretch, typically by tearing. *B*. A shell that dips too shallowly for the hinge radius must be compressed, typically by buckling into folds.

ing wall; second, the low density of water-saturated sediments makes them buoyant with respect to the surrounding rocks in a subduction zone, which affects their behavior; and third, the overriding block can deform during the process.

The mode of flow between the downgoing and overriding plates may depend largely on the relative rates of sediment supply and removal by underplating (Fig. 7.25). If the supply of sediment is less than or equal to the capacity of the subduction zone, then subduction of the sediment is a steady-state process (Fig. 7.25*A*). Sediment may leave the layer by underplating (Fig. 7.25*B*) if the shear stress on the hanging wall does not exceed a critical value. Conversely, erosion of the hanging wall can occur if the sediment layer is thin and the shear stress along the hanging wall exceeds the critical value. Underplating on the hanging wall causes thickening and uplift of the accretionary prism, whereas erosion of the hanging wall causes thinning and subsidence.

If the sediment supply exceeds the maximum capacity of some choke point along the subduction zone, part of the sediment that reaches the choke point must double back and flow up the subduction

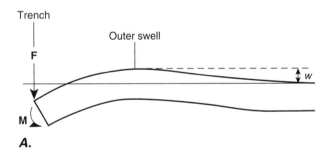

A.

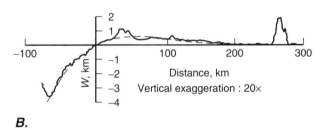

B.

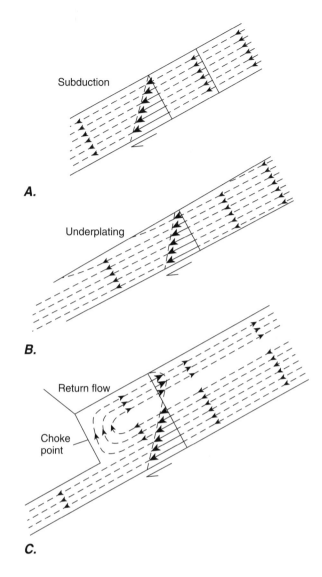

Figure 7.25 Modes of sediment flow in a subduction zone. Dashed lines with arrowheads show the streamlines; solid arrows indicate the velocity profile. *A.* Steady-state flow with sediment supply less than or equal to the capacity of the subduction zone. *B.* Underplating during which sediment is lost from the subduction zone by being incorporated into the overlying accretionary prism. *C.* Return flow occurs when the supply of sediment exceeds the capacity at some choke point along the subduction zone. *(After Shreve and Cloos, 1986)*

zone (Fig. 7.25C). This flow may reach the surface and expose high-pressure metamorphic rocks. If sufficient sediment is removed by underplating, however, the flow may die out before reaching the surface.

If deformation of the sediment involved in the doubling back and return flow is extreme, it may produce a tectonic melange. This process may explain the presence of blocks of rock metamorphosed at very high pressure and low temperature that are found mixed into some low-grade melanges.

The pattern of sediment flow at the inlet to the subduction zone significantly affects the evolution of the accretionary prism (Fig. 7.26). As sediment is dragged toward the subduction zone inlet, it encounters the resistance and shearing imposed by the overriding plate, and it shortens and thickens, largely by thrust faulting (see Fig. 7.10).

The different model flow patterns depend on the supply of sediment relative to the capacity of the subduction zone and on the patterns of offscraping and underplating. Sediment entering the subduction zone may be all subducted (Fig. 7.26A); all subducted and then partly underplated (Fig. 7.26B); or partly offscraped, partly subducted, and partly underplated (Fig. 7.26C). If a return flow is set up that reaches the surface, it can carry melange to the inlet. The melange may be added to the accretionary prism by offscraping or underplating, and part of it may be resubducted (Figs. 7.26D and 7.26E).

The structure of the accretionary prism could be characterized as a layer of sedimentary deposits covering layers of offscraped sediment, offscraped melange, underplated sediment, and underplated melange. In practice, the distinction between offscraping and underplating at the mouth of the subduction zone is difficult to make. Such a model of the accretionary prism ignores the deformation that occurs within the prism itself, which we discuss below. Material added to the accretionary prism by offscraping and underplating thickens and uplifts the prism and causes it to grow toward the trench. These processes alter the inlet capacity of the subduction zone, thereby affecting the flow geometry. Erosion along the hanging wall has the opposite effect.

The accretionary prism may be analogous to a large thrust sheet with approximately a wedge-shaped cross section (see Box 7.1), and the dynamics of such a wedge may explain some of the observed features of subduction complexes. The decrease in slope from the toe of the wedge (the inner trench wall) to the rear (the forearc basin) could be explained by one or a combination of processes, including progressive lithification and strengthening of the sediments in the wedge with increasing thickness of the wedge, a buildup of pore fluid pressure along the décollement with increasing depth of the fault, and a transition from frictional sliding to ductile deformation with increasing depth along the décollement.

Any offscraping at the front of the wedge tends to extend the length of the wedge and lower the angle of its frontal slope (Fig. 7.27A). The steady-state

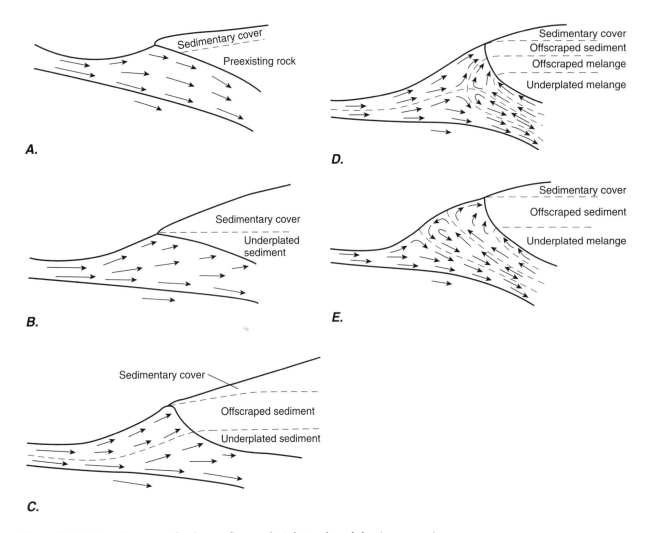

Figure 7.26 Inferred pattern of sediment flow at the inlet to the subduction zone. *A.* All sediment goes down subduction zone. *B.* Partial subduction and partial underplating. *C.* Sediment offscraped, as well as underplated. *D.* Return flow reaches the inlet. Incoming sediment is offscraped and subducted, and the melange is underplated, offscraped, and/or resubducted. *E.* Returning melange is offscraped, underplated, and/or resubducted. Incoming sediment is subducted. *(After Shreve and Cloos, 1986)*

Figure 7.27 Dynamic processes in the formation of an accretionary prism. *A.* At the toe of the accretionary wedge, if the slope is too low, shortening and thickening occur by formation of imbricate thrusts synthetic to the subduction zone, as well as some antithetic thrusts. Underplating occurs by duplex formation and folding at deeper levels. *B.* Underplating continues uplifting older rocks and steepening the slope at the rear of the wedge. The oversteepened slope drives gravitational collapse by listric normal faulting. Stippled pattern shows rocks that have been above 1000 MPa pressure. *C.* Growth of the accretionary prism through continued uplift by underplating from below, compensated for by gravitational collapse on listric normal faults above. Deeper rocks are uplifted, and the accretionary prism grows out over the subduction zone. *D.* Continuation of these processes forms complex normal fault geometry near the surface, nappes of high-pressure rocks on listric normal faults at depth, and extension of the toe of the accretionary prism on late thrust faults. *(After Platt, 1986)*

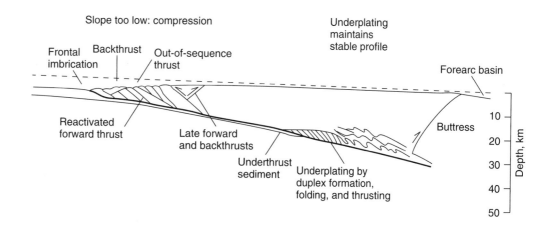

A.

Slope too low: compression

Frontal imbrication

Backthrust

Out-of-sequence thrust

Underplating maintains stable profile

Forearc basin

Reactivated forward thrust

Late forward and backthrusts

Underthrust sediment

Underplating by duplex formation, folding, and thrusting

Buttress

Depth, km — 10, 20, 30, 40, 50

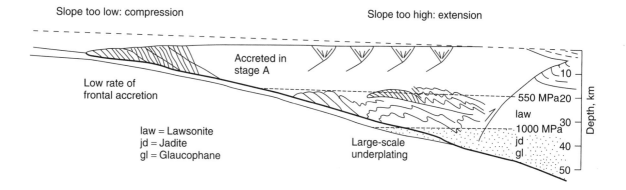

B.

Slope too low: compression

Slope too high: extension

Accreted in stage A

Low rate of frontal accretion

law = Lawsonite
jd = Jadite
gl = Glaucophane

Large-scale underplating

550 MPa
1000 MPa

law
jd
gl

Depth, km — 10, 20, 30, 40, 50

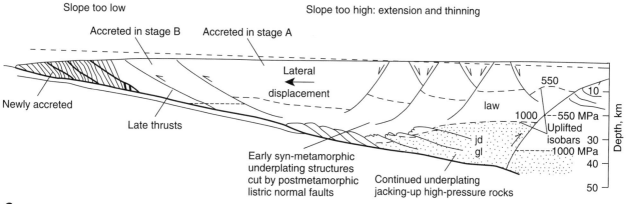

C.

Slope too low

Slope too high: extension and thinning

Accreted in stage B

Accreted in stage A

Lateral displacement

Newly accreted

Late thrusts

Early syn-metamorphic underplating structures cut by postmetamorphic listric normal faults

Continued underplating jacking-up high-pressure rocks

550
law
1000
jd
gl

550 MPa
Uplifted isobars
1000 MPa

Depth, km — 10, 30, 40, 50

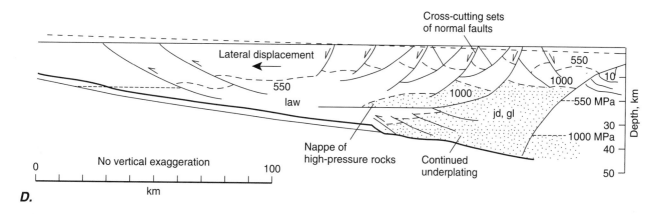

D.

Cross-cutting sets of normal faults

Lateral displacement

550
law

1000
1000
jd, gl

550 MPa
10

1000 MPa

Nappe of high-pressure rocks

Continued underplating

0   No vertical exaggeration   100
km

Depth, km — 10, 20, 30, 40, 50

value of the surface slope is maintained by internal thrusting and folding, which shortens and thickens the wedge, and/or by underplating, which thickens the wedge and uplifts the surface without requiring significant internal deformation. Both synthetic thrusting parallel to the subduction zone and antithetic backthrusting could occur (Fig. 7.27A).

With excessive underplating under the rear portion of the wedge, however, the surface slope becomes oversteepened (Figs. 7.27B and 7.27C), leading to gravitational collapse. The necessary extension and thinning of the wedge occur by listric normal faulting in the upper portion of the wedge. The net result of continued underplating at the bottom of the wedge and thinning and extension by listric normal faulting at the top is the uplift of the deeper rocks toward the surface (Figs. 7.27C and 7.27D). This process could account for the large areas of high-pressure, low-temperature metamorphism that are found at the rear of the accretionary wedges (see Fig. 7.12). Other models to account for the exposure of extensive blueschist terranes involve components of trench-parallel strike-slip faulting, trench-parallel extension, or collisional processes and are described in the next chapter.

The ages of blueschist terranes in such regions as Japan and California are concentrated in groups rather than being distributed continuously through time, which might be expected from a process of continuous subduction. Thus, acceptable models of accretionary prisms must account for the apparently episodic metamorphism, although there is little agreement at present about what the explanation might be.

The common association of oblique subduction and strike-slip faults in the forearc region has been modeled as shown in Figure 7.28. The model suggests that the displacement in the forearc region is partitioned into a component of strike-slip faulting parallel to the plate boundary and a component of dip-slip faulting down the subduction zone. In this way, a subduction zone can exhibit thrust and melange complexes, as described above, as well as strike-slip faults. The forearc sliver between the subduction zone and the strike-slip fault has been interpreted in some cases as a small plate. What happens at greater depth where ductile behavior becomes dominant and the earthquakes reflect deformation within the slab itself is not clear, however.

Thermal Structure and Melting at Subduction Zones

Heat flow measurements and calculations based on subduction rates and thermal diffusion rates indicate that a downgoing slab will be colder than the mantle into which it sinks. A model of the expected distribu-

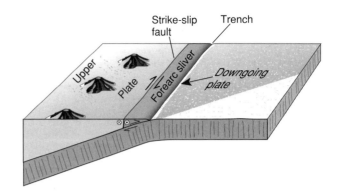

Figure 7.28 Block diagram showing model of processes in obliquely subducting zones. Partition of oblique displacement into strike-slip and dip-slip components leads to development of forearc sliver bounded by strike-slip fault and trench.

tion of isotherms is shown in Figure 7.29A for depths down to about 600 km. Figure 7.29B shows a more detailed view of the shallow thermal structure, including a schematic model of the structure in a continental arc.

It seems contradictory that melting should occur where a cold lithospheric plate descends into the mantle, and considerable controversy still exists about where and how the magmas form. One model proposes melting of the downgoing oceanic crust, but theoretical studies indicate that the temperature would be much too low at the required depths for melting to occur unless unusually large amounts of water were present. The heat produced by ductile shearing along the top of the downgoing plate could increase the temperature, but if this were a major factor, then higher rates of subduction should be associated with larger amounts of volcanism, a relationship that is not observed.

Other models rely on the fact that the presence of water considerably lowers the melting temperature of a rock. Thus, water contained in the subducted sediments and oceanic crust of the downgoing plate could migrate upward into the mantle wedge above the subducted slab and induce melting there (Fig. 7.29B). Experiments show, however, that such a mechanism should produce melts that are much more siliceous than the melts observed in island arcs.

A possible resolution of the problem is the hypothesis that the downgoing slab induces a corner flow in the mantle overlying the slab (Fig. 7.29C). This flow carries hot mantle into the corner of the overriding mantle wedge where it impinges on the

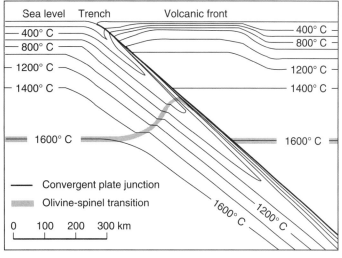

A.

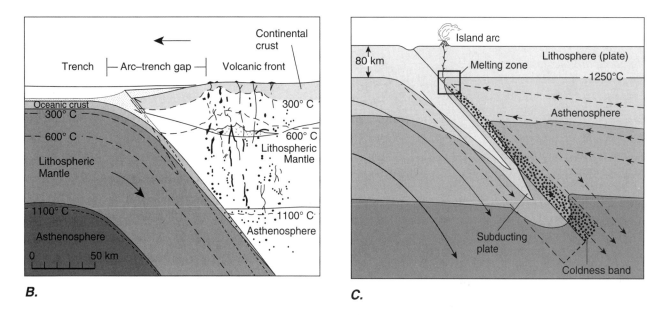

B.

C.

Figure 7.29 Models of the thermal structure of a consuming margin. The main feature of all the models is the subduction of a cool lithosphere that causes the depression of isotherms in the slab-trench region and their elevation in the arc region. *A.* Depression of isotherms in subduction zone and elevation in island arc. The olivine-spinel transition will be at shallower levels in the slab than in the mantle on either side. *B.* Upper portion of arc-trench system. Arc magma generation is assumed to occur within the mantle wedge overlying the subduction zone. The diagram also shows a region of partial melting in the continental crust, possibly giving rise to batholiths. Note that the isotherm distribution differs in detail from that in (*A*), owing to different assumptions and calculations. *C.* The downgoing slab sets up a corner flow in the overlying mantle wedge that brings hot mantle against the subducting slab, resulting in arc magma generation. *(A. after Oxburgh and Turcotte, 1970; B. after Ernst et al., 1970; C. after Marsh, 1979)*

Figure 7.30 The spacing of volcanic centers along the volcanic arc of a consuming margin. *A.* Magma is generated in a connected but restricted zone along the top of the downgoing slab. Fluid dynamic instability of a low-density magma in a higher density mantle causes magmas to form regularly spaced diapiric columns along which magmas persistently rise. *B.* The distribution of the volcanic centers at the surface is governed by the detailed geometry of the slab at depth. A buckled surface, a torn segment with different dip, or a change in slab dip at depth are all reflected in the distribution of volcanic centers in the volcanic arc.

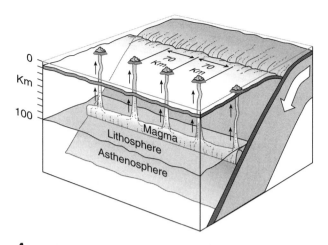

A.

downgoing slab and provides enough heat to overcome the coldness of the slab and cause local melting. Although there still seem to be problems in accounting for the detailed chemistry of the melts that we observe at the surface, this model seems the most likely explanation for the phenomenon at present.

Spacing of Volcanoes

As mentioned in Section 7.5 (see Figs. 7.2 and 7.14), volcanic centers are present at approximately 70-km intervals along a volcanic arc. One explanation for this spacing assumes that the magma source forms a more or less continuous linear zone at a constant depth along the subduction zone. If the magma is less dense than the surrounding rock, it is gravitationally unstable and tends to rise, forming buoyant diapirs at regular intervals. These intervals depend theoretically on the ratio of magma viscosity to mantle viscosity and on the thickness of the melt layer (Fig. 7.30A). Once a diapir has risen to the surface, the heated conduit remains a path of easy ascent for succeeding volumes of magma, thereby accounting for the longevity of individual volcanic centers. The irregularities in location and the segmented linear arrangement of the volcanic centers may reflect irregularities and breaks in the downgoing slab (Fig. 7.30B; compare Fig. 7.23).

This fluid dynamic model may also help explain the apparent episodicity of arc plutonic activity, in contrast to the more or less continuous record of mid-ocean spreading processes. The magnetic record from the ocean floor indicates that spreading, although discontinuous over time intervals of thousands of years, is continuous on the scale of millions of years or on the scale of magnetic reversals. In contrast, magma emplacement at convergent margins, like metamorphism, is episodic. For example, radiometric dating of plutons in the Sierra Nevada of California displays a pattern showing maxima at certain times, rather than a continuous distribution.

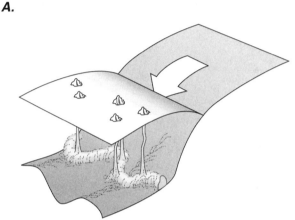

An undulating surface

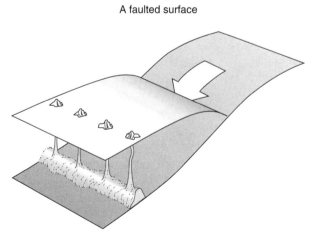

A faulted surface

A low-angle surface

B.

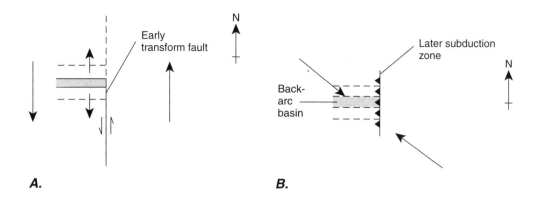

Figure 7.31 Diagrams of the formation of a marginal basin by entrapment of oceanic crust through a change in relative plate motion. *A.* Ridge-transform intersection, with north-south relative plate motion. *B.* Conversion of transform fault to subduction zone after change of relative plate motion to NW-SE. Dashed lines show magnetic anomalies about the now inactive spreading center.

Back-Arc Basins

Several origins of back-arc basins have been proposed. Some basins may result from the entrapment of preexisting oceanic crust if—during a change in plate motion, for example—an oceanic subduction zone develops at a preexisting fault zone. Figure 7.31 shows how a change in plate motion could convert a transform fault (Fig. 7.31A) into a subduction zone, trapping old oceanic crust behind the arc (Fig. 7.31B). In the Pacific region, the inactive back-arc basins with normal heat flow, such as the Aleutian Basin and the West Philippine Basin (see Fig. 7.17), may have formed in such a manner. Presumably the thrust fault forms by younger, less dense lithosphere moving over colder, denser lithosphere along the former transform faults (compare Figs. 6.5 and 6.10).

Other back-arc basins owe their existence to the formation of new crust behind an island arc. Three principal models have been proposed for such an origin: forceful injection of a diapir rising behind the arc from the subducting slab causes spreading to take place (Fig. 7.32A); a circulation in the asthenosphere overlying the subducting slab driven by the drag of the downgoing plate causes spreading behind the island arc (Fig. 7.32B); and the retreat of the overriding plate from the downgoing slab caused by global plate motion could cause spreading behind the arc (Fig. 7.32C). The geometry of the circulation above the downgoing slab, and whether it can account for both the island arc magmas and the back-arc spreading, is not clear.

The lack of clear magnetic anomalies in spreading back-arc basins may be accounted for in part by numerous closely spaced faults offsetting a spreading center (Fig. 7.33A), by asymmetrical or disordered spreading including small overlapping rifts (Fig. 7.33B), by diffuse spreading in a homogeneously extending region with no specific spreading axis (Fig. 7.33C), or by a complex combination of such magmatic and structural processes. The spreading process could account for the splitting of an island arc to form the remnant arcs that occur on the edges of back-arc basins.

Episodicity at Consuming Margins

Age information from metamorphic rocks, arc volcanic deposits, and back-arc basins suggest that metamorphism, volcanism, and back-arc basin spreading are episodic. For example, Figure 7.34 shows a plot of volcanic deposits from western North America, the southwest Pacific, and Central America that illustrates this episodicity. Some workers have argued that back-arc basin spreading alternates with arc activity. Others have argued just the opposite—that times of maximum arc volcanism are the same as times of back-arc spreading. The differences result from different data sets used by the various authors.

The evidence for episodic tectonic, magmatic, and metamorphic activity seems to contrast with the apparently continuous worldwide seafloor spreading.

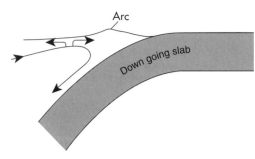

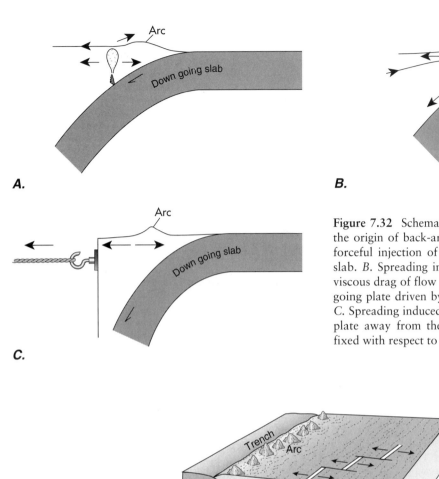

A.

B.

C.

Figure 7.32 Schematic cross sections depicting models of the origin of back-arc basins. *A.* Spreading caused by the forceful injection of a diapir rising from the downgoing slab. *B.* Spreading induced in the overriding plate by the viscous drag of flow in the mantle wedge above the downgoing plate driven by the motion of the downgoing plate. *C.* Spreading induced by the relative drift of the overriding plate away from the downgoing slab, whose location is fixed with respect to the deeper mantle.

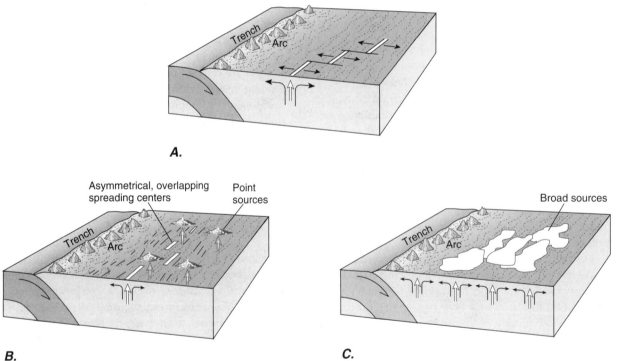

A.

B.

C.

Figure 7.33 Block diagrams illustrating possible mechanisms of spreading in back-arc basins. *A.* Symmetrical, ordered spreading of ridge-transform segments, similar to mid-ocean ridges, possibly as seen behind the Mariana, Izu-Bonin, and Tonga-Kermadec arcs. *B.* Disordered spreading, yielding asymmetrical or confused magnetic anomaly patterns, as possibly seen behind the New Hebrides, Solomon, and Bismarck arcs. *C.* Diffuse intrusion and extrusion from many point sources or broad sources, producing no discernible magnetic anomaly patterns, such as possibly seen in the Sea of Japan. *(A. after Lawver and Hawkins, 1978; B. after Weissel, 1981; C. after Uyeda, 1977)*

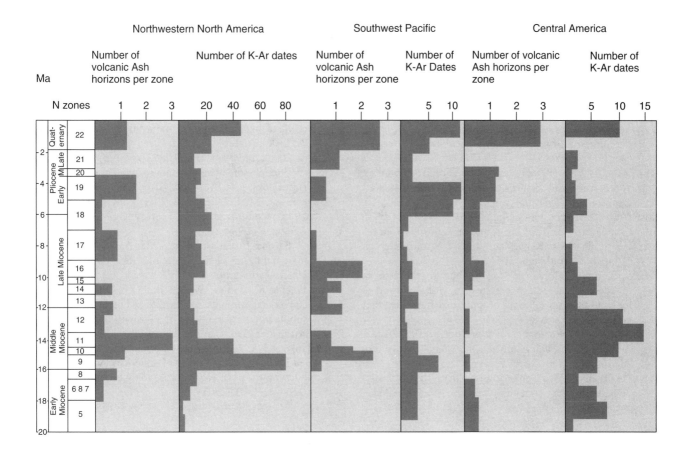

Figure 7.34 Correlation of volcanic activity with time for three areas: northwestern North America, southwest Pacific, and Central America. N-zones are foraminifera biostratigraphic zones. The first column shows the number of volcanic ash deposits, and the second column shows the number of radiometric dates for a given zone. Note peaks of activity in Miocene and Quaternary times. *(After Kennett et al., 1977)*

Seafloor spreading may also fluctuate, although the differences in the Tertiary are not as great as the changes in arc volcanism during the same period. Perhaps arc activity is a more sensitive indicator of global tectonic activity than seafloor spreading.

The Foreland Thrust Belt Problem

We have already discussed the discontinuity of the active volcanic arc of the Andes. In addition, active or recent thrusting toward the foreland is present along the entire length of the Andes. What is the relation-

ship between this foreland fold-and-thrust belt and subduction? Not all continental arcs possess such compressional features. Some workers have related the presence of antithetic thrust belts to compression of the margin of the overriding plate occasioned by the subduction of hot, buoyant lithosphere; others believe that there is an association between the fold-and-thrust belts and a shallow dip of the downgoing slab; and still others suggest that antithetic thrust belts are related to the subduction of an aseismic ridge. Are such foreland fold-and-thrust belts the consequence of subduction, or is there some other, more complex, interpretation? At present, we have no clear answer.

Additional Readings

Cowan, D. G. 1985. Structural styles in Mesozoic and Cenozoic melanges in the western Cordillera of North America. *GSA Bull.* 96:451–462.

Davis, D., J. Suppe, and F. A. Dahlen. 1983. Mechanics of fold-and-thrust belts and accretionary wedges. *Geophys. Res.* 88(B2):1153–1172.

Dickinson, W. R. 1970. Relations of andesites, granites, and derivative sandstones to arc-trench tectonics. *Rev. Geophys. and Space Phys.* 8:813–860.

Ernst, W. G., Y. Seki, H. Onuki, and M. C. Gilbert. 1970. Comparative study of low-grade metamorphism in the California Coast Ranges and outer metamorphic belt of Japan. *GSA Memoir* 124.

Isacks, B., and P. Molnar. 1971. Distribution of stresses in the descending lithosphere from a global survey of focal-mechanism solutions of mantle earthquakes. *Rev. Geophys. and Space Phys.* 9:103–174.

Karig, D. E. 1972. Remnant arcs. *GSA Bull.* 83:1057–1068.

Karig, D. E. 1974. Evolution of arc systems in the western Pacific. *Ann. Rev. Earth Planet Sci.* 2:51–76.

Lavarie, J. A. 1975. Geometry and lateral strain of subducted plates in island arcs. *Geology* 3(9):484–486.

Marsh, B. D. 1979. Island-arc volcanism. *Am. Sci.* 67:161–172.

Platt, J. P. 1986. Dynamics of orogenic wedges and the uplift of high-pressure metamorphic rocks. *GSA Bull.* 97:1106–1121.

Shreve, R. L., and M. Cloos. 1986. Dynamics of sediment subduction, melange formation and prism accretion. *J. Geophys. Res.* 91:10, 229–10, 245.

Silver, E. A., M. J. Ellis, N. A. Breen, and T. H. Shipley. 1985. Comments on the growth of accretionary wedges. *Geology* 13:6–9.

Taylor, B., and J. Natland, eds. 1995. Active margins and marginal basins of the western Pacific. *Amer. Geophys. Union.* Monograph 88.

von Huene, R., and D. W. Scholl. 1991. Observations at convergent margins concerning sediment subduction, subduction erosion, and the growth of continental crust. *Rev. Geophys.* 29:279–316.

Yamaoka, K., Y. Fukao, and M. Kumazawa. 1986. Spherical shell tectonics: Effects of sphericity and inextensibility on the geometry of descending lithosephere. *Rev. Geophys.* 24(1):27–55.

CHAPTER

8 Tectonics and Geology of Selected Triple Junctions

8.1 Introduction

In Chapter 4 (Sec. 4.3, Box 4.2), we describe the geometry and kinematics of triple junctions. In particular, we discuss criteria for the stability of a junction; that is, its ability to remain in an unvarying configuration over time, even though it might migrate with respect to one or more of the three plates surrounding the junction.

In this chapter, we examine briefly the deformation that triple junctions cause in rocks, their appearance as physical features on the Earth's surface, and the consequences of their evolution. We begin by outlining a refined representation of triple junctions that provides additional insight into the behavior of some types, particularly ones involving ridges and faults. We then apply our analysis to specific examples of triple junctions. Finally, we examine these examples to see if there are any generalizations tha209t we can make about the character of triple junctions.

8.2 Representation of Velocity Triangles in Physical Space[1]

In Chapter 4, the geometry and stability of triple junctions are represented, respectively, in physical space as maps (Fig. 8.1A, left) and in velocity space as a set of relative velocity vectors that describe the velocities of each plate relative to the others at the triple junction (Fig. 8.1A, right; compare Sec. 4.3, Figs. 4.7 through 4.9 and Fig. 4.2.4). A different method of representation (Fig. 8.1B) allows physical and velocity space for triple junctions involving ridges and transform faults to be combined on one diagram. In the Figure 8.1B diagram of a ridge-ridge-ridge (RRR) triple junction, the white areas are the lithosphere produced on each ridge in the time dt. The shaded triangle shows the

[1] This section is based substantially upon the work of Patriat and Courtillot, 1984.

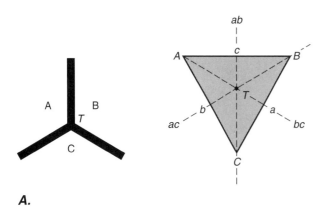

A.

Figure 8.1 Geometry of a ridge-ridge-ridge (RRR) triple junction. *A.* Conventional separate portrayal of junction in physical space (*left*) and velocity space (*right*). Compare with Figure 4.2.2. A, B, C, plates; *T*, triple junction; *ab*, *bc*, *ac*, reference frame lines for the boundaries between plates A and B, B and C, and A and C, respectively. *B.* Combined portrayal of physical and velocity space. White areas are new lithosphere produced in time *dt*. The shaded triangle shows new lithosphere produced by the interaction of the ridges at a triple junction. This triangle has the same geometry as the velocity triangle. White extensions of ridges into the triangle are the line segments *aT, bT,* and *cT,* which show how far each ridge has propagated in the increment of time *dt*. Arrows parallel to the ridges indicate the direction of propagation of the ridge. (*After Patriat and Courtillot, 1984*)

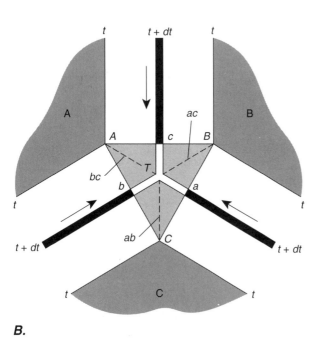

B.

area of crust generated by the interaction of the three ridges; it has the same geometry as the velocity triangle shown in Figure 8.1*A*. The white portions of the ridges in the shaded triangle in Figure 8.1*B*—labeled *aT, bT,* and *cT*—show how much each ridge has lengthened in the time *dt*. The line labeled *AB*, for example, represents the relative velocities between plates A and B and thus characterizes the ridge between these plates, which we can refer to as the *AB* ridge. Hence, the diagram shows velocity space superimposed on physical space.

Figure 8.1 shows three ridges that spread symmetrically and at the same rates; thus, the velocity triangle is equilateral. Most triple junctions do not

have such a perfect configuration, however, and the type of triple junction changes depending upon the relative orientations of the ridges and the spreading velocities.

Consider, for example, the RRR junction shown in Figure 8.2*A*, where two relatively fast spreading ridges (*AB* and *AC*) connect with a slow-spreading ridge (*BC*). In this particular case, the velocity triangle is a right triangle, and the triple junction *T* is located exactly on the hypotenuse. As shown in the diagram, the *AB* and *BC* ridges increase in length, but because the triple junction *T* lies on the boundary *AC* of the velocity triangle, the ridge *AC* does not lengthen. It is also possible for the triple junction between these three plates to be an RRF type, as shown in Figure 8.2*B*.

Another configuration for a triple junction is shown in Figure 8.3. A slow-spreading ridge (*BC*) intersects two faster spreading ones (*AB* and *AC*), so that the velocity triangle is an acute isosceles triangle. In this case, the junction can be either RRR type (Fig. 8.3*A*) or FFR type (Fig. 8.3*B*). In the RRR configuration, each ridge lengthens as in Fig. 8.1*B*, but in the FFR configuration, the slow-spreading ridge lengthens and the two faster spreading ridges remain the same length.

In the case of two slow-spreading ridges intersecting a fast-spreading one, the velocity triangle is obtuse, as illustrated in Figure 8.4. Again, two configurations are possible. If the junction is an RRR type (Fig. 8.4*A*), the two slow-spreading ridges (*AB* and *AC*) lengthen while the fast-spreading one (*BC*) recedes by an amount indicated by the dashed ridge segment. Alternatively, if the junction is an FFR type (Fig. 8.4*B*), the fast-spreading ridge (*BC*) propagates while the other two remain the same length.

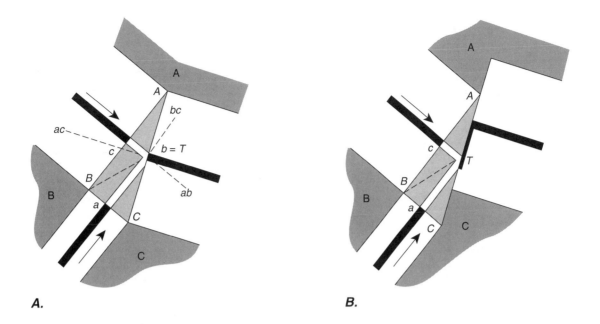

A.

B.

Figure 8.2 Combined portrayal of physical and velocity space for a triple junction between two faster spreading and one slower spreading ridge, where the velocity triangle has a right angle. Symbols are the same as in Figure 8.1. *A.* RRR configuration. The *AB* and *BC* ridges lengthen by amounts cT and aT, respectively; the *AC* ridge remains same length. *B.* RRF configuration of the same triple junction. The *AC* boundary moves away from the ridge by lengthening of the fault. *(After Patriat and Courtillot, 1984)*

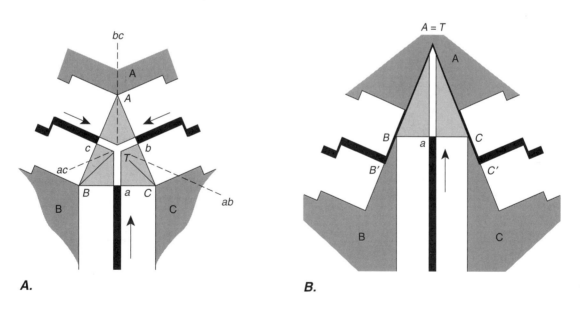

A.

B.

Figure 8.3 Triple junction between two faster spreading ridges and one slower spreading ridge, where the velocity triangle is an acute isosceles triangle. *A.* RRR configuration. The triple junction is at the centroid of the velocity triangle, and the *AB*, *BC*, and *AC* ridges lengthen by cT, aT, and bT, respectively. *B.* FFR configuration. The *BC* ridge lengthens by aT, and the faults lengthen by *AB'* and *AC'*, respectively. The triple junction is at *A*. *(After Patriat and Courtillot, 1984)*

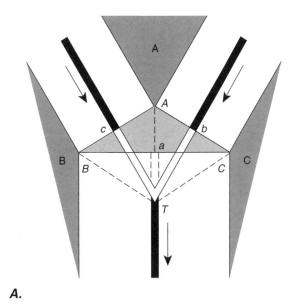

A.

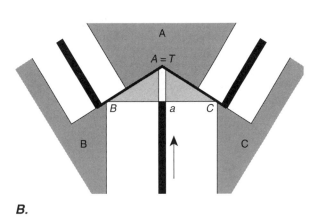

B.

Figure 8.4 Triple junction between two slower spreading and one faster spreading ridge, where the velocity triangle is an obtuse isosceles triangle. *A.* RRR configuration. The *AB* and *AC* ridges lengthen by *cT* and *bT*, respectively. The *BC* ridge recedes by *aT*. *B.* FFR configuration. The *BC* ridge lengthens by *aT*; faults lengthen. *(After Patriat and Courtillot, 1984)*

Observations of magnetic anomaly geometries suggest that a number of triple junctions alternate between two states, such as those illustrated in parts *A* and *B* of Figures 8.3 and 8.4. To account for this behavior, Patriat and Courtillot hypothesized that the triple junction would be in an RRR mode if the magma supply were sufficient to cause the ridge to propagate. If the magma supply to the propagating ridges

waned, however, then the FFR mode might develop, resulting in a lengthening of the transform faults. As the faults lengthen, the resistance along them increases. When a new pulse of magma arrives at the triple junction, the ridges propagate across the faults, and the junction switches back into an RRR mode.

8.3 Case Studies of Triple Junctions: The Bouvet Triple Junction

The Bouvet triple junction is located in the South Atlantic Ocean approximately 250 km west of Bouvet Island, at 1°W longitude and 55°S latitude. It is the point where the South American, African, and Antarctic plates meet. At the junction, the southern tip of the Mid-Atlantic Ridge meets a ridge-transform boundary between the South American and Antarctic plates and another one between the African and Antarctic plates. Two prominent fracture zones, the Conrad and Bouvet, help to determine the azimuth of relative plate motion. Magnetic surveys in the region have recognized anomalies 1, 2, and 2' (Fig. 8.5). The topography in the region is complex, with the Mid-Atlantic Ridge exhibiting no axial valley.

Figure 8.6 shows a velocity-space triangle for the Bouvet triple junction, illustrating azimuths and rates of spreading for the three plate boundaries. From the diagram, it is evident that the junction could be stable in either an FFR (Fig. 8.6*A*) or an RRR (Fig. 8.6*B*) mode. Figure 8.7 shows a model, based upon interpretation of magnetic anomaly records, for the evolution of the junction since about 20 Ma. According to the model, the junction was mostly in the FFR mode for the period 0–20 Ma, but at about 5 Ma, there was a brief period when it was essentially in the RRR mode. According to this model, the distance between the fracture zone labeled *X* and the triple junction labeled *T* should have increased between 20 and 10 Ma and should have decreased again since 10 Ma.

Based on the magnetic anomaly pattern, we can infer that the distance between fracture zone *X* and the triple junction *T* at the Bouvet triple junction was approximately 70 km at 20 Ma, increased to a maximum of approximately 110 km at 10 Ma, and subsequently decreased to its present 80 km. One might wonder whether the record of such a distance fluctuation would ever be preserved in subaerial exposures of oceanic crust in exemplary ophiolites such as the Troodos, Cyprus, or Semail, Oman, complexes. The distance of fluctuation is as large as most exposures, leading one to question how to recognize it in outcrop.

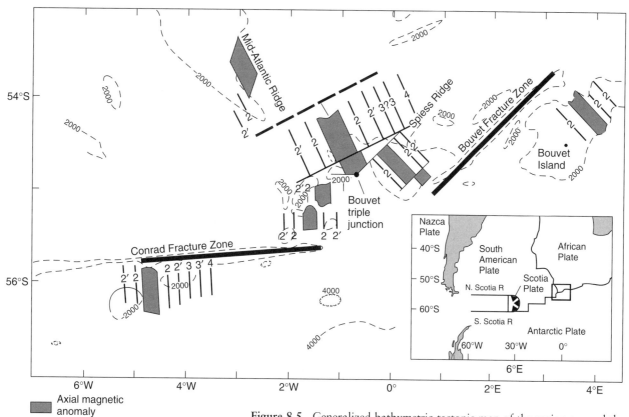

Figure 8.5 Generalized bathymetric-tectonic map of the region around the Bouvet triple junction in the South Atlantic. Note the position of Bouvet Island. Inset shows location. Map shows 2000m and 4000m contours, magnetic profiles, and anomaly interpretation. *(After Sclater et al., 1976)*

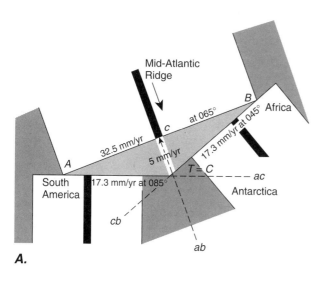

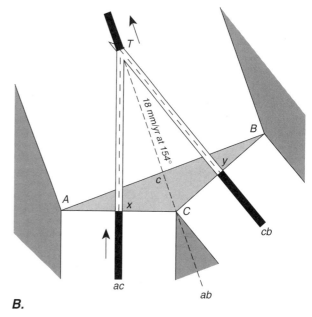

Figure 8.6 Two possible configurations of the Bouvet triple junction. *A.* FFR configuration. The Mid-Atlantic ridge lengthens by amount $cT = 168$ mm/yr. *B.* RRR configuration. The Mid-Atlantic ridge lengthens by $cT = 18$ mm/yr. The AC and BC ridges lengthen by $xT = 16$ mm/yr and $yT = 16$ mm/yr, respectively. *(After Patriat and Courtillot, 1984)*

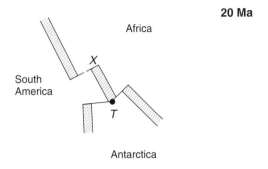

20 Ma

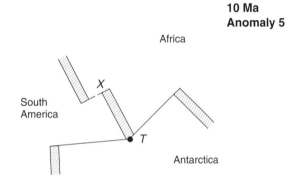

10 Ma
Anomaly 5

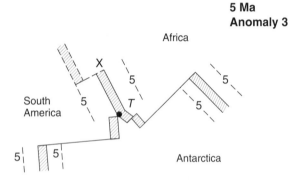

5 Ma
Anomaly 3

Present

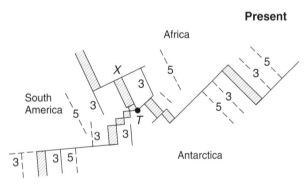

Figure 8.7   Model of the evolution of the Bouvet triple junction from 20 Ma to present. Note the shift in modes from FFR to RRR and back to FFR, as distance from fracture zone X to triple junction T first increases and then decreases. Current XT distance is 80 km. *(After Sclater et al., 1976)*

8.4   The Galapagos Triple Junction

This junction is an RRR triple junction that separates the Pacific, Cocos, and Nazca plates in the east-central Pacific Ocean (Fig. 8.8). The fast-spreading East Pacific Rise trends roughly north-south in this region and continues without offset past the triple junction. The east-trending, slower spreading Galapagos Ridge separates the Cocos and Nazca plates. Crust generated along the Pacific Ridge displays a smooth topography, whereas the Galapagos Ridge develops a rougher, more faulted topography. The boundary between the rough and smooth topography is approximately V-shaped, with the V opening to the east (dashed lines in Fig. 8.8) and enclosing a triangular area called the Galapagos Gore. Magnetic anomalies in this region indicate that the rough topography contains older crust in the east than in the west. Thus, the Galapagos Ridge appears to have propagated from east to west as it developed.

Figure 8.9 shows a combined diagram of physical and velocity space for the Galapagos triple junction. The diagram shows that the Pacific-Cocos and Pacific-Nazca rates of spreading are 137 mm/yr and 135 mm/yr, respectively. The Galapagos Ridge spreads at a rate of 41 mm/yr; its tip is propagating westward at the rate of 66 mm/yr.

As illustrated in Figure 8.10, the triple junction region contains an interplay of north-trending structures related to the East Pacific Rise and east-trending

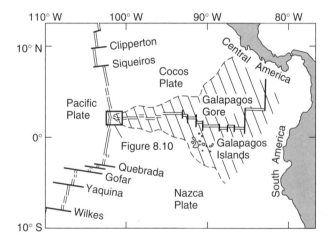

Figure 8.8   Tectonic map of equatorial eastern Pacific showing major plates, plate boundaries, and the location of the Galapagos triple junction. Dashed lines shows the rough-smooth boundary enclosing the ruled area—the Galapagos Gore. *(After Searle and Francheteau, 1986)*

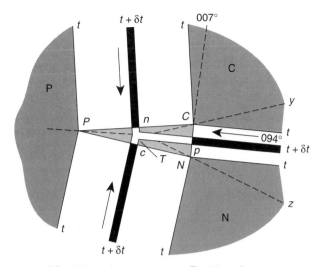

PC = 137 mm/yr   cT = 14 mm/yr
PN = 135 mm/yr   pT = 66 mm/yr
CN = 41 mm/yr   Azimuth of Ty = 077.5°
nT = 6 mm/yr   Azimuth of Tz = 112.5°

Figure 8.9 Combined velocity and physical space diagram for the Galapagos triple junction (T). P, Pacific Plate; C, Cocos Plate; N, Nazca Plate. Pacific-Cocos and Pacific-Nazca ridges propagate at 6 mm/yr and 14 mm/yr, respectively. Galapagos (Cocos-Nazca) Ridge propagates at 66 mm/yr. *(After Searle and Francheteau, 1986)*

Figure 8.10 Map of the Galapagos triple junction region showing major volcanic and tectonic features from SEA-BEAM and GLORIA surveys. The stippled line indicates the inferred spreading axis. The shaded area is the Hess Deep at the end of the propagating Galapagos Ridge. *(After Searle and Francheteau, 1986)*

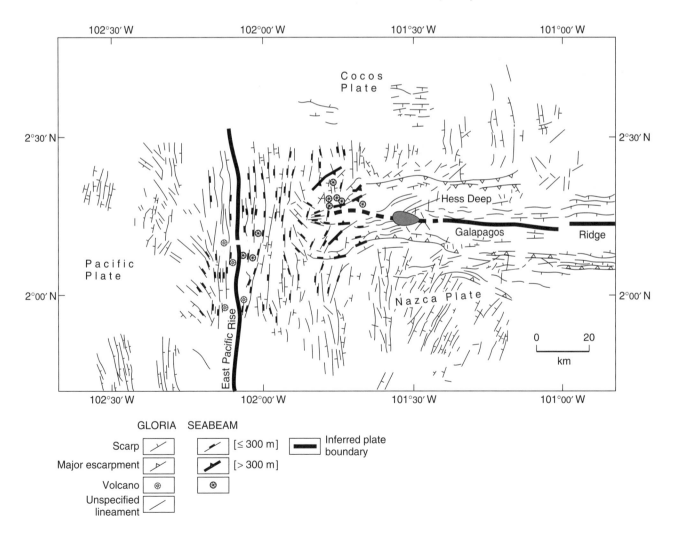

features related to the Galapagos Ridge. Close to the triple junction, the ridges change morphology and display more pronounced evidence of faulting. The East Pacific Rise changes from fast-spreading to intermediate-spreading morphology with the associated axial graben, whereas the Galapagos Ridge develops an axial valley that deepens to the west, finally reaching depths of more than 4000m in a basin called Hess Deep. West of Hess Deep, there is no clear expression of the Galapagos spreading center. Thus, there is a gap of approximately 25 km between the ridges. What might cause this gap? Is it related to episodic propagation of the Galapagos Ridge or to the availability of magma? Or is it an indication of the surface resolution of the geometry of plate processes at depth?

8.5 The Mendocino Triple Junction

This FFT junction marks the northwest tip of the San Andreas transform fault system (Fig. 8.11; see Sec. 6.4). It is one of two triple junctions between which the Pacific–North American plate boundary is a transform fault. As discussed in Section 6.4, the current thinking is that these two junctions formed as a result of the intersection of the East Pacific Rise with a subduction zone off the North American continent and the consequent disappearance of progressively larger portions of the Farallon Plate. The Juan de Fuca, Rivera, and Cocos plates constitute the remnants of the Farallon Plate in the northeast Pacific. The rise-trench intersection occurred about 30 Ma off southern California (Fig. 8.12). Subsequently, one triple junction moved northwestward to its present location at Cape Mendocino, and the other moved southeast to its present location at the Rivera triple junction. In between, the transform boundary grew in length at the expense of the trench, remnants of which still remain to the north off the coast of northern California, Oregon, and Washington, and to the south off the coast of southern Mexico and Central America. Some scientists have argued that prior to 5 Ma, the transform boundary was offshore, as

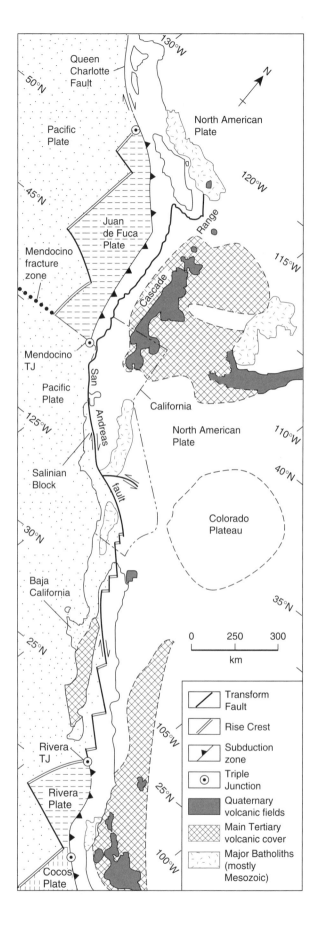

Figure 8.11 Tectonic map of portion of western North America showing current plate boundaries, including the positions of the Mendocino and Rivera triple junctions. Also shown are general areas of volcanic and plutonic rocks that relate to plate margin processes. *(After Dickinson and Snyder, 1979a)*

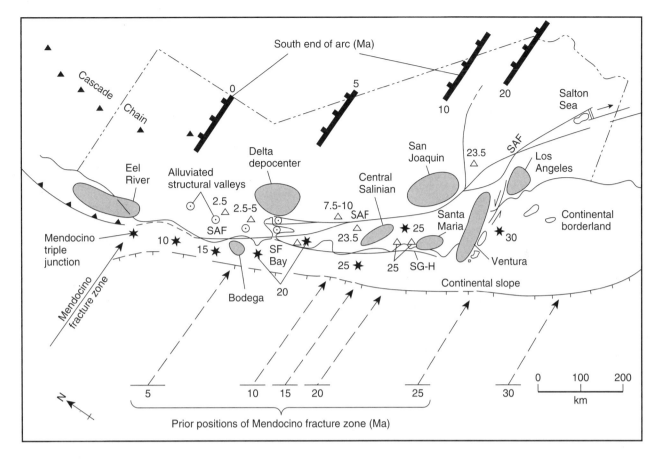

Figure 8.12 Diagram of San Andreas fault system showing associated tectonic features. The Mendocino fracture zone originally intersected the trench about 29–30 Ma. As it migrated northward, the southern limit of the subduction-related arc also migrated northward. Depositional basins developed along the fault system as the triple junction passed. Asterisks are inferred positions of the triple junction at different times, which have been displaced by the San Andreas fault. Open triangles with numbers are volcanic features and their approximate ages associated with passage of the triple junction. SAF, San Andreas fault; SG-H, San Gregorio–Hosgri fault. *(After Dickinson and Snyder, 1979a)*

shown in Figure 8.13, and the principal locus of transform faulting was west of the present San Andreas fault.

The stability and the position of the Mendocino triple junction have varied through time as the orientation of the plate boundary relative to the Farallon–North American plate motion has changed. As shown in Chapter 4 (Box 4.2, Fig. 4.2.4), a triple junction of FFT type can be stable only if the trench is colinear with one fault boundary, which does not apply to the present Mendocino triple junction configuration (Fig. 8.14). Figure 8.15 shows three possible configurations of a Mendocino-like FFT triple junction migration. In Figure 8.15*A*, the triple junction is unstable, and because it migrates, it changes from an FFT to a TTF

triple junction. In Figure 8.15*B* the *BC* (trench) and *AC* (fault) boundaries are colinear, and the junction is stable. In Figure 8.15*C*, the fault and trench boundaries are not parallel, and the junction is unstable. As the junction migrates to the northwest, spreading will take place behind it, and a complex trench boundary might develop northeast of it. The pattern shown in Figure 8.15*C* is similar to the present configuration of the Mendocino triple junction (Fig. 8.14), and seismic activity is present within the Juan de Fuca Plate northwest of the triple junction (Fig. 8.16). Fault-bounded sedimentary basins might form in regions of expected extension, as shown in Figure 8.15*C*.

As the San Andreas transform system developed, the unsubducted area of the Farallon Plate (that is, the

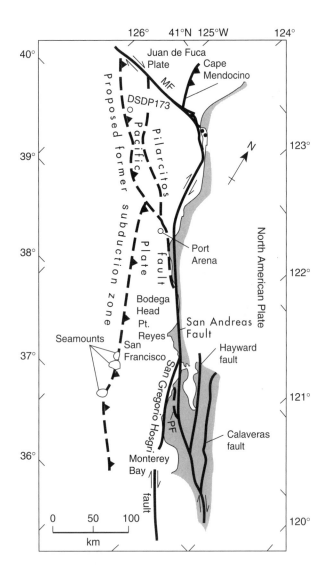

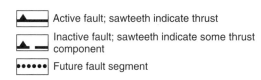

Figure 8.13 Tectonic map of northern San Andreas system and associated offshore features. The transform motion possibly moved progressively eastward over time, shifting from the Pilarcitos–San Gregorio–Hosgri fault to the San Andreas fault (SAF) approximately 3–5 Ma. At present, motion may be shifting farther eastward to the Hayward and Calaveras faults. *(After Griscom and Jachens, 1989)*

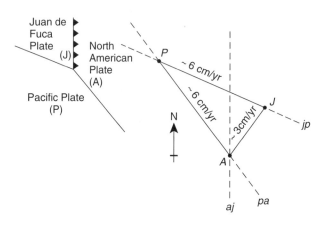

Figure 8.14 Physical and velocity space diagrams of Mendocino triple junction. Reference lines *aj* (North American–Juan de Fuca), *pa* (Pacific–North American), and *jp* (Juan de Fuca–Pacific) do not meet in a point, indicating that the junction is unstable. *(After Dickinson and Snyder, 1979a)*

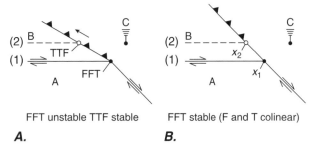

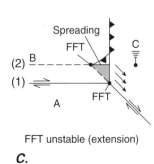

Figure 8.15 Diagrams illustrating varying configurations of the Mendocino triple junction depending upon relative orientation of the trench and relative Pacific–North American plate motion. Plate C is fixed. The triple junction migrates to northwest in all three diagrams. The boundary between plates A and B migrates from position 1 (solid line) to position 2 (dashed line). A. The triple junction migrates into a region where the trench trends more westerly than plate relative motion. The FFT triple junction converts to a TTF junction, where the trench southeast of the triple junction is a transpressive boundary. B. Trench and fault are colinear, and the FFT triple junction is stable. C. The triple junction migrates into a region where the trench trends more northerly than the transform fault. The junction is unstable and an extensional region develops, possibly with the formation of a deposition basin. *(After Dickinson and Snyder, 1979a)*

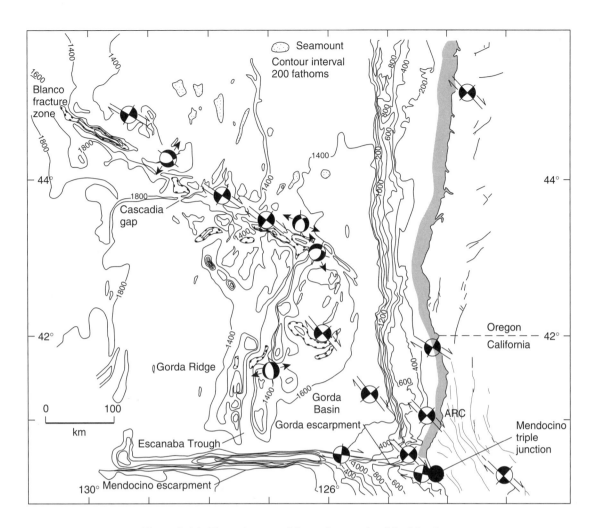

Figure 8.16 Tectonic map of the region north of the Mendocino triple junction showing extension of strike-slip faults into the Gorda Basin. First-motion diagrams show open quadrants (shortening quadrant; rarefaction first-motion) and filled quadrants (lengthening quadrant; compression first-motion). *(After Silver et al., 1971)*

Juan de Fuca and Cocos plates) diminished, and the subducted edges of the Farallon Plate occupied successive positions shown in Figure 8.17 as the lines $X_n Z'_n$ and $Y_n Z'_n$, where n increases with time. The triangular zone $X_n Z'_n Y_n$ represents a region within which the downgoing slab is absent because by time n, subduction has ceased between X_n and Z'_n. Within this expanding triangular slab window, hot asthenospheric material is in contact with the base of the North American Plate. The passage of the Mendocino triple junction northwestward along the California coast

has produced a NW-migrating boundary at which hot asthenospheric material abruptly contacts the base of the continent. This feature may have given rise to Neogene volcanic rocks within the California Coast Ranges, indicated by the open triangles in Figure 8.12, which show a progressive decrease in age toward the northwest. The increase in temperature associated with the impingement of hot asthenosphere on the base of the continent may have resulted in rapid uplift of the triple junction region. At present, Cape Mendocino is rising at a rate of 5–10 mm/yr.

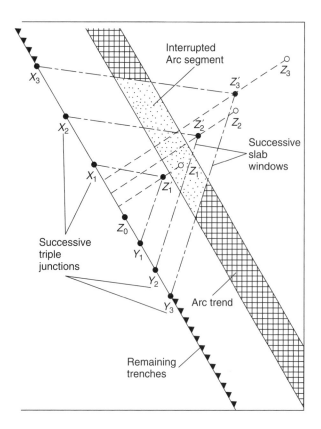

Figure 8.17 Diagram illustrating the development of a slab window as the San Andreas fault develops and the Mendocino and Rivera triple junctions migrate northwest and southeast, respectively. *(After Dickinson and Snyder, 1979b)*

nate it as currently part of the North American Plate (as indicated tentatively on Figs. 7.1, 8.18 and 8.19).

The Philippine–northeastern Japan (North American?) plate boundary is a complex region, the Sagami Trough, that represents a zone of highly oblique subduction extending west-northwest from the Boso triple junction (Fig. 8.19). Both thrust and strike-slip faults are present along this trough.

In Chapter 4 (Sec. 4.3, Box 4.2), we discuss the criteria for stability of TTT(a) triple junctions, such as the Japan junctions. In general, they are stable only if two of the plate boundaries are colinear or if the slip vector between the two plates being subducted beneath the third parallels their mutual boundary. Figure 8.19 shows maps and velocity triangles (diagrams in

8.6 The Japan Triple Junctions

Japan has two triple junctions. One, the Boso triple junction (named after the nearby Boso Peninsula) is off Japan along the Izu-Bonin-Japan Trench (Fig. 8.18). The other, a young and recently recognized one called the Fuji triple junction, is near Mt. Fuji along the southeastern coast of the island of Honshu. They constitute the only trench-trench-trench (TTT) triple junctions currently active in the world (see Fig. 7.1). Each is a TTT(a) type (see Fig. 4.2.4); that is, a triple junction where one subduction zone is of opposing polarity to the other two and one plate overrides the other two. The Boso triple junction joins the plate margins between the Pacific and northeastern Japan (North American?) plates, the Pacific and Philippine plates, and the Philippine and Eurasian plates. The Fuji triple junction joins the plate margins between the Eurasian and northeastern Japan (North American?) plates, the Philippine and northeastern Japan (North American?) plates, and the Philippine and Eurasian plates (Figs. 8.18 and 8.19).

The plate affinity of northeastern Japan is problematic. All workers agree that it was a part of the Eurasian Plate until recently, but many workers desig-

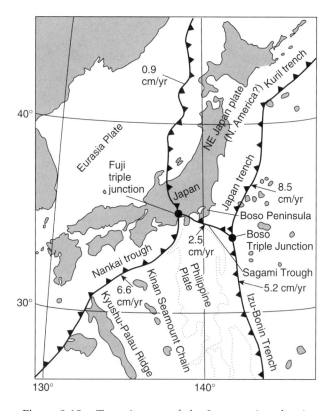

Figure 8.18 Tectonic map of the Japan region showing major plate boundaries and location of the Japan triple junctions. *(After Ogawa et al., 1989; Yamazaki and Okamura, 1989)*

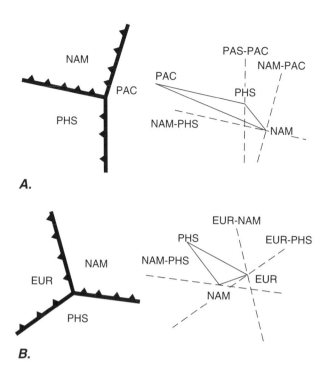

A.

B.

Figure 8.19 Physical and velocity space diagrams for the Japan triple junctions. EUR, Eurasian Plate; NAM, North American Plate; PAC, Pacific Plate; PHS, Philippine Plate. Note that reference lines for both triple junctions do not meet in a point, indicating that the junctions are unstable. *A.* Diagrams for the Boso triple junction. *B.* Diagrams for the Fuji triple junction. *(After Ogawa et al., 1989)*

the Philippine Plate is highly contorted, possibly because it does not have enough room as it is inserted between the Pacific and the North American plates (Fig. 8.21).

The Boso triple junction seems to have evolved rapidly in shape and configuration over the past few million years. Figure 8.22 shows a model of the evolution of the junction for the past 10 million years. From late Miocene through early Pliocene time, motion between Eurasia and the Philippine Plate was northward, with a pole of rotation near the triple junction (Figs. 8.22*A* and 8.22*B*). In early Pliocene time, the junction had migrated southward and reached the bend in the Izu-Bonin-Japan Trench, an unstable configuration. This migration corresponded

physical and velocity space) for the Boso triple junction (Fig. 8.19*A*) and the Fuji triple junction (Fig. 8.19*B*). Also shown are the reference frame lines (dashed lines) for both triple junctions. From the figure, it is clear that neither junction is stable at the present time because the reference frame lines do not meet in a point.

There is another condition attached to the existence of a TTT(a) triple junction. It stems from the fact that at each junction, one plate, the Northeastern Japan (North American?) Plate, overrides both the others—at both junctions. In addition, one plate (the Pacific Plate at the Boso junction and the Philippine Plate at the Fuji junction) descends beneath the other two. For each junction, there is an intermediate plate (the Philippine Plate at the Boso junction and the Eurasian Plate at the Fuji junction) that must be sandwiched between the other two plates. The attitude and motion of this intermediate plate must be such that everywhere it is above the bottom plate and below the top plate. Thus, at the Boso triple junction, the Philippine Plate must be sandwiched between the lower Pacific Plate and the overriding North American Plate; whereas at the Fuji triple junction, the Eurasian Plate must be sandwiched between the overriding North American Plate and the lower Philippine Plate.

Figure 8.20 shows a structural contour map of both the Philippine and Pacific plates based upon earthquake hypocenters. The map suggests that the Pacific Plate is relatively simple in structure, whereas

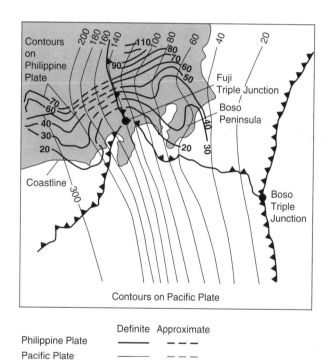

Figure 8.20 Map of the Japan triple junction area showing structural contours on the Pacific and Philippine plates. The Pacific Plate is relatively simple in three dimensions. The more complex pattern of structural contours on the Philippine Plate suggests contortion in the plate, possibly because it does not have enough room. *(After Huchon and Labaume, 1989)*

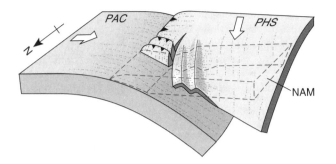

Figure 8.21 Block diagram showing a possible interaction between Pacific (PAC), Philippine (PHS), and North American (NAM) plates leading to contortions of the Philippine Plate because of a lack of space between the other two. The overriding North American Plate is shown only in ghost outline so as to reveal the Phillipine Plate. *(After Huchon and Labaume, 1989)*

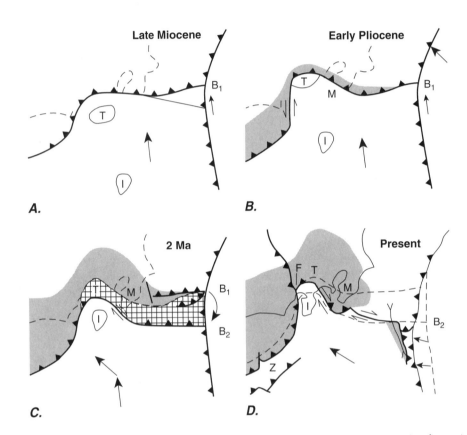

Figure 8.22 Schematic model of the evolution of the Japan triple junctions. *A.* Configuration in late Miocene time. T, volcanic island of Izu-Bonin arc approaching trench to form Tanzawa block; I, Izu block, a volcanic island; B_1, location of Boso triple junction. *B.* Configuration in early Pliocene time. Arrival of Tanzawa block and emplacement of Mineoka ophiolite (M). *C.* Configuration at 2 Ma. Shift of relative plate motion and southward shift of location of Boso triple junction to position B_2. The change of motion of the Philippine Plate between early Pliocene time and 2 Ma resulted in a shift of location of the junction southward, the development of transform motion along the Sagami Trough, and the initiation of the Fuji triple junction (F). *D.* Current configuration of triple junctions. Izu block (I) arriving at trench. *(After Huchon and Labaume, 1989)*

to the arrival at the subduction zone of the Tanzawa block (T in Fig. 8.22), an island in the Izu-Bonin arc system, and the emplacement of the Mineoka ophiolite (M in Fig. 8.22). As a result of a change in the relative motion of the Philippine–Eurasian plates at 2 Ma, a mostly strike-slip belt developed in the Sagami Trough, the plate boundary became highly curved, and a separate plate boundary developed west of northeastern Japan (see Fig. 8.18). Because of this change in relative plate motion, the Boso triple junction should evolve into a TTF or a TFF triple junction. Since the Philippine Plate is now moving northwest, its boundary with the Pacific Plate (the trench) will also move northwest relative to North America(?), and the location of the triple junction should migrate to the west along the Sagami Trough.

8.7 Discussion

This brief discussion of four selected triple junctions of different configurations, though certainly not a comprehensive survey of all triple junctions on Earth, permits us to make three generalizations:

1. Despite their diversity, triple junctions generally tend to obey the rules laid out by Dan McKenzie and Jason Morgan in 1969 (see Box 4.2).

2. Because plate boundaries are geologic regions up to tens of kilometers wide, any triple junction involving such boundaries must itself involve an area of tens to hundreds of square kilometers; that is, an area at least the size of an average field mapping area.

3. Each junction displays evidence of rapid evolution from different configurations in the very recent past.

A similar analysis of every triple junction on Earth would illustrate the principal kinematic processes for development of deformational structures at these plate boundary nodes. We would emphasize that, despite their overall simplicity and constancy, the motions associated with plate tectonics, on the scale of a map area, triple junctions undergo surprisingly rapid changes and produce a surprising complexity in deformational structures.

Additional Readings

Dickinson, W. R., and W. S. Snyder. 1979a. Geometry of triple junctions related to San Andreas transform. *J. Geophys. Res.* 84:561–572.

Dickinson, W. R., and W. S. Snyder. 1979b. Geometry of subducted slabs related to San Andreas transform. *J. Geol.* 87:609–627.

Ogawa, T. Seno, H. Akiyoshi, H. Tokuyama, K. Fujioka, and H. Taniguchi. 1989. Structure and development of the Sagami Trough and the Boso triple junction. *Tectonophysics* 160:135–150.

Patriat, P., and V. Courtillot. 1984. On the stability of triple junctions and its relation to episodicity in spreading. *Tectonics* 3:317–332.

Sclater, J. D., C. Bowin, R. Hey, H. Hoskins, J. Pierce, J. Phillips, and C. Tapscott. 1976. The Bouvet triple junction. *J. Geophys. Res.* 81:1857–1869.

Searle, R. C., and H. Francheteau. 1986. Morphology and tectonics of the Galapagos triple junction. *Marine Geophys. Res.* 8:95–129.

CHAPTER

9 Collisions

9.1 Introduction

As consuming margins evolve, eventually a downgoing plate will carry continental or island arc crust into a subduction zone. At that point, a collision zone replaces the consuming margin. The introduction of low-density crust into the subduction zone fundamentally alters the plate geometry and kinematics because its buoyancy counteracts the driving forces of subduction and thereby changes the motion of the plates.

There are several possible types of collision, as shown in Figure 9.1, depending in part on whether one or two subduction zones are involved. If only one subduction zone is present, a passive continental margin on the downgoing plate can collide with an active continental margin (Fig. 9.1A) or a forearc (Fig. 9.1B) on the overriding plate. If two subduction zones dip the same way, a back-arc on the downgoing plate can collide with an active continental margin (Fig. 9.1C) or a forearc (Fig. 9.1D) on the overriding plate. Finally, if the two subduction zones dip in opposite directions, two overriding plates can collide in three possible combinations of forearc and active continental margin (Figs. 9.1E, 9.1F, and 9.1G). Thus, we can envision two types of continent-continent collision (Figs. 9.1A and 9.1E), three types of arc-continent collision (Figs. 9.1B, 9.1C, and 9.1F), and two types of arc-arc collision (Figs. 9.1D and 9.1G).

Two fundamental considerations are important for understanding collisions. First, for several collision types, the buoyancy of continental or island arc crust tends to terminate subduction at the convergent margin. The result is either a shift in the location of subduction or an alteration of the plate tectonic geometry and kinematics. The right-hand diagrams in Figure 9.1 show how the collision zones might evolve.

At single subduction zones, subduction is shut off when the passive continental margin enters the subduction zone (Figs. 9.1A and 9.1B). For the continent-continent collision (Fig. 9.1A), the convergence must be accommodated elsewhere in the plate tectonic system, and this accommodation could lead to a worldwide reorganization of plate kinematics. For the arc–passive continent collision (Fig. 9.1B), the oceanic crust can begin to subduct under the collision zone, resulting in a reversal in the polarity of subduction.

At parallel double subduction zones (Figs. 9.1C and 9.1D), the subduction zone that remains active after the collision could accommodate the increase in subduction rate required by the elimination of one subduction zone.

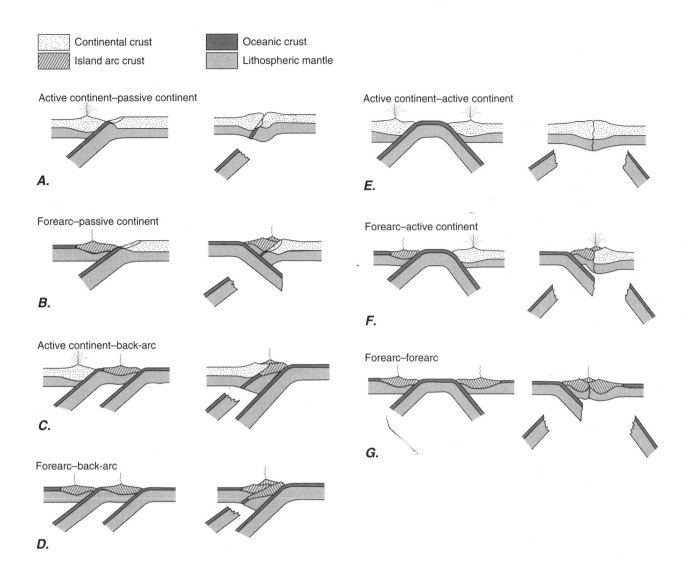

Continental crust
Island arc crust
Oceanic crust
Lithospheric mantle

Active continent–passive continent

A.

Forearc–passive continent

B.

Active continent–back-arc

C.

Forearc–back-arc

D.

Active continent–active continent

E.

Forearc–active continent

F.

Forearc–forearc

G.

Figure 9.1 Diagrams showing the theoretically possible types of collision involving one and two subduction zones. The left column shows the precollision geometry; the right column shows the postcollision geometry of the crustal blocks and the downgoing plate(s). In all cases, the original downgoing lithospheric plate is detached from the surface plate and the subduction geometry is reorganized. *A.* Collision of an active continental margin with a passive continental margin. After collision, subduction ceases and must be accommodated elsewhere in the plate tectonic system. *B.* Collision of a forearc margin with a passive continental margin. After collision, the subduction polarity reverses and the downgoing slab dips under the collision zone. *C.* Collision of a back-arc margin with an active continental margin. After collision, the subduction zone dipping under the island arc can accommodate the required increase in subduction rate. *D.* Collision of a forearc margin with a back-arc margin. After collision, the required increase in subduction rate can be taken up by the second subduction zone. *E.* Collision of two active continental margins as the intervening plate is subducted at opposite-dipping subduction zones. After collision, subduction ceases and must be accommodated elsewhere in the plate tectonic system. *F.* Collision of a forearc margin with an active continental margin with the subduction of the intervening plate under both margins. After collision, a new subduction zone could form, dipping under the collision zone. *G.* Collision of a forearc margin with a forearc margin with the subduction of the intervening plate under both margins. After collision, a new subduction zone could begin dipping in either direction underneath the collision zone.

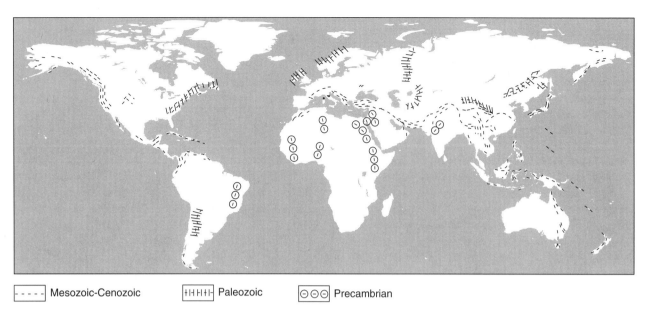

| - - - - | Mesozoic-Cenozoic | ʰʰʰʰ | Paleozoic | ⊖⊖⊖ | Precambrian |

Figure 9.2 World map showing ophiolite belts of various ages. *(After Gass, 1982)*

At opposing double subduction zones, a continent-continent collision (Fig. 9.1E) terminates convergence, which then must be accommodated by changes elsewhere in the plate tectonic system, possibly leading to a major worldwide shift in plate kinematics. If an island arc is involved in such a collision (Figs. 9.1F and 9.1G), collision could be followed by the formation of a new subduction zone carrying oceanic crust under the collision zone.

Another fundamental consideration is that ophiolite belts found *within* continents represent the remnants of the ocean basins that once separated the continental masses. Such ophiolites typically exhibit tectonic contacts with underlying shelf sediments. In many places, the contacts are characterized by subhorizontal thrust faults that have transported the ophiolites and underlying material large distances toward the interior of the continent. Such ophiolites can be emplaced in two ways: (1) most are emplaced by collision of a passive continental margin with an intraoceanic subduction zone and are preserved within continental crust by a subsequent continent-continent collision; (2) a minority are emplaced by incorporation of part of a downgoing oceanic plate into a subduction zone. The great variety and number of ophiolite belts in the world (Fig. 9.2) indicate that many collisions have taken place and that collision is an important tectonic process.

With collisions being so important, and with so many possible collision geometries, we can expect the interpretation of the geologic record to be quite difficult. Therefore, we first look at regions of the Earth

that are currently undergoing collision (Secs. 9.2 and 9.3) to examine the geological characteristics of areas where the geometry is fairly easy to determine. We then review several models of collisional tectonics and compare them using geological examples (Sec. 9.4). We discuss the formation of mountain roots in Section 9.5 and models of ophiolite emplacement in Section 9.6. Finally, we discuss sutures, sites of former oceans, which bear on the interpretation of ancient collision zones (Sec. 9.7).

9.2 Arc-Continent and Arc-Arc Collisions: The Southwest Pacific Region

The southwest Pacific (Fig. 9.3) is a region of complex interaction between three large plates: the Pacific, Australian-Indian, and Eurasian. These plates generally have consuming margins along their mutual boundaries. In addition, there is the medium-sized Philippine Plate, as well as many smaller plates. There are two major trench-trench-trench triple junctions in the area, involving the Australian-Indian–Pacific–Eurasian and the Eurasian-Philippine-Pacific plate boundaries. We describe some features of the island arc systems of this region in Chapter 7. In this section we look in more detail at the complexity and diversity of the subduction zones and the collision processes that are taking place here.

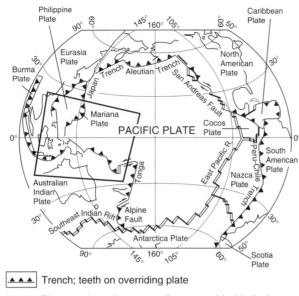

Figure 9.3 Plate tectonic maps of the southwest Pacific region. *A.* Major active plate boundaries and *B.* Detail of area in *A.* Boxes outline details shown in Figures 9.4, 9.5, 9.6*A* and 9.6*C.* *(After Addicott and Richards, 1982; McCaffrey et al., 1991)*

▲▲▲ Trench; teeth on overriding plate

Divergent boundary, spreading center (double line) and transform fault

A. Transform fault with (relative movement indicated)

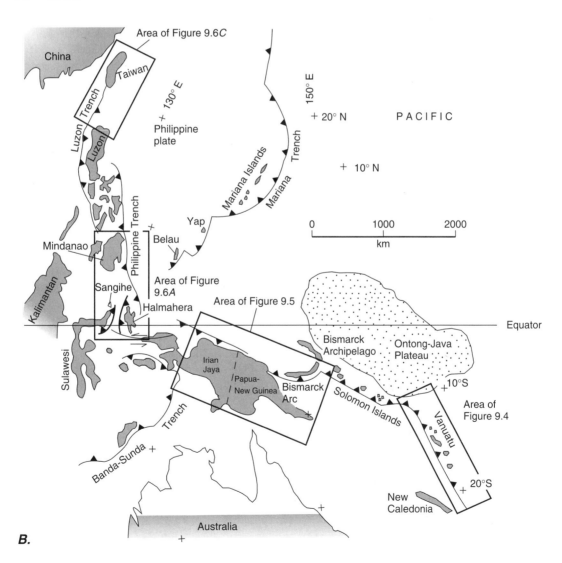

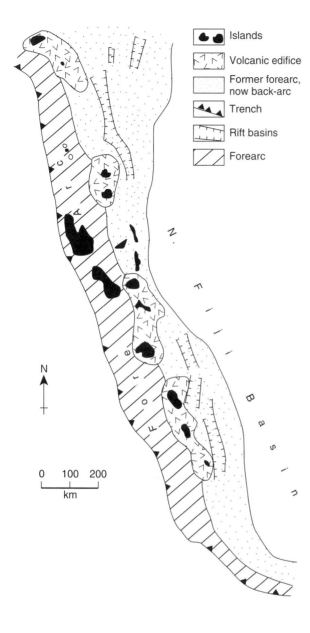

Islands

Volcanic edifice

Former forearc, now back-arc

Trench

Rift basins

Forearc

N
↑

0   100   200
km

Figure 9.4 Map of the Vanuatu arc, southwest Pacific, where the polarity of the subduction zone has reversed. *(After Carney and MacFarlane, 1982)*

Another collision is presently occurring along the northern edge of Australia in New Guinea between the Banda and Bismarck arcs (Figs. 9.3 and 9.5). In this case, however, the northern passive margin of the Australian continent is being carried under the Pacific Plate along a north-dipping subduction zone. To the west in Irian Jaya (western New Guinea), the process of collision is the most advanced, and the subduction polarity has reversed so that north of Irian Jaya subduction is north-vergent, that is, the downgoing plate, dips to the south under the collision zone (Fig. 9.5; compare Fig. 9.1B). Along the eastern part of the collision zone, an active south-vergent fold-and-thrust belt is growing southward towards the Australian shore (Fig. 9.5). The edge of the Australian continental mass, composed of metamorphic rocks and deformed passive margin sediments, is thrust back over the continent in a complex of imbricate thrust slices, and it is overthrust from the north by a complex of ophiolitic rocks. The ophiolite was part of the island arc that lay above the north-dipping subduction zone and was thrust over the Australian continent when the continent collided with the subduction zone. The original arc is still active east of the collision zone in the Bismarck arc of New Britain and New Ireland. Thus, the suture between the two plates lies approximately at the base of the ophiolite.

The Banda arc lies southwest of New Guinea, where the Australian passive continental margin is being subducted under the Asian Plate (see Fig. 9.3; compare Fig. 9.1B). At the northeast end of the Banda arc, the active subduction zone curves back on itself. This form may have resulted from folding of the plates at the zone of complex interaction as the continental margin of Australia collided with the subduction zone.

In Indonesia, between the islands of Sangihe and Halmahera on the equator northeast of Sulawesi, two forearc regions overlying opposite-dipping subduction zones are colliding (Fig. 9.6; compare Fig. 9.1G). The collision is eliminating an intervening plate and juxtaposing the Pacific and Asian plates. Figure 9.6A shows a map of the region with structure contours on the dipping seismic zones. The western one (Sangihe) dips west, and the eastern one (Halmahera) dips east. Figure 9.6B is a cross section showing an interpretation of the structure based upon seismicity, gravity measurements, and dredge sampling. Gravity model-

The Ontong-Java Plateau northeast of Australia is an oceanic plateau of continental crustal thickness that now rests on the overriding plate of the Vanuatu-Solomon Islands arc (Figs. 9.3 and 9.4; compare Fig. 7.1). Geologic evidence, however, indicates that before Pliocene time, the Pacific Plate carrying the plateau was being subducted toward the southwest. The plateau collided with the subduction zone, after which the polarity of subduction reversed (see Fig. 9.1B). The map of the Vanuatu arc (Fig. 9.4) shows that the former forearc region is now a back-arc in which small rift basins are developing; the trench is to the west. Seismicity under this arc is discontinuous, with a deep zone of earthquakes dipping southwest and a shallow zone of earthquakes dipping northeast (see Figs. 7.4 and Fig. 9.1B).

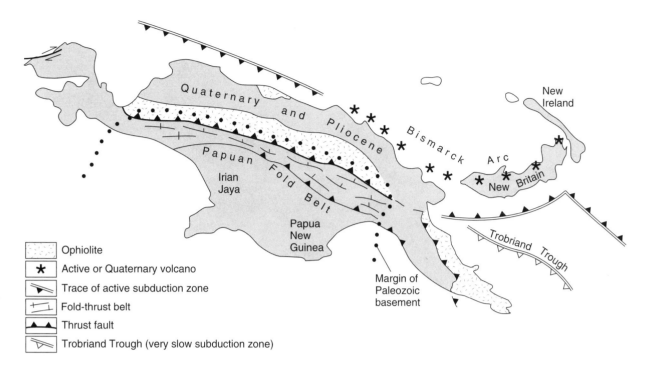

Ophiolite

* Active or Quaternary volcano

Trace of active subduction zone

Fold-thrust belt

Thrust fault

Trobriand Trough (very slow subduction zone)

Figure 9.5 Tectonic map of the New Guinea region showing major plate boundaries, Papuan fold belt, margin of Australia, ophiolite complexes, and Quaternary volcanoes. The ongoing collision is accompanied by a reversal in the polarity of subduction. New Guinea is divided politically into Irian Jaya in the west and Papua-New Guinea in the east, as indicated on Figure 9.3 *(After Bain, 1973; Pigram and Davies, 1987; Silver et al., 1991; McCaffrey et al., 1991)*

ing suggests that the western forearc is beginning to thrust over the eastern one.

In Taiwan, the passive continental margin of China is currently colliding with the east-dipping Manila subduction zone. This collision is marked by an actively deforming fold-and-thrust belt in western Taiwan where the sediments of the passive continental margin of China are being thrust westward over the advancing margin. An ophiolite complex, which is part of the volcanic arc over the subduction zone, has been emplaced structurally above the fold-and-thrust belt on eastern Taiwan (Fig. 9.6C). Note that the suture marking the boundary between the two plates is approximately at the thrust fault below the volcanic arc ophiolite and not at the physiographic trench axis.

In this brief summary of the active geology in this region of complex plate interactions, we have found striking evidence that fold-and-thrust belts form where passive continental margins collide with a subduction zone. The system of thrust faults can be interpreted as an imbricate set that splays off a major

décollement, which ultimately is the subduction zone itself. Thus, the vergence of the faults generally reflects the polarity of the subduction. We have also seen cases in which the emplacement of an ophiolite occurs when a passive continental margin is subducted under oceanic and island arc crust. The base of the ophiolite, therefore, must be close to the actual suture where the two collided plates are juxtaposed. The examples cited also provide evidence supporting our assumption that continental crust cannot be subducted very far and that reversals in the polarity of subduction zones are a common way of adjusting the plate tectonic system to a collision.

Finally, the southwest Pacific region illustrates the extreme complexity of geology that can occur where multiple plate collisions are taking place. These complexities may not be representative of every collision zone, but they serve as a warning against oversimplified interpretations of ancient collision zones where the process has evolved to completion and details of the geology are not nearly as well preserved.

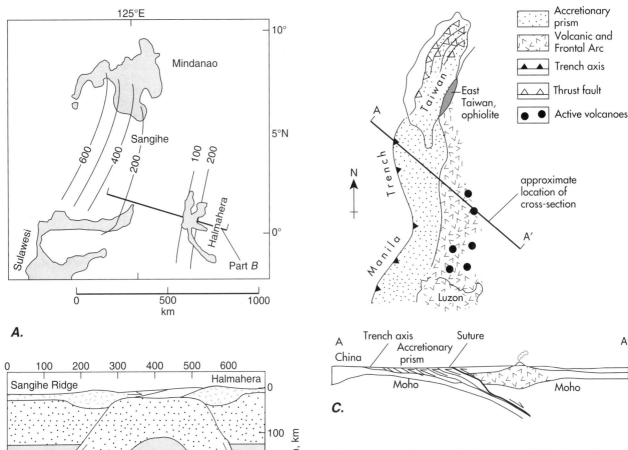

A.

B.

C.

Figure 9.6 Collisions in the Indonesia-Philippine region. *A.* Tectonic map of the Sangihe-Halmahera region, Indonesia, an arc-arc collision, showing structural contours on the seismic zone and the general location of the cross section. *B.* Crustal model of the collision zone. Note that the western arc is overriding the eastern one. C. Map and cross section of Taiwan region, Asia. Map shows western nonvolcanic and eastern volcanic arcs and the presently active major faults. *(A. after Cardwell, et al., 1980; B. after McCaffrey et al., 1980; C. after Liou et al., 1977)*

9.3 Continent-Continent Collisions: The Alpine-Himalayan System

The Alpine-Himalayan mountain system is one of the major structural features of the Earth, extending from Gibraltar in the west, past India in the east (Fig. 9.7). This system includes four large continent-bearing plates (the Eurasian, African, Arabian, and Australian-Indian), three inland seas containing oceanic crust (the Mediterranean, Black, and Caspian), the highest mountains on Earth (the Himalaya), and several high plateaus, (the Turkish-Iranian, Pamir, and Tibetan).

The active tectonics in this region provides the best evidence available of the behavior of continents when they collide. Even though these continents have already come in contact, they are still converging at an impressive rate, ranging from about 1 cm/yr in the west to about 5 cm/yr in the east, as indicated in Figure 9.7.

Despite the evidence for rapid convergence, practically all seismic events are relatively shallow. Earthquakes at depths greater than 100 km have

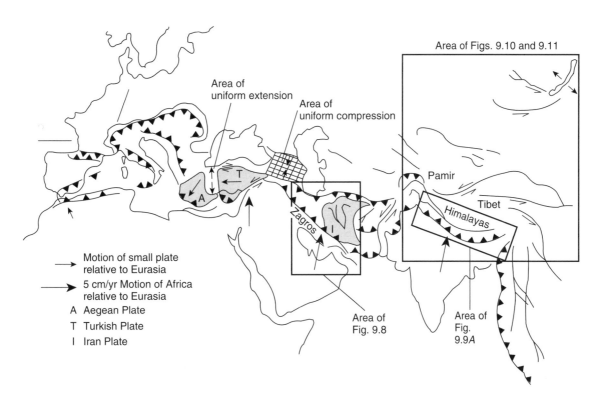

Figure 9.7 Map of Alpine-Himalayan system showing principal thrust regions and three mini plates: the Aegean, Iranian, and Turkish plates. Small arrows indicate relative motion along faults or motion of small plates. Large arrows indicate overall convergence rates of Eurasia with respect to the African, Arabian, and Indian plates. *(After Dewey, 1977; McKenzie, 1970)*

been located only beneath the eastern Mediterranean, Rumania, the Hindu Kush, and Burma (see Fig. 7.4). In contrast to the circum-Pacific subduction zones, there are no continuous seismic zones extending from the surface to depths of greater than 300 km are known.

Two modes of seismicity have been observed. In the first, seismic events are localized in a zone beneath the region of continent-continent collision, such as the Zagros region of Iran and the Himalaya. In the second, seismic events are widely distributed shallow earthquakes such as are found in central and eastern Asia. We consider each region briefly.

Localized Seismicity in Collision Zones:
The Zagros Region

In the Zagros region of southwestern Iran, a passive continental margin on the Arabian Plate is being subducted under an active continental margin on the Asian Plate (Fig. 9.8; compare Fig. 9.1*A*). The main

thrust zone, known as the Zagros Crush Zone (Figs. 9.8*A* and 9.8*B*), is a region of intense deformation and northeast-dipping (southwest vergent) thrusts. To the southwest of the crush zone is a wide foreland fold-and-thrust belt with actively developing folds and associated décollement-style thrusts.

Seismic events are concentrated in a broad area beneath the fold-and-thrust belt, including a significant part of the mantle lithosphere (Fig. 9.8*B*). A well-defined Wadati-Benioff zone does not occur. The focal-mechanism solutions for a number of these events are consistent with thrust faulting (Fig. 9.8*C*), indicating shortening of both crust and mantle lithosphere.

Seismicity of the Himalaya

The Himalayan Mountain Range, the highest on Earth, is actively rising as India continues to move northward beneath it. The major structures of the Himalaya include the south-vergent Main Boundary

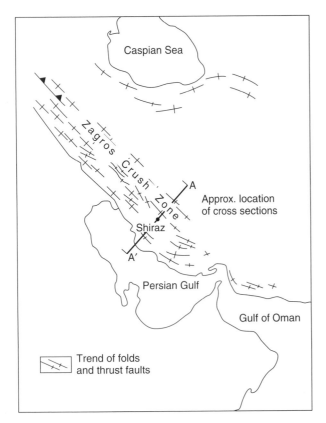

A.

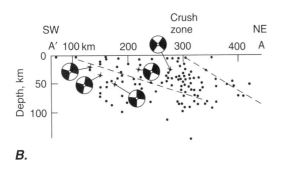

B.

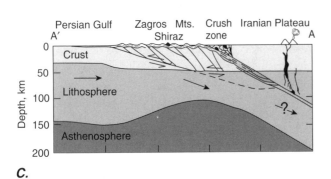

C.

Figure 9.8 The Zagros collision zone, Iran. *A.* Map of Iran showing Zagros Crush Zone and folded region of Zagros Mountains. *B.* Seismic cross section of southeastern Iran showing crush zone, subsidiary thrust, and focal mechanisms. The maximum shortening axis bisects the unshaded quadrants. Quadrant boundaries are potential shear planes. *C.* Tectonic interpretation seismic cross section. (*A. after Stocklin, 1974; C. after Bird, 1978*)

Thrust (MBT, Fig. 9.9*A*), along which Paleozoic and late Precambrian sedimentary rocks are emplaced over the Plio-Quaternary Siwalik sediments; the south-vergent Main Central Thrust (MCT, Fig. 9.9*A*), along which crystalline rocks of the main high Himalaya are emplaced over the Paleozoic and late Precambrian sedimentary sequences; and the Indus-Tsangpo suture zone, which represents the remnants of the ocean formerly separating India from the lands to the north. The Main Boundary Thrust is active today, and some geologists think that the Main Central Thrust is also active. In addition, faults in front of the Main Boundary Thrust emplace Plio-Quaternary Siwalik sediments southward over recent alluvium (Main Frontal Thrust, MFT, Fig. 9.9*A*).

All recorded earthquakes beneath and in front of the Himalaya are shallow; most are within the crust, and almost all focal mechanisms show thrust fault solutions (Fig. 9.9*B*). The locations of these earthquakes and their focal mechanisms suggest that India is being thrust beneath the Himalaya along the Main Boundary Thrust, the Main Frontal Thrust faults, one or more blind thrusts beneath the Ganges Plain, and possibly the Main Central Thrust (Fig. 9.9*A*). Because most seismicity is shallow and gives shallow-dipping focal mechanisms, the geometry of these faults at depth to the north is not clear. Three possibilities are shown in Figure 9.9*A*. Likewise, the extent of the Indus-Tsangpo suture at depth is unknown.

Distributed Seismicity in Collision Zones: The Tectonics of Central Asia and Tibet

Thrusting like that observed in the Himalaya and the Zagros is widely thought to be the main mode of deformation in collisional belts. In Tibet and central Asia, a significant amount of thrusting is suggested by the very high altitudes, especially of the Himalaya and the Tibetan Plateau (Fig. 9.10), which reflect greatly thickened crust (crustal thickness can be obtained approximately by multiplying the elevation by 10).

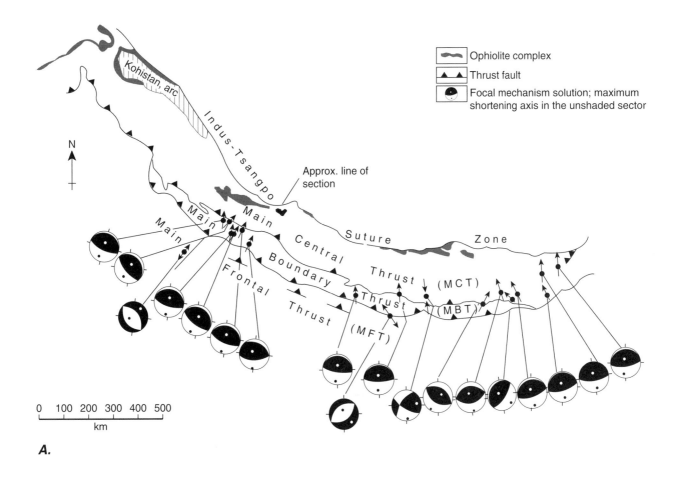

A.

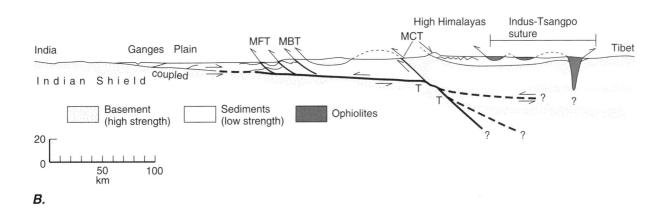

B.

Figure 9.9 The Himalaya. *A.* Tectonic map of the Himalaya showing the principal thrust faults and the Indus-Tsangpo suture zone, the Kohistan arc, an entrapped island arc system, principal ophiolites, and focal mechanisms. *B.* Cross section showing areas of active thrusting. The dip of thrusts to the north is unknown, and three possibilities are shown. The extent of the Indus-Tsangpo zone at depth is also unknown. *(A. after Gansser, 1980; and Ni and Barazangi, 1984; B. after Baranowski et al., 1984)*

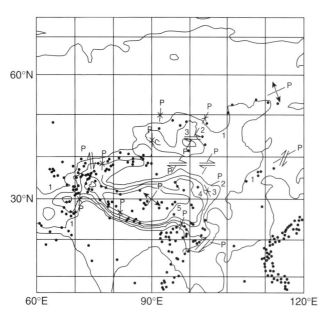

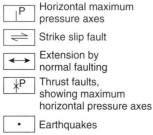

P | Horizontal maximum pressure axes

⇌ | Strike slip fault

↔ | Extension by normal faulting

✳P | Thrust faults, showing maximum horizontal pressure axes

• | Earthquakes

Figure 9.11 (*opposite page*) Tectonic map of central-east Asia showing major thrust faults, subduction zones, displacement directions of major continental blocks, regions of extension, and shear senses on strike-slip faults. The letters A, B, and C indicate the extending regions of the Andaman Sea, the South China Sea, and the Shansi region of China, respectively. (*After Tapponnier et al., 1982*)

Figure 9.10 Map of central-east Asia showing general contours of elevation in kilometers averaged over 1° by 1° elements and smoothed with a 2° by 2° filter (approximate thickness of the crust can be obtained by multiplying the elevation by 10). Also shown are earthquake focal mechanisms and the orientation of the horizontal P axes (axes of maximum shortening). Note that regardless of the nature of the earthquake, most first motions show a N or NE orientation to the principal axis of shortening. (*After Tapponnier and Molnar, 1976; England, 1982*)

The observations in Asia north of the Himalaya are somewhat surprising, however, because they indicate that strike-slip faulting is much more important than has been recognized in most collision zones. Figure 9.11 shows a simplified map of the major tectonic features of eastern Asia. The area is characterized by a complex system of predominantly strike-slip faults spanning the continent from north of the Himalaya to Siberia. Most NW-trending faults in the Tien Shan and Altai ranges north of India are right-lateral and have displacements that do not exceed 10 km. East- and southeast-trending faults in central and southern China are largely left-lateral, and a few of the major ones show displacements that may reach several hundred kilometers. Present displacement on the Red River fault is right-lateral, although this appears to be a fairly young reversal of large, predominantly left-lateral displacement.

North and east of the main Himalayan Range, the seismic activity is distributed for approximately 3000 km across Tibet and central Asia (see Fig. 9.10). The earthquakes show no consistent type of focal mechanism, with normal, thrust, and strike-slip mechanisms all being observed. In all cases, however, the principal horizontal axis of shortening has a fairly consistent orientation between N-S and NNE-SSW, regardless of the type of faulting. This fact implies that for the entire continent, the principal axis of shortening is homogeneous across a zone several thousand kilometers wide.

There are no discrete plate boundaries in most of central and eastern Asia. The widely distributed seismicity and deformation associated with continent-continent collisions suggest that the model of rigid plate motion is inappropriate for such regions. Some other model to account for the behavior of the rocks seems necessary, and we discuss this subject further in Section 9.4.

Miniplates of the Alpine-Himalayan System

West of the Himalaya, the pattern of distributed seismicity is broken in several discrete seismically quiet regions: the Aegean, Turkish, and Iran miniplates. Data from earthquakes surrounding the Aegean and Turkish plates suggest that they are moving southwest and west relative to Eurasia, respectively. Between these plates, zones of uniform north-south shortening are found in the Caucasus, and zones of uniform extension are found in western Anatolia.

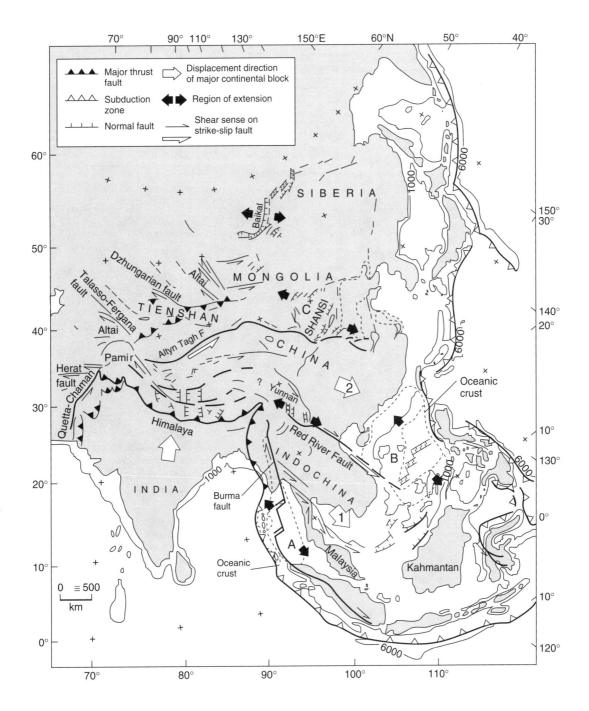

Displacement direction of major continental block

Region of extension

Shear sense on strike-slip fault

Major thrust fault

Subduction zone

Normal fault

9.4 Models of Collisional Deformation

The model of rigid plate motion cannot account for the distributed deformation associated with a continent-continent collision, and we assume, therefore, that the continents can be described as a continuously deformable material. To understand the deformation of such regions, we use continuum theory, scale models, and numerical models, and in this section, we describe the application of these approaches to the problem of understanding the defor-

mation associated with the collision along parts of the Alpine-Himalayan belt, particularly the collision of India with Asia.

Plastic Slip-Line Fields as Models of Asian Tectonics

The motion along a fault is characterized by a discontinuity in the velocity distribution across the fault. Comparable discontinuities can develop in a homogeneous, isotropic, rigid-plastic material undergoing plane strain, and this type of material is therefore

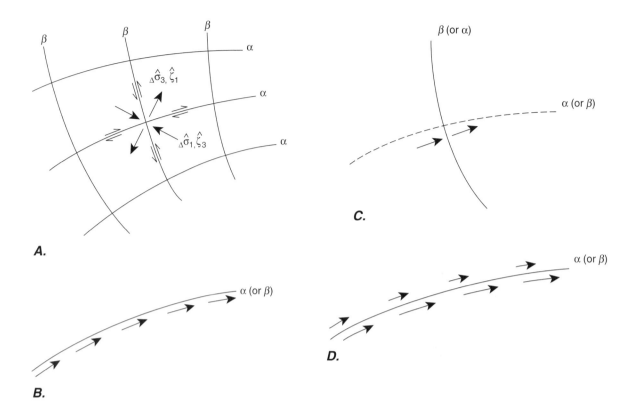

Figure 9.12 Characteristics of plastic slip-line fields. *A.* Slip lines form two sets of curves, α and β, that are everywhere mutually orthogonal. Each line is tangent to the orientation of maximum resolved shear stress and maximum resolved incremental shear strain, so the principal axes of both deviatoric stress $_\Delta\hat{\sigma}_k$ and incremental strain $\hat{\zeta}_k$ bisect the angles between the two sets of slip lines. *B.* Along any one slip line, the tangential component of the velocity and displacement must be constant. *C.* The component of velocity and displacement normal to a slip line must be the same on both sides of the slip line. *D.* The component of velocity and displacement parallel to a slip line can be discontinuous across the slip line, so the slip line can behave like a fault. *(After Backofen, 1972)*

useful in modeling the behavior of the Earth's lithosphere. It can be shown that the equations describing rigid-plastic behavior in plane strain (see Twiss and Moores, 1992; Eq. 18.3.5, Box 18.3), when combined with a form of Newton's second law called the equation of equilibrium (see Twiss and Moores, 1992; Eq. 20.2), result in a hyperbolic differential equation. The solution to this equation defines two families of lines called slip lines across which velocity discontinuities can occur. By convention these slip lines are designated α and β (Figs. 9.12A and 9.13). The mathematical development of the theory is beyond the scope of this book but can be found in a variety of references (see, for example, Johnson, 1970, and Odé, 1960). The results, however, are not difficult to understand. Each line in one family is

orthogonal to all the lines in the other family, and each line is everywhere tangent to the maximum shear stress and the maximum shear strain rate in the material. Thus, the maximum and minimum principal deviatoric stresses ($_\Delta\hat{\sigma}_1$, $_\Delta\hat{\sigma}_3$) and the principal axes of the incremental strain ellipse ($\hat{\zeta}_3$, $\hat{\zeta}_1$) bisect the right angles between any two intersecting slip lines (Fig. 9.12A).

The components of displacement and velocity parallel and normal to the slip lines conform to a few simple rules. Extensional strain parallel to any of the slip lines must be zero. As a result, the component of displacement or velocity parallel to a slip line must be constant along that slip line (Fig. 9.12B). Also, the component of displacement or velocity normal to any slip line must be constant across the slip line, because the slip lines are mutually orthogonal (Fig. 9.12C). The component of displacement or velocity parallel to a slip line, however, may be different on opposite sides of that slip line without violating these restrictions (Fig. 9.12D), and any such discontinuity must be the same along the entire length of that particular slip line. Such a discontinuity in velocity across a slip line is comparable to the motion on a fault, and it is this similarity that makes slip lines relevant to the interpretation of faults.

The geometry of the slip lines depends on the geometry of the deforming system. For example, consider the slip lines resulting from a rigid punch impinging on the edge of a plastic layer (Fig. 9.13). In Figure 9.13A, the plastic material is constrained between two rigid boundaries placed symmetrically on either side of the punch; in Figure 9.13B, the punch impinges on the straight edge of a plastic layer that is unconstrained on the sides. The heavily stippled area in each figure is a region of "dead" material that moves with the punch and does not deform. The discontinuity in slip that occurs between the punch and the plastic material can propagate along slip lines that emanate from the corners of the punch. These slip lines form patterns through the plastic material

that resemble some observed patterns of strike-slip faults in the Earth's crust. This similarity provides the rationale for using the slip lines as a model for understanding the faulting.

The slip-line model is appealing because it provides a mechanism by which velocity discontinuities across a surface can be described. The specific pattern formed by the slip lines depends strongly on the boundary conditions assumed; that is, the particular conditions of stress or velocity that must be specified along the boundaries of any mathematical model to find a solution. Although a variety of boundary conditions can result in model slip-line patterns that mimic a variety of observed fault patterns, it is difficult to evaluate the significance of the model boundary conditions in terms of how the Earth deforms because often there are no definite boundaries in the Earth on which specific conditions can be identified.

Nevertheless, we can interpret the faults of eastern Asia (see Fig. 9.11) as a system of slip lines associated with the collision of a rigid Indian continental mass with a deformable southern Asia. Unlike the patterns in Figure 9.13, the faults in Asia definitely are not symmetrical about India. Figure 9.14 portrays a model of southern Asia that incorporates a

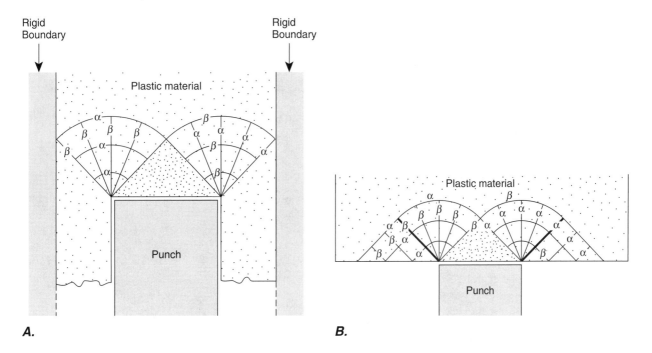

Figure 9.13 Geometry of plastic slip-line fields in an isotropic, homogeneous, rigid-plastic material deformed by a rigid punch. The stippled area represents the plastic material. The heavily stippled triangular area is the "dead zone" in which no deformation occurs. A. The plastic material is constrained by rigid boundaries placed symmetrically on either side of the punch. B. The punch impinges on the straight edge of the plastic material, which is unconstrained on either side of the punch. *(After Backofen, 1972)*

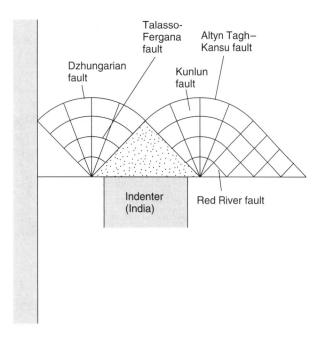

Figure 9.14 Slip-line field for a square indenter impinging on the planar face of a plastic plate with a rigid boundary located half the width of the indenter away from the indenter on the left, and an infinite plastic plate on the right. Some slip lines are labeled with the names of faults to which they might be compared (see Fig. 9.11). *(After Tapponnier and Molnar, 1976)*

combination of a rigid boundary on the west, an unconstrained boundary on the east, and an infinite extent of material to the north. The model produces a slip-line pattern reminiscent of a number of the faults around India, in particular the Talasso-Fergana, Dzhukgarian, Altyn Tagh–Kansu, Kunlun, and Red River faults of Figure 9.11. Figure 9.15 shows a different model consisting of a corner-shaped indenter impinging on the nonplanar edge of a plastic plate. This model produces a pattern reminiscent of the Herat, Talasso-Fergana, Quetta-Chaman, Karakoram, Altyn Tagh–Kansu, Kunlun, and Red River faults (see Fig. 9.11).

Instead of a plastic plate with no boundary opposite the indenter, we could consider one in which the boundary opposite the indenter is less than a few widths of the indenter away (Fig. 9.16). If this boundary is fixed and can support no shear stress, the pattern of the slip-line field is markedly altered. Moreover, the stress near the opposite boundary becomes tensile, as shown by the Mohr circles in Figure 9.16A. This feature suggests a possible origin of the Baikal rift system in Siberia and a number of faults in Mongolia, as indicated in Figure 9.16B

(compare Fig. 9.11), although the applicability of the imposed model boundary conditions in the Baikal region is not clear.

Despite similarities between the theoretical slip-line fields and the fault pattern in Asia, the correspondence is certainly not unique. Various simple boundary conditions seem to explain different aspects of the observed pattern, but the actual boundary conditions must be more complex than those for the models we have discussed. In particular, plastic slip-line field theory assumes plane strain; in fact, the slip lines are not a feature of three-dimensional plastic deformation. Thus, the theory cannot account for crustal thickening at all. Furthermore, the theoretical models apply only to the first small increment of deformation. After the boundary of the plastic layer has been deformed, the boundary conditions used in the theory no longer apply; so the evolution of the slip-line field for a finite deformation, such as certainly has occurred in Asia, cannot be determined from this approach. This problem can be overcome only with model experiments or with numerical calculations.

A Plasticine Model of Asian Tectonics

Asian tectonics also have been modeled experimentally using plasticine, an approximately plastic material (Box 9.1). The boundary conditions include a rigid boundary on the left (west), an unconstrained boundary on the right (east), and a plane-strain deformation imposed by lubricated plates above and below the plasticine layer. The indenter was advanced a long way into the layer to simulate the finite deformation (Fig. 9.17). The faults that develop to the left of the centerline of the punch are similar to the α slip lines

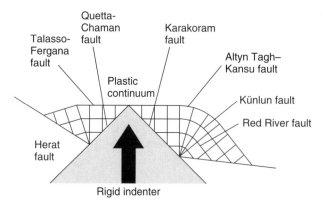

Figure 9.15 Slip-line fields for a corner indenter impinging on the nonplanar face of a plastic plate. Lines possibly corresponding to faults in Figure 9.11 are shown. *(After Tapponnier and Molnar, 1976)*

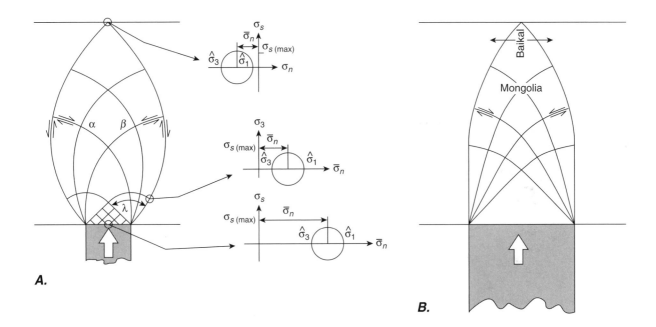

Figure 9.16 Slip lines from a rigid indenter impinging on a plastic plate with a boundary opposite the indenter that is rigid and can support no shear stress. *A.* Tensile stresses develop near the boundary opposite the indenter. The plate boundaries are 4.5 indenter widths apart. The Mohr circles show the stress magnitude at different points in the plate. *B.* The plate boundaries are less than 2 indenter widths apart. A possible correlation with extension along the Baikal Rift in central Asia is indicated. *(After Tapponnier and Molnar, 1976)*

shown in Figure 9.13*A*. The faults that develop to the right of the centerline are similar to the β slip lines shown in Figure 9.13*B*, except that in the experiment, some of the slip lines intersect the unconstrained right side of the layer rather than the bottom edge on which the indenter impinges.

The history of the development of the plasticine model faults is indicated by the three frames from successive stages in the experiment. Fault F_1 initiates from the discontinuity in displacement at the left corner of the indenter (Fig. 9.17*A*). The one-sided constraint and the plastic deformation of the layer result in rotation and flattening of the fault until it becomes unfavorably oriented for further slip, at which time a new fault, F_2, initiates from the left corner of the indenter (Fig. 9.17*B*). One can anticipate that if the deformation had continued, a similar fate would have befallen this second fault, and yet a third fault would have initiated at the left corner of the punch to accommodate the continued extrusion of the plasticine from in front of the advancing punch.

The slip-line field in Figure 9.17 shows striking similarities to the tectonic pattern in eastern Asia. We can make several comparisons between the features shown in Figure 9.17*C* and those shown on the map in Figure 9.11. Faults F_1 and F_2 compare respec-

tively with the Red River and the Altyn Tagh faults. Blocks 1 and 2 correspond respectively to Indochina and southeast China. The gaps that opened up in the model in Figures 9.17*A*, 9.17*B*, and 9.17*C* correspond respectively to the extensional tectonics observed in the Andaman Sea west of Indochina, the South China Sea between Indochina and southeast China, and the Shansi in northeastern China.

The correspondence in the geometry between model and prototype is impressive, but we must ask how relevant the model is to the behavior of the Earth. In Box 9.1 we show that it is kinematically, and possibly dynamically, similar to Earth and thus provides a reasonably good scale model. The model makes very specific predictions about the timing, history, and total displacement on the faults and thus provides a focus for further geologic study.

Although available geologic data are generally compatible with the plasticine model, this model does not resolve all tectonic questions. First, it assumes that all the deformation is two-dimensional. Much of the present deformation in the Himalayan region, however, involves thrust faulting, which presumably accounts for the thickening of the crust from the Himalaya northward (see Fig. 9.10). This deformation cannot be accounted for by the horizontal

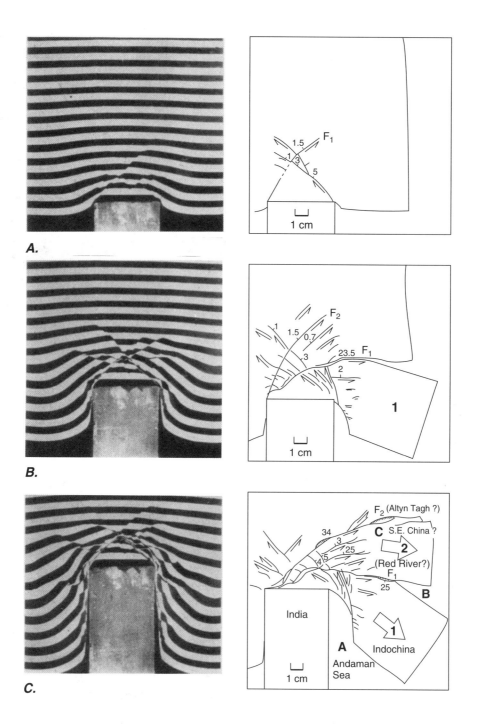

Figure 9.17 Experimentally developed slip-line field displayed in striped plasticine confined by a rigid boundary only on the left side, with a free boundary on the right side, and deformed by a rigid indenter. Deformation is constrained to plane strain by lubricated rigid plates above and below the sheet of plasticine. Line drawings emphasize the faults in the plasticine. *A.* Fault F_1 propagates out from the corner of the indenter. *B.* Block 1 extrudes to the right on fault F_1. Advance of the indenter rotates F_1 to a position unfavorable for further slip, whereupon fault F_2 propagates out from the corner of the indenter. *C.* Block 2 extrudes out from in front of the indenter by slip on fault F_2. Boldfaced numbers and letters correspond to numbers and letters in Figure 9.11. Faults F_1 and F_2 correspond to the Red River fault and the Altyn Tagh fault, respectively. Note pull-apart basins near faults F_1 and F_2. *(Tapponnier et al., 1982)*

We wish to examine whether the plasticine model of Asian tectonics shown in Figure 9.17 is a realistic scale model of the Earth. To describe completely the mechanical relationship between a scale model and the prototype that it is intended to model, we need three independent scale factors for length (λ), time (τ), and mass (μ). We define them by

$$\lambda = \frac{l_m}{l_p} \qquad \tau = \frac{t_m}{t_p} \qquad \mu = \frac{m_m}{m_p} \qquad (9.1.1)$$

where l is length, t is time, and m is the mass of material in a volume element and where subscripts m and p refer respectively to the model and the prototype. The prototype and model are said to be **geometrically similar** if all linear dimensions of the model are λ times the equivalent dimension in the prototype. The prototype and model are called **kinematically similar** if the time required for the model to undergo a change in size, shape, or position is τ times the time for the prototype to undergo a geometrically similar change. The terms m_m and m_p are the masses of material in geometrically similar volumes of the model and the prototype, respectively. In the model, the response of the material to a particular scaled force must be geometrically and kinematically similar to the response of the prototype to natural forces. These requirements are satisfied if each force in the model is related to the corresponding one in the prototype by the same scale factor, in which case the model and the prototype are said to be **dynamically similar.**

The scale factors for all the mechanical quantities of interest, such as forces, can be derived from the three scale factors defined in Equation (9.1.1), and these mechanical scale factors restrict the properties of an accurate scale model. Table 9.1.1 lists the mechanical quantities, their units, and the scale factors derived for them.

Two kinds of forces are important for this analysis: body forces and surface forces. Of the body forces that can act in mechanical systems, we consider only the gravitational force $\mathbf{F}_g$ and the inertial force $\mathbf{F}_i$, which are both defined by the mass times an acceleration. We ignore such other body forces as electrostatic and magnetic forces. Surface forces arise from the resistance of a material to an imposed deformation, and this resistance is expressed by a constitutive equation that relates, for example, the stress to the rate of incre-

mental strain for a viscous fluid or for a material deforming according to a power law (see Eq. 9.1). If the material is a linear viscous fluid, then the stress exponent in Equation (9.1) is $n = 1$, and the constant of proportionality is a material constant called the coefficient of viscosity η. Scale factors for material constants can be derived from the appropriate constitutive equation. For viscous fluids, for example, the coefficient of viscosity is given by the shear stress divided by the shear strain rate. Because strain is a dimensionless quantity, the unit of strain rate is inverse time. From these relationships, we derive the scale factor for viscosity shown in Table 9.1.1 on page 230.

Let us now apply these considerations to an analysis of whether the plasticine models of Asian tectonics can be considered good scale models for representing the behavior of the Earth. The length scale can be determined from the width of India in Figure 9.10 (about 2000 km) compared with the width of the indenter used in the experiments in Figure 9.14 (5 cm). The resulting length scale is

$$\lambda = 2.5 \times 10^{-8} \qquad (9.1.2)$$

The time scale is derived from the velocities of the prototype compared with those of the model (see Eq. 9.1.1; Table 9.1.1). India has been converging with Asia at the rate of about 5 cm/yr, or 5.7×10^{-4} cm/hr. In the model, the punch is driven into the plasticine at a rate of 2.5 cm/hr. We form the scale factor for velocity (see Table 9.1.1), from which we can derive the scale factor for time using the scale factor for length from Equation (9.1.2):

$$\frac{v_m}{v_p} = \frac{2.5 \text{ cm/hr}}{5.7 \times 10^{-4} \text{ cm/hr}} = \frac{\lambda}{\tau} \qquad (9.1.3)$$

$$\tau = \frac{(2.5 \times 10^{-8})(5.7 \times 10^{-4})}{2.5} = 5.7 \times 10^{-12} \qquad (9.1.4)$$

Choosing λ and τ as independent scale factors defines the scale factor for the acceleration, because acceleration scales as λ/τ^2 (see Table 9.1.1). Using the values above gives a scale factor for acceleration of 7.7×10^{14}. Clearly, we cannot choose the gravitational acceleration in the model to be 7.7×10^{14} times the gravitational acceleration in the Earth, so the model can represent the Earth only if the gravitational forces do not contribute significantly to the deformation both in the

Table 9.1.1 Scale Factors for Selected Variables in Mechanics

Quantity	Symbol	Units	Ratios	Scale factor
Area	A	l^2	$\dfrac{A_m}{A_p} = \dfrac{(l_m)^2}{(l_p)^2}$	λ^2
Volume	V	l^3	$\dfrac{V_m}{V_p} = \dfrac{(l_m)^3}{(l_p)^3}$	λ^3
Density	ρ	$\dfrac{m}{V}$	$\mathrm{P} = \dfrac{\rho_m}{\rho_p} = \dfrac{m_m}{m_p}\dfrac{(l_p)^3}{(l_m)^3}$	$\mathrm{P} = \dfrac{\mu}{\lambda^3}$
Velocity	v	$\dfrac{l}{t}$	$\dfrac{v_m}{v_p} = \dfrac{l_m/t_m}{l_p/t_p} = \dfrac{l_m}{l_p}\dfrac{t_p}{t_m}$	$\dfrac{\lambda}{\tau}$
Acceleration	a	$\dfrac{l}{t^2}$	$\dfrac{a_m}{a_p} = \dfrac{l_m/t_m^2}{l_p/t_p^2} = \dfrac{l_m}{l_p}\dfrac{t_p^2}{t_m^2}$	$\dfrac{\lambda}{\tau^2}$
Force	F	$\dfrac{ml}{t^2}$	$\dfrac{F_m}{F_p} = \dfrac{m_m(l_m/t_m^2)}{m_p(l_p/t_p^2)} = \dfrac{m_m l_m t_p^2}{m_p l_p t_m^2}$	$\dfrac{\mu\lambda}{\tau^2}$
Stress	σ	$\dfrac{\mathrm{F}}{A}$	$\Sigma = \dfrac{\sigma_m}{\sigma_p} = \dfrac{F_m/A_m}{F_p/A_p} = \dfrac{F_m l_p^2}{F_p l_m^2}$	$\Sigma = \dfrac{\mu}{\lambda\tau^2}$
Viscosity	η	$\dfrac{\mathrm{F}t}{A}$	$\dfrac{\eta_m}{\eta_p} = \dfrac{\sigma_s^{(m)}/\dot{\varepsilon}_s^{(m)}}{\sigma_s^{(p)}/\dot{\varepsilon}_s^{(p)}} = \dfrac{\sigma_s^{(m)}}{\sigma_s^{(p)}}\dfrac{\dot{\varepsilon}_s^{(p)}}{\dot{\varepsilon}_s^{(m)}}$	$\Sigma\tau = \dfrac{\mu}{\lambda\tau}$

* The subscript s indicates shear stress or sheer strain rate.

Earth and in the model. The scale of the vertical motions in Asia associated with thickening of the crust and isostatically driven vertical tectonics is small compared with that of the horizontal motions, and so to a first approximation we may be justified in ignoring gravity as a major factor in this particular deformation of the Earth. In the model, vertical deformation is eliminated by confining the plasticine between lubricated upper and lower plates. Furthermore, we can ignore inertial forces because accelerations are so small as to be negligible. By this reasoning, we argue that accelerations do not significantly affect this deformation, and we may safely ignore the associated scaling for the model.

Because gravitational forces are assumed to be negligible, the scaling of mass is not important in itself, because the force that depends on the mass is just the weight, which is the mass times the gravitational acceleration. Thus, for the third independent scale factor we are free to choose one other independent quantity as the significant one to be scaled. An obvious choice would be to scale the stress or the rheological characteristics of model and prototype, but to do this we must know the rheological properties of the prototype and the model materials. The scale factors for stress and viscosity are related by the scale factor for time (see Table 9.1.1), which we already know. In what follows, we make independent estimates of the scale factors for stress and viscosity from our knowledge of the Earth and the model materials, and we show that they give reasonably consistent results.

Plasticine deforms according to power-law rheology (Eq. 9.1). For plasticine the stress exponent is about $n = 7.5$ at 25°C. Its behavior, therefore, is very close to that of a perfectly plastic material. The yield stress at strain rates on the order of 10^{-7} s^{-1} is approximately 0.1 MPa. Using these numbers in equation 9.1 and solving for A yields

$$\left|{}_s\dot{\varepsilon}_n\right| = 3.2 \, {}_D\sigma^{7.5} \qquad (9.1.5)$$

where strain rate is in units of s^{-1} and stress is in units of MPa. The piston advances at about 25mm/hr, and if this shortening is distributed across a distance of about 70 mm in the plasticine (see Fig. 9.17B), then the approximate strain rate is 9.9×10^{-5} s^{-1}. The associated stress, from Equation (9.1.5), is then about 0.25 MPa. We can compare the rheologies of the model and prototype materials using the effective viscosity, which for the plasticine is given by

$$_{\text{eff}}\eta_m = \frac{{}_D\sigma}{\left|{}_s\dot{\varepsilon}_n\right|} = 2.5 \times 10^3 \text{ MPa s} = 2.5 \times 10^9 \text{ Pa s} \qquad (9.1.6)$$

To evaluate the scale factor for viscosity, we need to estimate the appropriate effective viscosity for the continental lithosphere. The appropriate rheology for the continental crust is far from clear. The mechanical properties of the rocks involved must vary from brittle near the surface to ductile at depths below about 15 km, judging from the maximum depth of earthquakes on the San Andreas fault, for example. If the entire continental crust to a depth of 40 to 80 km is involved in the deformation, it may well be controlled by the power law creep of the deeper parts of the crust. This conclusion would also hold if the upper mantle were involved in the deformation. Thus, to assign a single viscosity to the entire crust or lithosphere is a large simplification of the actual state of the material. Nevertheless, we can make a rough estimate of the effective viscosity of the lithosphere. Geologic strain rates tend to be on the order of 10^{-14} s^{-1}. If we calculate the strain rate for the collision of India in the same manner as we did for the model, we find a convergence rate of 50 mm/yr distributed over a distance of perhaps 2800 km, which gives a strain rate on the order of 5.7×10^{-16} s^{-1}. Stresses in the Earth are at most on the order of 1000 MPa, which we deduce from the dynamically recrystallized grain size in very strongly deformed ductile shear zones. Average stresses are probably one or two orders of magnitude smaller. Let us assume that a reasonable range for tectonic stresses is between 1 and 1000 MPa. Using these ranges for stress and strain rate gives us a range of effective viscosities for the continental lithosphere of

$$_{\text{eff}}\eta_p = 10^{20} \text{ Pa s to } 10^{25} \text{ Pa s}$$

On the basis of these estimates, the scale factor for the stress is

$$\Sigma = \frac{\sigma_m}{\sigma_p} = \frac{0.25}{(1 \text{ to } 1000)} = 0.25 \text{ to } 0.00025 \qquad (9.1.7)$$

and the scale factor for viscosity is

$$\frac{_{\text{eff}}\eta_m}{_{\text{eff}}\eta_p} = \frac{2.5 \times 10^9}{(10^{20} \text{ to } 10^{25})} = 2.5 \times (10^{-11} \text{ to } 10^{-16}) \qquad (9.1.8)$$

But from Table 9.1.1, the scale factor for viscosity equals $\Sigma\tau$. Using $\tau = 5.7 \times 10^{-12}$ from Equation (9.1.4), we find

$$\Sigma = 4.4 \text{ to } 0.0044 \qquad (9.1.9)$$

Thus, given the uncertainties in our knowledge, the stress scale factors derived on the one hand from estimates of stress in the model and in the Earth (Eq. 9.1.7), and on the other hand from estimates of effective viscosity in the model and in the Earth (Eq. 9.1.9), fall in a comparable range. We could reasonably conclude, therefore, that within the limits of our understanding, the plasticine models of collision provide a dynamically scaled model of the Earth's deformation.[1]

---

[1] Students interested in more discussion of scale models can find it in Twiss and Moores, 1992, Chapter 20.

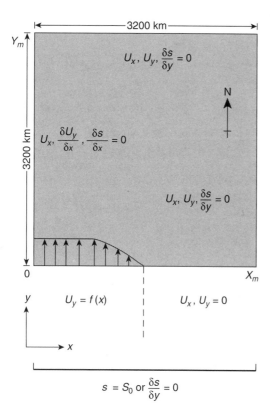

3200 km

Y_m

U_x, U_y, $\dfrac{\delta s}{\delta y} = 0$

N

3200 km

U_x, $\dfrac{\delta U_y}{\delta x}$, $\dfrac{\delta s}{\delta x} = 0$

U_x, U_y, $\dfrac{\delta s}{\delta y} = 0$

0

X_m

y

$U_y = f(x)$

U_x, $U_y = 0$

x

$s = S_0$ or $\dfrac{\delta s}{\delta y} = 0$

Figure 9.18 Boundary conditions for the numerical modeling of a plate of material with a power law rheology confined within a rigid square box with an influx of material at the bottom left side. The arrows indicate the velocity profile of the influx. For comparison with Asia, the box can be imagined to be 3200 km on a side. U_x, U_y: velocity in x and y directions; s, S_0: thickness and original thickness (35 km) of the layer. *(After England, 1982)*

plane-strain conditions imposed on the model. Second, it is not clear why one continent behaves as a rigid indenter and the other as a yielding plastic material. The answer may be that the continental margin characterized by a consuming plate boundary (Asia) is hotter and therefore weaker.

A Numerical Model of Asian Tectonics

Another approach to modeling continental collisional deformation in Asia uses numerical calculation of the deformation (see Twiss and Moores, 1992; Secs. 20.1 and 20.2). With this approach we assume that the lithosphere deforms according to a power law rheology (see Twiss and Moores, 1992; Eq. 18.4) such as

$$|_s\dot{\varepsilon}| = A_{D}\sigma^n \qquad (9.1)$$

that includes a changeable stress exponent n. Here $|_s\dot{\varepsilon}|$ is the absolute value of the steady-state extension rate, A is a material constant that depends exponentially on the temperature, and $_D\sigma$ is the differential stress (the maximum minus the minimum principal stress). The boundary conditions are defined by the requirement that the deformation occurs within a fixed box and that material enters the box along half of one side with a velocity profile as indicated in Figure 9.18. The material entering the box is comparable to the indenter in the plastic models.

The horizontal components of the velocity are calculated from the rheological equations, which require values for the constants defining the mechanical properties of the material. Obviously, the Earth's lithosphere is complexly layered, but for the sake of first-order calculation, these material properties are taken to be the average of the properties across the entire thickness of the lithosphere. The vertical strain rate is calculated from the horizontal strain rates using the assumption of constant volume deformation. The vertical deformation creates buoyant forces on the thickened crust that are included in the driving forces of deformation. Thus, if the effective viscosity of the lithosphere is sufficiently low, the gravitational collapse of an overthickened crust contributes significantly to the deformation.

The model is not intended to represent Asia except in a very general way. Figure 9.19, however, shows one model whose features are comparable to the tectonics of Asia. In this model, the stress exponent $n = 3$, and the rheology of the lithosphere was chosen so that gravitational collapse of overthickened crust contributes significantly to the deformation. Figure 9.19A shows the velocity field for the deformation, Figure 9.19B shows crustal thickness (compare Fig. 9.10), and Figure 9.19C shows the general style of deformation.

The velocity field (Fig. 9.19A) shows a general flow of material toward the east, away from the area of influx (compare Figs. 9.11 and 9.17). The area of maximum thickening (Fig. 9.19B) is adjacent to and in front of the area of influx, and thickening decreases to zero toward the east (compare Fig. 9.10). The style of deformation (Fig. 9.19C), determined from the magnitudes of the principal horizontal strain rates, includes zones of compression, extension, compression with a significant strike-slip component, and predominantly strike-slip deformation with components of compression (c) and extension (e). Figure 9.20 shows that similar zones are present in Asia (compare Fig. 9.11).

A.

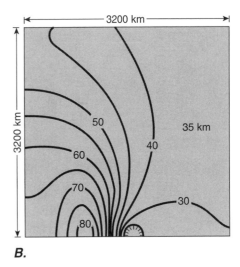

B.

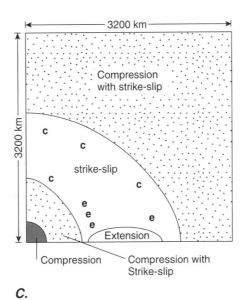

C.

Figure 9.19 Results from one of the numerical calculations for the deformation of the layer in Figure 9.18. The stress exponent $n = 3$ (see Eq. 9.1), and the lithosphere is assumed to have a sufficiently low yield strength that the buoyant forces caused by thickening of the crust result in a significant amount of deformation. *A.* Velocity field for the deformation. Note that the general form of the field is comparable to the plastic extrusion observed in the plasticine experiments (Fig. 9.17). *B.* Thickness of crust, starting at an initial 35 km. The thickness is greatest near the influx and near normal on the right side of the box (compare Fig. 9.10). *C.* The distribution of the dominant modes of deformation, including compression, compression with strike-slip, strike-slip with components of either compression (c) and extension (e), and extension. *(After England, 1982)*

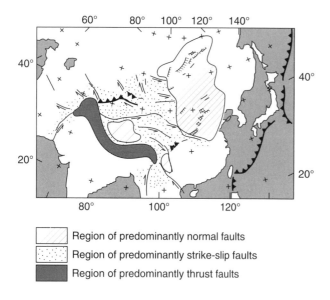

Region of predominantly normal faults

Region of predominantly strike-slip faults

Region of predominantly thrust faults

Figure 9.20 Tectonic map of Asia showing principal modes of deformation. Note areas of normal faulting within region of predominant strike-slip faulting (see Fig. 9.17C). *(After Tapponnier and Molnar, 1976)*

An advantage of the numerical models is that they take crustal thickening and topography into account. Because of the specific rheological equations assumed, and the three-dimensional deformation, however, they do not explicitly predict faults, and the boundary conditions are only rough approximations, at best, of the conditions in Asia. Thus, the deformation patterns of the model do not reproduce the observed patterns very closely (compare Figs. 9.19C and 9.20).

9.5 The Formation of Mountain Roots

Mountain roots are regions of greater than normal crustal thickness that underlie almost all continental mountain belts in the world. Indeed, the high topography of mountains exists because of the isostatic support provided by the thickened crust "floating" in a higher density mantle.

Two hypotheses have been proposed to explain the formation of mountain roots, one related to subduction processes and the other to collision processes. The subduction hypothesis, also called the Cordilleran hypothesis, holds that mountain roots form beneath an active continental margin by the rise of mafic magmas off the downgoing slab and intrusion of the magmas into the lower crust (Fig. 9.21A).

Cooling and crystallization of these magma bodies heat the surrounding crustal rocks to form a mobile core, including granitic magmas. The thickening of the crust causes an isostatic rise of the continental surface. Gravitational collapse of this thickened mobile crust causes thrusting toward the foreland and antithetic to the subduction zone.

Two famous regions to which this model is applied include the central and southern Sierra Nevada of California and the central Andes (Figs. 9.21B and 9.21C). In these regions, batholithic activity was more or less concentrated in regions now characterized by such a root. The currently active faults on the east side of the Andes (Fig. 9.21A; see Sec. 10.9) may correspond to foreland thrust faults that arise from gravitational collapse of the thickened and heated continental crust.

The collisional model of mountain roots holds that they result from the compressional thickening of the crustal rocks themselves. Such a root could be formed by the underthrusting of one continent by another (Fig. 9.22A), as in the Zagros Mountains discussed in Section 9.3, perhaps followed by antithetic thrusting in the other direction (Fig. 9.22B); it could form by symmetrical thrust faulting of the two colliding continental edges (Fig. 9.22C); or it could form simply by uniform ductile thickening in the collision zone (Fig. 9.22D).

A collisional origin of mountain roots, rather than magmatic intrusion, seems the more likely explanation in most cases. Neither the Andes nor the Sierra Nevada show convincing evidence along their entire length of a magmatic origin for crustal thickening. Each mountain chain shows clear evidence of collision along at least part of its length. Most continental arcs, as outlined in Chapter 8, do not show evidence of a root beneath the volcanic chain. For example, Japan was a part of eastern Asia until Neogene time, when the opening of the Sea of Japan separated it from the mainland. Apparently, eastern Asia (including Japan) has been an active subduction zone intermittently since late Paleozoic time, yet Japan shows little evidence of such a root with associated structures as implied in Figure 9.21A.

In young orogenic belts, the thickness of a mountain root should approximately reflect the amount of shortening that has taken place, as shown schematically in Figure 9.23. The total area of crust between points *a* and *b* remains the same. Figure 9.23A shows schematically two continents converging; Figure 9.23B shows the initiation of collision; and Figure 9.23C shows the final postcollisional state with a thickened crust forming a depressed root and an uplifted surface. To form a root, the rate of thickening or convergence should be rapid compared with

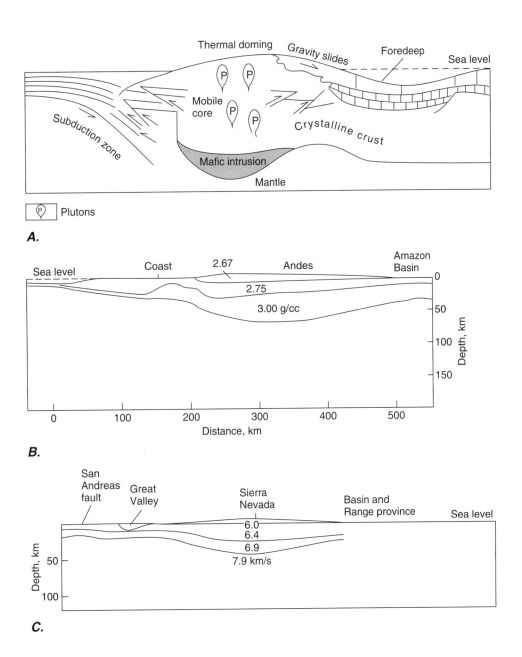

Figure 9.21 Mountain roots in continental arc regions. *A.* Hypothetical development of "orogenic welt" in continental arc region. Note difference in geometry from true-scale features. *B.* True-scale layered model cross section of central Andes showing crustal section, based upon gravity measurements. Numbers show model densities for individual layers. *C.* True-scale layered model cross section of crustal structure of Sierra Nevada, California, based on seismic refraction measurements. Numbers are P-wave velocities in km/s for individual layers. *(A. after Dewey and Bird, 1970; B. after Couch et al., 1981; C. after Eaton, 1966)*

the rate at which erosion or ductile collapse occurs. Of course, any real area of thickened crust with a mantle root and a topographic welt would be smoother than that shown in Figure 9.23C because of erosion of the uplift, thrusting out on either side of the uplift, and ductile deformation at depth. In addition, if strike-slip tectonics became important, as in central Asia, material would move out of the cross section, and the thickening of the crust would provide only a minimum estimate of the shortening.

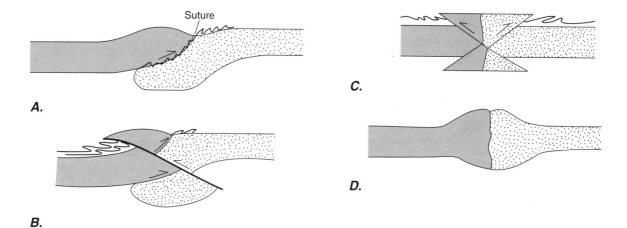

Figure 9.22 *Diagrams illustrating development of mountain root by collision. A. Development by simple thrusting of one continental margin over another, as in Zagros Mountains. B. Modification of (A) by thrusting in the antithetic direction produces a more complex structure, as in the Alps. C. Formation of root by symmetrical antithetic thrusting of both continents simultaneously. No real examples are known on Earth. D. Formation of root by homogeneous ductile deformation of collision region. No real examples are known of this exact scenario.*

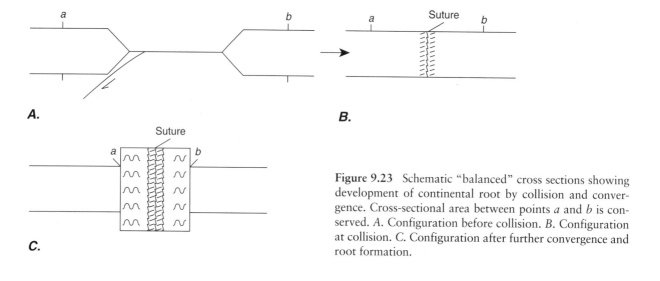

Figure 9.23 *Schematic "balanced" cross sections showing development of continental root by collision and convergence. Cross-sectional area between points a and b is conserved. A. Configuration before collision. B. Configuration at collision. C. Configuration after further convergence and root formation.*

In some mountain ranges, such as the Alps, the amount of shortening calculated from balanced cross sections exceeds that predicted by measuring the cross-sectional area of the root. In other words, the thickness of crust is not as great as it would be if a continental crust of normal thickness were shortened by the amount indicated by the sediments. Strike-slip motion carrying material out of the cross section may account for some of this discrepancy. Alternatively,

the colliding crustal margins may have been thinner than normal. They could have been tectonically thinned (probably by listric normal faulting) during earlier continental rifting. Figure 9.24 shows this situation schematically. If continental crust of full thickness (dashed lines) were shortened the full amount shown for points *a* and *b*, then the resulting continental thickness would be more like the dashed outline in Figure 9.24*B*, rather than the solid line.

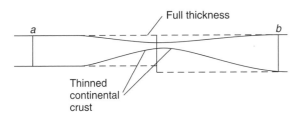

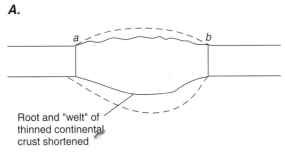

Figure 9.24 Schematic "balanced" cross section illustrating root-forming effect of collision of two thinned continental margins (solid lines) compared with ones of normal thickness (dashed lines). Approximately true scale (horizontal = vertical). Profile of margin taken from that of Red Sea margins (Fig. 5.11B). A. At collision. B. After collision and mountain root formation. Area of continental crust between points *a* and *b* remains constant. *(After Helwig, 1976)*

9.6 Models of Ophiolite Emplacement

As discussed in Section 9.1, ophiolite belts are thought to represent collisional sutures where former ocean basins were subducted and remnants of the oceanic crust have been thrust up onto downgoing continental crust. The fact that these belts are widespread (see Fig. 9.2) indicates that a great number of collisions have taken place and that collision is a primary tectonic process. Ophiolites are present both as intact, well-preserved complexes with a predictable sequence of rock types (Fig. 9.25) and as dismembered lithologic sequences in subduction complexes. Not all well-preserved ophiolite complexes contain a complete ophiolite sequence, a feature that may reflect tectonic dismemberment or original lithologic differences that developed during formation of the complex.

We consider here the question of ophiolite emplacement. Dismembered ophiolitic lithologies in subduction zones presumably were emplaced as tectonic slices or melanges during subduction of the oceanic lithosphere. The emplacement of well-preserved complexes, however, constitutes a tectonic problem of greater importance deserving further discussion.

Many ophiolites are tectonically emplaced over island arc or continental crust, typically with a thin layer of amphibolite at their bases. Complexes associated with continental crust are tectonically emplaced over continental shelf and slope rocks along low-angle faults that verge toward the continental interior

(Figs. 9.25B and 9.25C). On the ocean floor, these shelf and slope sediments are initially arranged as shown in Figure 9.25C. Beneath ophiolite complexes, they are stacked up by imbricate thrust slices, with the deep-water sequences consistently thrust over the shallow-water sequences (Fig. 9.25B). Also, many ophiolite complexes are overlain unconformably by shallow-water or even subaerial sediments (Fig. 9.25A). Ideally, the stratigraphy associated with ophiolite complexes can constrain the time of emplacement, because it must have occurred after the deposition of the youngest sediments involved in the basal thrust complex, but before the deposition of the shallow-water or terrestrial sediment unconformably on top.

There is considerable controversy over the mechanism by which ophiolites are emplaced onto continental crust. Four models have been proposed.

The first model holds that ophiolites are emplaced onto a passive continental margin during a collision of the passive continental margin with a mantle-rooted thrust, that is, a subduction zone. An incipient subduction zone will have no arc, whereas a mature one will have a substantial island arc developed above it (Fig. 9.26). In either case, the continent rides on the downgoing plate, which is subducted until the continent reaches the subduction zone. The passive margin of the continent begins to subduct until the buoyancy of the continental crust arrests the subduction (Fig. 9.26A). Isostatic adjustment then occurs, and the continent rises back up, lifting on its

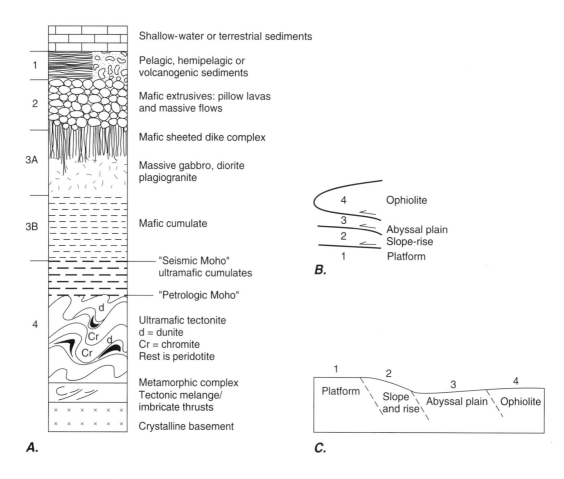

Shallow-water or terrestrial sediments

1 — Pelagic, hemipelagic or volcanogenic sediments

2 — Mafic extrusives: pillow lavas and massive flows

Mafic sheeted dike complex

3A — Massive gabbro, diorite plagiogranite

3B — Mafic cumulate

"Seismic Moho" ultramafic cumulates

"Petrologic Moho"

4 — Ultramafic tectonite
d = dunite
Cr = chromite
Rest is peridotite

Metamorphic complex Tectonic melange/ imbricate thrusts

Crystalline basement

A.

B.
4 Ophiolite
3 Abyssal plain
2 Slope-rise
1 Platform

C.
1 Platform
2 Slope and rise
3 Abyssal plain
4 Ophiolite

Figure 9.25 Ophiolites and their surroundings. *A.* Expanded ophiolitic assemblage showing crystalline substrate with platform sediments, deformed sedimentary sequence, melange, and metamorphic rocks at base, and an unconformity and shallow-water sediment on top. Also shown is the inferred correlation with an oceanic seismic model. The seismic Moho is the boundary between the seismic crust and mantle. The petrologic Moho is the boundary between mantle tectonite below and igneous rocks above. *B.* Typical tectonic stacking of thrust sheets of different provenance beneath ophiolites. *C.* Original paleogeographic positions of thrust sheets. *(After Twiss and Moores, 1992)*

margin a torn-off slice of the former overriding plate, an ophiolite (Fig. 9.26B).

The second model proposes that an ophiolite is emplaced onto an active continental margin, under which ocean floor is being subducted, by a process called *obduction*[2] (Fig. 9.27). In this model, part of a downgoing oceanic plate detaches, presumably along a preexisting discontinuity such as a fault, or along a

ridge axis (Fig. 9.27A), and is shoved onto the continental margin as the rest of the slab continues to subduct (Fig. 9.27B). Either the fault at the base of the ophiolite has to come back up to the surface of the seafloor, or ophiolite slices must accumulate in an imbricate thrust stack (Fig. 9.27C).

The third and fourth models hypothesize the emplacement of ophiolite complexes onto active convergent continental margins by two different processes of gravity sliding, neither of which seems adequate to account for most well-preserved ophiolite complexes (Fig. 9.28). Figure 9.28A shows a possible

[2] From the Latin *obducere*, to shove up.

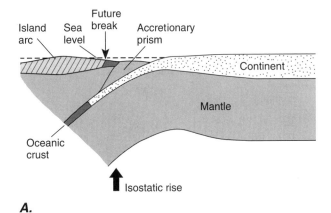

A.

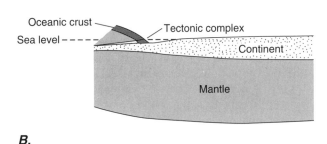

B.

Figure 9.26 Cross sections illustrating emplacement of ophiolites by collision of continental margin with subduction zone. Compare with Figure 9.1. *A.* Continental margin rides down into zone, chokes it, and then rises isostatically. *B.* Separated fragment of oceanic lithosphere now present as ophiolite complex. Accretionary prism remnant becomes tectonic complex. *(After Moores, 1982)*

model whereby crust from the downgoing plate detaches at the outer swell and slides across the trench and onto the continental margin. The alternative model (Fig. 9.28*B*) proposes that an uplift develops along the edge of an overriding plate, possibly resulting from diapiric activity. Subsequent gravitational collapse of the uplift detaches a piece of the oceanic crust and drives it up on the continental margin.

These models of ophiolite emplacement clearly have radically different tectonic implications. The collision model implies that because subduction is shut off by the collision, a major perturbation in the motion of the plates results. The other three models, however, imply no such perturbation. It is not clear, however, that gravity-driven emplacement could push the ophiolite sheet far enough over the continent. A gravity-sliding model can be convincing only if the uplift were large enough and if a sufficient downward slope toward the continent existed to drive the thrusting. Many ophiolite complexes are as much as 10 km thick and contain unserpentinized mantle peridotite; thus, they cannot have been emplaced simply by gravitational collapse. The diagrams shown in Figure 9.28 both suggest thrusting up a slope to an altitude that exceeds that of the collapsing uplift, which is impossible.

The collision model of ophiolite emplacement has the most clearly documented modern analogues: the localities in the southwest Pacific and eastern Mediterranean discussed in Section 9.2—New Guinea, Taiwan, Sulawesi, Sangihe-Halmahera, Cyprus—where oceanic crust is being emplaced or has just been emplaced over a continental margin by collision of the margin with a subduction zone. In these cases, the chemistry of the ophiolitic rocks also suggests derivation from a position over a subduction zone. The obduction model is possible if pre-existing oceanic crustal faults on a downgoing plate become thrust faults as the plate goes into the trench. Ophiolites thus emplaced could have chemical compositions indicating an origin at a mid-ocean ridge. Clear examples of ophiolites emplaced by this mechanism, however, are few and minor in extent. In fact, we know of only two examples—the Pliocene Taitao ophiolite, southern Chile, and the early Tertiary Resurrection Bay ophiolite, Alaska—that clearly seem to have been emplaced by this mechanism. Thus, according to our current understanding, ophiolite emplacement by collision seems to be the more common mechanism.

The collision model implies that the basal thrust is a fossil subduction zone. If so, the metamorphic

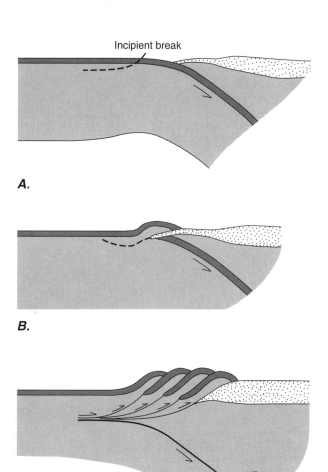

A.

B.

C.

Figure 9.27 Cross sections illustrating obduction of ophiolite complex on the continental margin of an overriding plate. *A.* Initial configuration. *B.* Obduction if fault comes back to surface. *C.* Sequence of thrusts if fault continues to dip into mantle, thereby emplacing successive slices. *(After Moores, 1982)*

China and Australia with subduction zones clearly do not involve continent-continent collisions. In other words, ophiolite emplacement histories in a given mountain belt can be expected to provide evidence of plate interactions prior to final continent-continent collision. All ophiolites thus emplaced have some history of arc activity, the extent of which must vary depending on the amount of subduction prior to collision.

What effects do ophiolite emplacement events have on the plate tectonic regime? As we discussed in Chapter 4, one of the principal driving forces of plate tectonics seems to be the sinking of downgoing slabs. If ophiolite emplacement is a collision of nonsubductable continental margins with subduction zones,

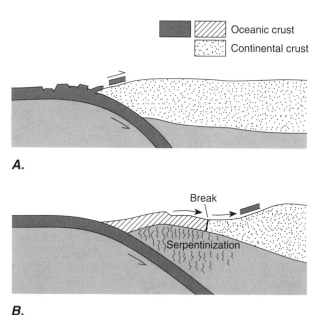

A.

B.

Figure 9.28 Two hypotheses for gravity-driven emplacement of ophiolite. *A.* Sliding onto continental margin from outer swell. *B.* Gravitational collapse of a topographic bulge formed by a diapiric rise above subduction zone. Note that ophiolites must slide up to an altitude that exceeds the altitude of the topographic high from which they originate, an impossibility. *(Redrawn after Gass et al., 1975)*

rocks present along the base of many complexes represent rocks formed or preserved in the subduction complex. Ophiolite basal thrusts are thus important tectonic features in any deformed belt. Their displacements must be very large but inherently indeterminate, because we cannot know how much subduction might have taken place before collision. In addition, because a substantial, but indeterminate, amount of downgoing plate with its overlying sediments would have disappeared down the subduction zone, *these ophiolitic thrusts intrinsically cannot be balanced.*

Collisions that result in ophiolite emplacement can occur between an island arc and a subduction zone (see Fig. 9.1D) or between a continental margin and a subduction zone (see Fig. 9.1B). These subduction zones may be out in the middle of an ocean away from the margin of any other continent. Thus, these collisions generally occur before the final continent-continent collision that completely closes the ocean basin. The modern collisions of the margins of

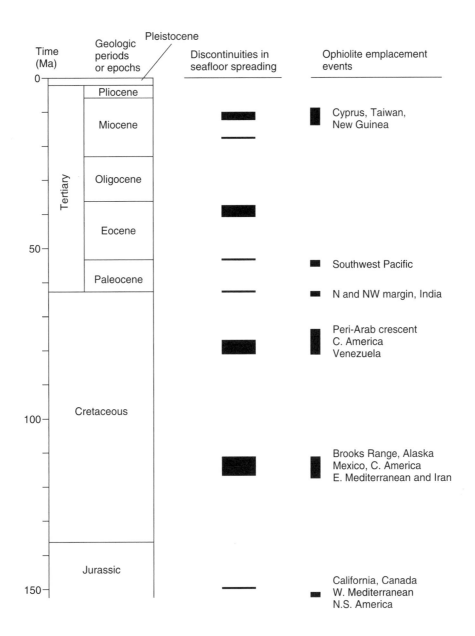

Figure 9.29 Diagram illustrating time correspondence between major periods of ophiolite emplacement and discontinuities in seafloor spreading. *(After Moores, 1982)*

such events may interrupt smooth plate motion. Thus, times of major ophiolite emplacement might correspond to times of change in plate tectonic behavior of the Earth. Figure 9.29 shows a correlation of major ophiolite emplacement events with times of change in seafloor spreading, as reflected in magnetic anomalies. To a first approximation, it appears that there may be a correspondence. The data on timing of ophiolite emplacement are incomplete and inexact, however, which makes it difficult to draw firm conclusions from this apparent correlation.

9.7 Sutures: Evidence of Ancient Plate Boundaries

Sutures are the boundaries between crustal blocks that were carried on two different tectonic plates and that have been juxtaposed by plate motion. The recognition of sutures in the continental geologic record is essential to our ability to reconstruct Earth's tectonic history for times older than the age of the ocean basins.

Generally, a suture is not a single fault or plane but a complex zone of deformation, several kilometers wide, that separates the crustal blocks. A **collisional suture** marks the site of a subducted ocean basin that separated the crustal blocks before collision. The crust on opposite sides of such a suture may have originated as any of the pairs of collided continental or island arc margins illustrated in Figure 9.1. Thrust faults are prominent in a collisional suture. A **transform suture** is the site of a fossil transform fault along which unrelated crustal blocks were juxtaposed. Most sutures show some evidence of both thrust and strike-slip motion.

Sutures can be recognized either by the characteristics of the boundary itself or by a major discontinuity across the boundary, such as marked changes in lithology, geologic history, structural style, paleomagnetic vectors, or faunal assemblage. Of course, a combination of such lines of evidence greatly strengthens the case for a suture's existence.

Inherent Characteristics

An ophiolite belt between two continental crustal blocks is a remnant of the former ocean basin that separated the blocks, and it is the most direct evidence for the existence of a collisional suture (see Sec. 9.6) Thus, the ophiolite belts illustrated in Figure 9.2 represent the locations of major sutures.

A suture in a continent is sometimes marked by a belt of tectonic melange, which may be part of a larger deformed belt. As discussed in Section 7.6, the distinction between a melange of tectonic origin and one of sedimentary origin may not be easy to recognize. If the melange is exclusively sedimentary, however, it might have developed by sedimentary processes such as slumping and therefore need not have developed at a suture.

Some melange zones, whether ophiolitic or not, contain sediments that have no counterpart on either side of the zone. These sediments were presumably deposited in regions far from the rocks on either side of the melange belt and were subsequently preserved by offscraping before being caught in a collision. Therefore, they imply the disappearance of a wide oceanic region and indicate a suture.

Some sutures are characterized by a mylonitic or ductile shear zone, although without other evidence, such a zone does not necessarily imply the presence of a suture.

The Indus-Tsangpo suture of Asia is an especially good example of a suture zone. It extends some 2500 km along the northern margin of the high Himalaya and separates rocks belonging to the Indian subcontinent from those of Tibet (see Fig. 9.9). The zone is marked by a melange containing numerous ophiolite fragments, the largest of which is some 3500 km² in area, as well as by a thick sequence of oceanic volcanic rocks with overlying pelagic sediments. Along its western end, the suture separates into two zones that enclose an exotic island arc sequence, the Kohistan arc (see Box 7.2), that was caught up in the Eurasian-Indian collision. Note that the suture is apparently folded, as illustrated in Figure 9.9B.

Geologic Discontinuities

If crustal blocks are separated by a boundary across which there is a markedly different stratigraphic or tectonic history or a discontinuity in the orientation or the style of structures, the boundary may be a suture, especially if it is highly sheared. The Kolar schist belt of southern India, for example, separates two Precambrian regions of markedly different ages and distinct trace element and isotopic signatures, as illustrated in Figure 9.30. The Kolar belt itself is a highly

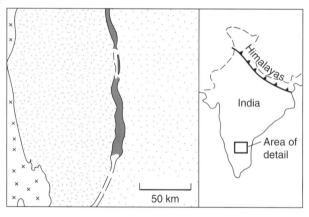

Kolar schist belt (suture)

Eastern gneiss, ~ 2530 Ma old, derived from mantle source

Western gneiss, 2553–2632 Ma old, derived from older crustal source

Younger granite

Figure 9.30 Map of portion of southern Precambrian shield of India, showing Kolar schist belt, a possible suture between two gneissic units. The eastern gneiss is about 2530 m.y. old, derived from a mantle source, whereas the western gneiss is 2553–2632 m.y. old, derived from an older continental crustal source. *(After Krogstad et al., 1989)*

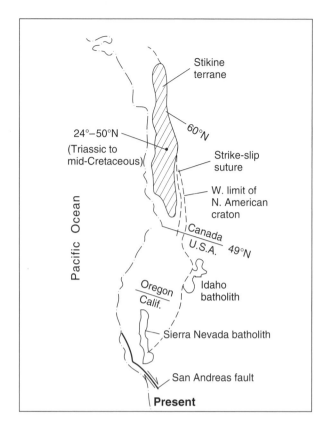

Figure 9.31 Relationships between the Stikine terrane, western Canada, and cratonic North America. Stikine Mesozoic paleolatitudes are sharply discordant with North American ones and indicate that the mid-Cretaceous position was approximately 1300 km south of the present position, as indicated, opposite Oregon and California. Figure ignores possible rotations. *(After Monger and Irving, 1978)*

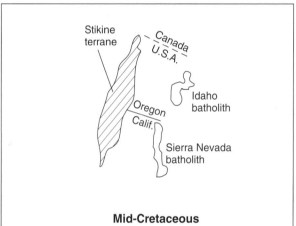

deformed zone with multiple fold episodes that may reflect both convergent and strike-slip motion. The different geochemical signatures of the amphibolites and gneisses on either side of the Kolar belt suggest that they had distinctly different origins. Rocks from east of the belt may have originated from mantle-derived high-Mg andesites, whereas those on the west may have originated by partial melting of older crustal rocks. Thus, the Kolar belt may represent a suture.

Discrepant orientations of paleomagnetic vectors determined from rocks of similar age across a shear zone may indicate that the crustal blocks formed in widely separated regions and were later juxtaposed along a suture. The Stikine terrane in western Canada provides an example of this situation (Fig. 9.31). Paleomagnetic measurements on Stikine rocks of Mesozoic age indicate that they were originally deposited at a latitude some 1300 km south of their present location. The northward movement may have been along a series of strike-slip faults that separate the Stikine rocks from North America. Thus, the Stikine–North America boundary may represent a transform suture.

In some cases, sutures can also be identified by dissimilarities in fossil assemblages across a boundary. This criterion must be used with caution, however, because many faunal or floral differences arise from environmental factors alone. The most useful types of fauna for this kind of analysis are shelf benthonic forms that display limited mobility of larval and adult stages. Using these forms as a guide, the present oceanic areas may be grouped into distinct faunal provinces (Fig. 9.32). Continental shelves, subduction zones, and midplate island chains are dispersal routes; oceanic ridge-transform systems, and to some extent climatic zones and large continental land masses, are faunal barriers. In the present world's oceans, the Indo-Pacific faunal province extends from East Africa to the Tuamotu Islands in the central Pacific, a distance of over 10,000 km. Other provinces, such as the one around the southern tip of South America, are much more limited in extent.

Applying this information to the past, we conclude that two distinctly different shelf benthonic faunas from a similar environment should have had widely separate locations at the time of deposition. If these same faunas are juxtaposed in a deformed region, there may be a suture between them. Two similar faunas, on the other hand, originally could have been either close or distant. Careful examination of the geology associated with the two faunas is a necessary adjunct to the fossil evidence.

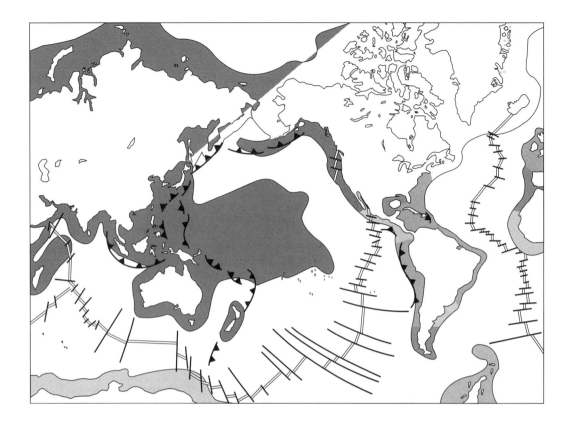

Figure 9.32 World map showing general plate boundaries and shelf benthonic zoogeographic provinces. Of particular note is the vast Indo-Pacific province extending from eastern Africa to the central Pacific. The figure illustrates the ease of spreading of fauna along chains of islands, particularly island arcs, and the difficulty of faunal spreading across rift-transform systems. *(After Valentine and Moores, 1974)*

The Cambrian–early Ordovician faunas of the Appalachian-Caledonide orogenic belt provide evidence for the existence of a suture (Fig. 9.33). The faunal assemblages of this age that occur on opposite sides of the orogenic belt are distinctly different. Late Ordovician faunas, however, do not display this distinction, suggesting that a suturing event occurred during Ordovician time. This event probably corresponds to the emplacement of ophiolite complexes along the length of the Appalachian orogen.

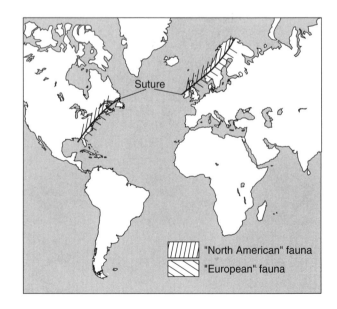

Figure 9.33 Distribution of "European" and "North American" Cambrian-Ordovician faunas about the North Atlantic Ocean. These distributions suggest the existence of a suture, as indicated. *(After Whittington, 1973)*

Additional Readings

Coward, M. P., and A. C. Ries, eds. 1986. *Collision Tectonics* GSA Special Publication 19.

Dewey, J. F. 1977. Suture zone complexities: A review. *Tectonophysics* 40:53–67.

Dewey, J. F., and J. M. Bird. 1970. Mountain belts and the new global tectonics. *J. Geophys. Res.* 75:2625–2647.

Eaton, J. P. 1966. Crustal structure in northern and central California from seismic evidence. *Calif. Div. Mines Geol. Bull.* 190:419–426.

England, P. 1982. Some numerical investigations of large-scale continental deformation. In *Mountain Building Processes.* K. J. Hsü, ed. New York: Academic Press.

Hilde, T. W. C., S. Uyeda, and L. Kroenke. 1977. Evolution of the western Pacific and its margin. *Tectonophysics* 38:145–166.

Johnson, A. 1970. *Physical Processes in Geology.* San Francisco: Freeman Cooper.

Moores, E. M. 1982. Origin and emplacement of ophiolites. *Rev. Geophys. and Space Phys.* 20:735–760.

Nur, A., and Z. Ben-Avrahem. 1982. Oceanic plateaus, the fragmentation of continents and mountain building. *J. Geophys. Res.* 87:3644–3661.

Odé, H. 1960. Faulting as a velocity discontinuty in plastic deformation. D. Griggs and J. Handin, eds. *Rock Deformation. Geol. Soc. Am. Memoir* 79:293–321.

Tapponnier, P. 1977. Evolution tectonique du systeme alpine en Mediterranee: Poinçoinnement et ecrasement rigid-plastique. *Bull. Soc. Geol. France* (no. 19):437–460.

Tapponnier, P., and P. Molnar. 1976. Slip-line field theory and large-scale continental tectonics. *Nature* 264:319.

Vink, G. E., W. J. Morgan, and W.-L. Zhao. 1984. Preferential rifting of continents: A source of displaced terranes. *J. Geophys. Res.* 89:10,072–10,076.

INTERLUDE The Scientific Method and the Plate Tectonic Revolution

In the preceding chapters we outline the tectonic processes currently active on Earth and something of their recent history. At this point, we would like to pause in our narrative and discuss the scientific method and, in particular, how the study of geology proceeds. Much of the material in Chapters 3–9 generally postdates the plate tectonic revolution in geology, which occurred in the 1960s and early 1970s. This material therefore provides insight into how scientific revolutions occur and what happens in the aftermath of such a revolution.

In titling this aside an interlude, we have unabashedly followed the example of one of the twentieth-century giants of structural geology and tectonics, Hans Cloos, who inserted an Interlude titled "To a Former Student" in his autobiographical Book *Conversation with the Earth*. Cloos's almost lyrical prose about the teacher-student relationship still makes good reading, although it was written well before the advent of plate tectonics, and he himself strongly doubted the validity of Wegener's hypothesis. Cloos wrote:

My third teacher was by far the wisest and most trustworthy, though he used neither books nor pictures, and was a poor lecturer. He was painfully silent, would talk only after repeated urging, and then often spoke almost unintelligibly in a language which had to be translated to be understood. But what he had to say was final and exhaustive, and could be passed on in good conscience to those who either would not or could not go directly to him for information. This teacher was nature itself. (Cloos, 1953)

I.1 The Scientific Method

The Science of Geology

Geology is the science of the history of Earth's evolution. It is based, first, on observations of the Earth itself and other planetary bodies, but the application of such sciences as biology, chemistry, physics, and materials science is also required to understand the processes we observe. Geology differs from these other sciences in at least three ways, however.

First, geology is fundamentally a historically oriented science dealing with processes that for the most part occur on a time scale that is immense compared with human lives. Thus, it is impossible to observe an entire process directly; we can see only what is happening at a single geological instant in time. Because of this constraint, the inference of geologic processes relies heavily upon the *fundamental assumption* that *spatial variation can be interpreted as temporal evolution.* In other words, we assume that the same process can be found in various stages of advancement in different places and that therefore we can piece together observations made in different places to infer a temporal evolution of that process.

Second, geology deals with large-scale and complex systems for which controlled experiments are difficult if not impossible to construct. Thus, the observation and description of natural features acquire proportionately more importance than they have in most other sciences.

Third, the fact that geologic evidence is fragmentary and incomplete makes many of the inferences drawn from the data nonunique and highly dependent upon our intuition and experience.

Despite these differences, the methods employed by geologists to investigate the Earth are philosophically much the same as those used in other realms of science. In the remainder of this section, we explore some of the philosophical underpinnings of our science that are usually understood implicitly and imparted at best by osmosis.

From Hypotheses to Laws

In discussions of structural and tectonic features, we refer repeatedly to *models* of such processes as the production of crystallographic preferred orientations, the formation of folds and faults, and the formation of rifted continental margins, or the evolution of triple junctions. This continual use of models prompts the question: What do scientific models have to do with reality? To answer the question, and to understand a scientific discussion, it is necessary to understand the meanings of the terms *fact, model, hypothesis, theory,* and *law* as they are used in science.

A **fact** is an objective observation or measurement, verifiable by any trained observer. The rock type that occurs at a given outcrop or the attitude of a dipping bed are two examples of objective facts. Facts are as near to "truth" as one gets in science, but because error always exists, the precision and accuracy[1] of a measurement or observation are never perfect.

A **hypothesis** is a proposed explanation of one or more observations. The term **model** is often synonymous with hypothesis. The mere proposal of a hypothesis or model does not, of course, imply that it is useful. For example, the ancient Egyptian belief that frogs formed from the mud of the Nile did not contribute much to the understanding of biology.

Perhaps the most straight-forward example of such a model in geology is a simple geologic map. Even in such a basic exercise as mapping, the scientist must exercise judgment. Two observers of the same field area will see and record different things depending upon the sharpness of their vision, their interests and knowledge, the goals of their search, and the models they wish to test. If geologic mapping were a completely objective and reproducible exercise, there would be no field work left to do once an area was carefully mapped. In fact, there is always more to do because new models such as plate tectonics spawn new questions that cannot be answered with existing information.[2] Old maps are useful in evaluating areas for further study or in posing new questions, but they rarely provide answers to the kinds of questions we pose today. In poorly known regions, it is now common practice to use regional images from aerial photography or satellite imagery to get an idea of what is on the ground before committing oneself to time-consuming and expensive fieldwork.

Thus, good models are important for understanding structural and tectonic features. When reading about the structure or tectonic development of a feature or an area, a tectonic or structural geologist should keep in mind such questions such as: How strongly do the data constrain the model? Are other models equally plausible? Are the models testable, and has the researcher provided any tests? Do the available data satisfy the tests? Only such a critical approach to scientific interpretation can lead to efficient advances in our understanding of the Earth and other planets.

In order to be useful, a model must satisfy three criteria: first, it must be *testable,* which means it must provide predictions that can in principle at least be verified by observation. Ideally, the predictions indicate some previously unrecognized or unexpected aspect or behavior of the process or system under investigation. Second, it must be *powerful,* which means it must be capable of explaining a large number of disparate observations.

---

[1] *Precision* is the closeness of one measurement to the mean value of many measurements of the same quantity. The standard deviation, for example, is a statistical measure of precision. *Accuracy* is the closeness of a measurement to the true value of the quantity. Thus, one can measure the attitude of a smooth, planar bedding surface more precisely than a rough, irregular one. If, in doing so, one used a compass with the wrong magnetic declination, the measurement would be inaccurate even though it may have been precise.

[2] The common lack of understanding of this aspect of the scientific method is illustrated by the story of the legislator who asked the state geologist, "Why do you need funds for geologic mapping? You have had a map of the state since 1910."

Third, it must be *parsimonious,* which means it must use a minimum of assumptions compared to the amount of data that it explains. In other words, the model should satisfy the principle of *Occam's razor,* which requires the elimination of all unessential features of a model.[3]

These principles provide only a qualitative standard, because there is no objective way of quantifying power or parsimony. Nevertheless, different models can usually be compared against one another on the basis of these principles. The best models are useful, simple, and elegant, but they necessarily are simplifications and idealizations of reality and should not be confused with "Truth," with a capital "T."

A useful model need not be right or "True," it need only be testable, which means it must include the potential for being disproven. Disproof of a model can be absolute, because in principle, a single observation can show that the model does not work. Proof, on the other hand, is never absolute, because regardless of the extent to which a model has successfully accounted for the characteristics of the observed world, it is impossible to guarantee that there will never be an inconsistent observation. In practice, of course, scientists tend to rationalize away one inconsistent observation by questioning the methods or ability of the observer, or by simply ignoring it as an unexplained anomaly. So disproof of a hypothesis typically is accepted only after the buildup of a significant quantity of anomalous observations, and the more entrenched the hypothesis initially is, the more resistant scientists typically are as a group to accepting its demise.

The restrictions on what constitute an acceptable model are quite severe, and they impose a stringent limit on the sorts of questions that scientists can legitimately pose and that science can be expected to answer. Thus, science is not an approach that can be used to answer all questions about everything. For example, most moral, religious, artistic, and legal questions simply are not within the realm of science. Investigating such issues requires a different system of inquiry and thought, which should be understood to coexist with the scientific philosophy rather than to compete with it.

A **theory** is a model or hypothesis that has gained more general acceptance through repeated verification. A **law** is a basic theory that has been shown so often to be consistent with observation that its validity is no longer in serious question. To be regarded as a

---

[3] After the English philosopher William of Occam (or Ockham), 1280–1349, who expressed it thus: "multiplicity ought not to be posited without necessity."

law, moreover, a proposition must generally deal with the very foundations of our scientific understanding. Even laws, however, can change. Although Newton's laws of motion, for example, are still useful for many everyday applications, they do not work at velocities approaching the speed of light; for these conditions they have been supplanted by Einstein's theory of relativity.

Deductive and Inductive Science

Scientific inquiry proceeds by two distinct methods, the **inductive** (or **Baconian**) **method,** named for the English philosopher Sir Francis Bacon (1561– 1626), who first espoused it, and the **model-deductive** (or **Darwinian**) **method,** named after the English naturalist Charles Darwin (1809–1892), who championed and practiced it. The Baconian method involves the collection of observations without regard to theory, in the expectation that explanations and natural laws will become apparent from the organization and synthesis of large amounts of data. The Darwinian method presumes that the scientist can devise a model that accounts for a set of observations. The model is then used to make predictions about nature, which then must be tested by comparing the predictions against objective observation of nature. By a continual iterative process of devising models, testing them through prediction and observation, modifying them to eliminate inconsistencies between prediction and observation, and further testing, a model emerges that accounts well for the observations.

The Baconian method is intended to provide a systematic and objective approach to scientific study. Although its application results in the amassing of large quantities of data, as a system of scientific investigation it is ultimately inefficient, as it lacks a means of deciding what data are important to gather. The data that could be gathered are in principle unlimited, and we could easily miss a critical observation simply because it never occurred to us that it might be useful. This problem is embodied in the common expression, "The eye seldom sees what the mind does not anticipate." The historian and philosopher of science, Thomas H. Kuhn, expressed the problem with the Baconian method as follows:

Since any description must be partial, the typical natural history often omits from its immensely circumstantial accounts just those details that later scientists will find sources of important illumination. (Kuhn, 1970)

Thus, it is impossible to make a set of observations of natural phenomena that are sufficiently complete to serve all the needs of future investigators. In addition, despite the best efforts to the contrary, every scientist's work is affected to some extent by the prejudices of the society, including its scientists, the limits of current knowledge, and the brain's limitations in dealing with the unfamiliar.

Nevertheless, the Baconian method is useful in the early stages of study of a subject when data gathering helps define the field and provides the basis on which preliminary models can be formulated. Examples of the application of this method in geology include the reconnaissance mapping of a poorly known area, and the accumulation of basic information that generally follows the application of a new technology to the study of some aspect of the earth.

The strength of the Darwinian method is that it provides a system for focusing the search for data. The method works best if one can invent more than one model to account for a given set of observations because one thereby avoids the common pitfall of becoming psychologically wedded to a particular model and ignoring inconsistent data that would tend to refute it. We call this technique the **method of multiple working hypotheses**. The predictions of a model, or contradictory predictions of two different models, direct the scientist's attention toward gathering specific data that are critical for the evaluation of the models.

Even if an individual scientist does not succeed in devising multiple hypotheses, however, others in the scientific community will, and fierce arguments in the scientific literature between proponents of different models are not unusual. The net result is that this method is more efficient at focusing our investigations and improving our understanding.

It is a common but egregious misunderstanding of scientists and the scientific method to believe that scientists are strictly rational and unintuitive individuals, and that science progresses only by the application of strict principles of logic and rational thought. The Baconian method, for example, does not specify how to carry out the process of induction that leads from the amassed data to natural laws and an understanding of nature; and the Darwinian method also provides no method for the development of hypotheses or models that are required for the process of prediction and testing against observation. In the end, we must admit that both processes rely on the non-rational and non-logical creativity, imagination, and intuition of the scientist, and that therefore these "scientific methods" are not entirely an objective and rational approach to understanding our surroundings.

In fact, this non-rational and non-logical part of the scientific process is *crucial to the progress of science*. This aspect of science can never be taught by a teacher to a student, because we cannot teach how to get good ideas or how to find a flash of insight. The importance of this aspect of science also means that science can never totally be systematized in any epistemology.[4] The drive of curiosity, the flash of insight, the leap of intuition, are the expressions of a process that we must accept as a part of our human nature, but one that we do not as yet understand and cannot predict or control. It is the factor that unites the work of scientists with all the creative endeavors of human beings, whether the result of the creativity be a scientific hypothesis, a piece of music, a dance, a painting or a piece of sculpture, a poem, a novel, or a play, an invention, or any result of the myriad of creative human activities that lead us in new directions and to new expression and understanding.

For scientists, however, the results of this creative process must be passed through the filter of rational objective analysis before they can become an accepted part of the body of scientific knowledge and understanding. The filtering process requires that predictions be derived from the models and that those predictions be tested by comparing them against the objective observation of nature. Those hypotheses that fail the tests are ultimately discarded. This is the part of the scientific process that is rational and logical and that usually receives the most attention. It is the part that teachers try to teach, that students try to learn, and that philosophers try to systematize. But the irony is that this part of the scientific process is only the part that comes *after* the fundamental insight, and it is really the insight that is the core of the process.

In the final analysis, the practice of science is the practice of phenomenology: it is a practice of finding what works as an explanation of nature, not of finding what is "True," whatever that may be. If we find a model that works consistently, we use it to improve our ability to predict and control our surroundings; if the model does not work, we discard it and look for a better one. It is ultimately a totally practical and utilitarian endeavor. Questions about how any scientific endeavor should or should not be accomplished, or how it was or was not accomplished, or what the significance of the results might be in terms of "Truth" are philosophical in nature and thus fall within a different realm of inquiry.

---

[4] Epistemology is the branch of philosophy that considers how we know what we know and what the limits of our knowing are.

I.2 The Normal Progress of Science and Scientific Revolution

Scientific disciplines, according to T. S. Kuhn, undergo a recognizable pattern of evolution. Most natural sciences begin with an initial Baconian phase of development in which observations are collected essentially at random without much regard to their significance or their relationship to one another. We might call this the **immature stage** of development of a science (Kuhn calls it the pre-historic stage). As the number of observations increases, eventually someone synthesizes them into a first comprehensive model, which Kuhn calls a **paradigm**. With the creation of a paradigm, the discipline enters its **mature stage** (Kuhn calls it the historic stage), and further progress occurs primarily by the application of the Darwinian, or model-deductive, method of investigation.

The paradigm is a framework of hypotheses and methods that constrains the questions that may be posed and that is assumed to be correct when observations are interpreted or other models are proposed. If observations contradict the predictions of a model, then the model is modified within the constraints of the accepted paradigm.

The normal progress of science can be interrupted either by the amassing of many facts that cannot be explained by the reigning paradigm or by the application of a new technology to an existing field. The first situation generally arises as investigation proceeds in greater and greater detail and more and more instances of incompatibility with the existing paradigm accumulate. At first, the inconsistencies are few and are regarded as either errors or anomalies. Eventually, however, so many inconsistent observations accumulate that the old paradigm becomes progressively less credible and a crisis develops in the discipline.

The second situation arises when a new technology is developed that provides information of a kind not available before. This happened, for example, when techniques for the radiometric dating of rocks were developed, and when new geophysical instruments and computers permitted investigation of the ocean floor in great detail. The first applications of a new technology often resemble an immature phase of scientific development. Data are collected according to the Baconian method, and practitioners of the technique measure just about everything they can get their hands on. This process continues until the data thus amassed are synthesized and compared with the currently accepted paradigm. If the new data do not fit well with the paradigm, the usefulness of the paradigm is impaired and again a crisis develops.

When a crisis arises in a scientific field, the inadequacy of the old paradigm becomes clear, but no new idea is present to explain the contradictions. Eventually a new synthesis occurs, often initiated by an intuitive leap of understanding or a flash of insight. Scientists erect a new paradigm radically different from the old one that explains the contradictions as well as many additional observations. This process leads to a whole host of new corollaries, new ways of looking at natural phenomena, and the posing of new questions.

At first, many scientists resist the new model, but as its success in explaining the observations increases, the resistance crumbles. Such an event constitutes a **scientific revolution**. It is followed by a "mopping-up" phase of normal scientific progress during which old facts are fit into the new paradigm, new observations are made to test its corollaries, and new hypotheses are erected and tested within the constraints of the new paradigm.

I.3 The Development of Plate Tectonics: A Scientific Revolution

The development of the theory of plate tectonics provides an excellent example of a scientific revolution as Kuhn meant the term. Although many books have been written about the subject (see Menard, 1985; Uyeda, 1978; Takeuchi et al., 1967; Hallam, 1973; Marvin, 1973; Glen, 1982), it is nevertheless worth discussing briefly here, for it illustrates the stages of the development of a science that Kuhn has outlined, and it also serves to introduce our discussion of tectonic history.

The accompanying table shows the development through time of many of the major ideas that led to the plate tectonic revolution. Time progresses from oldest events at the bottom to youngest events at the top, with the dates shown in the leftmost column. The lower part of the table contains five columns, each showing the major discoveries within a particular field of study. The fields are labeled across the bottom of the table. Gray areas across the columns indicate syntheses involving two or more fields. The last column, labeled Development Status, indicates the stages of development of the science of geology before, during, and after the revolution, which occurred between 1960 and 1970.

CHRONOLOGY OF DEVELOPMENT OF PLATE TECTONIC MODEL
Possible New Tectonic Subdivisions
(EACH RECOMBINING PARTS OF OLD ONES BELOW)

Date	Geometry of plate tectonics (Ch. 4)	Tectonics of plate margins (Chs. 5–8)	Collisions (Ch. 9)	Plate tectonics and geologic history (Chs. 10–12)	Plate tectonics on other planets (Ch. 13)	DEVELOPMENT STATUS	
1995 1990 1985 1980 1975						New paradigm	MATURE (HISTORIC)
1970	Asilomar conference: Plate tectonics and orogeny					Geologic revolution	
	Seismology and new global tectonics/Tectonics on a sphere/Spreading confirmed worldwide					Seismological revolution Marine geological revolution	REVOLUTION
1965	Melanges	Computer Atlantic Fit / Vine Matthews Morley Hypothesis			Seismic map Lithosphere Transform faults		
		Hess's history of ocean basins		Mag. Reversal time scale	Focal mechanisms	Synthesis	
1960		Fracture zones	Continental drift and an expanding earth	Pacific magnetic stripes	WWSSN		
1955	Deep water origin of turbidites	World rift system		Polar wander paths	Low velocity zones	Crisis	
		Thin oceanic crust			World—dipping seismic zones		
		Thin sediments					
1950		Oceanic exploration					
		W. Pacific topography					
		Guyots					
1945	North American geosynclines				Mountain roots		IMMATURE (PREHISTORIC)
1940		WW II - radar sonar				Early synthesis	
1935	Hess's early synthesis attempt						
		Tectogene					
1930	More Steinmann's trinity	Negative gravity over trenches	AAPG rejection of Wegener	Pole reversals	Dipping seismic zone—Japan	Early problems	
1925							
1920							
1915	Verschluchung (sub-duction) in Alps		Wegener's continental drift hypothesis		Core-mantle recognition		
1910				Magnetically reversed lavas	Mohorovicic discontinuity		
1905	Steinmann's trinity—deep sea sediments					Old view	
1900	Geosynclinal mode						
	OROGENIC BELTS & GEOSYNCLINES 1	MARINE GEOLOGY & GEO-PHYSICS 2	CONTINENTAL DRIFT 3	MAGNETISM & PALEO-MAGNETISM 4	SEISMOLOGY 5	DEVELOPMENT STATUS 6	

OLD GEOLOGIC / GEOPHYSICAL SUBDIVISIONS

The Old Paradigm

It is convenient to begin our account around 1900. At the turn of the century, geology was exclusively the study of rocks on land. It was a mature discipline unified by the laws of evolution, uniformitarianism, and stratigraphy. In structure and tectonics, the principal efforts were to understand the characteristics and causes of deformation observed in the world's mountain systems, principally the Appalachian and Cordilleran systems of North America, the Alpine belt through southern Europe, and the Caledonide belt of the British Isles and Scandinavia.

Despite the deformation evident in these orogenic belts, the continents were believed to be essentially stationary on the Earth's surface, the so-called fixist hypothesis. Accepted motions of the crust were predominantly vertical, as evidenced by marine rocks in high mountains and covering large areas of continents. The driving force for such motions was thought to be isostasy. This theory—originally proposed in 1865 by George Airy, the British Astronomer Royal, to explain gravity measurements in India—had received considerable subsequent support from the studies of Clarence Dutton in the western United States.

Orogenic belts (column 1) were explained according to the geosynclinal hypothesis, which was formulated by the American James Hall in 1857, and finally was accepted by skeptics 35 years later. The geosynclinal hypothesis supplanted the previously held contraction hypothesis, which proposed that the mountains of the Earth were in effect the result of the wrinkling of the Earth's crust as the radius of the Earth decreased. The geosynclinal hypothesis ascribed mountain building or orogeny to the development of a subsiding linear trough containing thick, shallow-water sedimentary rocks along both sides of the trough (miogeosynclines) and deeper water sedimentary and volcanic rocks in the center (eugeosynclines). Following the deposition of these thick sequences, deformation occurred, with the central rocks of the eugeosyncline becoming the most highly deformed and metamorphosed, and the flanking rocks of the miogeosyncline being more mildly deformed and metamorphosed. Deformation was believed to be symmetrical, with thrusting of all the rocks in both directions away from the center over the flanking undeformed continental platforms.

The development of such thick deposits of shallow-water sediments clearly required some form of isostatic compensation. Continents, however, were thought to be fundamentally fixed, and although very little was known about the oceanic crust, it was considered to be as ancient as the continents when it was

thought about at all. A major weakness of this model was that there was no explanation for what caused geosynclines to develop and subsequently to deform, a situation that was later described as a theory of mountain building with the mountain building left out!

At this time, marine geology (column 2) and geophysics (columns 4 and 5) were immature disciplines. The broad outlines of the topography of the oceanic floor were known as a result of wire-line soundings during oceanographic expeditions of the late nineteenth century, but many features of the oceanic crust were simply unknown. Studies in geophysics (magnetism, paleomagnetism, and seismology) were in their infancy. The French geophysicist B. Brunhes discovered reversely magnetized basalts in 1909. The discovery was confirmed in 1928 by the Japanese geophysicist M. Matuyama, who found that older lavas in Japan were reversely magnetized, and he proposed that the Earth's magnetic field had been reversed in polarity during the Quaternary. In 1909, the Yugoslav seismologist Mohorovičić discovered the sharp increase in P-wave velocity at the base of the crust (the M discontinuity or Moho named for him), and in 1914 the German seismologist B. Gutenberg first recognized the core-mantle structure of the Earth.

There was no unifying model of the Earth in existence at this time. Initially, marine geology and geophysics developed with very little influence from, or effect upon, the problems in continental geology.

Early Problems

Soon after the turn of the century, problems with Hall's fixist geosyncline paradigm began to crop up. In 1905, the German geologist Gustav Steinmann recognized the ubiquitous association of serpentinites, pillow lavas, and radiolarian cherts in the Alpine belt. This association became known as Steinmann's trinity in the literature published in the English language, but Europeans recognized it as the ophiolite suite. Steinmann proposed that the sediments were laid down in deep water over rapidly subsiding continental platforms. The serpentinites and pillow lavas were thought to be magmatic intrusions and extrusions, respectively, in these deep-water regions. This recognition led to the European view that geosynclines could form in deep water between continents, rather than in shallow water on continents, as Hall proposed. Steinmann's observations received little attention in the English-speaking world until three decades later, when they played a role in an early attempt by the American geologist H. H. Hess to synthesize land and marine geology, as described below.

Another problem with the geosyncline model was that it could not explain the very large amounts of shortening exhibited by the nappes in the Alps (hundreds of kilometers by some calculations). This led the Austrian geologists Ampferer and Hammer in 1911 to formulate the hypothesis of **Verschluchung,** the downsucking or **subduction,** of the European platform beneath the Alpine geosyncline. This idea was one of the first signs that the fixist hypothesis was not adequate, although it had little influence outside the German-speaking geologic community.

A direct challenge to the fixist model came in 1915 with Wegener's hypothesis of continental drift (column 3). He maintained that all the continents were originally joined in a single continent (Pangea) and had drifted apart to give the present distribution. Wegener argued that continental drift not only explained the fit of coastlines around the Atlantic but also was superior to the fixist model in explaining the distribution of ancient coal and glacial deposits, as well as faunal and floral provinces around the Atlantic Ocean.

Wegener's hypothesis created a great deal of interest worldwide, but it was received with skepticism by many American geologists. The climax of the controversy came in 1926 in at a meeting of the American Association of Petroleum Geologists in New York, at which continental drift was the focus of debate and Wegener himself was present. Although a number of speakers either favored drift or advocated an open mind, the consensus was to reject the hypothesis on several grounds, some of which seem quite spurious in retrospect. Wegener was a meteorologist and therefore not a part of the geologic establishment (a scientifically poor reason, but sociologically not an uncommon reaction to outsiders); he could come up with no convincing driving mechanism to explain how continents moved through the oceanic crust; his hypothesis was considered nonuniformitarian; and the geosynclinal theory was considered satisfactory (despite the fact that no convincing driving mechanism had been proposed for this process either). Furthermore, the British geophysicist H. Jeffreys had supposedly proved that the transmission of elastic seismic waves through the Earth meant that the slow inelastic motion required by drift was impossible.

After the New York meeting, continental drift as a viable model was essentially dead in North America and much of Europe. It was still popular in the southern hemisphere, however, espoused most vigorously by the South African geologist A. DuToit in his book *Our Wandering Continents,* published in 1937. Continental drift would not be revived as a major force until the late 1950s and early 1960s, when it received support from paleomagnetism (column 4).

In the early 1930s, the Dutch geophysicist F. A. Vening-Meinesz began a study of gravity at sea[5] that became one of the first marine geophysical studies to have repercussions for models of continental tectonics. Vening-Meinesz and his team made the surprising discovery that deep-sea trenches in the Caribbean and in Indonesia were associated with negative gravity anomalies. It seemed that isostasy was violated in these areas and that something was holding the crust down. These observations gave rise to the concept of a downbuckling of the crust into the mantle, subsequently called a **tectogene,** that developed in response to horizontal compression. The British geologist Arthur Holmes proposed that a tectogene formed in response to a downgoing convection current in the mantle. The American geophysicist D. L. Griggs modeled this mechanism experimentally in 1939. Vening-Meinesz and Holmes were among the first to suggest that forces other than isostasy were important in tectonics, and the idea that convection occurred in the mantle was not at all consistent with a fixist notion of the Earth.

At about the same time, in 1928, the Japanese seismologist K. Wadati recognized that earthquake sources beneath Japan are located along an inclined planar zone that extends from the trench east of Japan and dips westward under the islands, providing yet another problem that had no simple explanation in the fixist model.

Early Synthesis

In an early attempt at synthesis, Hess proposed at the 1937 Moscow International Geological Congress the hypothesis (later published in 1939) that the tectogene was related to geosynclines and to orogeny. In a study of Appalachian peridotites, Hess thought he could see two belts of peridotite approximately 120 km apart. He proposed that these peridotites were magmas intruded at the margins of a tectogene (geosyncline) at the onset of the orogenic phase of geosynclinal development but not during later stages in the orogeny.

---

[5]The measurements had to be done below wave base because the accelerations from normal ocean swells far exceed the size of most gravity anomalies, and for his work in the Caribbean, Vening-Meinesz borrowed a surplus submarine from the U.S. Navy. On several cruises, Vening-Meinesz took with him the young American geologist Harry Hess. Hess purportedly received a reserve commission in the U.S. Navy to satisfy Navy regulations that only an officer could give orders to crew members, thus enabling the scientific team to give directional instructions directly to the helmsmen.

Hess's ideas received little attention largely because at that time he had not yet built up a reputation and was in a junior position in the geologic community (again a scientifically poor excuse, but sociologically a common phenomenon[6]), but also possibly because the geologic community was not yet prepared to accept radical ideas. He published an update of the same article in 1955, after his reputation was firmly established, and it was widely influential.

Marine Geology and Geophysics and the Developing Crisis

Although World War II interrupted much of the normal research of the scientific community, two inventions early in the war years proved to be crucial to subsequent oceanographic investigation: sonar, which allowed the determination of ocean depths by timing the echo of sound waves off the bottom of the ocean; and radar, which eventually enabled ships to determine their position at sea more accurately. During the war, millions of miles of bathymetric profiles of unprecedented accuracy were collected, particularly in the Pacific by U.S. Navy ships.[7] Hess compiled and synthesized these results in two landmark papers that appeared in 1946 and 1948 and in which many of the detailed topographic characteristics of the western Pacific Ocean basin were evident for the first time, including such features as the true dimensions of the trenches off Japan, the Marianas, and the Philippines.

After World War II, marine geologic work began in earnest in the United States and Great Britain. U.S. efforts centered around the oceanographic institutions of Scripps in La Jolla, California; Woods Hole in Woods Hole, Massachusetts; and the Lamont (later Lamont-Doherty) Geological Observatory of Columbia University in New York. British efforts centered around the team of E. O. Bullard and M. N. Hill at Cambridge. Although most work concentrated in the North Atlantic and North Pacific, it was worldwide in coverage. This new work coincided with the development of a whole array of sophisticated new equipment, including computers, that enabled the rapid analysis of data to provide information such as gravity and magnetic anomalies and earthquake locations and focal mechanisms.

With burgeoning knowledge about the ocean basins, more and more data seemed to have no relationship to the fixist and geosyncline paradigm. This situation is not surprising, for the paradigm was based solely on continental geology, and the new information was coming from the heretofore unknown ocean basins. The explosion in knowledge as a result of the advent of marine geophysics is an excellent example of the Baconian method of inquiry that commonly follows the introduction of a new technology. That the new knowledge caused problems for the reigning paradigm is also a classic example of a crisis brought on by the application of a new technology.

The Scripps seismologist R. W. Raitt soon found that the oceanic crust was uniformly thin, approximately 5 km everywhere. In addition, he and his colleagues found that the sediment cover was much thinner than expected, if the oceans were as old and permanent as the fixist model implied. Workers at Lamont-Doherty, led by Maurice Ewing, soon confirmed the seismic measurements in the Atlantic. In the late 1950s, they also reported the existence of a rift valley along the axis of a world-girdling mountain range, or mid-ocean ridge system, that was located in the middle of the world's oceans and stood 2 to 2.5 km above the ocean floor. The rift valley, they argued, was a continuation of the well-known East African Rift on land, and it indicated that the ocean basins were undergoing extension.

In the late 1950s and early 1960s, the pace of new discoveries and problems accelerated. Scientists at Scripps, such as H. W. Menard, discovered fracture zones thousands of kilometers long in the crust of the Pacific Ocean, and similar zones were found in the Atlantic Ocean. A. D. Raff and R. G. Mason, and V. Vacquier, published results of magnetic surveys off the west coast of the United States that revealed a series of puzzling striplike magnetic anomalies that trended north-south and were offset tens to hundreds of kilometers along the newly discovered fracture zones. Structural models to explain these anomalies involved complex folding and faulting, but no evidence for such complex oceanic crustal structure was forthcoming from marine seismic results. Scripps workers also discovered that the ridge axes exhibited heat flow two or more times higher than the average for the rest of the oceanic crust and that trenches showed low heat flow.

In the 1950s, paleomagnetism ended its isolated development and contributed to the growing crisis when P. M. S. Blackett and his colleagues E. Irving and S. K. Runcorn in England recognized that Europe

---

[6] Hess observed many years later that it was a tactical error for a young upstart scientist to attempt a major synthesis without first establishing his or her reputation in a more conventional field.

[7] During the war, Hess served in the Navy, eventually rising to the command of an attack transport, the U.S.S. *Cape Johnston*. One of the many stories that circulated about Hess was that he ran the depth sounder continuously, even in deep water, without regard for possible nearby hostile submarines.

and North America had different polar wander paths. The simplest explanation was that the continents had moved relative to one another, a major blow to the fixist paradigm.

Meanwhile, seismology was progressing relatively independently. But its discoveries about the structure of the Earth's interior nevertheless laid important groundwork for the field's subsequent major contributions to the crisis. In 1943, Gutenberg discovered the existence of mountain roots, confirming the theory of isostasy; in the late 1940s and early 1950s, the American seismologist Hugo Benioff recognized the presence of dipping seismic zones around the Pacific extending deep into the mantle, confirming Wadati's earlier discovery under Japan; and in the late 1950s, Gutenberg discovered a zone of low seismic velocity at approximately 100-km depth in the mantle.

Thus, the late 1950s was a time of growing crisis in Earth science. Remarking on the new and unexplained data coming in from the ocean basin research, Hess wrote in 1955 that "some vital ingredient is lacking."

The Australian tectonicist S. W. Carey tried to provide this missing ingredient in 1958 with an impressive worldwide synthesis. He accepted drift away from the mid-ocean ridges and proposed many basic tectonic relationships that have subsequently been confirmed. Carey carefully constructed maps of the Earth on a spherical table and tried to show that the continents fit together much as Wegener had proposed. He argued forcefully for Earth expansion (an increase in the Earth's diameter) as the driving mechanism, an unpopular concept that hampered widespread acceptance of his views.

Synthesis and Revolution

The revolution began in 1960 with the circulation by Hess of a manuscript entitled "Evolution of Ocean Basins" (published in 1962 as *History of Ocean Basins*). In it, he argued that the ocean crust was young and was created over a rising limb of a mantle convection cell. He accepted the paleomagnetic evidence for polar wander, thereby accepting Wegener's hypothesis (shown in the table as the end of the boundary between column 2, marine geology and geophysics, and column 3, continental drift). The preprint caused considerable excitement and spawned a series of companion papers by other workers, in one of which R. S. Dietz coined the term *seafloor spreading*.

Meanwhile, work by A. Cox, R. Doell, and G. B. Dalrymple on volcanic rocks in the western United States produced the startling result that Earth's magnetic field had reversed not once but several times in

the past several million years, and they published a magnetic reversal time scale. Thus Brunhes's and Matuyama's findings were confirmed.

In 1963, the next important synthesis came independently from L. W. Morley in Canada and from F. J. Vine and D. H. Matthews in Cambridge, England. They proposed that seafloor spreading, by acting essentially like a tape recorder, preserved a continuous record of the polarity reversals in Earth's magnetic field and thereby accounted for the puzzling strips of alternately normally and reversely magnetized oceanic crust that are symmetrically arrayed about the mid-ocean ridges.[8]

The years 1965 and 1966 were climactic for marine geology and geophysics. The Vine-Matthews-Morley hypothesis was confirmed by data from all the oceans, much of which had been amassed by the team at Lamont-Doherty under the direction of Maurice Ewing. It became clear that the magnetic anomalies off the west coast of the United States were produced by spreading ridges located near the North American continental margin.

At this time the first results also became available from the computerized Worldwide Standardized Seismic Network (WWSSN), which had begun operation in 1960. These records showed that earthquakes along the fracture zones in the Atlantic Ocean were confined to the segment between the offset ridge crests. Thus, fracture zones were divided into active and inactive portions. Furthermore, first-motion studies of the earthquakes showed that the shear sense of the fractures is opposite to the sense of offset of the ridges. In 1965, the active parts between ridge crests were christened **transform faults** by the Canadian geophysicist J. T. Wilson. Also in 1965, the Cambridge group published a computer-generated fit of the continental margins bordering the Atlantic Ocean showing that the fit was very good indeed.[9]

---

[8] Morley's article was rejected by two international journals; the reviewer for one of them stated that the idea was suitable for cocktail party conversation but hardly publishable as a scientific paper. One of the journals that rejected Morley's paper subsequently published the Vine and Matthews article. Morley's version did not appear until a year later, in 1964, in a less prominent Canadian publication. Parts of his original communication appear in *Glen*, 1982, p. 299.

[9] One story contends that Bullard organized his computer analysis of the Atlantic fit in response to H. Jeffreys's comment, after a lecture by S. W. Carey, objecting that Carey had not proved mathematically that a fit existed. Jeffreys argued that there was a 15° misfit, but it turned out that he was talking about a fit of the coastlines, rather than the edges of the continents (approximated by the 500-fathom contour).

As 1966 drew to a close, the Geological Society of America held its annual convention in San Francisco. Cox, Doell, Dalrymple, and Vine gave keynote addresses. The meeting marked the general acceptance of seafloor spreading and continental drift by most American marine geologists and geophysicists, although not yet by considerable numbers of Earth scientists whose research was land-based. The old paradigm, however, was crumbling.

Meanwhile, events were moving rapidly in seismology. The American seismologists B. Isacks, L. Sykes, and J. Oliver of Columbia University were studying the transmission of earthquake waves in the Tonga-Kermadec region of the South Pacific. They discovered a 100-km-thick slab of mantle having an abnormally high seismic velocity that started in front of the arc near the trench and extended down into the mantle under the arc along the zone of earthquakes. Transmission of seismic energy was particularly efficient within the slab, and this property was correlated with the high strength and density, and therefore the low temperature, of the material. Subsequently, they found similar dipping slabs beneath all western Pacific island arcs. All the first motions of earthquakes in the dipping seismic zones were consistent with these slabs moving downward beneath the island arcs. This was the first recognition of the role of subduction as complementary to the spreading at ocean ridges, and it provided a plausible explanation for the concentration of earthquakes beneath island arcs reported earlier by Wadati and Benioff.

The map of the world's seismicity from 1960 to 1968 published by the American seismologists M. Barazangi and L. Dorman in 1969 showed a pronounced concentration of seismicity in shallow zones along the mid-ocean ridges and in planar zones dipping beneath island arcs to depths of up to 700 km. This map made it clear how tectonically active the Earth is and how the preponderance of the deformation is concentrated along very narrow belts. The map stunningly confirmed the idea that tectonics on Earth is characterized by the rigid-body motion of large plates on a sphere and that deformation and seismic activity are concentrated along the boundaries of the plates where they interact with one another.

The relative motion of nondeforming plates on a sphere became the focus of studies by D. McKenzie and R. L. Parker and by J. Morgan. They showed that the relative motion between two segments of the Earth's crust could be accounted for by rotation about a pole common to the two segments. Transform faults should be small circles about the pole of rotation, and the rate of motion should be proportional to the distance from the pole of rotation. Morgan's article was the more general of the two. He noted that the great fracture zones of the Pacific were not straight but curved, and he hypothesized that they could result from motion about a pole of rotation. He showed that the rate of new crust generation in the Atlantic was consistent with constant angular velocity about a pole of rotation, which implies a linear velocity that increases from zero at the pole to a maximum at 90° to the pole, and that the pole for spreading on the mid-Atlantic ridge was near Iceland.

Morgan's article was published in March 1968, together with a number of articles by the Lamont-Doherty group, notably J. R. Hiertzler, W. Pitman, and others; these articles extended marine magnetic anomalies to all oceans and extended the magnetic reversal time scale to the late Cretaceous. Later in 1968, two landmark articles appeared, one by B. Isacks, J. E. Oliver, and L. R. Sykes, entitled "Seismology and the New Global Tectonics," and the other by X. Le Pichon, entitled "Sea Floor Spreading and Continental Drift." Isacks et al. summarized the evidence for the existence of segments of the outer portion of the Earth in relative motion with respect to one another, based upon the location of earthquakes and the transmission of seismic waves. This article brought seismology (column 5) fully into the plate-tectonic revolution. Le Pichon used the magnetic anomaly stripes on the ocean floor and the magnetic reversal history to give a more quantitative estimate of the relative positions of the continents at past times.

Thus, by the end of 1968, practically the entire North American and British geophysical and marine geological communities were converted to the new view of the Earth. Although the application of this new view to land geology was not yet apparent, geologists rapidly began to see the implications of seafloor spreading and plate tectonics for continental geology. The climax of this new insight perhaps came at the Geological Society of America Penrose Conference at Asilomar, California, organized by William R. Dickinson in December 1969. At this meeting, geologists and geophysicists from around the world convened to explore the application of plate tectonics to continental geology and orogeny. During the discussions, it became evident that plate tectonic processes could account not only for recent tectonic activity but also for ancient continental geology. Thus, the historical record preserved in the continents was incorporated into the synthesis. For example, all ophiolites were recognized as fragments of oceanic crust and mantle, some formed at spreading centers; geosynclines were reinterpreted in terms of modern plate tectonic settings, with the miogeosynclines equivalent to the sediment accumulations along passive continental margins and the eugeosynclines largely equivalent to the sediments and volcanics of the deep ocean basins and island arcs; and the convergence of plates and the collision of continental crustal fragments accounted for orogenic structures, the tectonic juxtaposition of shallow-water and deep-water deposits, and the production of igneous and metamorphic rocks in volcanic arcs and orogenic zones.

It soon became clear that continental drift was not a single episode in Earth's history but an inherent and continuous process and that Wegener's Pangea had been assembled from an earlier group of dispersed continental fragments. Thus, Wegener's hypothesis was only the last act in a long play, and continental drift was shown to be a uniformitarian process, contrary to the assertions of Wegener's critics in 1926. The barriers separating the various fields of geological investigation finally broke down, and plate tectonic theory became the new paradigm for all of geology.

Geology as a Mature Science

As indicated on the right side of the table, geology as a whole has evolved from a fragmented to a unified mature science as a result of the plate tectonic revolution. Research in tectonics is now largely in a phase of "mopping-up." The new paradigm is generally accepted, and progress in our understanding of the Earth's tectonics is achieved through the process of normal science.

The acceptance of the new paradigm has opened new questions for research in tectonics, and research efforts have crystallized around topics that are markedly different from those that preceded the revolution. Major divisions involve the geometry and kinematics of plate motion, the history of plate motions, the tectonics and geology of plate margins, the tectonics and geology of plate interiors, the processes of collision, the plate tectonic interpretation of orogenic belts, and the mantle-core processes affecting the driving mechanism of plate tectonics. Several of these divisions are the subjects of succeeding chapters, as indicated at the top of the table.

The development of the plate tectonic revolution shows many features of a classic scientific revolution. For land geology, the revolution consisted of a progress from old ideas through increasing problems to crisis and then new synthesis. For marine geology and geophysics, however, the revolution was a combination of exploration in previously unknown regions and the introduction of new technology. The field progressed from an immature to a mature phase in its history, with an associated change in the method of investigation from Baconian to Darwinian. Because the marine world makes up most of Earth's surface and because so much of the tectonic action is there, for geology as a whole the revolution must be regarded as proceeding from an immature to a mature stage of development.

The geologic community is still exploring all the ramifications of plate tectonics and incorporating new ideas as problems arise. Will there be another revolution in Earth science? Of course we cannot answer that question. Two of the questions at the frontiers of research, however, are whether the plate tectonic process exists on other planets and whether the process itself evolved and changed through time. We discuss these questions briefly in the final chapters.

Additional Readings

Historical Accounts

Cloos, H. 1953. *Conversation with the Earth*. New York: Knopf.

Cox, A., ed. 1972. *Plate Tectonics and Geomagnetic Reversals*. San Francisco: W. H. Freeman and Company.

Glen, W. 1982. *The Road to Jaramillo*. Stanford: Stanford University Press.

Hallam, A. 1973. *A Revolution in the Earth Sciences*. Oxford: Clarendon Press.

Hsü, K. J. 1986. *The Great Dying*. New York: Harcourt, Brace, Jovanovic.

Marvin, U. B. 1973. *Continental Drift: The Evolution of a Concept*. Washington, D.C.: Smithsonian Institution Press.

Menard, H. W. 1986. *The Ocean of Truth: A Personal History of Global Tectonics*. Princeton, N.J.: Princeton University Press.

Schwarzbach, M. 1986. *Alfred Wegener: The Father of Continental Drift*. Madison, Wis.: Science Tech Inc.

Takeuchi, H., S. Uyeda, and H. Kanamori. 1967. *Debate about the Earth*. San Francisco: Freeman, Cooper, & Co.

Vine. F. J. 1977. The continental drift debate. *Nature* 266:19–22.

Pre-Revolution Articles

Continental Drift

Carey, S. W. 1958. A tectonic approach to continental drift. S. W. Carey, ed. *Continental Drift*. A Symposium of the Geological Department, University of Tasmania.

Van Waterschoot Van der Gracht, W. A. J. M., ed. 1928. *Theory of Continental Drift: A Symposium*. Tulsa: American Association of Petroleum Geologists.

Wegener, A. 1966. *The Origin of Continents and Oceans*. Translated from the 4th ed. by John Biram. New York: Dover.

Orogeny and Geosynclines

Ampferer, O., and W. Hammer. 1911. Geologischer Querschnitt durch de Ostalpen vom Allgäu zum Gardasee. *Jahrbuch der Geologische Reichsanstalt* 61 (no. 3–4):531–710.

Dott, R. J., Jr. 1974. The geosynclinal concept. In *Modern and Ancient Geosynclinal Sedimentation*. R. H. Dott, Jr., and Robert H. Shaver, eds. Society of Economic Paleontologists and Mineralogists. Special Publication No. 19.

Hess, H. H. 1939. Island arcs, gravity anomalies, and serpentine intrusions: A contribution to the ophiolite problem. *International Geological Congress, Moscow, 1937* 2:262–283.

Kay, M. 1951. *North American Geosynclines*. GSA Memoir 48.

Steinmann, G. 1905. Die Geologische Bedeutung der Tiefseeabsætze und der ophiolitischen Massengesteine. *Naturforschung Gesellschaft Freiburg. Bericht.* 16:44–65.

Steinmann, G. 1927. Die Ophiolitzonen der Mediterranean Kettengebirge. Int. Geological Congress, Madrid.

Marine Geology and Geophysics

Hamilton, E. L. 1959. Thickness and consolidation of deep-sea sediments. GSA Bull. Geological Society (London) Memoir 70:1399–1424.

Heezen, B. C. 1960. The rift in the ocean floor. *Scientific American* 203:98–110.

Hess, H. H. 1946. Drowned ancient islands of the Pacific Basin. *American Journal of Science*. 244:779–791.

Hess, H. H. 1948. Major structural features of the western North Pacific. *GSA Bull.* 59:417–446.

Menard, H. W. 1964. *Marine Geology of the Pacific*. New York: McGraw-Hill.

Raitt, R. W. 1956. Seismic-refraction studies of the Pacific Ocean Basin. *GSA Bull.* 67:1623–1640.

Vening Meinesz, F. A. 1934. *Gravity Expeditions at Sea, 1923–1932*. Vol. 2. Interpretation of the results. Publications, Netherlands, Geodetic Commission. Delft: Waltman.

Seismology

Benioff, H. 1954. Orogenesis and deep crustal structure: Additional evidence from seismology. *GSA Bull.* 64: 385–400.

Gutenberg, B. 1959. *Physics of the Earth's Interior*. New York: Academic Press.

Magnetism and Paleomagnetism

Cox, A., R. R. Doell, and G. B. Dalrymple. 1964. Reversals in the earth's magnetic field. *Science* 144:1537–1543.

Raff, A. D., and R. G. Mason. 1961. Magnetic survey off the west coast of North America. *GSA Bull.* 72: 1267–1270.

Runcorn, S. K. 1956. Paleomagnetic comparisons between Europe and North America. *Proc. Geol. Assoc. Can.* 8: 77–85.

Articles of the Revolution: 1960–1968

Barazangi, M., and J. Dorman. 1968. World seismicity map of ESSA Coast and Geodetic Survey epicenter data for 1961–1967. Bull. Seismol. Soc. Amer. 59: 369–380.

Bullard, E. C., J. E. Everett, and A. G. Smith. 1965. The fit of the continents around the Atlantic. *Phil. Trans. Roc. Soc.* 258: 41–51.

Heirtzler, J. R., G. O. Dickson, E. M. Herron, W. C. Pitman III, and X. Le Pichon. 1968. Marine magnetic anomalies, geomagnetic field reversals, and motions of the ocean floor and continents. *J. Geophys. Res.* 73: 2119–2136.

Hess. H. H. 1962. History of ocean basins. In *Petrologic Studies: A Volume to Honor A. F. Buddington*. A. E. J. Engel, H. L. James, and B. F. Leonard, eds. New York: Geological Society of America.

Isacks, B. L., J. E. Oliver, and L. R. Sykes. 1968. Seismology and the new global tectonics. *J. Geophys. Res.* 78: 5855–5899.

Le Pichon, X. 1968. Sea-floor spreading and continental drift. *J. Geophys. Res.* 73:3661–3705.

McKenzie, D. P., and R. L. Parker. 1967. The North Pacific: An example of tectonics on a sphere. *Nature* 216: 1267–1280.

Morgan, W. J. 1968. Rises, trenches, great faults, and crustal blocks. *J. Geophys. Res.* 73:1959–1982.

Morley, L. W., and A. Larochelle. 1964. Paleomagnetism as a means of dating geologic events. *Geochronology in Canada* (Roy. Soc. Canada Spec. Pub. 8). F. F. Osborne, ed. Toronto: University of Toronto Press.

Oliver, J. E., and B. Isacks. 1967. Deep earthquake zones, anomalous structures in the upper mantle, and the lithosphere. *J. Geophys. Res.* 72:4259.

Sykes, L. R. 1966. Seismicity and deep structure of island arcs. *J. Geophys. Res.* 71:2981.

Sykes, L. R. 1967. Mechanism of earthquakes and the nature of faulting on the mid-oceanic ridges. *J. Geophys. Res.* 72:2131.

Vine, F. J. 1966. Spreading of the ocean floor: New evidence. *Science* 154:1405–1415.

Vine, F. J., and D. H. Matthews. 1963. Magnetic anomalies over oceanic ridges. *Nature* 199:947–949.

Wilson, J. T. 1965. A new class of faults and their bearing on continental drift. *Nature* 207:343–347.

Moores, E. M., and F. J. Vine. 1971. The Troodos massif, Cyprus, and other ophiolites as oceanic crust: Evolution and implications. *Trans. Roy. Soc. London* 278A: 443–466.

Articles Published or Submitted Just Before Asilomar Conference

Atwater, T. 1970. Implications of plate tectonics for the Cenozoic tectonic evolution of western North America. *GSA Bull.* 81:3513–3536.

Dickinson, W. R. 1970. Relation of andesites, granites and derivative sandstones to arc-trench tectonics. *Rev. Geophys.* 8:813–860.

Dickinson, W. R., and T. Hatherton. 1967. Andesitic volcanism and seismicity around the Pacific. *Science* 1557: 801–803.

Hamilton, W. B. 1969. Mesozoic California and the underflow of the Pacific mantle. *GSA Bull.* 80:2409–2430.

Karig, D. E. 1970. Ridges and trenches of the Tonga-Kermadec island arc system. *J. Geophys. Res.* 75:239–254.

Articles Published Immediately After Asilomar Conference

Dewey, J. F., and J. M. Bird. 1970. Mountain belts and the new global tectonics. *J. Geophys. Res.* 75:2625–2647.

Dewey, J. F., and J. M. Bird. 1971. Origin and emplacement of the ophiolite suite: Appalachian ophiolites in Newfoundland. *J. Geophys. Res.* 76:3179–3206.

Dickinson, W. R. 1971. Plate tectonic models of geosynclines. *Earth and Planet. Sci. Lett.* 10:165–174.

Hsü, K. J. 1971. Franciscan melanges as a model for eugeosynclinal sedimentation and underthrusting tectonics. *J. Geophys. Res.* 76:1162–1170.

Moores, E. M. 1970. Ultramafics and orogeny, with models of the U.S. Cordillera and the Tethys. *Nature* 228: 837–842.

Oxburgh, E. R., and D. L. Turcotte. 1971. Origin of paired metamorphic belts and crustal dilation in island arc regions. *J. Geophys. Res.* 76:1315–1327.

White, D. A., D. H. Roeder, T. H. Nelson, and J. C. Crowell. 1970. Subduction. *GSA Bull.* 81:3431–3432.

PART

III   Tectonic History

IN the preceding chapters we covered the principles of modern plate movements and examined how these movements have affected the geologic record. With this background, we can consider the historical record of plate movements over geologic time. Tectonic history is the study of the development of a feature, such as a basin or an orogenic belt, through various stages to its present form. The challenge is to interpret the geologic record in terms of the ongoing plate processes, starting from the current plate and seismic situation.

One perspective on tectonic history is shown in the accompanying illustration. The diagram includes two types of phenomena: oscillatory and secular. Oscillatory phenomena are events that occur in cycles, such as earthquakes and tides. Secular phenomena are gradual permanent changes in response to tectonic activity. Both types of phenomena can be divided into several distinctive sectors.

The seismological sector is especially important. By examining the worldwide pattern of earthquakes over several decades geologists have discovered that most earthquakes occur at plate boundaries. Thus, seismology has provided one of the essential keys to the development of plate tectonic theory.

In our discussion of the geological consequences of plate tectonics (Chapters 5–9), we rely on features formed within roughly the past 200 million years, a period for which some oceanic record is still available.

The largest part of the Earth's history unfolded in pre-ocean basin time, that is, before the oldest rocks in the present ocean basins were even formed. The record of this older tectonic activity is recorded in continental rocks, and much of the tectonic history that we can find is recorded in the Earth's orogenic belts. Thus, in Chapter 10 we set the stage by describing the features that are typical of orogenic belts. From this type of information, we must learn to infer most of the Earth's tectonic history.

Typical geologic analyses rely on techniques that have a precision of around a million years, so in effect our preceding description of plate tectonics focuses on features that date from 1 Ma or earlier. The study of geologic features in the time interval between the past few decades and more than 1 Ma is the realm of neotectonics. We begin our investigation of orogenic activity, therefore, in Chapter 11, by discussing the principles of the emerging field of neotectonics and what it tells us about present-day orogenic processes.

In Chapter 12, we examine the characteristics of several orogenic belts and consider the plate tectonic events that led to their development. For orogenic belts younger than about 200 Ma, we can infer the plate movements associated with their origins partly from the magnetic anomalies of the ocean basins. For older orogenic belts, we must rely solely on the paleomagnetic evidence for continental drift and the indirect *petrotectonic*[1] indicators of plate tectonic activity preserved in the mountain belts themselves.

Chapter 13 focuses on the tectonics of the terrestrial (Earth-like) planets and satellites of our solar system. The field of comparative planetary tectonics, although very new, provides valuable insights into the principles of tectonics on Earth.

When discussing tectonics, it is useful to divide Earth's history into intervals defined by the record of tectonic activity (see the illustration below):

Neotectonic time (0–5 Ma): The time between current tectonic activity and the conventional geological record. Alternative methods of study are necessary.

Ocean basin time (1–200 Ma): The time for which direct evidence of seafloor spreading is still preserved in the oceans. Orogenic events on the continents can be compared directly with the spreading record for clues about cause and effect.

Plate tectonic time (200–950 Ma): The time dominated by plate tectonics that predates the record preserved in the oldest oceanic crust. The geologic record of this time is preserved in continental orogenic belts. This time includes the late Proterozoic and the Phanerozoic.

Middle to lower Proterozoic time (950–2500 ± 200 Ma): Some features reminiscent of modern plate tectonics are present, but others are lacking.

Archean time (2500–3800 ± 200 Ma): Tectonic conditions seem to have been substantially different from those of today.

---

[1]As used here, *petrotectonic* refers to a distinctive suite of rocks and/or structures that characterizes a certain tectonic setting. Examples include blueschists or calc-alkaline volcanic rocks that indicate a subduction zone and continental marginal sequences or ophiolite complexes that indicate a rifted margin.

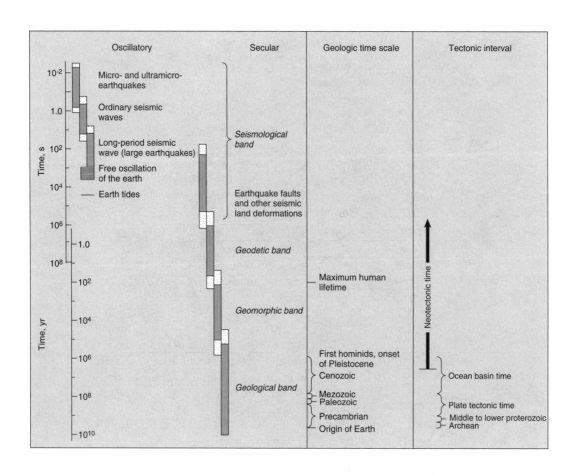

CHAPTER

CHAPTER
10 Anatomy of Orogenic Belts

10.1 Introduction

With an understanding of the major tectonic processes presented in the preceding chapters as a background, we can turn to an investigation of orogenic belts. Not accidentally, orogenic belts coincide with some of Earth's great mountain chains. Indeed, the terms *orogen, orogenic belt,* and *mountain belt* are used more or less synonymously in the literature. We follow that tradition, noting, however, that other tectonic processes, such as continental rifting, can also produce uplift and the formation of topographic mountains.

The solitude and majesty of the Earth's mountains have been as much a source of inspiration to geologists as to poets and philosophers, and for geologists they have also long been an object of intellectual curiosity. Only recently, however, have we come to recognize orogenic belts as the principal continental record of plate tectonic activity. Orogenic belts record the subduction of one plate beneath another, as well as collisions between crustal masses, such as two continents, a continent and an island arc, or a continent and an oceanic plateau. Because all oceanic crust older than about 200 Ma (early Jurassic) has

been subducted, orogenic belts are the prime repository of information about plate tectonic interactions for the first 95 percent of Earth's history. By studying these features, we have a chance to decipher part of the tectonic history of the Earth that is preserved nowhere else.

In this chapter, we discuss the structural characteristics of the major parts of orogenic belts, illustrating them with examples from the major mountain belts of the world, chiefly the North American Cordillera, the Appalachian-Caledonide, and the Alpine-Himalayan systems (Fig. 10.1).

Figure 10.2 shows general maps of the North American Cordillera, the Alpine-Iranian portion of the Alpine-Himalayan orogen, and the Appalachian-Caledonide orogen on a predrift reconstruction. Of these three orogens, only the last is inactive. It formed by the collision of Africa and Europe with North America in late Paleozoic time, and it was fragmented into its separate parts by the subsequent opening of the Atlantic Ocean in Mesozoic-Cenozoic time. The North American Cordillera and the Alpine-Himalayan orogens include plate margins that are still active. The major plates involved in the North American orogen are the North American, Pacific, and Juan de Fuca plates; those in the Alpine-Himalayan orogen

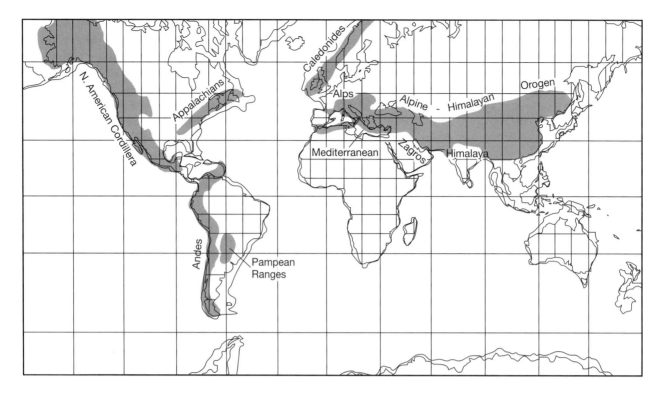

Figure 10.1 World map showing the locations of major orogenic belts referred to in this chapter: North American Cordillera, Appalachian-Caledonide, Alpine-Himalayan, and Andean.

are the Eurasian, African, Arabian, and Australian-Indian plates. Other, smaller plates are also included in each orogen.

From the arrangements of the various tectonic zones and associated structures shown in Figure 10.2, it should be clear that no single map or cross section can provide a universal model of an orogenic belt. Nevertheless, these and other orogens possess a number of features in common: a rough bilateral structural symmetry; a foreland or undeformed plate on either side; outer foredeeps; fold-and-thrust belts; sutures marked by ophiolitic rocks; one or more slate belts; and an internal crystalline core zone of metamorphosed and deformed sedimentary and volcanic rocks, mafic-ultramafic complexes, and granitic plutons. Thus, we can discuss these belts in terms of their common features.

The cross section of any portion of an orogen differs from that of another part of the same orogen and from that of other orogens. For example, orogens such as the North American Cordillera have an undeformed ocean on one side and a continent on the other, whereas orogens such as the Alps have a continental platform or foreland on each side. These contrasts have led to the classification of orogenic belts into two main types—(1) Cordilleran or circum-Pacific and (2) Alpine-Himalayan—based upon whether a given belt is bordered by an ocean and a continent or by two continents. These contrasting orogens display enough similarity in their structures, however, that we can discuss them in terms of a composite cross section, as shown in Figure 10.3.

10.2 The Outer Foredeep or Foreland Basin

Between the main orogenic belt and the undeformed continental platform lies a thick series of clastic sediments derived from a rising source area in the adjacent mountains. These sediments were deposited in a **foredeep** or **foreland basin** (see Figs. 10.2 and 10.3) (a **molasse basin**[1] in the terminology of Alpine geology. The sediments reach thicknesses of as much as 8 to 10 km near the mountain front, and the coarseness of the basin fill usually decreases away from the

---

[1] From the French word *molle*, soft, weak, a reference to the fact that these sediments are often unconsolidated.

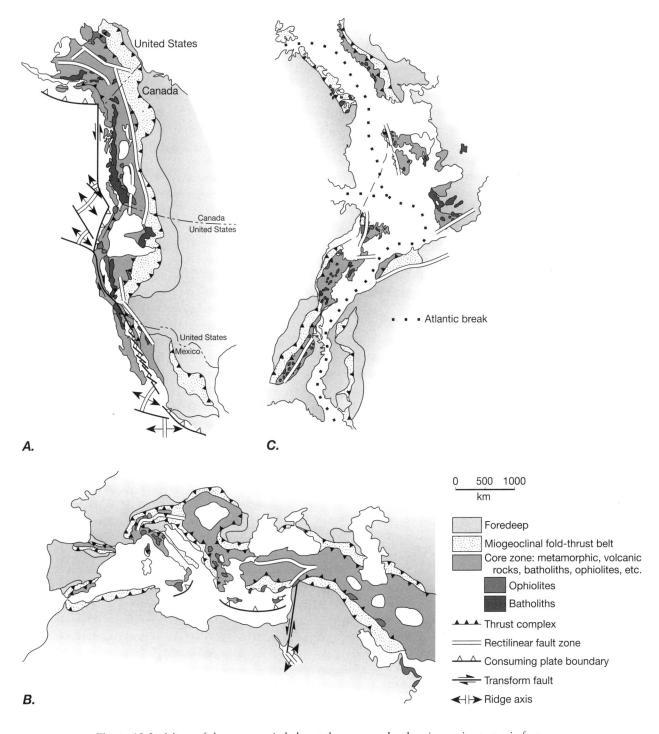

Figure 10.2 Maps of three orogenic belts at the same scale, showing major tectonic features to be compared with the model cross section in Figure 10.3. Note also location of other figures. *A.* North American Cordillera. *B.* Alpine-Iranian, or western, segment of the Alpine-Himalayan Orogen. *C.* Appalachian-Caledonide orogenic belt (including the West African orogen), on a predrift reconstruction of the continents around the Atlantic Ocean. *(A. after King, 1977; B. after Dewey, 1977; C. after Williams, 1984)*

mountain front. Conglomerates pass into sandstones and shales, which in turn may pass into carbonate marine shelf sediments (Fig. 10.4).

Environmental indicators from various basins suggest two possible modes of basin development. Along the eastern side of the Cordilleran belt in the

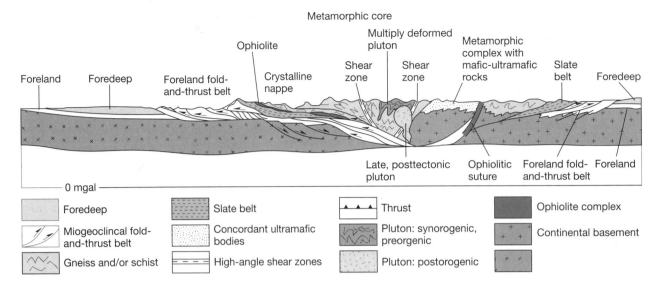

Figure 10.3 Cross section across a model composite orogenic belt. *(After Hatcher and Williams, 1986)*

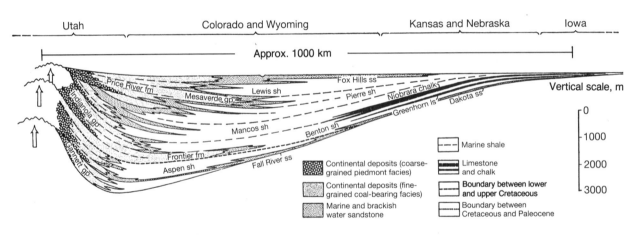

Figure 10.4 Cross section of a Cretaceous foredeep on the east side of the Cordillera from Utah to Iowa. Such basins are also called molasse basins or foreland basins. Note the thickening and coarsening of sediments toward the western source area. *(After King, 1977)*

western United States, for example, the Cretaceous stratigraphy begins with deep-water sediments that pass upward into shallow-water deposits, suggesting that the basin formed relatively abruptly and was filled up gradually. This gradual shallowing of the basin is indicated by the expansion through time of the shallow-water sandstones and continental deposits from west to east across the basin (see Fig. 10.4). By

contrast, the Devonian foredeep trough of the northern Appalachians, exposed in New York, formed slowly enough that sedimentation kept pace with subsidence, thereby producing a thick sequence of shallow-water sediments.

In some foredeeps, the clasts in the conglomerates and sandstones reveal an **unroofing sequence** in which stratigraphically younger deposits contain debris from successively deeper levels in the mountains. Such a sequence reflects the progressive uplift and erosion, or unroofing, of the adjacent mountains.

Typically, the foredeep rocks are only slightly deformed, which indicates either that they were deposited mostly after the main phase of deformation in the interior of the orogenic belt or that they were far enough away not to be affected by the deformation. Near the mountain front, however, the foredeep deposits are commonly deformed by folds and thrust faults. The folds are generally open, multilayer, class 1B folds, having folding angles of less than about 90°. The wavelengths, which can exceed 1 km, are controlled by the thick competent layers in the stratigraphy. Folding increases in intensity toward the mountain front, as evidenced by higher folding angles; larger aspect ratios; a change toward multilayer class 1C geometry; and, in some instances, inclined, possibly overturned folds with a vergence toward the stable platform. Allochthonous masses of the sediments are found in a few places in the foredeep basins, presumably emplaced by gravity sliding of surficial sheets of sediment.

In some regions, such as the Central Rocky Mountains of the United States or the Pampean Ranges of the southern Andes (Fig. 10.5), low-to-moderately dipping thrust faults bring basement to the surface over shelf or outer foredeep deposits. Seismic reflection profiles over one such uplift, the Wind River Mountains of Wyoming, indicate that the thrust penetrates to the base of the continental crust (Fig. 10.5B).

10.3 The Foreland Fold- and-Thrust Belt

Behind the foredeep basins, toward the center of the orogenic belt, lies a foreland fold-and-thrust belt. This belt consists predominantly of folded and thrust-faulted miogeoclinal sedimentary rocks that have been pushed away from the orogenic core and out over the stable foreland (Fig. 10.6). The foredeep deposits are overthrust by the miogeoclinal rocks at the front of the fold-and-thrust belt, and some foredeep deposits may be incorporated into the hanging-wall blocks of some of the thrusts. The miogeoclinal sediments usually thicken toward the core of the orogenic belt.

Figures 10.6 and 10.7 show typical examples of these structural regions from the Appalachian Valley and Ridge province (Figs. 10.6A and 10.7A); the Cordilleran overthrust belt, especially north of the Basin and Range province (Figs. 10.6B and 10.7B);

and the southern Himalaya (Fig. 10.6C and 10.7C). Other examples of such regions include the Jura Mountains north of the Alps, the Papuan fold belt (see Fig. 9.5), the west Taiwan fold belt (see Fig. 9.6C), and the Zagros Mountains of Iran (see Fig. 9.8).

A fundamental structural characteristic of fold-thrust belts is the presence of a sole fault that separates the deformed rocks of the thrust sheet from the underlying undeformed basement. These faults rise through the stratigraphic section toward the foreland, giving a wedge-shaped geometry to the thrust sheet. Above the sole fault, there may be several décollements, all of which ultimately are branches off the main sole fault and, like the sole fault, tend to rise through the stratigraphic section toward the foreland (see Fig. 10.7). Each fault characteristically adopts a ramp-flat geometry in which it cuts steeply up through competent layers such as sandstone or limestone and forms bedding-parallel faults in incompetent layers such as shale, gypsum, or salt. Duplex structures are common (Fig. 10.8).

Movement on such faults accommodates shortening and thickening of the thrust wedge and creates fault-ramp folds, which may tighten to accommodate further deformation. Other folds may form above a flat décollement to accommodate shortening and thickening of the thrust wedge. Some thrust faults may develop when folding becomes too tight to accommodate more shortening and faults cut up from the décollement through the steep or overturned limb of a fold. The dominant fold style is class IB to IC, which of geometrical necessity must be associated with a décollement because the radius of curvature of the anticlinal folds decreases downward, requiring a discontinuity at depth. Many folds are asymmetrical, in most cases having a vergence away from the orogenic core.

In many fold-and-thrust belts, age relationships consistently show that thrusts and folds near the orogenic core and those shallower in the thrust stack are older than those near the foreland and deeper in the thrust stack. This decrease in the age of deformation toward the foreland is sometimes called a **prograding deformation.** It is not uncommon, however, to find **out-of-sequence thrusts,** which are younger thrust faults that form behind the older frontal thrust.

Prograding deformation can be accounted for by the dynamic wedge model of a thrust sheet (see Chapter 7, Box 7.1). The wedge-shaped cross section of the thrust sheet is the geometry that is required if the entire sheet moves and if the rock in the wedge is everywhere at the critical point for fracture, which is the maximum possible stress. On any vertical plane through the wedge perpendicular to the direction of

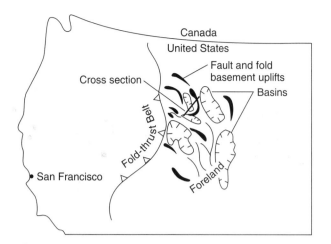

Figure 10.5 Basement uplifts in front of mountain belts: the Central Rocky Mountains, United States, and the Pampean Ranges, Argentina. *A.* Map of Central Rocky Mountains showing fold-and-thrust belt, basins, and basement uplifts in foreland region. *B.* Cross section of the Wind River Thrust, Wyoming, United States, showing the structure at depths including and below the crust-mantle boundary. *C.* Map of South America showing Andean thrust belt, the Pampean Ranges, Argentina. *D.* Cross section of a Pampean Range, the Sierra de Valle Fertil, showing involvement of Precambrian basement in the thrusting. *(A. after King, 1977; B. after Smithson et al., 1979; Jordan et al., 1983; C. after Dalziel and Forsythe, 1985; D. after Jordan et al., 1983)*

A.

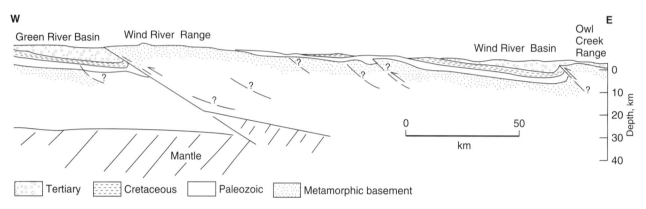

B.

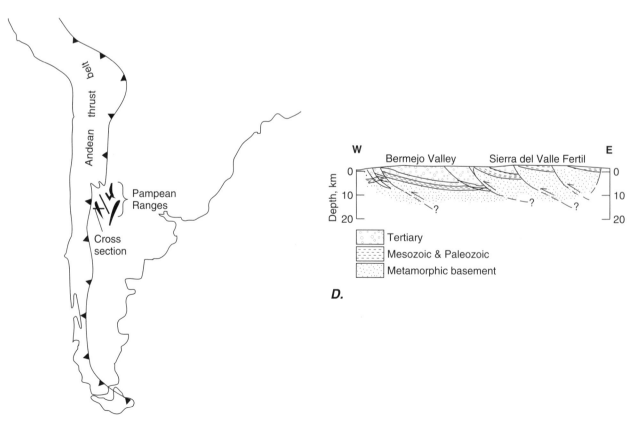

C.

D.

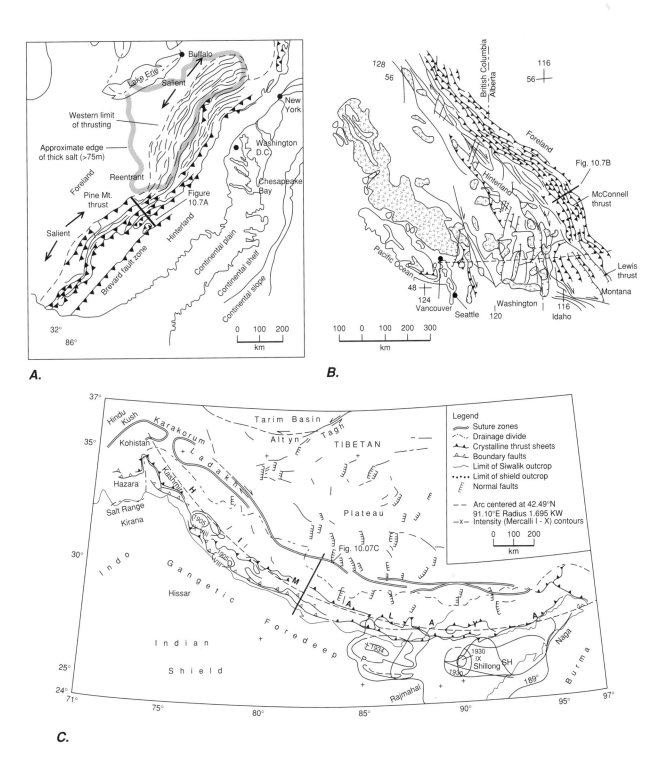

A.

B.

C.

Figure 10.6 Major thrust systems showing the foreland, hinterland, salient or virgation, and reentrant or syntaxis relative to the direction of movement for each fold-and-thrust belt. *A.* Map of the Appalachians. Solid lines are fold hinges; barbed lines are thrust faults with barbs on the hanging-wall side. *B.* Map of the Cordillera. *C.* Map of Himalaya, showing main thrusts, normal faults, the Indus-Tsangpo suture (northern boundary of the Hamalaya proper), and regions of historic earthquakes. *(A. after Harris and Bayer, 1979; B. after Price and Hatcher, 1983; C. after Seeber et al., 1981)*

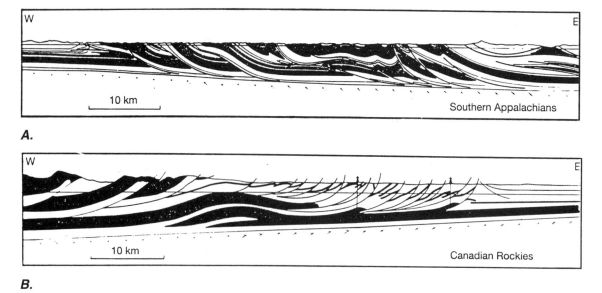

A.

B.

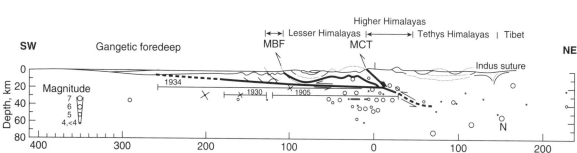

C.

Figure 10.7 Cross sections of the major fold-and-thrust belts shown in Figure 10.6. *A. Southern Appalachians. B. Cordillera. C. Himalaya. Note difference in scale compared with (A) and (B). (A. after Davis et al., 1983; Roeder et al., 1978; B. after Davis et al., 1983; Bally et al., 1966; C. after Seeber et al., 1981)*

motion (Fig. 10.9), the horizontal driving force must balance the frictional resistance on the part of the sole fault ahead of that vertical plane. With increasing distance from the front of the thrust sheet, the area of the sole fault ahead of such a vertical plane increases, and therefore so does the frictional resistance. This relationship holds because the resistance is given by the product of the area of the fault and the frictional shear stress σ_f on the base of the thrust sheet. To compensate for the increased resistance, the driving force must increase. This increase can develop only if the area of the vertical plane increases, because the total driving force depends on the area of the vertical plane multiplied by the horizontal normal stress σ_t, and because σ_t is already at the fracture strength and therefore cannot increase further. Thus, the thickness

of the sheet must increase with distance from the toe, leading to the wedge shape.

The surface slope of the wedge is determined by details of the resistance to sliding on the sole fault and the strength of the rock in the wedge, both of which are strongly affected by the pore fluid pressure, as well as the slope of the sole fault. Internal folding and thrusting of the rocks in the thrust sheet shorten and thicken it and thereby steepen its surface slope (gray line, Fig. 10.9). If such deformation creates an oversteepened surface slope, then the propagation of the sole fault deeper and farther out into the unfaulted foreland causes the slope to decrease and reestablishes a steady state (dashed lines, Fig. 10.9). This is the situation that typically leads to prograding deformation.

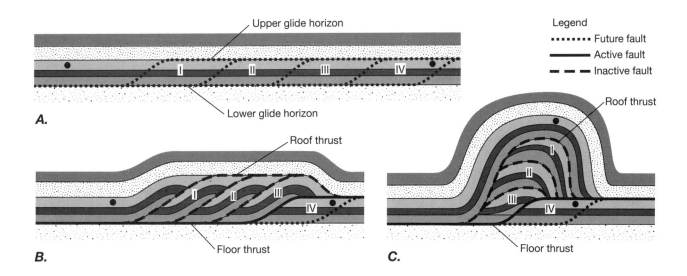

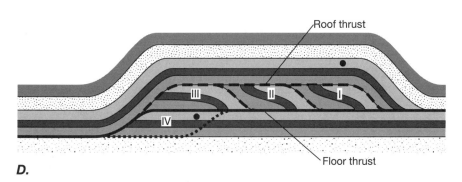

Figure 10.8 Geometry of duplex structures resulting from the progressive cutting of the thrust fault into the footwall block. Thrust faults are marked by heavy lines: short dashed lines indicate future faults; solid lines indicate active parts of the fault; long dashed lines indicate inactive parts of the fault on which displacement has occurred. The large black dots in the upper layer mark the same two points in each diagram. The roman numerals mark the same horses in each diagram. *A.* Undeformed section. *B.* A hinterland-dipping duplex. *C.* An antiformal stack. *D.* A foreland-dipping duplex.

Faults that form early in the deformation are commonly folded when a new décollement develops. Folds that develop above the new, deeper décollement must fold all preexisting overlying structures, including older décollements. This folding of the older faults makes continued slip on them increasingly difficult, and they eventually become inactive. Such deformed faults may be cut by later out-of-sequence faults.

In some cases, imbricate thrust faults dominate the deformation of the thrust wedge, as in the southern Appalachian Valley and Ridge province (see Figs. 10.6A and 10.7A). In other cases, the formation of folds characterizes the deformation, as in the northern Valley and Ridge province.

Most major sole faults remain above the strong crystalline basement and within the miogeoclinal sed-

imentary sequence, which characteristically contains abundant layers of weak rocks such as shales, gypsiferous layers, or salt. This style of deformation, in which the basement remains undeformed by the thrusting, is known as **thin-skinned tectonics.**

The sedimentary sections involved in foreland fold-and-thrust belts are chiefly miogeoclinal sections that become thinner with increasing distance from the orogenic core, consistent with the model. Thus, there is a gentle but significant upward slope to the basement of the sedimentary section from the hinterland toward the foreland. This slope reflects the rifting and thinning of the continental basement at a passive continental margin (see, for example, Figs. 5.13 and 5.14). Where these sections are deformed, they exhibit transport from the margins toward the

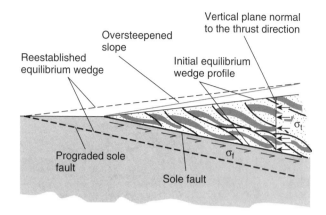

Figure 10.9 Diagram of imbricate thrust wedge showing the interplay of horizontal and frictional stresses (σ_t and σ_f, respectively). A push from behind on the wedge with an initial equilibrium profile causes it to be modified first to an oversteepened slope (gray line) and then to the reestablished equilibrium profile (dashed lines). See text for discussion.

foreland, and the basement slope helps direct the thrust faults upward in this direction.

Most rocks in foreland fold-thrust belts are unmetamorphosed, but low-grade metamorphism is present in some regions. Clay minerals, for example, are recrystallized to chlorites and micas, coal is anthracite-grade, and the magnetic vectors of the rocks typically are reset. Slaty cleavage is characteristic of argillaceous sediments and may be cut by a second generation of spaced foliation formed by solution during deformation. Despite these metamorphic effects, fossils are fairly common, the stratigraphy is relatively easy to work out, and the rocks can therefore be correlated from one thrust block to another.

In the inner parts of foreland fold-and-thrust belts, however—that is, the parts nearer to the orogenic core—fault slices of crystalline rocks become incorporated into the thrust sheets. In some cases, such as the western Alps, these rocks originally formed the basement on which the miogeoclinal rocks were deposited; they are called **external massifs** (Figs. 10.10A and 10.10B). In other cases, the crystalline rocks bear no obvious relation to the sediments in the belt of thin-skinned deformation. Regardless of their origin, these basement crystalline rocks are variably deformed and in some cases contain mylonite zones and folds that reflect faults and folds in the associated cover.

As we move from the front of the fold-and-thrust belt farther toward the interior of the orogenic belt, several changes take place. The sediments involved in the deformation change from shallow-water miogeoclinal sediments to deep-water eugeoclinal sediments. In some areas, called **slate belts**, the rocks are characterized by a monotonous predomi-nance of relatively unfossiliferous shales and slates. The monotony of the stratigraphy, the lack of fossils, and the poor exposure of the easily eroded shales and slates make stratigraphic analysis and correlations difficult and imprecise. The rocks in these regions were apparently laid down beyond the edges of continental margins as continental rise or abyssal deposits, or in offshore volcanic environments. It is possible to distinguish between these provenances by stratigraphic and petrologic analysis of the sediments.

Because slate belts are closer to the orogenic core, the grade of metamorphism is higher, reaching high zeolite or low greenschist facies. Ductile deformation becomes increasingly prominent, and the style of folding becomes class IC to II, reflecting a more ductile sedimentary pile and a decrease in the influence of competent layers on the folding of the sedimentary section. In places, multiple generations of folding are present. Folds become more inclined, even recumbent, and they can form huge fold nappes. This change in fold style is attributable in part to the higher temperature indicated by the increase in metamorphic grade and in part to a change in lithology from predominantly sandstone and limestone to the more ductile shales and slates.

As the rocks are more highly deformed and recrystallized, they tend to exhibit pervasive continuous foliation such as slaty cleavage and phyllitic foliation, which in places may be overprinted by later spaced foliations. Faults are present, but in many cases the lack of distinctive markers or piercing points makes them difficult to interpret.

The slate belt of Wales, Britain, is one of the best studied examples of an orogenic slate belt (Fig. 10.11A). There, a thick sequence of late Precambrian

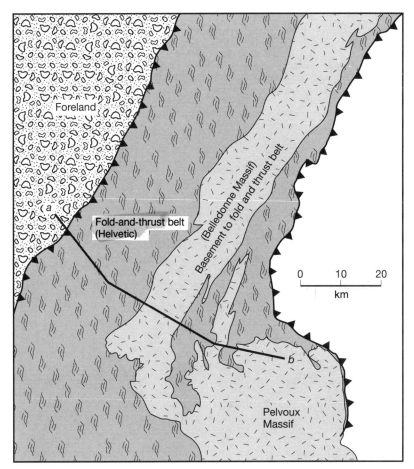

Figure 10.10 External masiffs of the western Alps, resulting from the involvement of basement rock in a fold-and-thrust belt. *A.* Map of the Belledone and Pelvoux massifs, France. *B.* Cross section along line *ab*. *C.* Index map, reduced version of Figure 10.2*B* showing the location of *(A)*. *(A. after Ramsay, 1963)*

A.

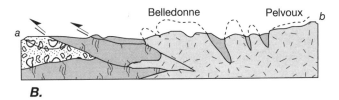

B.

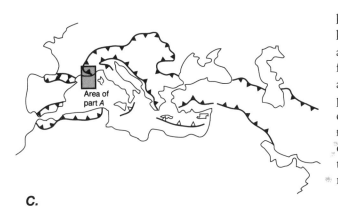

C.

and lower Paleozoic deep-sea sediments and associated volcanic rocks and melange display a pervasive sequence of upright folds and penetrative axial cleavage with down-dip extension. Thrust faults are recognized in a few areas, but rarely can they be traced for long distances. As shown in the cross section (Fig. 10.11*B*), the structures are thought to be part of a series of imbricate thrusts above a basal décollement, but the question marks on the section emphasize the uncertainty of this interpretation.

In many orogenic belts, ophiolites are present as large subhorizontal thrust sheets hundreds of square kilometers in area. Many of these complexes occupy a fairly internal position, over the inner parts of the fold-thrust belt. The best preserved complexes, such as many in the Alpine-Iranian belt (see Fig. 10.2), are present as thrust masses over the slate belts or the décollement-style fold-thrust belt, as illustrated schematically in Figure 10.3. However, in a few cases, ophiolites are preserved even out in front of the fold-thrust belt, tectonically overlying rocks of the continental platform.

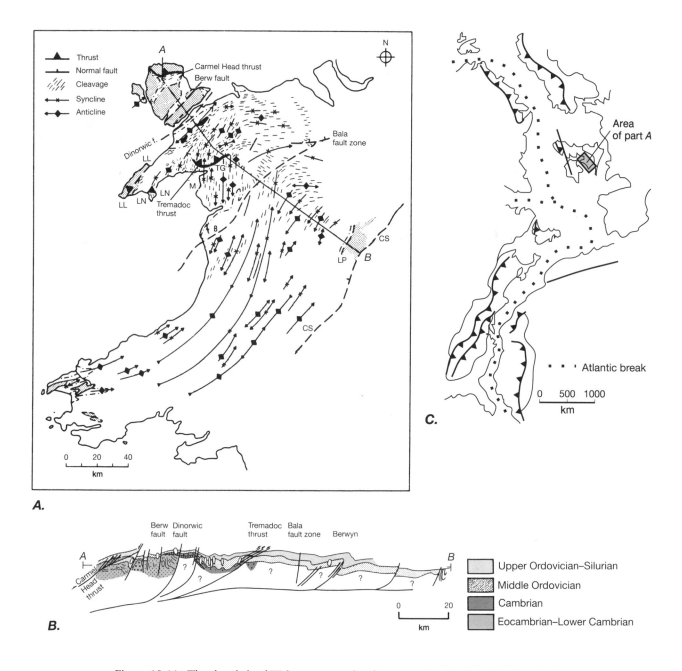

Figure 10.11 The slate belt of Wales, an example of an orogenic slate belt. *A*. Map. *B*. Cross section. *C*. Index map, a reduced version of Figure 10.2C showing location of *(A)*. *(B. after Coward and Siddans, 1979)*

10.4 The Crystalline Core Zone

The crystalline center or core of an orogenic belt contains metamorphic and plutonic rocks that have deformed extensively by ductile flow. The resulting structures include large thrust or fold nappes and complex multiple deformation structures. The core is invariably thrust out over the rocks of the foreland fold-and-thrust belt (see Fig. 10.3). Some of the nappes are very large (100,000 to 250,000 km²).

Multiple generations of folding in core zones produce a variety of fold interference structures. A structural sequence that often emerges from the geometrical analysis of such areas includes one or more generations of recumbent isoclinal folds; refolded by a generation of upright, more open folds; and finally deformed by a generation of either smaller scale kink and chevron folds or ductile shear zones (Fig. 10.12). Some studies have indicated, however, that the deformational events associated with particular styles or generations of folds do not necessarily correlate over large distances, and the significance of such sequences of fold generations remains a major problem in the interpretation of orogenic core-zone structure.

In rocks that preserve both fine-scale layering and large-scale stratigraphy, small-scale (high-order) folds mimic the orientations and styles of the larger (lower order) folds of the same generation. Detailed studies of high-order folds in critical outcrops, therefore, can be used to infer the basic geometry of regional deformation. In plutonic igneous rocks where no original layering is present, the deformed rocks become foliated, and deformation may become concentrated in large-scale anastomosing mylonite zones of high ductile shear strain that enclose volumes in which deformation has been less intense.

The rocks in core zones of mountain belts have diverse origins, but we discuss next the principal components that are common to most orogenic core zones.

Sedimentary Rocks and Their Basement

In some cases, such as in some Alpine crystalline nappes (Fig. 10.13), the crystalline rocks are meta-morphosed deep-water sedimentary rocks and their thinned continental crystalline basement, and these rocks appear now as amphibolites or gneisses. In such regions, former basement rocks commonly form the cores of nappe structures and are surrounded by an envelope or sheath of metasedimentary rocks. Figure 10.13 shows an example of such structures: the Adularia, Tambo, and Suretta nappes of the Penninic zone of the Alps.

In the deeper structural levels of an orogenic belt, metamorphic temperatures can approach or even exceed the granite solidus. The resulting highly ductile rocks are gravitationally unstable, and they rise diapirically, forming huge mantled gneiss domes that may contain a core of intrusive granite, as well as gneiss mantled by a metasedimentary envelope (Fig. 10.14)

The processes that form gneiss domes may be gradational into those that form domes by multiple generations of folding. Indeed, it is possible in some cases that little or no piercement—that is, diapiric intrusion—of the gneiss through the overlying rock has taken place. The contact between the gneiss and metasedimentary rocks generally is a ductile shear zone, however.

Volcanic and Igneous Rocks and Associated Sediments

Orogenic core zones typically contain large areas of rocks characterized by a lack of pronounced or continuous layered stratigraphy and by an abundance of intrusive rocks. Metamorphism is generally intense in these rocks, reaching amphibolite or even granulite conditions. Thus, such core zones tend to form vast areas of massive or banded amphibolite or granulite in which the original stratigraphic and intrusive relationships become next to impossible to determine.

The Appalachians in New England and southern Canada provide some especially well documented examples of such rocks. As shown in Figure 10.15, the rocks include metasedimentary rocks, metavolcanic rocks, and a number of gneiss domes. The rocks display complex, multiply folded structures generally with patterns and numbers of deformational phases similar to those outlined for the miogeoclinal or shelf sequences discussed above.

Such metavolcanic rocks were originally interpreted as "eugeosynclinal" sequences (see Sec. 3.5), but they are now understood to be a juxtaposition of continental rise-slope deposits with volcanic lls and volcanogenic sediments that formed in volcanic arcs, oceanic arcs, or mid-oce complexes. The juxtaposition is a resu at one or more subduction zones.

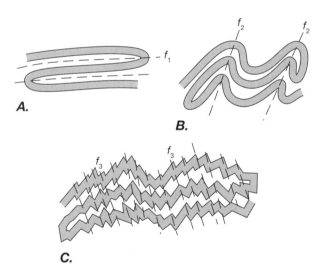

Figure 10.12 Diagrammatic cross sections illustrating progressive sequence of folding proceeding from *(A)* first-generation isoclinal folds, through *(B)* more upright second-generation folds superposed on the earlier deformation, to *(C)* a third-generation kinking superposed on all earlier foldings.

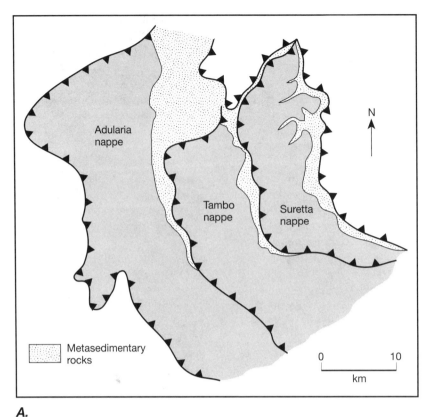

A.

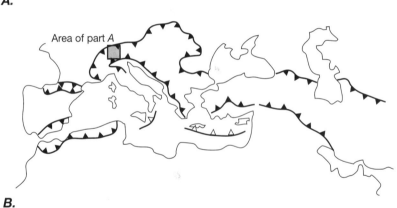

B.

Figure 10.13 Nappes in an orogenic core zone. *A*. Map of part of the Penninic (core) zone of the Alps, showing map view of three major subhorizontal crystalline nappes—the Adularia, Tambo, and Suretta—separated by thin septia of metasedimentary rocks. Compare with Figure 10.16. *B*. Index map, a reduced version of Figure 10.2*B* showing the location of *(A)*. *(After Spicher, 1980)*

Metamorphosed Ophiolite Sequences

Ophiolite belts are a feature of many orogenic core zones, and in many cases they have been affected by regional deformation and metamorphism. Where this has happened, the pseudostratigraphy of the ophiolites records complex structures including large-scale recumbent or multiply refolded folds. At high grades of metamorphism and large amounts of deformation, pillow lavas become massive greenschists or amphibolites; mafic dikes and plutonic complexes become massive or banded amphibolites; peridotites become serpentinized at lower grades of metamorphism and are dehydrated back into peridotite at higher grades, with the original mantle fabric overprinted or even obliterated by the later deformation.

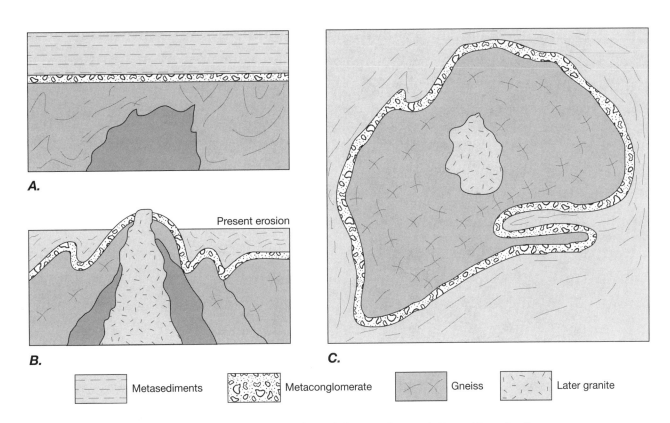

A.

B.

Present erosion

C.

| Metasediments | Metaconglomerate | Gneiss | Later granite |

Figure 10.14 Development of mantled gneiss domes. *A.* Deposition of sedimentary sequence unconformably on metamorphosed sediments and intrusive rocks. *B.* Deformation of entire sequence in *(A)* during a new deformation, followed by intrusion of a new granatic body. *C.* Map of the resulting gneiss dome after exposure by erosion. *(After Eskola, 1949)*

Lower Continental Crust and Mantle

In some regions of collisional orogenic belts where overthrusting and subsequent uplift has been extreme, highly metamorphosed quartzofeldspathic gneisses are found overlying peridotite. The Ivrea zone of the Alps is a good example of this situation (Fig. 10.16). These regions may represent the contact between the lower continental crust and the underlying mantle; that is, the original Moho. Like ophiolitic peridotites, both the peridotites and the gneisses of continental assemblages typically display an old metamorphic assemblage and deformation fabric overprinted by a younger one. Some subcontinental peridotites contain irregular layers and veinlets of gabbroic rocks that pass into bodies or dikes, suggesting that partial melting has occurred. The pressure-temperature relationships inferred from the gabbro

and peridotite mineralogy suggest that this melting took place during emplacement of the complex.

Gneissic Terranes with Abundant Ultramafic Bodies

Some orogenic core zones include terranes characterized by amphibolite or by granulite facies gneisses and schists in which are included numerous small, discontinuous ultramafic bodies, most less than 1 km long, consisting of fresh peridotite, pyroxenite, dunite, or their serpentinized equivalents. The southern Appalachians provide an especially good example of such bodies, as shown in Figure 10.17 on page 280, but similar terranes are also present in the Alps and the Caledonides. Characteristically, the bodies are elongate parallel to the regional structural grain

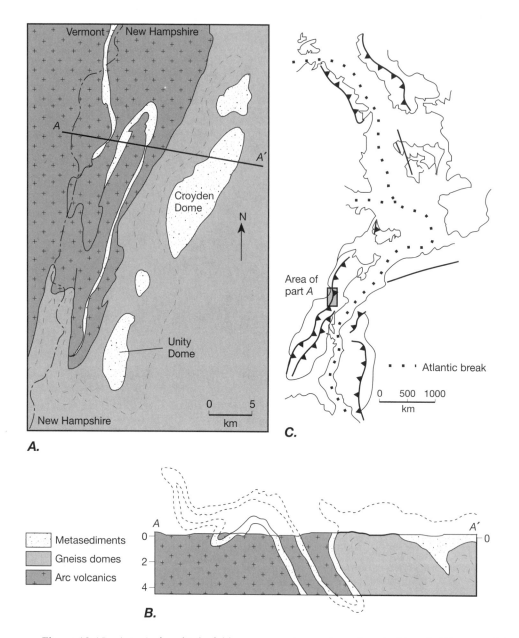

Figure 10.15 A typical multiply fold core zone, the northern Appalachians in Vermont. Of particular interest is the metavolcanic sequence, possibly an early Paleozoic island arc complex. *A.* Map. *B.* Cross section. *C.* Index map, a reduced version of Figure 10.2*C* showing the location of *(A). (After Thompson, 1968)*

and display an internal fabric that is concordant with the regional fabric.

The interpretation of these terranes is one of the great unsolved tectonic problems of mountain belts. We understand neither the mechanism by which the ultramafic rocks are incorporated into the metamorphic terranes nor the protolith[2] of the metamorphic rocks. We can suggest three possible origins for these enigmatic regions. They may be (1) exposures of deep crustal levels where fragments of the subcontinental or subisland arc mantle were somehow incorporated into the crust during deformation; (2) remnants of ophiolitic or mantle slabs that were completely disrupted after emplacement by the extreme deformation and metamorphism that produced the gneissic terranes; or (3) mafic and ultramafic igneous rocks intruded into continental margin sediments during rifting.

---

[2] After the Greek protos, first, and lithos, rock.

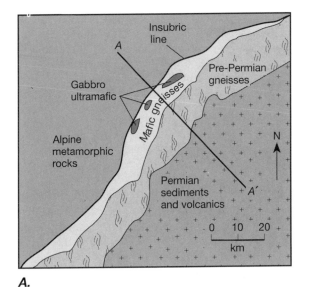

A.

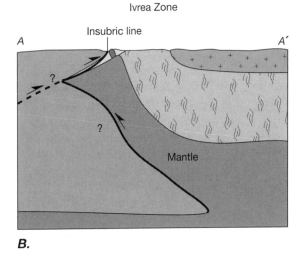

B.

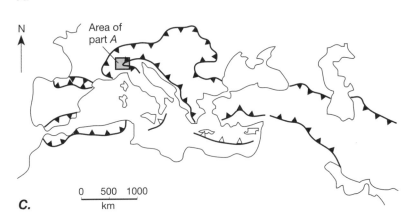

C.

Figure 10.16 Exposure of lower continental crust and mantle, Ivrea zone, southern Alps. *A.* Map. *B.* Cross section based upon geology and geophysics showing the mantle thrust over the northern continental edge, as well as backthrust in the opposite direction. *C.* Index map, a reduced version of Figure 10.2*B* showing the location of *(A)*. *(B. after Zingg and Schmid, 1979)*

Granitic Batholiths

Granitic batholiths are plutonic igneous rocks, generally of dioritic to granitic composition, that occupy vast areas in many mountain belts, in some cases dominating the area of the orogenic core. In the North American Cordillera, for example, batholithic rocks extend discontinuously from northern Alaska to northern Mexico (see Fig. 10.2*A*). Most batholiths are not a single body, but comprise tens to hundreds of individual plutons, each having an exposed area of a few tens to hundreds of square kilometers. They exhibit wide differences in rock type, degree of deformation, and apparent depth of emplacement.

Granitic rocks are either I-type or S-type granites, the I and S implying derivation by partial melting of an originally igneous or sedimentary source, respectively. Characteristically, I-type granites are composed of hornblende-biotite quartz diorites. They are thought to have formed by partial melting of a hy-

drous mantle or a previous crystallized igneous rock. S-type granites are richer in K and typically contain both biotite and muscovite and, less commonly, garnet. They may be derived from the partial melting of sedimentary rocks.

The timing of batholithic activity relative to deformation in the orogen is also variable. In a number of mountain belts, the age of granitic rocks overlaps with periods of deformation. Older granitic plutons commonly exhibit evidence of this deformation, such as the development of foliation and/or folds parallel to regional trends, whereas younger intrusive rocks do not. Thus, granitic plutons can be considered preorogenic if they intruded before the main deformation, synorogenic if they intruded during deformation, and postorogenic if they intruded after deformation.

The predominant rock type in batholiths varies in different mountain ranges. Quartz diorite or granodiorite predominates in the western North American Cordillera and the Andes, whereas granite is most common in the Appalachians and Caledonides. The Alpine belt contains few granitic bodies at all. The composition of the granitic rock displays a crude correlation with the type of country rock and the timing of intrusion. Both preorogenic batholiths and those intruding oceanic sedimentary or volcanic rocks tend to be poorer in K than postorogenic batholiths and those invading continental rocks.

The Sierra Nevada of California and Nevada provide a good example of the variability of batholithic history. In this region, the granitic rocks have an intrusive history ranging in age from approximately 70 to 200 Ma (Figs. 10.18A and 10.18B). Some plutons in the Sierra Nevada are deformed; others are not. Granitic rocks in the Sierra Nevada cluster into three age groups: early-middle Jurassic, late Jurassic, and middle-late Cretaceous (Fig. 10.18B). Because the principal deformation in this region was middle-late Jurassic (175 to 140 Ma), these groups may generally correspond to preorogenic, synorogenic, and postorogenic intrusives.

We can gain some insight into the variations observed in granitic batholithic terranes by considering the possibilities offered by the model of plate tectonics. Preorogenic I-type granitic bodies may be the

Figure 10.18 (*At right*) Batholiths in the North America Cordillera. A. Map of Sierra Nevada, California. B. More detailed map of part of the central Sierra Nevada, showing radiometric age relationships of granitic plutonic rock. C. Index map, a reduced version of Figure 10.2A showing the location of (A). (A. after Bateman, 1981)

product of normal igneous activity at a noncollisional consuming margin. These granitic bodies are then deformed when a collision occurs at the margin. The granites are Na-rich and are associated with volcanic rocks of similar composition that are also deformed in the later collision. The synorogenic and postorogenic S-type granites may result from partial melting of the lower part of the continental crust that has been thickened by formation of a mountain root during collision. They are K-rich, and possibly even Al-rich, granitic rocks. Postorogenic alkalic granites form after all orogenic phases have ceased and may reflect early stages of an episode of subsequent continental rifting.

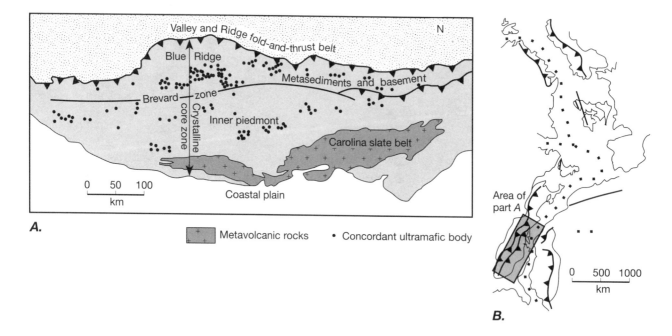

Figure 10.17   A. Concordant ultramafic bodies in the crystalline core zone of the southern Appalachians. B. Index map, a reduced version of Figure 10.2C showing the location of (A). (A. after Misra and Keller, 1978)

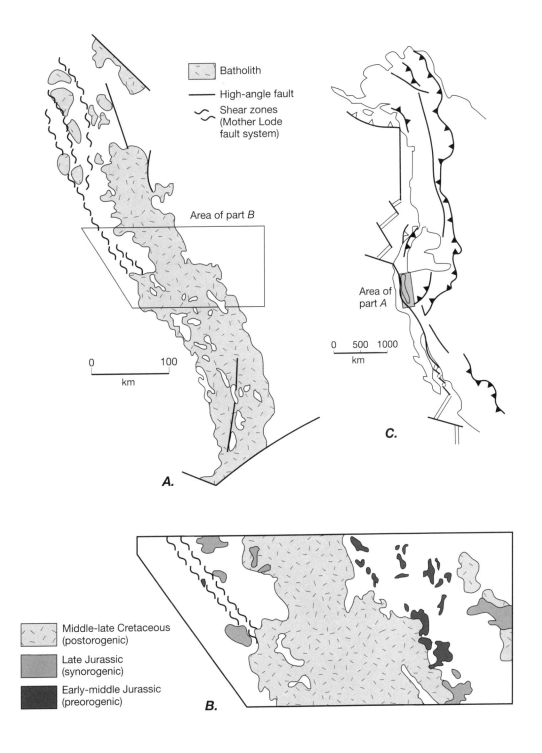

Batholith

High-angle fault

Shear zones
(Mother Lode
fault system)

Area of part *B*

0   100
km

A.

Area of
part *A*

0   500 1000
km

C.

Middle-late Cretaceous
(postorogenic)

Late Jurassic
(synorogenic)

Early-middle Jurassic
(preorogenic)

B.

10.5 The Deep Structure of Core Zones

The core zones of orogenic belts generally overlie the roots of mountain belts where the crust is thickest. Thus, their deep structure is related to the formation of mountain roots. What is the structure of these core zones at depth? How far down do the structures that we observe at the surface extend?

For many orogenic belts, a cross-sectional view shows that the major folds are recumbent and faults dip at a low angle. In places, however, these features dip more steeply. Folds are upright or vertical to steeply reclined; thrust faults are nearly vertical; pronounced down-dip lineations occur; and, in places, ductile strains become very large. Some of these areas are also the sites of major shear zones. Generally, structures and tectonic units are not continuous across the region.

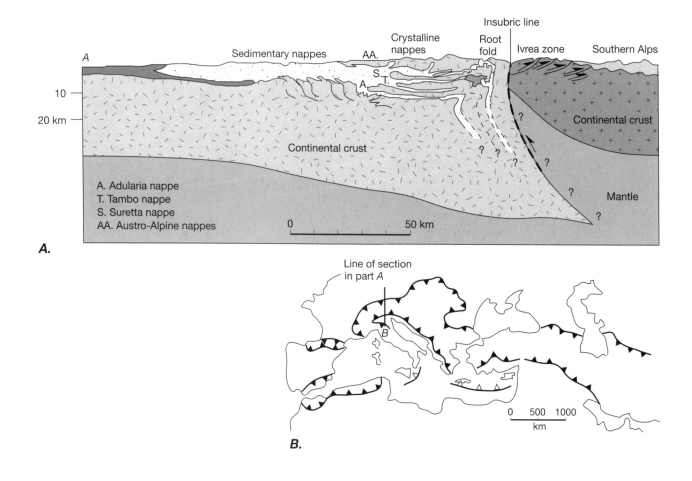

Figure 10.19 *A.* Cross section of the Swiss Alps showing recumbent nappes and root fold in the crystalline core zone of the Alps. Note the offset in the Moho, comparable but not identical to that in Figure 10.16. Compare also with Figure 10.10, which is located just west of the cross section. *B.* Index map, a reduced version of Figure 10.2B showing the location of *(A).* *(A. after Laubscher, 1982; Miller et al., 1983)*

Recent work in the Alps has shown that the steep-dip region is in fact the limb of a huge second-generation fold that deforms originally subhorizontal nappes and thrust sheets (Fig. 10.19). Because these folds have a vergence that is opposite the vergence of the nappe structures, they are referred to as **backfolds**. Formerly, such regions were incorrectly thought to be the source area, or root zone, of large-scale nappes and folds, a zone of extreme horizontal shortening out of which the nappes were squeezed.

Backfold structures may develop as a consequence of collision and subsequent change in the direction of dip of a subduction zone, as illustrated schematically in Figure 10.20A. Such changes in regional-scale deformational geometry may be reflected on a small scale (Fig. 10.20B) by the superposition of structures related to backfolding (S_2) on earlier

structures (S_1). Alternatively, backfolds and related minor structures may form as the result of a differential isostatic uplift of the core zone relative to the flanks of the orogen, on which is superposed a continued horizontal shortening.

The depth to which surface structures descend seems to vary from one mountain belt to another. In the Alps, seismic reflection and refraction evidence suggests that the structures involved in the backfold extend to deep levels. In contrast, seismic reflection work in the southern Appalachians by the Consortium for Continental Reflection Profiling (COCORP) suggests that the crystalline core zone is allochthonous and tectonically overlies a series of flat-lying reflectors that may be the little-deformed equivalents to the sediments of the Valley and Ridge province (Fig. 10.21). This result implies that the entire deformed

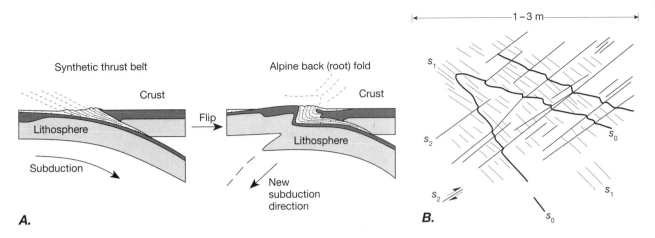

A.

B.

Figure 10.20 Relationship between thrust belts, backfolds, and subduction. *A.* Model for formation of multiply deformed structures such as the Alpine backfold by creation of synthetic thrust faults during collision of continents and subsequent backfolding after a reversal of subduction polarity. *B.* Outcrop-scale structure from within the root zone showing two episodes of deformation, marked by folded layering S_0, cleavage S_1, and axial surfaces of kink bands S_2. S_1 and S_2 are thought to represent deformation during synthetic thrusting and backfolding, respectively. *(After Roeder, 1973)*

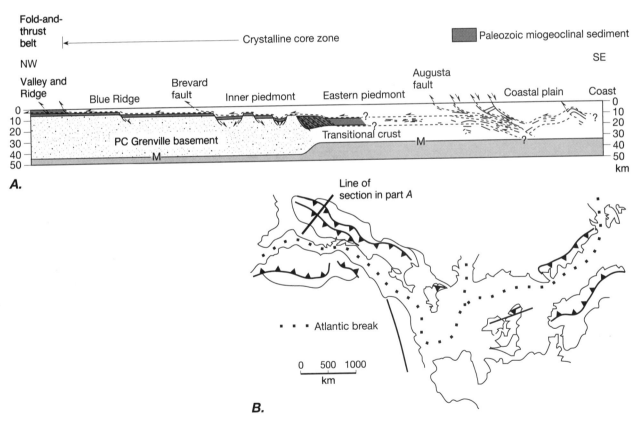

A.

B.

Figure 10.21 *A.* Cross section of southern Appalachians showing interpretation of COCORP seismic reflection results. Continental basement is believed to extend beneath the crystalline core of the mountain belt, including the Blue Ridge and Piedmont provinces. The basal décollement may extend to beneath the Coastal plain or "root" beneath the eastern Piedmont. *B.* Index map, a reduced version of Figure 10.2C showing the location of *(A)*. *(After Cook et al., 1981)*

belt may be allochthonous and may be displaced hundreds of kilometers over an autochthonous continental basement. Similar seismic results from elsewhere in this orogenic belt suggest that along much of its length, the entire core zone is allochthonous. Such a thrust feature involving an entire orogenic belt has profound implications for the character of the movements and forces that caused it.

10.6 Rectilinear (High-Angle) Fault Zones

Rectilinear, or high-angle, fault zones, which are usually moderately to steeply dipping and transect all the other features, are present in nearly every mountain belt (for example, see Fig. 10.2). They vary in width from a few hundred meters to 10 km, although most are relatively narrow and are marked by a band of well-developed mylonite. Typically, they extend for tens or hundreds of kilometers roughly parallel to the axis of the deformed belt, forming a major structural boundary along the orogenic belt.

Some shear zones offset geologic features identifiable on both sides of the fault, making estimates of the displacement possible. In such regions, both dip-slip faults and strike-slip faults have been found. The lineations observed in such zones tend to be approximately parallel to the direction of demonstrable displacement.

Other fault zones separate regions of distinctly different geologic history. For example, the predominant metamorphic ages on either side of the fault may be radically different, or the geologic or paleogeographic history may be different. In these cases, no correlation can be made across the fault, and such zones probably represent sutures (see Sec. 9.7).

Rocks within such shear zones are highly recrystallized and even mylonitic, and they often possess a metamorphic grade generally lower, but in places higher, than that of the surrounding region. In most places, a variety of diverse lithologies is present in discontinuous lenses. Mineral lineations are common, as are minor folds with their axes parallel to the mineral lineations. These lineations can be horizontal or steep, depending on the geometry of deformation within the fault zone.

The timing of the deformation in these faults is variable. Some zones are late faults that sharply transect the preexisting structures. Others grade into the structure of the surrounding rocks and are correspondingly difficult to interpret. Near the fault zones,

axial surfaces of folds tend to be deflected from the regional attitude into attitudes parallel to the zones.

Most well-documented zones display a complex history of repeated movement with several different senses. Three examples illustrate this characteristic. The Brevard zone of the southern Appalachians (see Figs. 10.17 and 10.21) extends for more than 500 km along strike. It appears to have an early dip-slip history, then a subsequent dextral strike-slip motion. The Insubric line of the Alps (see Figs. 10.16 and 10.19) is a steep fault zone that separates the Ivrea zone from the Penninic zone. It also appears to have an early dip-slip history on which is superposed a subsequent dextral strike-slip deformation. The Mother Lode fault system of the Sierra Nevada in California (see Fig. 10.18) comprises steeply east-dipping faults that may represent the traces of highly deformed thrust faults that once dipped westward and preserve predominantly west-over-east dip-slip motion. The faults were later folded and reactivated during an episode of east-over-west backfolding, and they were again reactivated as recent dextral strike-slip and normal faults. The Mother Lode fault system and the Insubric line represent remnants of sutures in their respective mountain systems.

10.7 Metamorphism and Tectonics

The rocks in the central portions of all orogenic belts are metamorphic. The distribution of metamorphic zones in most deformed belts is roughly symmetrical, so that the highest grade rocks constitute the central portions of the belt and unmetamorphosed rocks occur on the flanks. Although a comprehensive discussion of metamorphism is beyond the scope of this book, a few points are of structural and tectonic interest.

The series of metamorphic assemblages developed in metamorphic zones are a reflection of the temperature-depth profile at the time of metamorphism. Based on the mineral assemblages, we can distinguish metamorphism that has occurred under conditions of high temperature and low pressure (Buchan type), normal pressure and temperature (Barrovian type), and high pressure and low temperature (blueschist type) (Fig. 10.22). The temperature-depth profiles implied by these types of metamorphism are the result of different tectonic conditions.

High temperature, low pressure metamorphism implies that temperatures are elevated above a normal geothermal gradient. This condition develops in contact aureoles around shallow igneous intrusions in

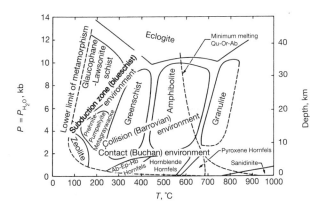

Figure 10.22 Metamorphism in mountain belts. Pressure-temperature diagram showing major metamorphic facies and their locations for conditions of high pressure, low temperature (blueschist) metamorphism characteristic of a subduction zone environment; intermediate pressure and temperature (Barrovian) metamorphism characteristic of normal temperature gradients in a collisional environment; and low pressure, high temperature (Buchan) metamorphism in a shallow contact aureole environment. *(After Turner, 1968)*

a volcanic arc environment. The metamorphism can be preorogenic, in which case it may be overprinted by subsequent synorogenic Barrovian metamorphism, or it can be postorogenic, in which case it overprints all earlier metamorphic phases.

Blueschist metamorphism implies that the temperatures of the rocks are significantly lower than they would be in a normal geothermal gradient. This situation occurs in subduction zones where cold shallow rocks are carried to large depths faster than normal temperatures can be reestablished. Such metamorphism tends to occur relatively early in the orogeny and is rare in most mountain belts. Where present, it is commonly overprinted by younger Barrovian metamorphism. Buchan (high temperature, low pressure) and blueschist (high pressure, low temperature) metamorphism, where present together, may constitute a paired-metamorphic belt (see Chapter 7, Sec.7.5).

Barrovian metamorphism indicates a normal geothermal gradient and is widespread in all mountain belts. For this reason, it is also known as classic regional metamorphism.

The Alps provide a good example of the relationships among the various metamorphic types (Fig. 10.23). The blueschist metamorphism of 70–85 Ma age is overprinted in places by the 15–25 Ma Barrovian regional metamorphism. Elsewhere, only Barrovian metamorphism is reflected in the rocks. Evidence of any previous metamorphic event has been completely obliterated.

In areas of good exposure, one can obtain an idea of the shape of isograd surfaces in three dimensions. Where they are undeformed or only mildly deformed, they appear to be gently curved surfaces that intersect the Earth's surface at small angles. For example, Figure 10.24 shows maps of the structure of part of the northern Appalachians (Fig. 10.24A) and of the Barrovian metamorphic zones in the same region (Fig. 10.24B). In Figure 10.24A, the upper, central, and lower nappes form a subhorizontal stack along which there is a north-trending antiform-synform pair and a concentration of gneiss domes. These folds in turn are warped into a series of culminations and depressions about west-northwest–trending axes, giving rise to type 1 and type 2 interference patterns that are particularly evident in the boundary between the upper and central nappes. The metamorphic grade generally decreases from the lower to the upper nappes (Fig. 10.24B), and the metamorphic isograds are gently folded about the same two directions as the nappes, giving the interference patterns on the isograd map (Fig. 10.24B). Thus, although the isograds do cut across the nappe boundaries, they share in the two gentle foldings of the tectonic units, suggesting that deformation of the nappes ceased before or during peak metamorphism and that later the nappes and isograds were both gently folded.

In other regions, metamorphic zones clearly have been displaced or even inverted. In the central Himalaya, for example (Fig. 10.25), the metamorphic grade increases from the chlorite zone continuously up through the sillimanite zone with progressively higher positions in the structure. Such an occurrence could indicate a primary inversion of isotherms, as would develop at a subduction zone above a downgoing cold slab (Fig. 10.26A), or it could indicate a tectonic inversion of the isotherms following metamorphism, as would occur on the inverted limb of a recumbent fold (Figs. 10.26B and 10.26C).

The interpretation of metamorphic zones in terms of plate tectonics is complex because the rates of downwarping and uplift of the rocks are similar to the rate at which the rocks heat up or cool off. The rate of erosion must be considerably slower than the rate of tectonic thickening to account for the thickened crust in orogenic belts such as the Zagros and Himalaya-Tibet areas. Thus, a given volume of rock might be expected to be buried quickly and then exhumed more slowly. The temperatures achieved during the process depend upon the heat flux into the rock and the rate of uplift and denudation.

Figure 10.27 illustrates burial and uplift histories for a volume of rock as a set of pressure-temperature paths along which time increases nonlinearly in

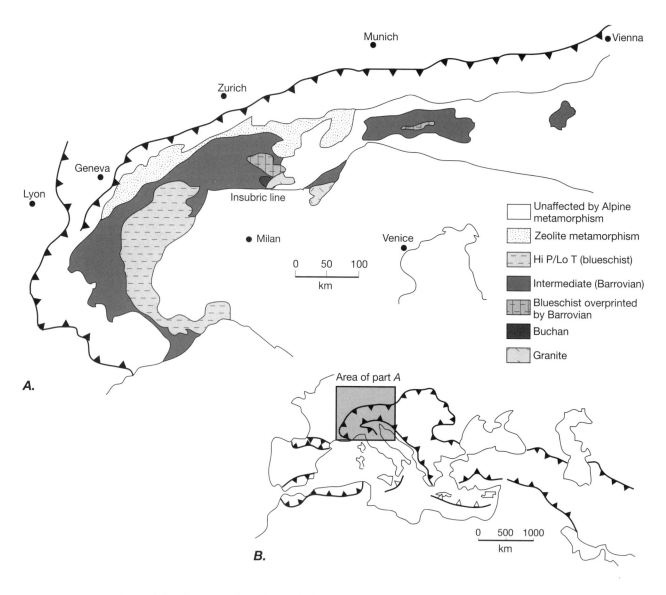

Figure 10.23 Metamorphism in a typical orogenic belt, the Alps. *A.* Metamorphic map of the Alps. Note the scarcity of granitic rocks. Compare with Figure 6.23. *B.* Index map, a reduced version of Figure 10.2B showing the location of *(A)*. *(A. after Frey et al., 1974)*

the direction of the arrows—so-called **PTt (pressure, temperature, time) paths.** A rock starting out at the surface (upper left-hand corner) is rapidly buried along path A to its maximum depth at B. If it is uplifted immediately, it might follow a path such as C. If the rock is uplifted more slowly, it might follow paths D or E. In case E, considerable time elapses before uplift, so that the temperature of the rock approaches the steady-state geotherm.

Note that in all cases, the maximum temperature is not reached at the maximum pressure. Thus, the peak temperature of metamorphism must occur after

maximum burial. If the mineral assemblages most likely to be preserved are those formed at peak temperatures of metamorphism, the metamorphic events recorded in the rocks must have occurred after collision. Numerical calculations for such models applied to the Alps suggest that the time lag between the end of the collision and the peak of metamorphism may be 10 m.y. or more.

The metamorphic mineral assemblages exhibited by a rock vary depending upon the PTt path and the relative rate of chemical reaction. Thus, path A-B-C might give rise to surface exposure of blueschist metamorphism, path A-B-D to blueschist overprinted by Barrovian metamorphism, and A-B-E to Barrovian metamorphism.

The radiometric age of a rock or an individual mineral is a measure of the time since the radiometric clock within the rock was isolated. Thus, it is the time at which the rock or mineral cooled through the "closure temperature" for a particular radioactive element and its decay products. That temperature is different for different decay schemes. For example, the closure temperature for the K-Ar decay scheme is lower than that for the U-Pb or Rb-Sr schemes, so K-Ar dates are nearly always slightly to significantly younger than the others. Thus, discordant ages from different decay schemes in the same rock reflect the rock's thermal history.

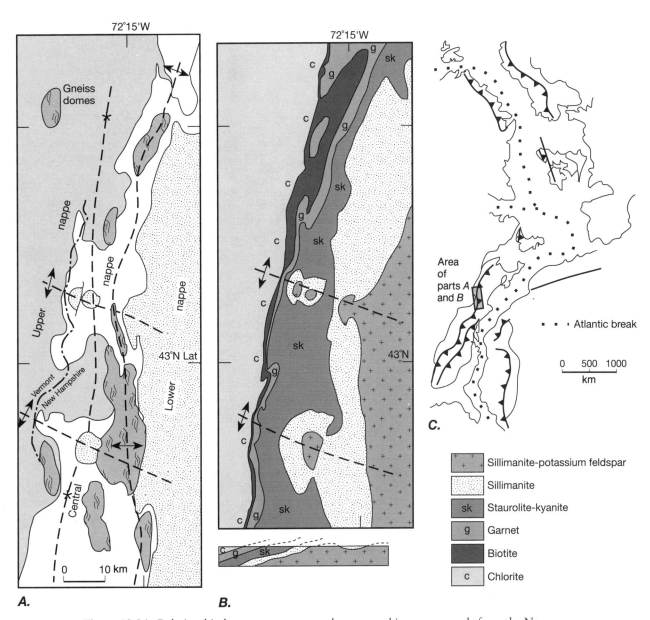

Figure 10.24 Relationship between structure and metamorphism, an example from the New England Appalachians. *A.* Tectonic map showing distribution of major nappes and gneiss domes. *B.* Map and schematic cross section of mineral isograds of Barrovian metamorphism. Note the general increase of grade with tectonic level. *C.* Index map, a reduced version of Figure 10.2C showing the location of *(A)* and *(B)*. *(B. after Thompson et al., 1968)*

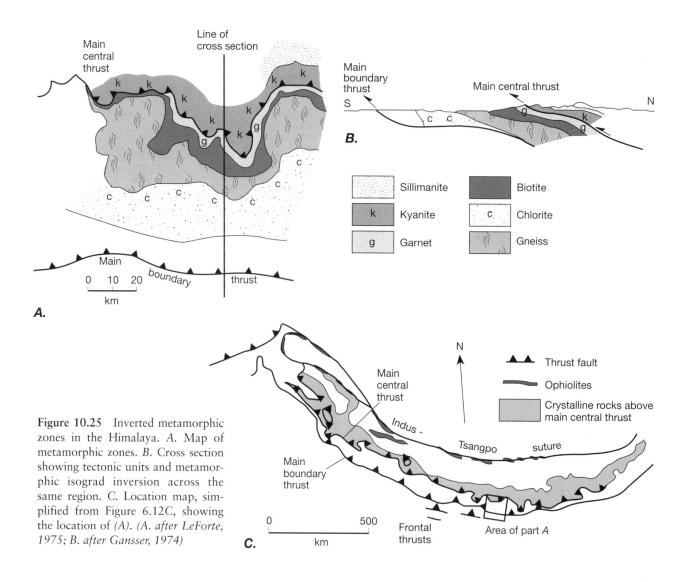

Figure 10.25 Inverted metamorphic zones in the Himalaya. *A.* Map of metamorphic zones. *B.* Cross section showing tectonic units and metamorphic isograd inversion across the same region. *C.* Location map, simplified from Figure 6.12C, showing the location of *(A)*. *(A. after LeForte, 1975; B. after Gansser, 1974)*

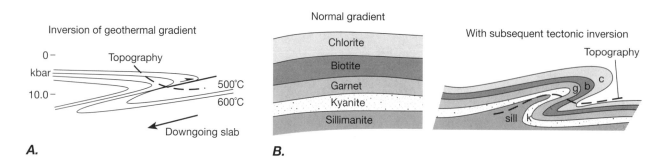

Figure 10.26 Possible origin of inverted metamorphic zones. *A.* Inversion of isotherms during thrusting, showing possible temperature distribution during metamorphism. *B.* Development of a recumbent nappe of metamorphic rocks after metamorphism. *(A. after LeForte, 1975)*

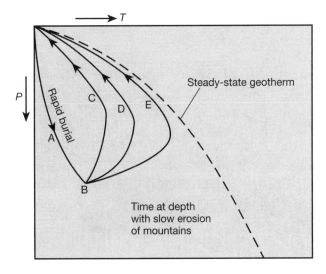

Figure 10.27 Diagram showing various pressure-temperature-time trajectories of rocks undergoing metamorphism. See text for discussion. *(After England and Thompson, 1984)*

10.8 Minor Structures and Strain in the Interpretation of Orogenic Zones

Our efforts to understand the origin of orogenic cores and their relationship to plate tectonic events prompts several questions. In what directions were the rocks transported during the deformation? What is the distribution of strain through the rocks? What large-scale pattern of flow gave rise to the observed structures and strain distributions, and how can these features be explained in terms of tectonic processes? These questions bear upon the way that orogenic cores evolve, and the answers ultimately may aid in the correlation of core zone evolution with plate kinematics.

In this section, we present examples of how the analysis of minor structures is applied to orogenic core zones. Our focus is the kinematic and strain analysis of folds, foliations, mineral fibers, and crystallographic preferred orientations. We assume that the reader has basic familiarity with these features. They are described in many standard textbooks, such as our book *Structural Geology* (Twiss and Moores, 1992), and we adopt the terminology and notation of that book. We cannot present a comprehensive account of the worldwide significance of such analyses, a formidable if not impossible task. Rather, we wish to give some idea of how kinematic analyses might contribute to understanding the formation of a given orogenic belt.

Kinematic Analysis of Folds

Application of the Hansen method for determining the slip direction (Twiss and Moores, 1992; Sec. 12.8) to deformed rocks in the Alps shows that in the outer areas of the thrust nappes, the shear direction is transverse to the orogenic belt, as is generally expected. In the central region of the orogenic core, however, the shear direction tends to be parallel to the axis of the orogenic belt (Fig. 10.28). Similar results have been found in the Norwegian Caledonides and the northern Appalachians. These longitudinal shear directions could be accounted for by flow models in which localized collision results in lateral flow away from the collision zone. Such models have been proposed for the Himalaya-Tibet region (see Sec. 9.4) and the Alps (see Box 12.1)

Kinematic Interpretation of Foliations

Because foliations are typically parallel to the plane of flattening of the finite-strain ellipsoid, they can be used to infer a shear sense in a fault zone if the foliation is a result of the shearing and if the orientation of the shear zone is also known. The intersection of the foliation plane and the shear plane is a line approximately perpendicular to the direction of shear, and the acute angle between the foliation and the shear plane points in the direction of relative motion of the material on the opposite side of the shear plane (Fig. 10.29A).

This relationship is confirmed by studies of the foliation developed in sediments in southern Alaska that have been deformed in the accretionary prism above the subduction zone (Fig. 10.29B). We can determine the relative plate motion independently from earthquake focal mechanisms and Pacific Plate reconstructions, and this motion indeed lies perpendicular to the intersection of the foliation with the thrust plane.

Strain Analysis

The regional distribution of strain in the Morcles nappe of the Swiss Alps southeast of the Lake of Geneva is shown in Figure 10.30. The map (Fig. 10.30A) shows the distribution of the maximum and intermediate extension axes ($\hat{s}_1$ and $\hat{s}_2$ with the trend of $\hat{s}_1$ shown correctly. Note that $\hat{s}_1$ is usually perpendicular to the fold axes and subparallel parallel to the direction of displacement of the nappe, although in places it is parallel to the fold axes and perpendicular to the direction of displacement. Examination of the cross section (Fig. 10.30B) shows that the strains are highest near the base of the nappe where the $\hat{s}_1$

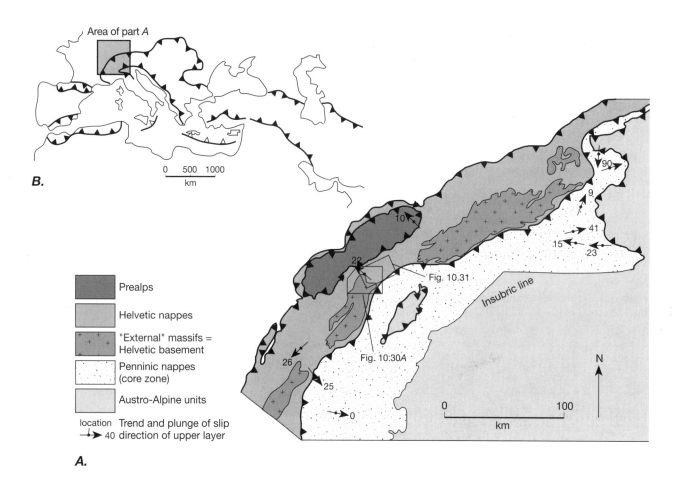

Figure 10.28 *A.* Tectonic map of the Alps showing the orientation of the slip lines deduced by Hansen's method from various parts of the orogenic belt. Arrows point in the direction of movement of the upper layers. The reason for some arrows pointing in toward the core is not understood. Note locations of Figures 10.30*A* and 10.31. *B.* Index map, a reduced version of Figure 10.2*B* showing the location of *(A).* *(After Hansen et al., 1967)*

axis is oriented at small angles to the subhorizontal thrust. Higher in the nappe, the strain magnitude decreases progressively, and the orientation of the $\hat{s}_1$ axes becomes steeper.

The total deformation which is recorded in the finite strain ellipsoid, includes a component of initial flattening associated with sedimentary compaction, followed by multilayer buckle folding during which the limestones behaved as the competent members, followed by inhomogeneous simple shearing associated with the emplacement of the nappe. Thus, the finite-strain ellipsoids do not record just the process of nappe emplacement, and the results of such a complex history of deformation are often difficult to interpret.

Kinematic Interpretation of Mineral Fibers

Mineral fibers in oriented overgrowths on pyrite crystals have been used to infer the extension history over

a region of the Swiss Alps (Fig. 10.31) that overlaps with Figure 10.30. The lines are everywhere parallel to the axis of maximum incremental extension $\hat{\zeta}_1$, and the length of any segment of the line is proportional to the magnitude of the extension in that direction. Note that the lines do *not* indicate the direction and amount of *displacement* of material points in the rocks. The small triangles indicate the locations of the measurements and plot at the youngest end of the extension history line.

The Morcles nappe is shown in both Figures 10.30 and 10.31. In the west, lower in the nappe, the extension history recorded by the fibers is simple, and the maximum extension direction of the finite-strain ellipsoids $\hat{s}_1$ is approximately parallel to the fiber extension directions $\hat{\zeta}_1$. In the east, however, which is structurally high in the nappe, the fiber extension histories are curved, and the orientation of $\hat{\zeta}_1$ is less regular. The strain ellipsoids, of course, probably

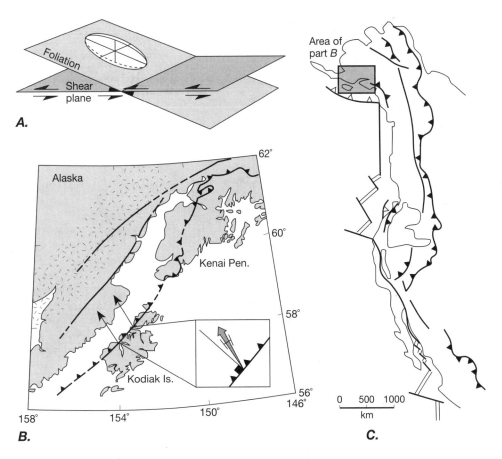

Figure 10.29 Relationship between foliation and shear plane. *A.* The foliation and the shear plane intersect in a line perpendicular to the direction of shearing. The acute angles of the intersection (shaded) point in the direction of motion of the material on the opposite side of the shear plane. *B.* An example of the application of this relationship in the field: Kodiak Island, Alaska. The generalized map shows arrows indicating the relative plate motion deduced from an analysis of Pacific Plate motions. The inset shows the relative motion (thick arrow) and the shear direction deduced from the foliation–shear plate relationship (thin line). *C.* Index map, a reduced version of Figure 10.2A, showing the location of *(B)*. *(B. after Moore, 1978)*

record more of the deformation than do the mineral fibers. Nevertheless, the information deduced from the fibers about the history of the strain accumulation provides more stringent constraints on any model that attempts to account for the emplacement of these rock masses.

In the lower and central parts of the Morcles nappe, the extension directions are relatively constant to the northwest throughout the period recorded by the fiber growth. A notable feature of much of the Wildhorn nappe and of the upper parts of the Morcles nappe is the change from roughly north-south extension approximately normal to the orogenic core zone early in the deformation to significant components of east-west extension roughly parallel to the core zone later in the deformation. This consistent pattern over a large area suggests a fundamental change in the geometry of the deformation. Such details of deformation history have not yet been incorporated into a unified model of the emplacement of these nappes during orogeny.

Kinematic Analysis of Crystallographic Preferred Orientations

Along the basal thrusts of ophiolites, the ultramafic rocks or underlying metamorphic rocks exhibit a foliation oriented approximately parallel to the thrust surface. In some cases, preferred orientation fabrics

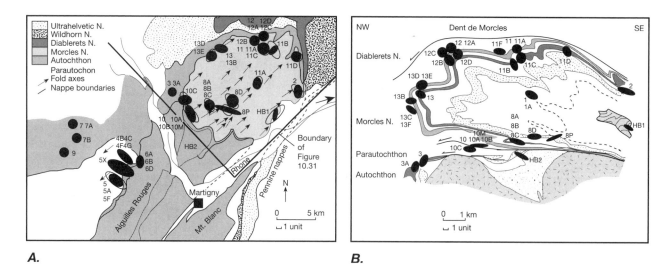

A.   **B.**

Figure 10.30 Distribution of strain in the Morcle nappe in the Alps of western Switzerland. *A.* Geologic map of the Morcles nappe and its surroundings showing horizontal sections through the strain ellipsoid (solid black ellipses) and the $\hat{s}_1 - \hat{s}_2$ section of the finite-strain ellipsoids (open ellipses) with $\hat{s}_1$ oriented parallel to its correct bearing. *B.* Composite down-plunge projection through the Morcles nappe showing the distribution of the $\hat{s}_1 - \hat{s}_3$ section through the finite-strain ellipsoids. *(After Siddans, 1983)*

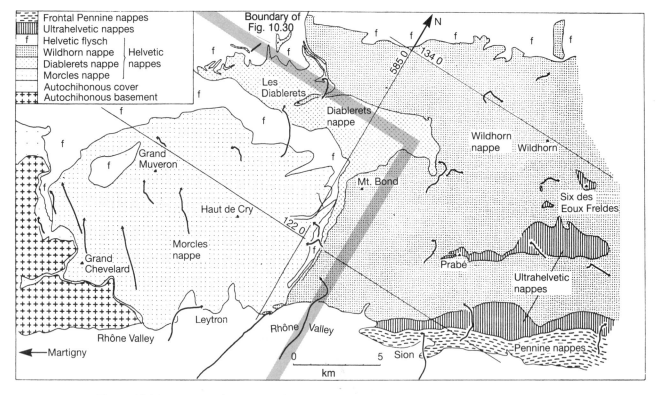

Figure 10.31 History of incremental extension in the western Helvetic nappes of the Swiss Alps as deduced from fibrous overgrowths on pyrite crystals. Lines are parallel to the directions of maximum incremental extension $\hat{\zeta}_1$ and line lengths in any given direction indicate magnitudes of the extension in that direction. The lines do not indicate the *displacement* of material points in the rock. Small triangles are at the youngest end of the line and are plotted at the location where the data were measured. See Figure 10.28 for location. *(After Durney and Ramsay, 1973)*

for crystallographic axes of olivine and orthopyroxene in the ultramafic rocks and of quartz in the underlying metamorphic rocks can be related to the sense of shear. If these basal thrust contacts represent fossil plate boundaries, as suggested in Section 9.6, these fabrics may indicate the relative plate motions along the boundaries.

10.9 Models of Orogenic Deformation

One aim of studying the structure of an area is to understand the relationship between local small-scale structures and large-scale tectonic processes, up to and including plate tectonics. The significance of the local structures can emerge only from the integration of detailed studies over large areas, which requires much time-consuming field work and analysis. Despite decades of study, however, a general model of orogenic deformation has not emerged, and we can only describe pieces of a puzzle that have not yet been assembled into a complete picture. Relating local minor structures to regional tectonics remains a fascinating problem of orogenic belts.

It is easiest to interpret results from areas of good exposure. The models of regional tectonics derived from such areas can then provide the frame-

work for interpreting the structure of less well exposed regions. Because the Alps and the Caledonide Mountains have recently been scraped clean by continental glaciation, exposure is exceptionally good, and much of the work on these problems has come from studies in these areas.

Simple Patterns of Ductile Flow

The pattern of rock flow during ductile deformation determines the distribution of strain in the rocks and the types and orientations of the structures that form. It is best described by the **streamlines** of the flow, which are everywhere tangent to the velocity vectors of material points. We consider three idealizations of the flow patterns that could occur: convergent flow, divergent flow, and shear flow. It is important to recognize, however, that the structures in the rocks are formed as a result of inhomogeneities in the flow, which are not considered explicitly in these models.

In convergent flow (Fig. 10.32A), all streamlines converge in the downstream direction, and velocity must increase in that direction if the material maintains a constant volume. Any volume of rock is subjected to a coaxial deformation, and the strain path plots in the constrictional strain field of the Flinn diagram (Twiss and Moores, 1992; Fig. 15.19). The axis of maximum extension $\hat{s}_1$ of the finite-strain ellipsoid

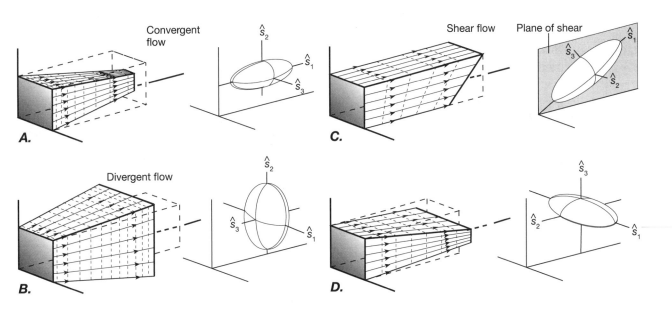

Figure 10.32 Streamlines for various flow geometries. *A.* Convergent flow. The streamlines all converge in the downstream direction, and velocity increases downstream. *B.* Divergent flow. The streamlines all diverge in the downstream direction, and velocity decreases downstream. *C.* Shear flow. The streamlines are all parallel, and velocity does not change in the downstream direction. The velocity of the upper surface is the highest, and of the lower surface the lowest. *D.* Combined convergent and divergent flow. *(After Hansen, 1971)*

is oriented essentially parallel to the streamlines, and the two other principal axes are directions of shortening (Fig. 10.32A). We therefore expect fold hinges and stretching lineations formed during the deformation to be parallel to the streamlines. Because material lines tend to rotate toward the $\hat{s}_1$ direction during a coaxial deformation (Twiss and Moores, 1992; Fig. 15.14), we also expect older lineations defined by material lines to rotate toward parallelism with the streamlines.

In divergent flow, all streamlines diverge from one another downstream, and the material velocity decreases in the downstream direction (Fig. 10.32B). The deformation is again coaxial, but the strain path in this case lies in the flattening strain field of the Flinn diagram (Twiss and Moores, 1992; Fig. 15.19). The axis of maximum shortening $\hat{s}_3$ of the finite-strain ellipsoid is essentially parallel to the streamlines, and the other two principal axes are directions of extension. Stretching lineations and fold axes formed during the deformation are perpendicular to the streamlines, and lineations defined by material lines are rotated toward the $\hat{s}_1$ direction and therefore toward being perpendicular to the streamlines.

In shear flow, the streamlines are parallel, and the velocity of the material does not change in the downstream direction. The velocity does change, however, in a direction perpendicular to the streamlines (Fig. 10.32C). The deformation is noncoaxial. During a shear flow, the $\hat{s}_1$ direction and material lines rotate progressively toward the streamlines. In principle, however, neither $\hat{s}_1$ nor rotated material lines ever become exactly parallel with the streamlines. Progressive simple shear is one example of such a shear flow (Twiss and Moores, 1992, Fig. 15.15).

We can imagine these simple types of flow to be combined to give more complex types of flows. For example, the streamlines may converge in one plane but diverge in a plane normal to the first (Fig. 10.32D). Moreover, a shear flow may be combined with any of the flows involving convergent or divergent flow.

The orientation of the streamlines is equivalent to what is often called the **direction of tectonic transport**. The foregoing discussion shows that there is no simple relationship between streamlines and the principal axes of finite strain, the fold axes, or other material lineations. If the regional distribution of principal strain axes can be determined, however (see Sec. 10.8), they constrain the possible flow patterns (Fig. 10.32). Alternatively, local inhomogeneities in the flow may produce particular structures from which it is possible to deduce the orientation of the streamlines (see Sec. 10.8). It is important to remember, however, that because rigid translation causes no

deformation, it cannot be recorded by any structures in the rock. Thus, the orientations of structures are not associated with streamlines or the direction of tectonic transport, but only with variations—that is, gradients—in the velocity along or across the streamlines. Therefore, the direction of tectonic transport cannot be deduced from deformational structures.

Models of Crystalline Nappe Emplacement

The processes by which thrust sheets are emplaced are complex, and they are an on-going subject of research. Particular attention has centered on foreland fold-and-thrust belts in which high pore fluid pressure plays an important role in reducing frictional resistance at the base of the thrust wedge. Crystalline nappes constitute a different problem, however, because even though shearing may be concentrated at the base of a nappe, the deformation is mainly ductile, and so friction is irrelevant to the mechanism of nappe emplacement.

We consider three simple models to explain the emplacement of ductile fold or thrust nappes: gravity glide, horizontal compression, and gravitational collapse. Each of these models predicts a different distribution of the principal finite-strain axes, and so in principle, we should be able to distinguish the different mechanisms of emplacement.

The critical differences show up on a cross section through the thrust sheet parallel to the streamlines. We represent a portion of such a cross section in two dimensions by a rectangular block resting on a base (Fig. 10.33A). The gravity glide model assumes that a nappe may be emplaced by gravitational forces that cause it to glide down a gently inclined base by ductile shearing within the nappe (Fig. 10.33B). In the simplest case, the nappe neither shortens nor extends parallel to the movement direction. The resulting flow is an example of inhomogeneous progressive simple shear flow (see Fig. 10.32C). The base of the nappe is a zone of intense shear strain. With progressively shallower depths in the nappe, the shear strain decreases, reaching zero near the surface (Fig. 10.33B). Thus, at the top of the nappe, the $\hat{s}_1$ axis of the finite-strain ellipse is oriented at 45° to the shear plane (see Fig. 10.32B). With increasing depth in the nappe, the orientation of the $\hat{s}_1$ axis rotates toward increasingly lower angles with the shear plane, reaching the minimum angle at the base of the nappe.

The second model assumes that the nappe is emplaced by a horizontal compression applied to the rear of the nappe, the so-called push from behind (Fig. 10.33C). The result is a ductile shearing over the base, similar to that described for the first model, on which is superimposed a shortening of the nappe par-

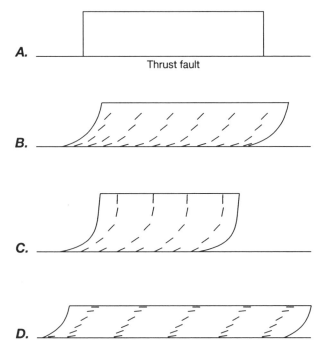

A.

Thrust fault

B.

C.

D.

Figure 10.33 Idealized distributions of $\hat{s}_1$ in thrust nappes deformed according to various possible mechanisms of emplacement. *A.* A portion of a nappe before deformation. *B.* Gravitational glide. The nappe deforms by simple shear parallel to the basal fault. *C.* Horizontal compression, or a "push from behind." The nappe deforms by a combination of simple shear and shortening parallel to the basal fault. *D.* Gravitational collapse. The nappe deforms by a combination of simple shear and extension parallel to the basal fault. *(After Sanderson, 1982)*

allel to the streamlines. For simplicity, we assume that the shortening is homogeneous through the nappe. The flow is then a combination of shear flow (see Fig. 10.32C) and divergent flow (see Fig. 10.32B). Thus, at the top of the nappe where the shear strain is zero, the shortening causes the $\hat{s}_1$ axes to be vertical. With increasing depth approaching the basal shear zone, the shortening component of the strain added to the simple shear causes the $\hat{s}_1$ axes to be oriented at a higher angle to the shear plane than for the case of simple shear alone.

The third model assumes that the nappe is emplaced by a process of gravitational collapse (Fig. 10.33D), which involves the ductile spreading and thinning of the nappe similar to the flow of a continental ice sheet. In this model, for both the nappe and the ice sheet, the flow is actually driven by the slope of the top surface of the body, which tends to decrease as a result of the flow. For simplicity, we assume that the surface slope is zero and that the flattening of the nappe is homogeneous within the illustrated portion of the cross section. The resulting deformation is a combination of shear flow (see Fig. 10.32C) and convergent flow (see Fig. 10.32A). At the top of the thrust sheet where the shear strain is zero, the flattening of the nappe causes the $\hat{s}_1$ axes to be horizontal and parallel to the streamlines (see Fig. 10.33D). Where the shear strain is nonzero, the extensional strain tends to rotate the $\hat{s}_1$ axes toward lower angles with the shear plane than is the case for simple shear alone. The net result is a sigmoidal pattern of $\hat{s}_1$ orientations with

the smallest angles between $\hat{s}_1$ and the shear plane near the top and bottom of the nappe and the largest angles near the center (Fig. 10.33D).

Figure 10.34 shows a plasticine model of gravitational collapse that includes a nonzero surface slope, vertical thinning, horizontal extension, and shear along the base, with a complex rolling under of the top of the nappe at the front. Such a process could explain the major fold noses and inverted limbs of fold nappes, as well as superposed crenulation cleavage like that found in the inverted limb of the Morcles nappe.

These models are highly simplified, of course, particularly in ignoring end effects, in the assumed geometry of the nappes, and in the assumed deformation that the nappes undergo. Nevertheless, they provide a basis upon which to begin to build an interpretation of field data. The strain in the Morcles nappe, for example (see Figs. 10.30 and 10.31), is reasonably consistent with the models for gravity glide or possibly gravitational collapse, but not with the push-from-behind model (see Fig. 10.33). Nevertheless, the inhomogeneity of the strain and the complicated history of the deformation (Sec. 10.8) make comparison with such simple models very imprecise.

More complex models of nappe formation have been investigated using scale model experiments. One hypothesis proposes that the emplacement of some nappes is driven by convective overturn of crustal rocks in an orogen. Models of the process produced by centrifuging layered blocks of various puttylike

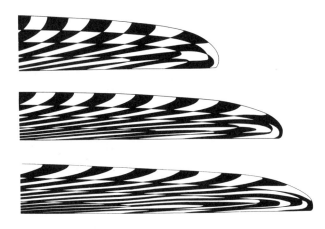

Figure 10.34 Strain distribution in a plasticine model of a nappe undergoing gravitational collapse with shear along the base. Note that the tip of the nappe at the front gets rolled under the advancing front of the nappe, a process that would explain the development of recumbent folds at the fronts of nappes. *(After Merle, 1986)*

materials to induce density-driven flow show striking similarities to cross sections of some deformed orogens (for example, see Figs. 10.15 and 10.19). Although the similarity does not prove the hypothesis, it is sufficient to indicate that this model must also be considered as a possible way to interpret the field data.

Plate Tectonic Models of Orogenic Core Zones

As mentioned in Section 10.5, the core zones of many collisional orogens show evidence of several generations of deformation, typically including an early isoclinal folding and subsequent upright and/or kink folding. Attempts to relate such structural features to plate tectonics must assess the possibility that the deformation reflects the relative plate motions.

Many proposals have been made to account for the generations of deformation that occur in orogenic core zones, although the data are not adequate to support any of them clearly. The multiple generations could correspond to shearing during subduction, followed by shortening and thickening during collision, and finally by isostatic collapse of the orogenic welt. Alternatively, they could relate simply to changes in plate motion during subduction and removal of the rocks in question from the active part of the accretionary zone at the plate margin.

Because there is usually no evidence that minerals defining foliations have recrystallized after defor-

mation, we infer that the deformation was more or less synchronous with the peak of metamorphism, which probably postdated any collision by a few tens of millions of years (see Sec. 10.7). Thus, it is still not clear whether any of the structures observed in the core zones reflect original subduction directions; rather, they may result from internal deformation within the collision zone during isostatic adjustment and gravity collapse of the orogenically thickened crust.

Despite decades of assiduous study of orogenic core zones by hundreds of geologists, it has been extraordinarily difficult to make precise associations between particular core zone structures and tectonic events, or to make general models that account for the observed structural characteristics of core zones in terms of plate tectonics. As absolute dating techniques improve, it may become possible to relate the age of formation of a single structure or fabric to the inferred relative plate motion for the same time, but for now the problem remains unresolved.

10.10 The Wilson Cycle and Plate Tectonics

Observations over the past century or so of the development of orogenic belts suggest a pattern that was termed the orogenic cycle. In this discussion, the term *cycle* is used rather loosely, because rarely, if ever, is there evidence of exactly the same sequence of events repeating itself in the same orogen. Although the advent of the plate tectonic theory has swept away many of the old concepts, the observations still must be accounted for by any new model. The characteristics of an orogenic cycle include:

1. Accumulation in separate areas of thick deposits of both shallow-water (miogeoclinal) and deep-water (eugeoclinal) marine sediments, the latter in association with intrusions or extrusions of mafic or intermediate magmatic rocks.
2. Commencement of deformation in the foreland fold-and-thrust belt together with the emplacement of ophiolitic rocks and the subsequent isostatic rise of the ophiolite and the deformed sediments beneath it.
3. Continued deformation in the fold-and-thrust belt—and metamorphism, deformation, and intrusion of granitic batholiths in the core zone—together with deposition of synorogenic sediments.

4. Further isostatic rise of the orogenic region and the deposition and partial deformation of post-orogenic continental sediments in the outer fore-deep (the *molasse* of Alpine geology).
5. Block faulting, the development of fault-bounded basins, and the intrusion of scattered alkalic dikes or intrusive bodies.

The Wilson cycle, a hypothesis proposed by the Canadian geophysicist J. T. Wilson, holds that the opening and closing of the oceans gave rise to the tectonic patterns observed on Earth. The Wilson cycle can accommodate many events leading to orogeny. Figure 10.35 on page 298 shows one possible scenario, by no means the only one, that might explain the orogenic events outlined above:

1. The rifting of a continent and the opening of a new ocean basin produce gradually subsiding passive continental margins on which thick deposits of shallow-water (miogeoclinal) sediments accumulate. Offshore, the ocean basin is formed by basaltic volcanism and deposition of deep-water sediments in the abyssal plains and continental rises (Figs. 10.35A and 10.35B). Eventually, with a shift in the pattern of plate motions, the spreading pattern changes and a subduction zone develops in the ocean basin, probably along preexisting fractures such as transform faults, oceanic fracture zones, or ridge-parallel faults (Fig. 10.35C) The eugeoclinal suite of deposits is completed by island arc volcanic and volcanogenic sedimentary rocks.
2. A passive continental margin on the downgoing plate collides with the oceanic subduction zone (Fig. 10.35D), emplacing a piece of oceanic crust and island arc crust (an ophiolite) onto the continental margin, juxtaposing the eugeoclinal and miogeoclinal suites, and initiating the formation of a foreland fold-and-thrust belt in the sediments of the passive margin. This collision is probably not synchronous along the entire margin but migrates along strike over time, depending on the relative geometries of the subduction zone and the continental margin.
3. Following the first collision, the polarity of the subduction zone reverses, producing a continental arc or Andean-style continental margin along the now-deformed former Atlantic-style margin (Fig. 10.35E) and initiating subduction of the remainder of the ocean basin. The consuming continental margin depicted in Figure 10.35E *need not* be a part of the originally rifted continent, because the convergent relative velocities need

not be the exact opposite of the original divergent relative velocities. Preorogenic plutons intrude the deformed continental margin. Deep basins in the trench along the continental margin, or deep-sea fans derived from the continental margin in the narrowing ocean, constitute the deep-water orogenic sediments. Eventually, the second continental margin arrives at the subduction zone and begins a continent-continent collision (Figs. 10.35F and 10.35G), which sutures the formerly separate pieces of continent. All the rocks in the suture region are deformed, the thrusting of the continental crust over continental crust creates a deep root, and partial melting at the base of the root produces late-stage intrusive or extrusive rocks. Strike-slip faulting takes place as the irregular edges of the continents adjust to each other.
4. Isostatic rise of the collision zone creates a source of postorogenic sediments and deforms earlier sediments. Strike-slip faults continue as the zone accommodates further postcollisional convergence (Fig. 10.35G).
5. Finally, the suture zone may be torn apart by another rifting event, which causes normal faulting and the intrusion and extrusion of alkalic to basaltic magmas.

Although the Wilson cycle is based largely on Appalachian-Caledonide tectonic history, recent investigations have shown that history to be much more complex. The cycle is a very broad generalizaton that does not provide a great deal of insight into the details of orogeny. Moreover, it is highly unlikely that two rifted continental margins will come back together in exactly the same place from which they rifted. Rather, given the changes in plate motion observed as a normal consequence of multiplate tectonic evolution, it is much more probable that two rifted continents will reapproach each other along different parts of the margin and that a misfit will occur. Indeed, it is probable that many continental collisions have occurred between two continental margins that were never previously close to each other. Misfits along colliding continental margins will adjust by lateral motion of crustal blocks, possibly by plastic slip-line processes (see Figs. 9.15, 9.16, 12.1.1, and 12.1.2). Thus, strike-slip motion in orogenic regions is probably as important as the more obvious contractional motion. Finally, as discussed in the next section, we find from examining several orogenic belts that the agglomeration of exotic terranes is a common feature of orogenies for which the Wilson cycle does not account.

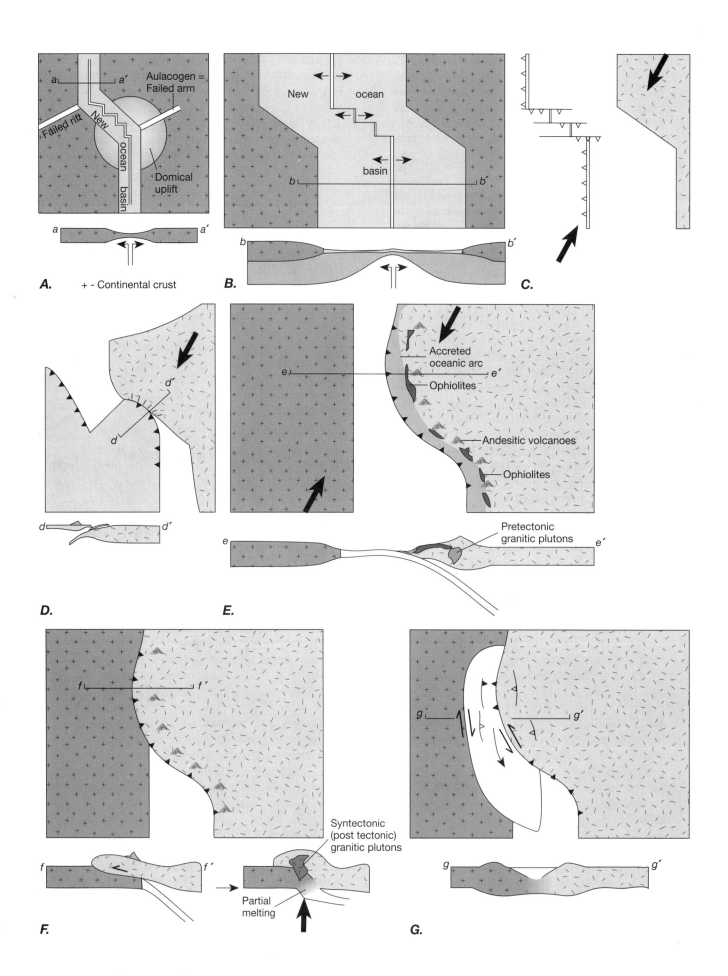

A.   + - Continental crust

Aulacogen = Failed arm

Failed rift

New ocean basin

Domical uplift

B.   New ocean basin

C.

D.

E.   Accreted oceanic arc

Ophiolites

Andesitic volcanoes

Ophiolites

Pretectonic granitic plutons

F.   Syntectonic (post tectonic) granitic plutons

Partial melting

G.

Figure 10.35 (*At left*) Sketch maps and coss sections illustrating possible development of a plate tectonic Wilson cycle to account for traditional observations of mountain belts. *A*. Rifting of continental margin; formation of domical uplifts and aulacogens along failed arms. *B*. Development of mature Atlantic-style ocean with passive margins. *C*. Change of relative plate motion; commencement of closure of ocean; and development of subduction zones along preexisting fault zones in oceans, along fracture zones, ridge-transform fault intersections, or ridge parallel faults. *D*. Collision of hypothetical intraoceanic arc system with continental margin; emplacement of ophiolites. *E*. Reversal of subduction direction; development of Andean-style margin on continent to right; cessation of subduction on continent to left; continued convergence with other continent. *F*. Collision of two continents with Andean- and Atlantic-style continental margins; formation of mountain root. *G*. Adjustment of continental margin collision zone by strike-slip movement and/or renewed rifting.

10.11 Terrane Analysis

Many, if not most, of the world's orogenic belts include a composite of distinct terranes that originate not only from the continent(s) or arc(s) involved in the final collision but also from areas that are clearly exotic to the main crustal blocks, or whose relationship to those blocks is suspect. We call such terranes **exotic** or **suspect terranes.**

A terrane is an area surrounded by sutures and characterized by rocks having a stratigraphy, petrology, or paleolatitude that is distinctly different from that of neighboring terranes or continents. In a collisional mountain belt, a terrane is a remnant of crust that has had a history different from that of the subducted oceanic crust or that of the main crustal blocks that have collided. Some mountain belts, such as the North American Cordillera, are characterized by collisions between a single continent and numerous exotic terranes rather than between two continents. Some of these terranes seem similar to the areas of anomalous oceanic crust discussed in Section 3.2.

To understand the role of these exotic or suspect terranes in a mountain belt requires a different kind of analysis, and their recognition has begun to shed new light on the history of these complex regions. The object of terrane analysis is to work out when adjacent terranes were apart and when they came together, as revealed by a detailed comparison of the geologic histories. Figure 10.36 illustrates the method of analysis with a hypothetical map (Fig. 10.36A) and a stratigraphic-tectonic diagram (Fig. 10.36B). Four terranes, numbered 1 through 4, are separated from one another and from continents

A and D by sutures. By plotting the stratigraphy from each terrane in a column, with adjacent columns for adjacent terranes (Fig. 10.36B), it is possible to determine when the terranes began a common history and thereby to determine when they collided, or "docked." The time of collision or docking of two terranes or of a terrane and a continent is given by the maximum age either of intrusives that cross-cut the suture or of sediments that unconformably overly both terranes. Apparent polar wander paths (see Secs. 2.4 and 9.7) of two separate terranes should be different until the docking, after which they should exhibit a single APW path.

Hypothetical terranes 3 and 4 contain rocks of Carboniferous through Triassic and Silurian through Triassic age, respectively. They share a common history from Jurassic time onward, as indicated by a date on the pluton (crosses) that intrudes the suture. In mid-Jurassic time, they collided with continent D, as indicated by the sedimentary unit (vertically ruled) that overlies that suture. In terrane 1 the rocks are Jurassic in age, and in terrane 2 they are early Triassic to late Jurassic. The diagram shows that terranes 1 and 2 share a common history beginning in the late Jurassic, when a rock unit (pattern with irregular blobs) was deposited across the suture between them. Terranes 1 and 2 docked or collided with the already amalgamated terranes 3 and 4 and continent D in early Cretaceous time, as indicated by the sedimentary unit overlying all the sutures between terranes 2 and 3, 2 and 4, 3 and 4, and 4 and D (horizontally ruled). Continent A collided with the composite of the terranes and continent D in Cenozoic time, when the sedimentary deposits covering all the previously accreted terranes also began to cover continent A.

The recognition of exotic terranes in mountain belts around the world requires a major modification of the Wilson cycle and of our ideas about how mountain belts develop. Given that areas of anomalously thick oceanic crust can develop in the ocean basins (see Sec. 3.2), the process of subduction must sweep them into the subduction zone ahead of any continent that rides on the same plate. Before collision, these terranes have their own individual geologic history, but after collision, they become part of the overriding crustal block. Thus, the argument about whether orogeny is an ongoing or highly episodic event may be settled in favor of both sides, in that the docking of exotic terranes may be a quasi-continual process during subduction, but the collision of major continents is only episodic. The motions of the major crustal blocks in a collision may represent only a part of the activity that constructs an orogenic belt, and many other relative plate motions may be represented by the numerous sutures among different exotic terranes.

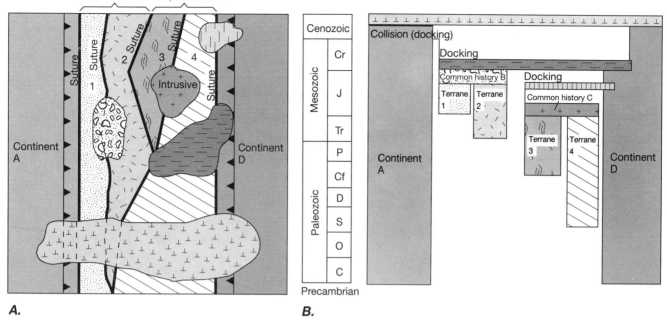

Figure 10.36 Analysis of exotic terranes. *A.* Schematic map of two continents separated by several exotic terranes. *B.* Stratigraphic-tectonic diagram illustrating ages of rocks in individual terranes, ages of "docking," and ages of common histories. See text for discussion.

Additional Readings

Bally, A. W. 1980. *Basins and Subsidence: A Summary.* AGU-GSA Geodynamic Series 1.

Bally, A., et al. 1979. Continental margins: Geological and geophysical research needs and problems. Washington, D.C.: National Academy of Sciences.

Burchfiel, B. C. 1983. The continental crust. In *The Dynamic Earth*. R. Siever, ed. Special issue of *Scientific American*. September.

Burchfiel, B. C., C. Zhiliang, K. V. Hodges, L. Yuping, L. H. Royden, D. Changrong, and X. Jiene. 1991. The South Tibetan detachment system, Himalayan orogen: Extension contemporaneous with and parallel to shortening in a collisional mountain belt. GSA Special Paper 269.

Clark, S. P., Jr., B. C. Burchfiel, and J. Suppe, Eds. 1988. Processes in continental lithospheric deformation. GSA Special Paper 218.

Francheteau, J. 1983. The oceanic crust. In *The Dynamic Earth*. R. Siever, ed. Special issue of *Scientific American*. September.

Hoffman, P. 1988. United Plates of America. *Ann. Rev. Earth and Plan. Sci.* 16:543–603.

King, P. B. 1977. *Evolution of North America.* 2nd ed. Princeton: Princeton University Press.

Kröner, A., and Greiling, eds. *Precambrian Tectonics Illustrated.* Stuttgart: Schweizerbartsche.

National Academy of Sciences-National Research Council. 1980. *Continental Tectonics.* Washington, D.C.

Nisbet, E. G. 1987. *The Young Earth: An Introduction to Archaean Geology.* Boston: Allen and Unwin.

Schaer, J.-P., and J. Rodgers, eds. 1987. *The Anatomy of Mountain Ranges.* Princeton: Princeton University Press.

Twiss, R. J., and E. M. Moores, 1992. *Structural Geology.* New York: W. H. Freeman and Company.

Uyeda, S. 1978. *The New View of the Earth.* New York: W. H. Freeman and Company.

Windley, B. F. 1984. *The Evolving Continents.* New York: Wiley.

CHAPTER

11

Neotectonics

11.1 Introduction

An important goal of any science is the discovery of new information that benefits society. In tectonics, we search for ways to predict the time and location of such damaging tectonic events as earthquakes and tectonically induced landslides. To be socially useful, the predictions must be precise enough to alleviate the loss of life and property. This quest for prediction techniques involves (1) the investigation of active tectonics, and (2) the investigation of neotectonics.

Active tectonics is defined formally as the "tectonic movements that are expected to occur within a future time span of concern to society" (Geophysics Study Committee, 1986). *Neotectonics* is the study of phenomena over time spans ranging from thousands to a few million years before present. Both fields are essential to the study of tectonic hazards because geologic events happen so infrequently. We must look at the history of local deformation and plate motions over the past several thousand to few million years to gain a perspective on present and future movements and to determine where we stand at the present with respect to the cycle of episodic or recurring hazardous geologic events.

In this chapter, we outline briefly the techniques and results of studying active tectonics and neotec-

tonics. We discuss the direct measurement of ongoing movements (Sec. 11.2), special dating methods that are appropriate for neotectonic studies (Sec. 11.3), geologic-geomorphic features that reflect plate movements (Sec. 11.4), and the study of fault and fold movements (Sec. 11.5).

11.2 Direct Measurements of Tectonic Movements

Three main techniques are used to study ongoing tectonic motions: direct measurements of geodetic networks that span areas of tectonic activity, sometimes called near-field geodesy[1]; determination of relative positions of a few sites on a regional or continental scale by triangulating them with reference to satellites or astronomically distant radiative sources; and the recently developed Global Positioning System (GPS) of satellites, which permits the accurate location of geographic features by determining their distances from orbiting satellites whose locations are well known.

---

[1]After the Greek *ge*, earth, and *dainein*, to divide.

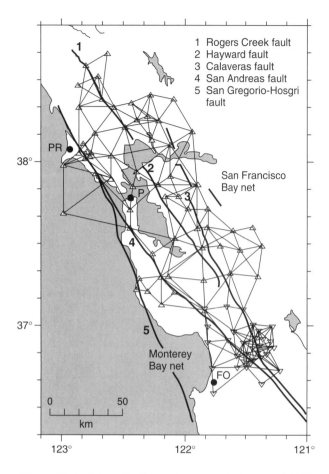

Figure 11.1 Map of trilateration network in central California. Large dots indicate VLBI stations, used as reference. FO, Fort Ord; P, Presidio; PR, Point Reyes. Curved lines are principal faults; heavy line is San Andreas fault. *(After Lisowski et al., 1991)*

Near-field geodesy employs leveling studies, where the vertical distance between points is carefully measured, and highly accurate surveys of the distances between pairs of reference points in a group of points that constitute a **trilateration network** (Fig. 11.1). By repeating the same survey at regular intervals, it is possible to monitor any changes that may have taken place in the length of the lines. The detection of systematic displacements across a structure is evidence of active tectonics. Trilateration networks are analogous to more traditional triangulation networks, but they differ in that they use modern laser-based instruments to measure distance precisely, instead of just measuring angles from any one station to two adjacent stations. Figure 11.1 shows a representative network for central California, which straddles the principal faults of the San Andreas system.

The principal aim of repeated trilateration surveys is to determine the displacement across the vari-

ous faults. Changes in the length of each line are converted by standard geodetic techniques into displacements of each station relative to a common reference such as a specific station or the centroid of the network as a whole. Accumulation of displacement data over a number of years can then be interpreted in terms of active fault motions. Figure 11.2 shows the results for 15 years of measurements across the network shown in Figure 11.1. The diagram indicates the calculated displacements of each station. The large jumps in displacement across several of the faults in the region shows that the total displacement is distributed across several faults of the system, rather than concentrated on a single fault. In this sense, the plate boundary is not the San Andreas fault itself, but the entire set of faults from the San Gregorio fault on the west to the Calaveras–Green Valley fault on the east.

Regional position determinations are of two types—satellite laser ranging (SLR) measurements and very long baseline interferometry (VLBI), both of which can resolve distances between stations to within a few millimeters. SLR involves measuring the distance from the station to an orbiting satellite whose position is very precisely known. The time required

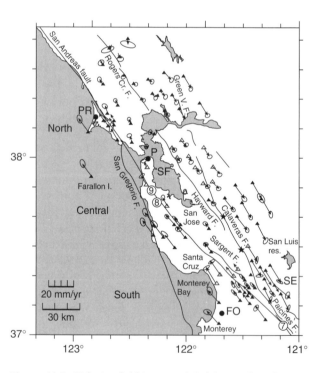

Figure 11.2 Velocity field in central California, based upon the trilateration network shown in Figure 11.1. FO site is effectively on Pacific Plate. Although the network indicates an abrupt change in velocity across the San Andreas fault, no surface creep has been observed along much of the portion of the fault shown. *(After Lisowski et al., 1991)*

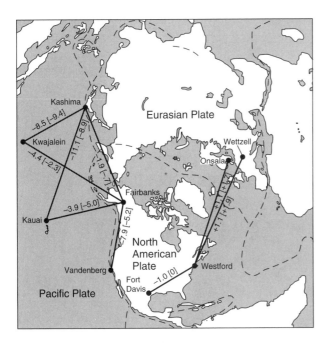

Figure 11.3 Relative motions of points on plates in the northern hemisphere, determined by VLBI using radio signals from quasars. Relative motions are in centimeters per year. Numbers in brackets are the theoretical motions according to the model of Minster and Jordan, 1978. *(After Carter and Robertson, 1986)*

for a laser pulse aimed at the satellite to reach the satellite and be reflected back is the basic measure that is converted to distance using the known speed of light. VLBI utilizes radio telescopes at two stations that simultaneously record the radio-wavelength emissions from two or more quasars billions of light years away. The distance between the stations results in different arrival times for the signals, so when the signals from the two stations are combined, they produce an interference pattern of "beats" that changes as the distance between the stations changes. Thus, repeated measurements can reveal small relative motions of the stations. Figure 11.3 shows a map of the northern hemisphere comparing recent length change measurements in cm/yr using this technique with the theoretical changes (in brackets) predicted by the plate motions proposed by Bernard Minster and Thomas Jordan. The agreement between the theoretical and measured amounts is reasonably good, with lines across and around the Pacific Ocean showing the largest inconsistencies. The VLBI measurement between Westford and Fort Davis inexplicably is shortening, even though it is within the same plate. The differences between these measurements and the theoretical motions remain to be reconciled.

The *Global Positioning System* provides a relatively new technique for measuring absolute position on the globe. Under the best of conditions, using a dish receiver that records data in one place for several hours at a time, the system can determine locations to within a few millimeters in horizontal directions and a few tens of millimeters in the vertical direction. Even relatively cheap hand-held receivers can deliver instantaneous locations to within a few tens of meters. The greatly increased speed, ease, and accuracy provided by this system is revolutionizing the use of geodetic measurements in tectonics.

The position of the reciever on the ground is determined by measuring the travel times of radio signals broadcast from overhead by satellites of the Global Positioning System whose orbits are very precisely known. These data are converted to distances from the satellites using the precisely known locations of these satellites, the known speed of light, and sophsiticated computer analysis. The distances from at least four satellites provide a unique location on the Earth's surface. The system, operated by the U. S. Defense Department, was developed for use by the military and by civilian ships at sea, but it has been adapted for use in precise and relatively inexpensive location measurement. With this system, it is no longer necessary that stations in a geodetic network be visible to one another, as is the case for traditional trilateration networks.

Although in some places discrepancies between traditional trilateration measurements and GPS determinations have appeared, in other areas there is such good agreement between the two sources of data that they have been combined to give a more complete picture of the deformation. The technique has already demonstrated its power, not only in directly measuring tectonic motions, but also in simplifying enormously the surveying requirements for gravity and magnetic surveys. We can expect a continual stream of important contributions to our knowledge of tectonics from this geodetic technique.

11.3 Dating Methods

Because of the short time span involved in many neotectonic movements, the traditional geological techniques used to determine the age of a structure and its rate of formation in older rocks are of limited use. As a result, a whole new set of dating techniques has been devised to augment the traditional methods. A few of these dating methods are summarized in Table 11.1, and we discuss selected methods in this section. In general, these methods are divisible into two broad categories: numerical (or absolute) and relative. Numerical methods well-suited to the short time spans

Table 11-1 Some Important Neotectonic Dating Methods

Numerical Methods	Most useful age range (yr)
Annual	
1. Historical records	0–3000
2. Dendochronology	0–3000
3. Varves	0–10,000
Radiogenic	
1. Carbon 14	0–50,000
2. Uranium series (based on decay of intermediate daughter products)	2000–200,000
3. Potassium-argon	10,000–1,000,000
4. Fission Track	10,000–1,000,000
5. Thermoluminiscence	1000–1,000,000
6. Cosmogenic isotopes other than ^{14}C (^{10}Be, ^{36}Cl, ^{26}Al, etc.)	Variable
Relative	
Chemical or biological changes	
1. Amino-acid raceminization	100–500,000
2. Obsidian hydration	100–500,000
3. Tephra hydration	1000–1,000,000
4. Lichenometry	100–1000
5. Soil development	1000–1,000,000
6. Rock and mineral weathering	1000–500,000
7. Progressive landform modification	5000–100,000
8. Rate of deposition	Variable
9. Geomorphic incision rate	Variable
10. Rate of deformation	Variable
Correlation	
1. Tephrochronology	
2. Paleomagnetism	Depends on age
3. Stratigraphy	of feature and
4. Fossils and artifacts	and accuracy
5. Stable isotopes	of recognition
6. Tektites and microtektites	

involved in neotectonic processes include counting techniques, which can be used if the effects of yearly changes are directly observable, and radiogenic techniques, for situations in which appropriate radioactive elements are present. Relative dating techniques include the measurement of the extent of chemical or biological changes in deposits and the stratigraphic correlation of different features.

Numerical or absolute dating techniques include the use of historical records, dendrochronology, and varves. Historical records are of value chiefly in regions of the world with a long written history— mostly such centers of ancient civilization as the Mediterranean area, central and western Asia, India, and China. Some records in these regions extend back several thousand years. Since uplift and subsidence characteristically proceed at rates of a few millimeters per year (a few meters per thousand years), significant changes can occur over several hundred or a few thousand years.

Dendrochronology is the technique of determining the age of a tree by counting the annual tree rings. In some cases, such as in the U.S. Southwest, the age of archeological sites has been determined by comparing the tree ring pattern of wood used in buildings with that of living trees. The oldest trees, however, are only a few thousand years old, and the technique is not applicable in areas where the trees are not old enough or do not show sufficiently distinct annual growth patterns.

Varves are annual sediment layers that accumulate in lakes that freeze in winter and thaw in summer.

The thickness and grain size of the sediments vary with the season and so can be used as a stratigraphic indicator of age. Errors can occur, however, because of small unconformities and miscorrelation of layers from one site to another.

In principle, radiogenic methods can yield accurate numerical ages. Suitable material is not always present, however, and errors can creep into the analysis if there is contamination of the samples or if the geologic context of the dated material is not clear. Thus, the manner in which the ages have been obtained should always be examined carefully before placing much reliance on the resulting numbers.

Carbon 14 (^{14}C) is perhaps the most widespread radioisotope used to date young deposits. It is produced by the interaction of cosmic rays with ^{14}N in the upper atmosphere. It is then incorporated into rocks, plants, and animals by such processes as photosynthesis and shell formation. ^{14}C decays with a half-life of 5570 years. If the rate of production of ^{14}C in the atmosphere is known, then the total amount in the atmosphere and the biosphere can be calculated, and the age of a sample can be determined, using standard radioactive decay equations. Early ^{14}C age determinations assumed that the cosmic ray flux was constant, an assumption now known to be incorrect. In the 1970s and 1980s, a fairly accurate evaluation of ^{14}C production through time was obtained by close correlation with tree ring, historical, and other independent sources, so that reasonably accurate age determinations are now possible. The functional limit of ages that can be analyzed by this method, however, is about 75,000 years. In samples of this age, the amount of ^{14}C remaining is only 0.0001 of its original concentration. For such old samples, the problems of accurate measurement and of contamination with younger material become severe.

Uranium series dating depends upon the measurement of concentrations of the intermediate and final products in the decay of ^{238}U, ^{232}Th, and ^{235}U to lead. The most common reaction used in these measurements is

$$^{234}U \rightarrow \,^{230}Th + \,^{4}He$$

^{234}U has a half-life of 75,000 years. Materials that can be dated by this method include calcite deposits; deposits of minerals such as anhydrite, halite, and silica; bones; and young volcanic rocks.

The potassium-argon dating method depends upon the decay of ^{40}K to ^{40}Ar; ^{40}K has a half-life of 4.55×10^9 years. This method is especially useful for igneous volcanic rocks and deposits that contain K-rich minerals. The method either compares the relative amounts of K and Ar or measures the ratio of radiogenic ^{40}Ar to nonradiogenic ^{39}Ar. In spite of the very long half-life, the usefulness of this method can be extended to dating relatively young samples, but generally no younger than about 75,000 years. Contamination from atmospheric argon can also be a problem.

Fission-track dating uses the fact that when an atom of uranium decays by spontaneous fission, the resulting *isotopes* passing through the adjacent crystal leave a track of damage behind them. These tracks can be observed in an electron microscope or with a standard petrographic microscopic if they are enhanced by etching. Because ^{238}U fissions spontaneously at a constant rate, the age of a mineral or glass can be determined from the number of tracks present, if the amount of uranium is known. Because the tracks anneal out at elevated temperature, the dating technique provides estimates for the length of time since the mineral passed through the temperature above which annealing occurs rapidly. This annealing temperature is relatively low, roughly 120°C or so, depending on the mineral, so the technique is often used to determine rates of uplift and associated cooling.

Thermoluminescence dating depends upon the effect of ionizing radiation on minerals. Minerals that grow in a sediment or a lava flow may be subjected to alpha, beta, or gamma radiation from the decay of radioactive nuclei of K, U, or Th. This radiation causes ionization of the atoms in the crystals. When the sample is heated, the electrons re-attach to the atoms, producing a characteristic glow at a temperature that varies depending on the mineral. If the rate of production of such ionizing radiation is known, the age can be calculated by comparing the amount of light produced during heating with that produced after a known dose of laboratory radiation. This technique has proved useful in dating loess deposits, eolian sands, and some water-laid sediments and soils. The ages accessible with this technique range from 1000 to 1,000,000 years.

Relative dating techniques cannot yield absolute ages but can provide the age of one deposit relative to that of another. In this sense, they are reminiscent of traditional stratigraphic techniques.

When an animal dies, the living amino acids in the animal's proteins, which are in the so-called L configuration, convert slowly to the D configuration; this process is called amino acid racemization. The rate of racemization depends chiefly on the temperature history after death, the genus of the animal, the type of amino acid, and the type of postdepositional diagenetic processes. Despite these complex factors, it has been possible in some cases to develop an idea of the relative ages of deposits having a similar character, and these correlations may be calibrated with other techniques, such as radiocarbon dating. The technique has been used most successfully on mollusks from Pacific coastal deposits, on lacustrine deposits in the Basin and Range province, and on terrestrial gastropods taken from glacially related sediments in the midcontinent.

As volcanic glass weathers in the natural environment, it picks up water. Hydration begins at the surface of a deposit and proceeds inward at a rate proportional to the square root of time. The obsidian hydration dating technique has been calibrated with K-Ar dates in the Yellowstone region and has proved to be a good indicator of relative ages.

Tephrochronology is based upon the fact that an individual volcanic ash deposit represents a widespread time line. If the deposit possesses unique characteristics by which it differs from other tephra layers of different ages, then identification of this layer provides an identifiable stratigraphic time horizon that can be used for relative dating. The identification of a deposit is based upon a number of criteria, such as the petrography and chemistry of the glass and its phenocrysts, paleomagnetism, and biostratigraphy. So far, this technique has been applied chiefly in the western United States and has resulted in significant revision of the dating of stratigraphic and faulting events that have occurred in the past 1 million years.

Magnetostratigraphy and paleomagnetism together provide a useful dating technique in many instances. The paleomagnetic polarity reversal profile from a given deposit can be correlated with the established paleomagnetic polarity reversal time scale to provide a time period for the deposition. The most recent reversal between the Matuyama reversed-polarity interval and the present Brunhes normal-polarity period occurred about 730,000 years ago. This change provides a worldwide benchmark for the comparison of deposits. Since that reversal, a number of partial changes in the Earth's magnetic field, or excursions, have occurred that may become useful as correlation tools.

11.4 Geologic-Geomorphic Features

Offsets in many geomorphic features—such as streams, terrace deposits, moraines, and soils—provide direct evidence of tectonic movement at time scales greater than one human lifetime. Evidence of recent tectonic movements also includes deformation of sediment or soil along active faults resulting from past earthquakes (paleoseismic indicators), changes in the elevation of coastlines, changes in stream channels, and patterns of slope erosion.

Paleoseismic Indicators

Trenching along active fault zones usually reveals features that provide evidence of a fault's movement history. Figure 11.4 shows a series of profiles across the main San Andreas fault and subsidiary faults. Fault movements caused by specific earthquakes can be identified and dated from an analysis of these profiles. Paleoseismic indicators include the following features.

1. A fault that breaks older strata but is overlain by another unfaulted deposit (Fig. 11.5A). Dating of the broken and unbroken strata brackets the time of the faulting.
2. "Sand blows" or injected sandstone dikes or sills that result from soil liquefaction during an earthquake and that are overlain by younger, unaffected deposits (Fig. 11.5B). They may be present in the surficial deposits whether the fault is exposed or not. Such features seem to be the only remaining identifiable deposits resulting from the 1811–1812 earthquake near New Madrid, Missouri, and from the 1886 earthquake near Charleston, South Carolina.
3. Convolute or disturbed bedding in recently deposited lacustrine or marine sediment (Fig. 11.5C). Such beds are disrupted during an earthquake while in an unconsolidated, water-saturated state. The disturbed beds are then covered by new, undeformed deposits. Dating of the disrupted and overlying deposits brackets the date of an earthquake. Disturbed bedding in lacustrine deposits has been used to infer paleoseismic events in the U.S. Basin and Range province.
4. A fault that formed a topographic scarp that was then covered over by younger sediments (Fig. 11.5D).
5. A stream that is associated with a series of channels each offset from the next by a distance on the order of only feet or meters. Such small offsets suggest that the different channels might have formed between events in a series of earthquakes (Fig. 11.5E), whereas larger offsets are characteristic of channels that have remained occupied by the same stream through many earthquakes.

Changes in Elevation of Coastlines

A large portion of the Earth's coastlines are in regions of tectonic activity, and in these areas marine terraces are widespread. Generally, only a few terraces are developed in any one area, although in a few places they

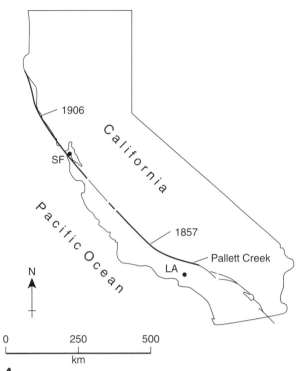

are numerous (Fig. 11.6). Extensive coastal terraces preserve a record of the interaction between land and sea that results from changes in sea level and from uplift or subsidence of the land. The principal task in the neotectonic study of coastal regions is to separate the various effects.

A coastline record of the *relative sea-level changes* shows the changes in sea level relative to the land surface. It includes both *eustatic sea-level changes*, which are worldwide changes that reflect real changes in the volume of the ocean basins or in the volume of water in the oceans, and *apparent sea level changes*, which are the inverse of the true vertical displacements of the land surface:

$$relative\ sea\text{-}level\ change =$$
$$eustatic\ sea\text{-}level\ change + apparent\ sea\text{-}level\ change,$$

or

$$relative\ sea\text{-}level\ change =$$
$$eustatic\ sea\text{-}level - true\ vertical\ land\ displacement.$$

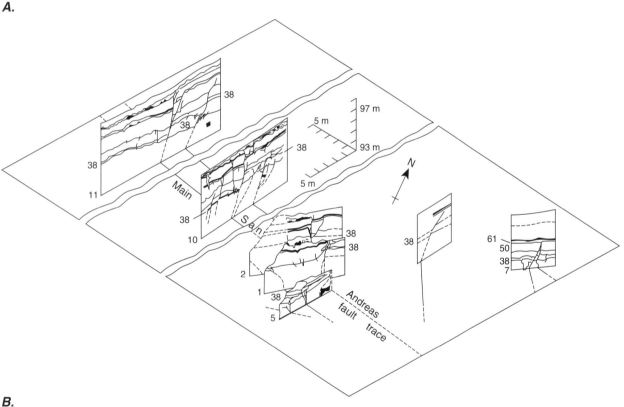

Figure 11.4 An example of cross sections determined from trench walls that cross the San Andreas fault near Pallet Creek, California, at the southeast end of the 1857 earthquake break. *A.* Map slowing location of Pallett Creek. *B.* Fence diagram showing generalized cross sections along trenches. Dark lines are peat layers. Other lines are prominent bedding markers. *(After Sieh, 1978)*

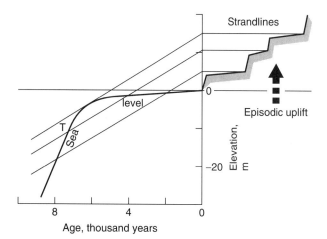

Strandlines

Sea level

T

0

Episodic uplift

Elevation, m

−20

Age, thousand years
8   4   0

Figure 11.8 (*At left*) Relationship between relatively constant sea level during the last 6000 years of Holocene time and successive episodic uplift of shoreline. (*After Lajoie, 1986*)

crease from left to right. As the gradient and velocity increase, a channel tends to change from straight to meandering to braided. This tendency has been verified experimentally. Figure 11.11*B* shows a curve derived from experiments in a laboratory flume that illustrates typical changes. For a given discharge, the shape of the channel will change from straight to increasingly meandering to braided with increase in slope.

about 6000 years ago. The profile of Holocene marine terraces is plotted on the right side as elevation versus distance. The highest marine terrace is associated with the bend in the Holocene sea-level curve, which was the time when eustatic sea-level stopped rising and became essentially constant. During the formation of lower and younger terraces there were no eustatic sea-level fluctuations; thus the terraces must record individual tectonic uplift events.

Figure 11.9 shows the *average* rates of Holocene and Pleistocene uplift for a number of coastal regions around the world. Both Pleistocene and Holocene rates range from nearly zero to a maximum of between 8 and 16 m/1000 yr (these units are the same as units of mm/yr). Comparison of these rates with historical uplift rates indicates that in the regions specified in Figure 11.9, uplift, tilting, and folding appear to be coseismic, rather than aseismic.

Stream Response to Tectonic Movements

Streams are active erosional and depositional agents that tend to maintain a dynamic equilibrium among such factors as slope, sediment load, bed erodibility, and discharge rate. The equilibrium or graded profile of any stream is characterized by a progressive decrease in slope from headwater to mouth (Fig. 11.10). The actual elevation at a given point along a stream is related to the elevation of its mouth, called the **base level.** If something acts to change the elevation of a point along the river's course so that it is no longer in equilibrium, the stream tends to erode its channel or to deposit sediment to regain equilibrium.

As the slope of a stream changes, its channel changes as well. Figure 11.11*A* illustrates the change of stream channel with changes in channel parameters. In the diagram, the stream gradient increases from top to bottom, and the velocity and power in-

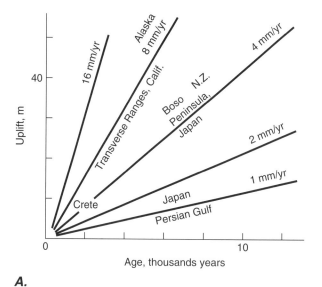

A.

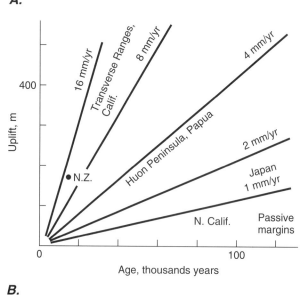

B.

Figure 11.9 Rates of uplift for selected (*A*) Holocene and (*B*) Pleistocene coasts. (*After Lajoie, 1986*)

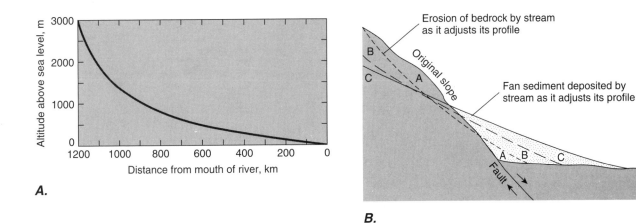

Figure 11.10 Adjustment of stream profile to fault movement. *A.* Profile of graded stream, the Platte and South Platte rivers. *B.* Diagram showing adjustment of stream profile to repeated movement—A, B, and C—by erosion of bedrock and deposition of fan. *(A.* Generalized from data of H. Gannett, in *Profiles of Rivers in the U.S.,* U.S. Geologic Survey Water Supply Paper 44, 1901. *A.* and *B.* in Press and Siever, 1986)

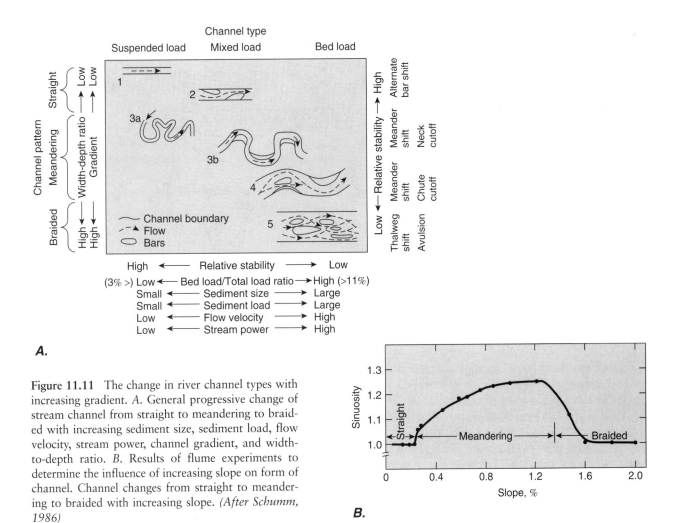

Figure 11.11 The change in river channel types with increasing gradient. *A.* General progressive change of stream channel from straight to meandering to braided with increasing sediment size, sediment load, flow velocity, stream power, channel gradient, and width-to-depth ratio. *B.* Results of flume experiments to determine the influence of increasing slope on form of channel. Channel changes from straight to meandering to braided with increasing slope. *(After Schumm, 1986)*

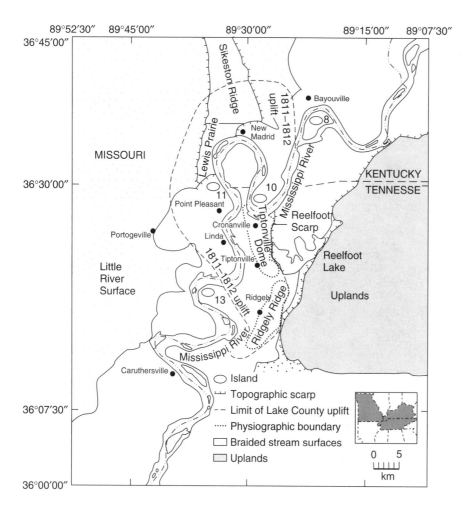

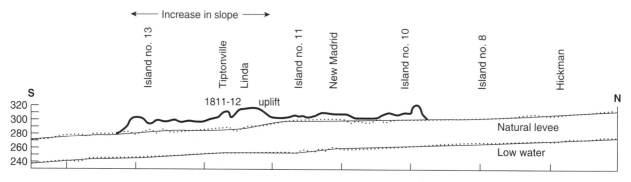

Figure 11.12 Map and profile of the Mississippi River showing increased sinuosity of the river in the region of Lake County, an area of uplift thought to have resulted from repeated movement on the Reelfoot Rift, the fault that produced the 1811–1812 earthquakes. *(After Schumm, 1986)*

The channel of the Mississippi River in the New Madrid region may be an example of this effect. Figure 11.12 shows a map of the region, which was the site of very large earthquakes in 1811–1812. An uplift (labeled the 1811–1812 uplift in Fig. 11.12*A*) was associated with these earthquakes. North of the uplift, between Cairo and Hickman, the river is relatively straight. Between New Madrid and Caruthersville, the river is markedly sinuous, but it becomes straighter again to the south. The area of increased sinuosity corresponds directly to the increase in slope of the river as measured on profiles.

Slope Retreat

Slopes formed by uplift along normal fault scarps tend to retreat either by parallel retreat of the slope, so that the angle remains the same, or by softening and rounding of the slope, so that the declivity of the slope decreases with time (Figure 11.13). The two

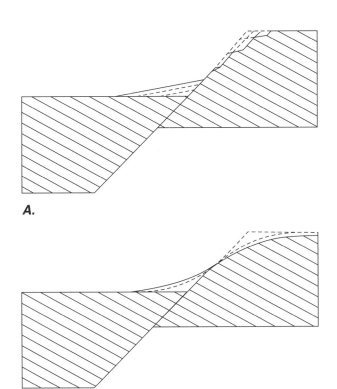

A.

B.

Figure 11.13 *A.* Loosening-limited slope. All material loosened from bedrock is transported away, resulting in parallel retreat of slope. *B.* Transport-limited slope development. All material loosened from bedrock is not transported away, resulting in rounding of slope and a stationary inflection point. *(After Nash, 1986)*

types are called **loosening-limited slopes** and **transport-limited slopes** because of the two rate-limiting factors. In the first instance, the transport mechanisms can remove all the material produced by weathering. In the second instance, more material is produced than the transport processes can remove, and the material tends to accumulate.

Figure 11.14 illustrates an example of the first process, where multiple faceted spurs formed on the western front of the Wasatch Range, Utah, by repeated uplift of the range caused by movement on the fault. This pattern implies that the slope underwent parallel retreat between discrete events (see Fig. 11.13A). In transport-limited slopes, new movement on a scarp shows up in the middle of an older rounded scarp (see Fig. 11.13B) formed by downslope movement of loose material.

Mountain-front sinuosity, defined as the length of the mountain front (assuming a measuring device of a given length) divided by the straight-line length, is another measure of the activity along a fault. The rationale for this measurement is that as a scarp erodes, it becomes more sinuous, whereas new scarps are relatively straight. If the uplift rates for a given area can be correlated with a specific sinuosity, this index can provide a quantitative indication of the rate of uplift. If not, this index at least provides an indication of relative uplift rates. Figure 11.15 shows this relationship for a desert region in southeastern California. The west side of the Panamint Range (Fig. 11.15B), an area of rapid uplift, shows a straight, steep mountain front, with little penetration of alluvial fans into the range itself. The east side of the Slate Range, an area of moderately rapid uplift, shows some penetration of the range by the alluvial fan deposits, and the contour shown in Figure 11.15A is more sinuous. An area of slow uplift, south of the Granite Mountains, shows a landscape dominated by alluvial fans, with isolated upland regions; the contours in this region are highly contorted.

Figure 11.14 Flight of faceted spurs along the Wasatch Range west front, Utah, thought to have been formed by parallel slope retreat in response to multiple movement on the Wasatch fault. *(Photograph courtesy of W. K. Hamblin)*

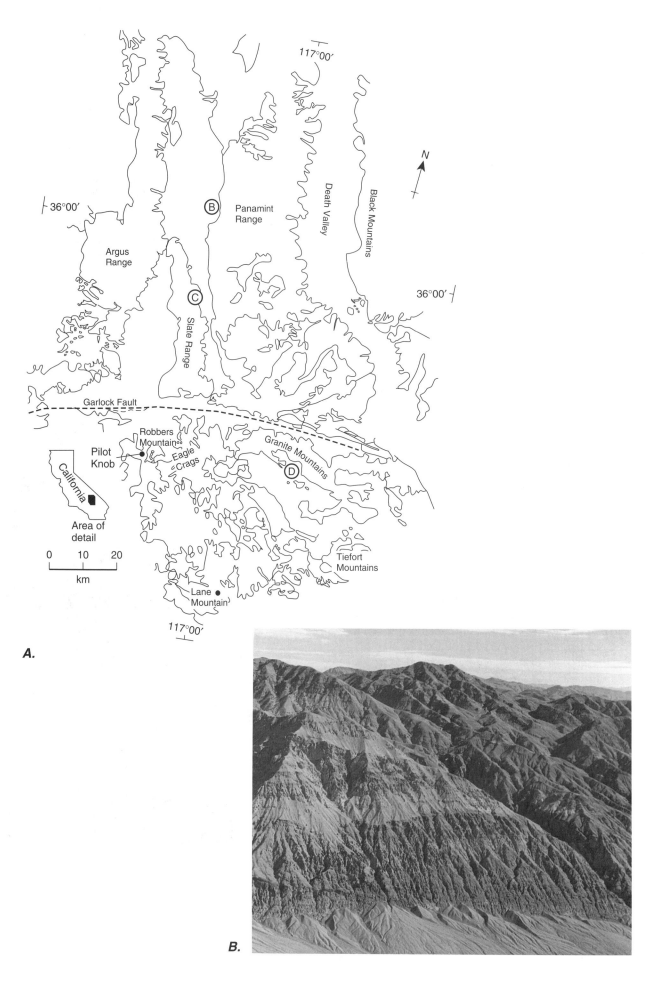

A.

B.

C.

Figure 11.15 Variations in sinuousity of mountain fronts, southeastern California. Classes are related to rate of uplift. Class 1, 1–5 mm/yr; class 2, 0.5 mm/yr; class 3, 0.05 mm/yr. *A.* Map showing region and approximately location of parts *B, C,* and *D* (circled letters). Note the differences in the shapes of the contours. *B.* Class 1 mountain front along the east side of Saline Valley, California. Mountain-piedmont junction coincides with a normal fault. Mountain slope displays slightly eroded faceted spurs. *C.* Class 2 mountain front along the east side of the Slate Range. *D.* Class 3 mountain front on southern margin, Granite Mountains, California. Only remnants of the Granite Mountains remain because uplift ceased long ago and erosion is removing the mountains. *(A. after Bull and McFadden, 1977; photographs courtesy of W. B. Bull)*

D.

11.5 Neotectonic Behavior of Faults and Folds

From analyses of active folds and faults, a pattern of fault and fold behavior has emerged. Exposed active faults appear to be separable into segments, with each segment moving semi-independently of the others. Each segment has a characteristic size of earthquake and an average recurrence time, which together result in the average slip on the fault over a long period of time. Segment boundaries are breaks in the fault trace; stepovers, bends, or offsets along the fault; or transverse boundaries of diverse origin. Figure 11.16 is a diagram of a typical fault trace, showing possible segments along the fault's length. Note that individual segments may be creeping or stuck; stuck segments are pinned (P) if they occur at restraining offset, unpinned (U) if they occur at releasing offset, or bent (B) if they occur at a change in strike.

Figure 11.17 shows segment models for the Wasatch and San Andreas faults. The Wasatch fault (Fig. 11.17A) is divisible into eight segments with different histories of movement. Segment 2 has moved twice in the past 1500 years; segment 1 has not moved in Holocene time; segments 3 and 4 have moved repeatedly in the Holocene; and segment 6 has moved three times in the past 4500 years, the last time about 300 years ago.

The San Andreas fault (Figs. 11.17B and 11.18) can be divided into a northern segment, which last moved in 1906; a central segment that is actively creeping; a south-central segment that has had a complex movement history over the past 1500 years and last moved in 1857; and a southern segment that has not generated large earthquakes in historical time, although a number of earthquakes occurred nearby in the early 1990s.

Many seismologists speak of a characteristic earthquake size, as well as a recurrence interval for such a characteristic earthquake. It seems reasonable to expect a relationship between the average rate of motion on a fault and the size and frequency of earthquakes along that fault. The relationship is a complex one, however, because of such factors as rock type, fault type, and fluid pressure.

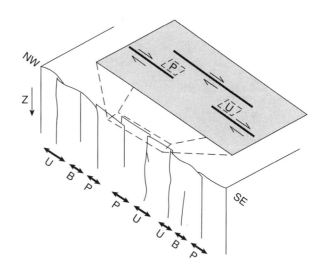

Figure 11.16 Model of segmented fault. Fault is either creeping or stuck. Sticking points are: U, unpinned at a right-stepping offset; P, pinned at a left-stepping offset; or B, bent at a change in strike. *(After Slemmons and Depolo, 1986)*

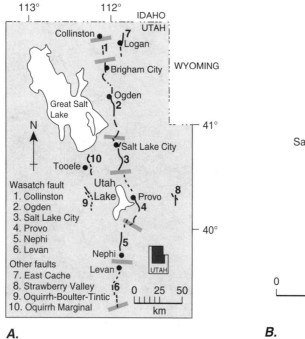

A.

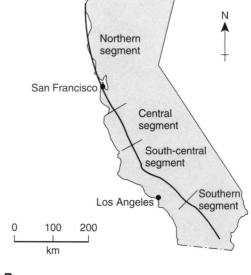

B.

Figure 11.17 Examples of segmented faults. A. Wasatch fault, Utah. B. San Andreas fault, California. *(After Schwartz and Coppersmith, 1986)*

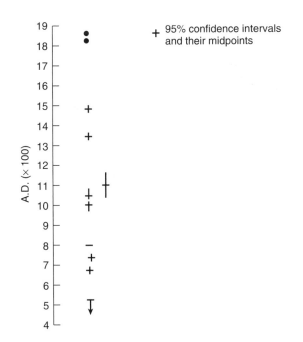

Figure 11.18 Movement history of the San Andreas fault, 55 km northeast of Los Angeles, California, in south-central segment of fault. *(Data from Sieh et al., 1989)*

Table 11.2 shows the definitions of four commonly used earthquake magnitudes: the local magnitude M_L, the surface-wave magnitude M_S, the body-wave magnitude M_B, and the moment magnitude M_W. M_L, M_S, and M_B are empirically derived relationships; they "saturate" at certain values listed in the table. In other words, above these values, the magnitude does not get much larger and does not represent the total energy released by the earthquake. Because the moment magnitude M_W is related to specific geometric and kinematic properties of the fault, it is unlimited in size. There are approximately linear relationships between M_L, M_S, and M_B. An empirical relationship between M_W and M_L determined by D. H. Chung and D. L. Bernreuter is

$$M_W = 0.63 + 0.887 M_L$$

The moment magnitude expresses the relationship between earthquake size and fault displacement. Figure 11.19 shows a possible relationship among earthquake magnitude (given as surface-wave magnitude M_S), recurrence interval, and slip rate on the faults.

The relationship between active folding and seismicity is unclear. The important question is whether folding is aseismic or coseismic; that is, whether folding occurs without earthquakes or in increments accompanied by seismic events. Active folds associated with faults generally form above a décollement or a detachment fault, or at the tip of a propagating dip-slip fault. These folds grow as slip occurs on the fault. Thus, most workers seem to agree that folding is a coseismic process related to faulting on a blind décollement or other dip-slip fault at depth. Folds may also generate seismicity because of flexural slip during folding. A key challenge for the future is to evaluate the seismic recurrence rate and its relationship to fault slip or fold growth rate.

Table 11.2 Earthquake Magnitudes

Symbol	Type	Comments
M_L	Local magnitude	The original Richter scale. Developed in southern California for earthquakes with epicentral distances of less than 600 km and focal depths of less than 15 km. Uses waves with periods of about 1 s. Saturates at $M = 7.25$.
M_S	Surface-wave magnitude	Suitable for global distances. Uses waves with 20 s periods. Saturates at about $M = 8.6$.
M_B	Body-wave magnitude	Suitable for global distances. Uses waves with periods of about 1 s. Saturates at $M = 7.25$.
M_W	Moment magnitude	Based on seismic moment ($M_0 = \mu AD$), where μ = shear modulus, A = area of fault rupture, and D = fault displacement. $M_W = 2/3 \log M_0 - 10.7$. Does not saturate.

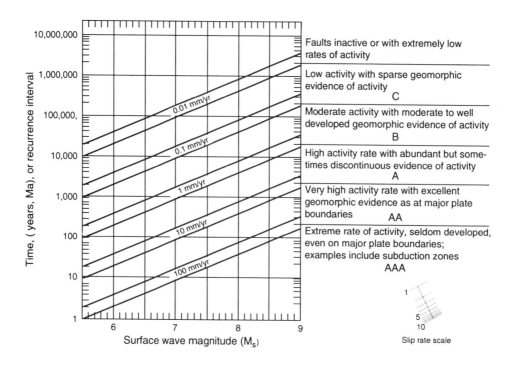

Figure 11.19 Possible relationship among recurrence interval of earthquakes, earthquake magnitude, and slip rate on a fault. The figure assumes that all energy is released in seismic events. *(After Slemmons and Depolo, 1986)*

Additional Readings

Carter, W. E., and D. S. Robertson. 1986. Studying the earth by very-long-baseline interferometry. *Scientific American*:46–54. November.

Dixon, T. H. 1991. An introduction to the global positioning system. *Rev. Geophys.* 29:249–276.

Ellis, M. A., and D. J. Merritts, assoc. eds. 1994. Tectonics and Topography. *J. Geophs. Res.* 99: Special Section, p. 12,133–12,315; 13,869–14,050; 20,061–20,321.

Geophysics Study Committee. 1986. *Active Tectonics.* Washington, D.C.: *National Academy Press.*

Hager, B. H., R. W. King, and M. H. Murray. 1991. Measurement of crustal deformation using the global positioning system. *Ann. Rev. Earth and Planetary Sci.* 19:351–382.

Lisowski, M., J. C. Savage, and W. H. Prescott. 1991. The velocity field along the San Andreas fault in central and southern California. *J. Geophys. Res.* 96:8369–8389.

Sieh, K. E. 1978. Prehistoric large earthquakes produced by slip on the San Andreas fault at Pallett Creek, California. *J. Geophys. Res.* 83:3907–3939.

Sieh, K. E., M. Stiover, and D. Brillinger. 1989. A more precise chronology of earthquakes produced by the San Andreas fault in southern California. *J. Geophys. Res.* 94:603–623.

Wallace, R. E., ed. 1990. The San Andreas fault system, California. U. S. Geol. Surv. Prof. Paper 1515:283.

CHAPTER

12

Case Studies of Orogenic Belts

12.1 Introduction

We look now at the characteristics of several orogenic belts and consider the plate tectonic events that led to their development. We can infer the plate movements associated with the origins of orogenic belts younger than about 200 Ma at least in part from the magnetic anomalies in the ocean basins. For older orogenic belts, however, we must rely solely on the paleomagnetic evidence for continental drift and the indirect petrotectonic indicators of plate tectonic activity preserved in the mountain belts themselves. Figure 12.1 illustrates the distribution of the orogenic belts and basement rocks that were formed during the major tectonic subdivisions of Earth's history. Figure 12.2 shows the distribution through Earth's history of the times these regions were formed.

In this chapter, we consider five mountain systems: the North American Cordillera, the Andes, two portions of the Alpine-Himalayan system—the western or Alpine-Iranian and the eastern or Himalayan-Tibetan—and the Appalachian-Caledonide system. We have chosen these belts because they illustrate two relationships: first, the division of orogenic belts into Cordilleran and Alpine type; second, the development of orogenic belts in ocean basin versus plate tectonic time. Cordilleran-type orogenic belts were

formed chiefly by the interaction between an oceanic and a continental region; Alpine-type belts were formed by an oceanic-continental interaction that ultimately became an interaction between two continental masses. The overall dimension of the orogenic belts, however, is approximately the same. We discuss the two Cordilleran-type examples first, followed by the Alpine-type examples.

Although the development of the North American Cordillera and the Andes (co and an, respectively, in Fig. 12.2) spanned much of ocean basin time and part of plate tectonic time, we concentrate on the younger, ocean basin time history. The Alpine-Himalayan system (a-h in Fig. 12.2) developed chiefly during ocean basin time. The Appalachian-Caledonide system (ap in Fig. 12.2) developed during plate tectonic time. In all cases, we find the Wilson cycle to be a major oversimplification of the process that produces orogenic belts.

We take the model of the composite orogenic belt outlined in Chapter 10 (see Fig. 10.3) as a reference point in our discussions. However, we use the model simply as a mechanism for organizing observations of real mountain belts and comparing them with one another. Basically, the model assumes that the orogen is the final result of the collision of two continents, and to that extent, the Appalachian-Caledonide system is most like the model. The Alpine-Himalayan

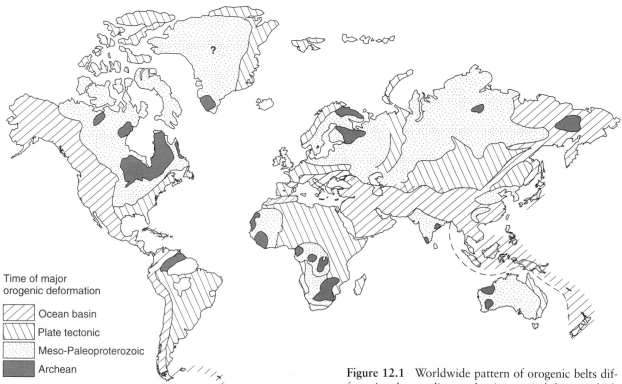

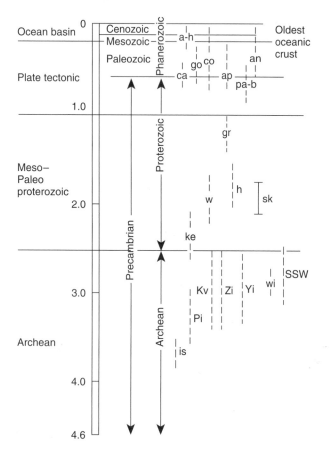

Figure 12.1 Worldwide pattern of orogenic belts differentiated according to the time period during which the deformation occurred. Belts are grouped according to their ages in ocean basin time (1–200 Ma), plate tectonic time (200–1000 Ma), Meso-Paleoproterozoic (1000–2500 Ma), and Archean (2500–3800 Ma). *(After B. C. Burchfiel, 1983)*

Figure 12.2 Bar graph showing times of formation of orogenic belts, separated into ocean basin, plate tectonic, Meso-Paleoproterozoic, and Archean times.

ca	Calidonide	a-h	Alpine-Himalayan
is	Isua, Greenland	go	Gondwanide
pi	Pilbara, Australia	co	Cordilleran-circum-Pacific
Kv	Kapvaal-South Africa	an	Andean
Zi	Zimbabwe	ap	Appalachian
Yi	Yilgarn, Availia	pa-b	Pan African–Brazilian
SSW	Superior Slave,	gr	Grenville
	Wyoming; N. America	h	Hudsonian
sk	Svecokarelian	w	Wopmay
		ke	Kenoran
		wi	Witwatersrand sediments

belt, the North American Cordillera, and the Andes show similarities to the model but differ in important aspects partly because they are still being formed.

Finally, we discuss the changes in tectonic style that may distinguish the middle to early Proterozoic and Archean times from plate tectonic and ocean basin times.

12.2 The North American Cordillera

Regional Characteristics

The North American Cordillera is one of several orogenic belts marginal to the Pacific Ocean that form the circum-Pacific orogenic system. As shown in Figure 12.3, the Cordillera extends from Alaska to central Mexico. The limits of the chain are somewhat arbitrary because it is part of an essentially continuous belt of deformation that extends from the Verkhoyansk Mountains in eastern Siberia to southern South America (see Fig. 12.1).

The mountain belt is part of the currently active plate tectonic margin of western North America, and therefore it is still evolving. The margin is a consuming margin beneath Alaska, the U.S. Pacific Northwest and southern-most Canada, and Central America; it is also a transform margin connecting the trenches in Alaska and the Pacific Northwest and connecting the Mendocino triple junction to the Gulf of California, which comprises the San Andreas fault system. Extension is also still active in the Basin and Range province.

The belt exhibits a long history of orogeny extending back into the late Precambrian and thus spanning all of ocean basin time and much of plate tectonic time (co in Fig. 12.2). Thus, we can compare only the more recent evolution of the system with the seafloor record of plate motions. For an analysis of the entire tectonic evolution, we must rely in part on terrane analysis and the paleomagnetic record from various blocks in the orogenic belt.

Although structures in the orogenic belt are west-vergent on the western side and east-vergent on the eastern side, the belt is not as bilaterally symmetrical as the composite orogen model shown in Fig. 10.3. On the western side the underthrust plate is oceanic, whereas on the eastern side the underthrust plate is continental. The foredeeps and associated structures are well developed in the United States and southern Canada, but only on the eastern side of the orogen (see Fig. 12.3). Modern deep-sea fans are the counterparts to foredeeps on the Pacific side. The foredeep in the western United States (see Fig. 10.4) is principally of Cretaceous age and is deformed by the thrust faults of the foreland fold-and-thrust belt and by the foreland basement uplifts (see Fig. 12.3).

The foreland basement uplifts of the western United States have no counterpart in the composite model. They are chiefly of late Cretaceous to early Tertiary age and consist of fault and fold structures in shelf sediments and underlying Precambrian crystalline basement (see Fig. 10.5). They are an anomalous feature of the North American Cordillera and are not easy to explain using current plate tectonic models.

As in the composite orogen model, most contractional deformation events can be attributed to collision, although many workers have advocated noncollisional processes. The fold-and-thrust belt is well developed along the whole eastern side of the orogen from western Alaska to the southwestern United States, and it continues into northeastern Mexico (Fig. 12.3). On the west side of the orogen, active folds and thrusts are present in the California Coast Ranges; active to inactive folds and thrusts are also present in accretionary prism rocks more or less continuously along the Pacific coast.

Rocks deformed in the eastern fold-and-thrust belt include Paleozoic to Mesozoic miogeoclinal rocks. This belt was deformed principally in the middle to late Mesozoic, with much of the deformation restricted to Cretaceous time. Middle and late Paleozoic thrusts, however, have been described from the boundary between eugeoclinal and miogeoclinal rocks in Nevada (see Figs. 12.3 and 12.4C), and deformations of the same age are present in rocks of the Sierra Nevada (approximately the latitude of cross section C–C'). The late Mesozoic structures exhibit classic décollement-style features, and they are best preserved in Canada. The fold-and-thrust belt in the United States is disrupted by superposition of the extensional Basin and Range deformation (not shown in Fig. 12.4C; but see Figs. 12.4B, 5.4, and 5.5).

The eugeoclinal portion of the Cordillera consists of a series of volcanic, volcanogenic, and deep-water terrigenous sedimentary rocks, as well as some ophiolitic sequences. Volcanic and sedimentary rocks range chiefly from early Paleozoic to Cretaceous in age, but Tertiary rocks are also involved in the deformation. Ages of ophiolitic igneous rocks are late Precambrian to early Paleozoic, late Paleozoic to early Mesozoic, and Jurassic, and the emplacement ages are similar, for the most part. Both well-preserved ophiolite complexes and disrupted sequences in ophiolitic melanges are present.

Volcanic rocks continue to be intruded and extruded right up to Holocene (Recent) time, and they comprise the modern oceanic arc volcanoes of the

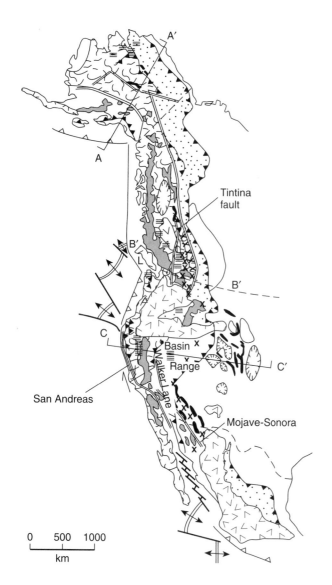

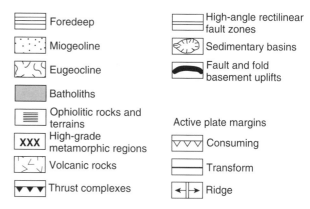

Figure 12.3 Map of U.S. Cordillera showing major tectonic features discussed in the text. Section lines A–A' through C–C', indicate locations of the cross sections shown in Figure 12.4. *(After King, 1977)*

Aleutians, the continental arc volcanics of the northwestern United States, and the related calc-alkaline volcanics in the southwestern United States and in Mexico. The chiefly basaltic plateaus in southern Canada and the northwestern United States are probably associated with the passage of the Yellowstone hot spot. Extensive Cenozoic silicic volcanic outpourings in the United States and Mexico associated with extension in the Basin and Range province are features not included in the composite orogen model. This volcanic cover masks much of the underlying geology, making regional correlations uncertain.

The Cordilleran crystalline core is not as uniformly developed as implied in the model. Batholithic rocks are well developed but discontinuous along strike. The predominant age of such rocks is late Mesozoic, although a few older (middle Paleozoic,

early Mesozoic) and younger (Tertiary) plutons are present. Coarse crystalline (high-grade) regional metamorphic rocks are present chiefly as "metamorphic core complexes" that have been exposed by Cenozoic extension and tectonic denudation, and the role of collisional processes in their development is a subject of controversy.

Cordilleran metamorphic core complexes display both Mesozoic and Tertiary radiometric ages. The metamorphism in these complexes affects thick sequences of miogeoclinal rocks and the underlying basement in the northern United States and southern Canada; in the southern United States and northern Mexico, it affects thinner platform sedimentary rocks and the underlying crystalline basement. This metamorphism of platform and miogeoclinal deposits contrasts strongly with the position of metamorphic rocks

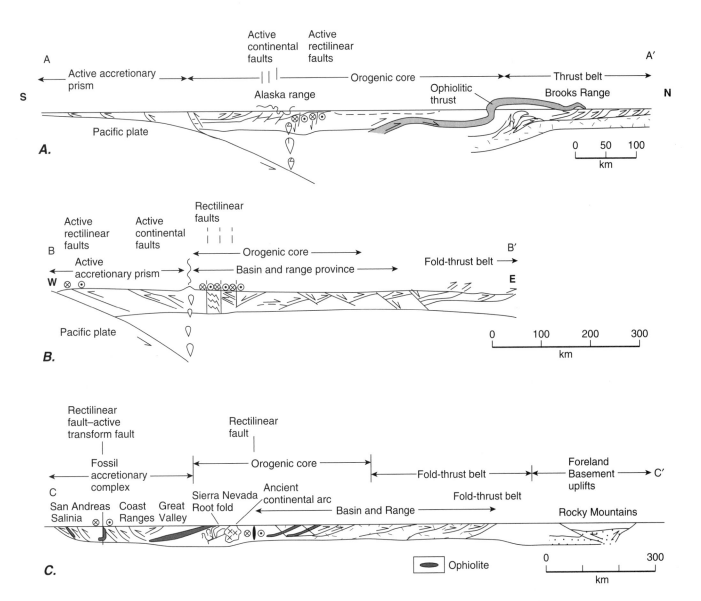

Figure 12.4 Cross sections of the North American Cordillera. Locations indicated in Figure 12.3. *A*. Cross section A–A′ across Alaska. *B*. Cross section B–B′ approximately along the 49th parallel. *C*. Cross section C–C′ of the Cordillera through the Basin and Range province. *(A. after Csejtey et al., 1982; Roeder and Mull, 1978; B. after Potter et al., 1986; C. after Maxwell, 1974; Allmendinger et al., 1986)*

in the composite orogenic model, which occur in the orogenic core region.[1]

---

[1]There is a possibility that similar metamorphic complexes may occur in other mountain ranges. At least two examples have been proposed for the Alpine-Mediterranean system (the Aegean area and the Tauern window in the Alps; see Box 12.1), which at least raises the question of whether many of the exposures of multiply deformed and highly metamorphosed rocks in core regions of other mountains belts could result from similar unrecognized extensional events.

Elsewhere in the Cordillera, metamorphism is lower grade. Blueschist metamorphism, generally of late Mesozoic age, is prominent along the western continental margins of the United States, Mexico, and Alaska, as well as in the Seward Peninsula of west-central Alaska.

High-angle (mostly strike-slip) fault zones are numerous throughout the Cordillera and display a variety of ages. Large faults in Alaska are currently active, as is the San Andreas system. The long Tintina fault in Canada and its companion, the Straight Creek system, were dominantly active with dextral displacement during late Mesozoic time. Faults in the United States and Mexico include the dextral Walker Lane system of Mesozoic to Recent age and the Jurassic sinistral Mojave-Sonora fault.

Extensional structures are widespread in the Cordillera, most obviously in the Basin and Range province, which is thought to be an incipient continental rift zone. In addition to the modern extension (see Sec. 5.2), an earlier phase of extension occupied much of middle Tertiary time and was associated with widespread silicic volcanism. A similar episode of extension, though without the obvious volcanism, characterized the high-grade metamorphic region straddling the United States–Canada boundary. Some writers have argued that extension in the forearc region (principally the Franciscan complex) gave rise to the exposure of blueschist metamorphic rocks.

Cross Sections of the Cordillera

The structure of the Cordillera is represented in the three schematic cross sections shown in Fig. 12.4. The cross sections are generalized from existing data including seismic reflection profiles along portions of cross sections B–B' and C–C'. All three cross sections exhibit a crude bilateral symmetry of thrust vergence, which is toward the continent on the east (or north in Alaska) side of the orogen and toward the Pacific Ocean on the west side (or south side in Alaska). Only the east (or north) side, however, is characterized by a classic thin-skinned fold-and-thrust belt involving the usual miogeoclinal rocks.

Cross section A–A' (Fig. 12.4A) shows the structure across Alaska from the presently active subduction zone in the south to the Brooks Range in the north. Proceeding from south to north, south-vergent structures associated with, and synthetic to, the present subduction zone apparently truncate earlier north-vergent structures in the Alaska Range. Neither of these thrust complexes could be called thin-skinned.

Ophiolitic rocks are present in the Brooks Range and around the metamorphic rocks of the Seward Peninsula. A vergence reversal or backfold is present in the southern Brooks Range. A thin-skinned fold-and-thrust belt is present along the northern edge of the section.

Cross section B–B' (Fig. 12.4B) shows the structure across southern Canada and the northern United States. In the west is a continental arc complex associated with present east-dipping subduction and characterized by active west-vergent structures in the accretionary prism and an active volcanic arc, the Cascade volcanoes. A series of dextral strike-slip faults cuts the west-vergent structures where transform faulting has replaced subduction. To the east is a metamorphic orogenic core characterized by a series of distinct high-grade metamorphic terranes that form a region of east-vergent recumbent nappes cut by normal faults that dip in both directions. To the east of the orogenic core lies the east-vergent décollement-style foreland fold-and-thrust belt. Ophiolitic rocks are present in the section but are not shown in the figure. The outcrops are small, and their tectonic significance is disputed.

Cross section C–C' (Fig. 12.4C) shows the structure across the central United States from northern California to southern Wyoming. In the California Coast Ranges on the west side of the orogen, the structures are chiefly west-vergent thrust faults and folds synthetic to the east-dipping subduction that preceded initiation of transform faulting. East-vergent structures are also present, however. The San Andreas transform fault system cuts the west-vergent structures but may itself bottom out in a midcrustal décollement. East of the Coast Ranges, the Great Valley sequence accumulated in a forearc basin atop ophiolitic basement.

East of the Great Valley, the Sierra Nevada consists of several distinct low-grade metamorphic terranes separated by major faults that are marked in places by ophiolites. The core is pervasively intruded by plutons that range in age from Jurassic to Cretaceous. Major west-vergent structures in the Sierra Nevada may represent a backfold. West-dipping seismic reflectors on a COCORP line in this region suggest that east-vergent thrust faults are present at depth; these may correlate with a few east-vergent thrusts in the eastern Sierra Nevada and western Nevada. Continental arc volcanics presumably related to pretransform east-dipping subduction off the coast occur near the Sierra crest and to the east.

The steep eastern scarp of the Sierra Nevada marks the beginning of the Basin and Range province, much of which is separated from the Sierra Nevada by the Walker Lane, a dextral high-angle fault zone of possibly large displacement. Its presence between the metamorphic batholithic core and the fold-and-thrust belt to the east is not consonant with the composite orogen model. The older structures of the western Basin and Range consist of east-vergent thrust-faulted eugeoclinal rocks, including some large ophiolites. These eugeoclinal and ophiolitic thrust sheets are in an unusual location between the batholithic belt and the foreland fold-and-thrust belts. East of these eugeoclinal rocks, major metamorphic core complexes, mentioned above, affect the miogeoclinal rocks. Still farther east, east-vergent thrusts define a foreland fold-and-thrust belt in miogeoclinal rocks. The earlier

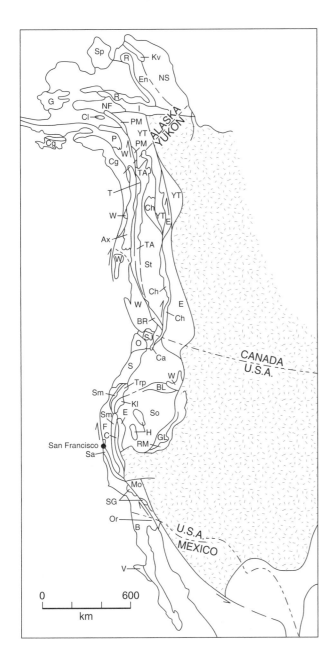

Principal terranes

Alaska
NS	North Slope
Kv	Kagvik
En	Endicott
R	Ruby
Sp	Seward Peninsula
I	Innoko
NF	Nixon Fork
PM	Pingston and McKinley
YT	Yukon-Tanana
CI	Chulitna
P	Peninsular
W	Wrangellia
Cg	Chugach and Prince William
TA	Tracy Arm
T	Taku
Ax	Alexander
G	Goodnews

Canada
Ch	Cache Creek
St	Stikine
BR	Bridge River
E	Eastern assemblages

Washington, Oregon, and California
Ca	Northern Cascades
SJ	San Juan
O	Olympic
S	Siletzia
BL	Blue Mountains
Trp	Western Triassic and Paleozoic of Klamath Mountains
KI	Klamath Mountains
Sm	Smartville (California) and Josephine (Oregon) belts
F	Franciscan and Great Valley
C	Calaveras
E	Northern Sierra, eastern belt
SG	San Gabriel
Mo	Mohave
Sa	Salinia
Or	Orocopia

Nevada
So	Sonomia
RM	Roberts Mountains
GL	Golconda
H	Humboldt

Mexico
B	Baja
V	Vizcaino

Figure 12.5 Map showing distribution of principal tectonostratigraphic terranes in North America. Extent of craton shown by pattern. *(After Coney et al., 1980)*

structures are all severely disrupted by extensional normal faulting.

Foredeep basins at the eastern edge of the fold-and-thrust belt were involved in the younger thrusting and in anomalous foreland thrusts that uplifted Precambrian basement and its platform cover.

Suspect Terranes and the Origin of the Orogen

Much recent work on the North American Cordillera has centered on the role of suspect terranes in the various orogenies. Throughout the Paleozoic, much of the western margin of the North American continent was a passive continental margin accumulating a typical sequence of miogeoclinal sediments. This passive margin was affected by deformation in Devonian-Mississippian time and in Permian-Triassic time. Neither of these deformations caused extensive deformation in the miogeocline, however. The miogeoclinal fold-and-thrust belt did not develop until the main Cordilleran orogeny in Jurassic to early Tertiary time.

West of the miogeoclinal fold-and-thrust belt, the Cordillera apparently is composed of a bewildering agglomeration of allochthonous or suspect terranes of various ages, rock types, and provenances. Figure 12.5 above shows a map of the principal terranes that have been identified. Some of these terranes seem to

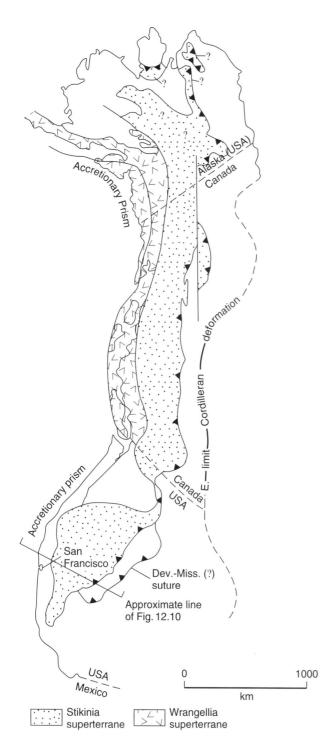

Figure 12.6 Map of western North America showing the generalized superterranes of Stikinia and Wrangellia. *(After Monger, 1984)*

The timing of docking of several terranes is similar to that of formation of the fold-and-thrust belt, leading some workers to suggest a cause-and-effect relationship. Specifically, these workers argue that the fold-and-thrust belt formed by collision of the terranes and partial subduction of the continental margin along westward-dipping (or southward-dipping, in Alaska) subduction zones. Subsequent large-scale strike-slip faulting has greatly modified the original geometry of the orogen. Other workers, however, hold that the collisions are of minor importance and that the orogen evolved above a subduction zone that dipped eastward under the continent, above which an antithetic fold-and-thrust belt developed in Jurassic to early Tertiary time along the old migeocline-platform boundary.

The number, complexity, and diversity of terranes exhibited in Fig. 12.5 make suspect any sweeping conclusions about the tectonic history of the orogen itself. In particular, there is still a great deal of controversy about the significance of individual terranes. In Canada and southern Alaska, however, most terranes can be clustered into two superterranes that display a distinctive history and timing of collision with North America: The Stikinia superterrane (named for its principal component, the Stikine terrane in Canada) and the Wrangellia superterrane (named for its major component, the Wrangell terrane of Canada and Alaska) (Fig. 12.6), are themselves composites of different terranes. Their amalgamation from various diverse sequences is summarized in Figure 12.7 (see Sec. 10.10, Fig. 10.36 for the principles involved in this analysis).

The Stikinia superterrane is the more complex. It includes Paleozoic and early Mesozoic deep-sea sediment, intraoceanic arc rocks, and ophiolitic rocks, as well as a distinctive shallow-water fusulinid-bearing limestone that is present chiefly as exotic blocks in melanges and olistostromes. All these rocks form a complex basement to a large early Mesozoic arc. The Stikinia superterrane extends from the southwestern United States northward through Canada and possibly into western Alaska; southward, the terrane may extend into Mexico and Central America. This superterrane collided with North America in middle Jurassic time in the United States, Canada, and possibly Mexico, and perhaps in early Cretaceous time in western Alaska and central America. Structural and stratigraphic evidence from Canada, the Sierra

have formed at or near their present positions, but others are clearly exotic. The docking of these terranes with the North American continental margin completed the assembly of what we now recognize as the Cordilleran orogen.

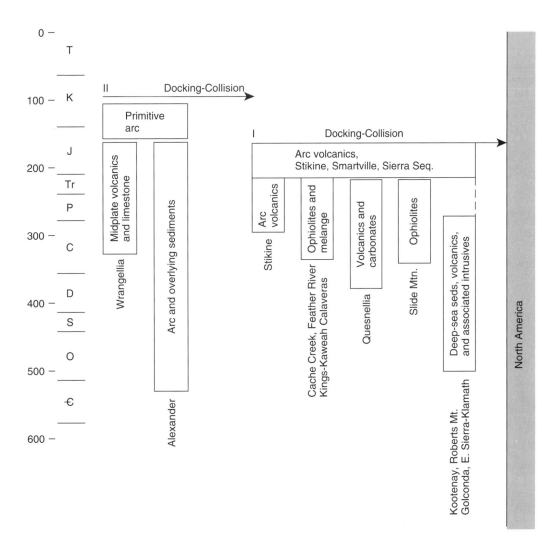

Figure 12.7 Terrane diagram showing development of two North American superterranes and their accretion to (or docking with) North America. Canadian terranes and their tentative correlations with U.S. terranes are shown. (See Figure 10.36B for principles of interpreting such diagrams.) *(After Monger, 1984)*

Nevada of California, and western Nevada indicate that the arc was active, but whether it was above a west- or east-dipping subduction zone is still ambiguous. This event apparently gave rise to the early décollement fold-and-thrust structures present along the miogeoclinal margin. If the subduction zone dipped west beneath the arc, the collision may have generated the fold-and-thrust belt and then been followed by a reversal in the subduction polarity.

The Wrangellia superterrane is composed principally of two fragments: the Alexander terrane, composed of lower Paleozoic oceanic arc deposits; and the Wrangellia terrane, composed of upper Paleozoic oceanic arc rocks overlain by lower Mesozoic oceanic

plateau rocks. These two terranes were amalgamated by late Jurassic time, when they formed the basement of an oceanic island arc that was active until it collided with North America in middle to late Cretaceous time. The polarity of the subduction zone beneath the Wrangellia superterrane is unclear.

Paleomagnetic data from each superterrane are incomplete. Some workers argue that the Stikinia superterrane may not have shifted greatly in latitude or may have arrived considerably to the south, off Baja California. Most agree, however, that the Wrangellia superterrane underwent large-scale shift in latitude (Fig. 12.8). It is not clear from the paleomagnetic data whether Wrangellia was derived from north

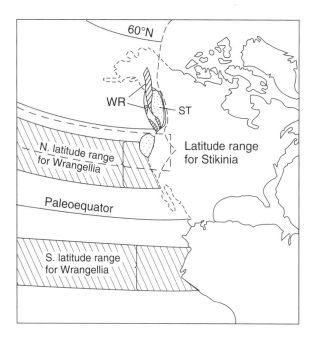

Figure 12.8 Paleolatitudes from Wrangellia (WR) and Stikinia (ST) superterranes for Triassic time. The Wrangellia superterrane could have been either north or south of the equator.

or south of the equator; hence, the two bands for Wrangellia. Indeed, some workers have suggested that the Alexander terrane may have originated in Australia.

Using oceanic magnetic anomaly data and the positions of continents, it is possible to obtain an idea of the plate configuration in the Pacific Basin in the past as well as the relative motion of the major oceanic plates with respect to North America. We have already presented generalized maps of the world plate configuration for times since the middle Jurassic, about 160 Ma (see Fig. 4.21). The magnetic anomaly and paleomagnetic data suggest the migration path for Wrangellia shown in Figure 12.9. The Stikine superterrane apparently collided at approximately 165 to 170 Ma. The presence of volcanic and plutonic rocks of this age in Stikinia implies the presence of an active subduction zone, shown in Figure 12.9 as a west-dipping zone, that gave rise to early phases of the east-vergent fold-thrust belt. The position of

Wrangellia at the time of the Stikinia collision is shown as the region with opposing thrust faults.

The hypothetical reconstruction shown in Figure 12.9 suggests that during Mesozoic time, western North America was bordered by a series of complex island arcs and small ocean basins reminiscent of those present today in the western Pacific. The migration and intersection of these offshore oceanic features with the North American continent gave rise to at least some of the features we see there. It is unclear whether all major tectonic events of the western Cordillera relate to the collision of allochthonous terranes. In particular, late Cretaceous–early Tertiary deformation of the platform and underlying basement

Figure 12.9 Trajectories of Wrangellia position versus time as indicated by reconstruction of the spreading record in the Pacific Basin and correlation with paleomagnetic data. Two trajectories are shown for exposures in southern Alaska and Vancouver Island, respectively. Approximate position of the Wrangellia superterrane at the time of its amalgamation is shown. This time corresponds with that of the collision of the Stikinia superterrane with North America. The southern hemisphere position for the Wrangellia superterrane in Triassic time shown in Figure 12.8 is more likely. *(After Debich et al., 1987)*

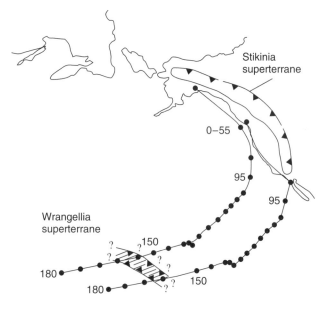

in the U.S. Rockies and Colorado Plateau seems difficult to relate to any large-scale collisional event. Many workers ascribe this deformation to an increase in the absolute motion of North America westward, which caused it to override a fringing east-dipping but nearly flat subduction zone.

Plate Tectonic Cross-Sectional Model of the Cordilleran Orogen

Three principal plate tectonic models have been proposed for the Cordilleran orogen. These models reflect the controversies over the number and types of collisions and the polarity of subduction zones at various times. Two are noncollisional models developed before the widespread recognition of suspect terranes. The first model proposes that the angle of subduction during late Mesozoic and early Tertiary time was so shallow that the overriding continental plate remained coupled to the downgoing oceanic plate as much as 1500 km from the coast. The model proposes that the late Mesozoic–early Tertiary décollement-style fold-and-thrust belt and the basement-involved deformation of the U.S. Rocky Mountains resulted from subduction of this flat oceanic slab. The second model, also a noncollisional model, proposes that a pronounced increase in the rate of subduction caused deformation in the overriding plate at great distances from the subduction zone. Both models invoke only one subduction zone, which dips eastward beneath the continent. The east-vergent fold-and-thrust belt thus developed antithetic to this west-vergent (east-dipping) subduction regime.

The third model is a collisional one; it was also originally proposed before the widespread recognition of suspect terranes but is compatible with their existence. This model holds that the successive episodes of orogenic deformation recorded in the rocks resulted from successive collisions of several different subduction zones with the western margin of North America.

Figure 12.10 shows a cross section of this model for the general latitude of section C–C′ in Figure 12.3. As indicated in the model, North America was characterized by a passive margin in early Paleozoic time (Fig. 12.10A). Approach and collision of an exotic island arc in late Devonian and Mississippian time resulted in the Antler orogeny (Figs. 12.10B and 12.10C). Polarity reversal in middle Mississippian time and back-arc spreading produced a Japan Sea–style continental margin in latest Paleozoic time (Fig. 12.10D). Apparent reversal of that polarity (for reasons still obscure), consumption of the back-arc basin, and subsequent collision resulted in the Sonoma orogeny (Fig. 12.10E). After a polarity reversal following the collision, subduction occurred beneath the de-

formed continental margin in early Meso? (Fig. 12.10F). The approach of an exotic isl? southern continuation of Stikinia?—Fig. 12.? sumed the oceanic crust under two opposite ?? subduction zones, leading to a collision of the arc and the continent. The resulting deformation of the continental margin produced a mountain root and developed the early phases of the miogeoclinal fold-and-thrust belt. Continued subduction beneath the continent produced the late Mesozoic Franciscan–Great Valley–Sierra Nevada trench-arc sequence, possibly with one or more dextral strike-slip faults. Thrusting antithetic to the active subduction zone may have continued in the fold-thrust belt (Fig. 12.10H). The arrival of Wrangellia at the North American continental margin at approximately 90 Ma (Fig. 12.10I) may have deformed the continental margin again and ultimately may have resulted in the final stages of fold-and-thrust belt development and the basement uplifts in the U.S. Rocky Mountains (east end of cross section C–C′). Subsequent continuation of subduction beneath the continental margin, intersection of the trench with the East Pacific Rise, and development of the Basin and Range province, as outlined in Section 4.4 and Figure 4.21, complete the plate tectonic scenario.

We favor this model as a working hypothesis because it has greater explanatory power than the other two and because it is more compatible with a number of principles outlined in the previous chapters:

1. It preserves the parallelism between the polarity of subduction zones and that of fold-and-thrust belts, as outlined in Sections 7.5, 9.4, and 10.3. Both noncollisional models contradict this relationship.
2. It is consistent with the tectonic significance of the emplacement of ophiolites, as outlined in Section 9.6. Neither of the noncollisional models takes into account the implications of ophiolite emplacement.
3. It proposes a number of collisions and subsequent polarity reversals in subduction zones, not unlike those observed in the present-day southwest Pacific, as outlined in Sections 9.1 and 9.2. Neither noncollisional model proposes any such scenario.
4. It provides a tectonic role for the arrival and docking (collision) of the suspect terranes. Neither noncollisional model ascribes any role to these bodies.

This model still leaves unresolved the origin of the Rocky Mountains, except as a possible result of stressing of the entire crust resulting from the collision with Wrangellia.

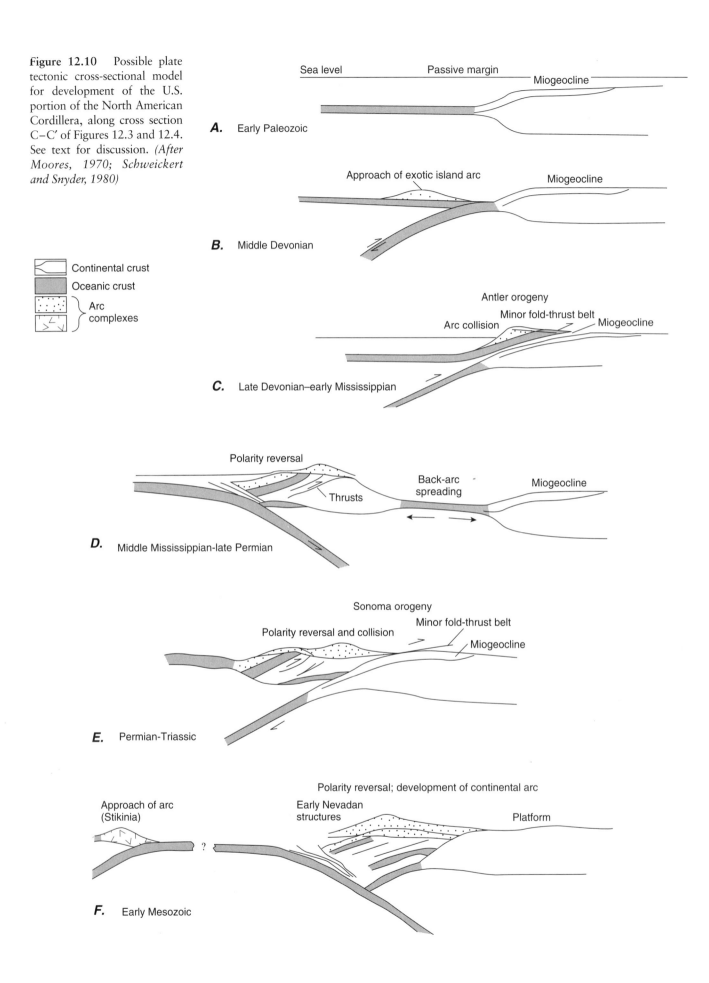

Figure 12.10 Possible plate tectonic cross-sectional model for development of the U.S. portion of the North American Cordillera, along cross section C–C' of Figures 12.3 and 12.4. See text for discussion. *(After Moores, 1970; Schweickert and Snyder, 1980)*

Continental crust
Oceanic crust
Arc complexes

Sea level   Passive margin
Miogeocline

A. Early Paleozoic

Approach of exotic island arc   Miogeocline

B. Middle Devonian

Antler orogeny
Minor fold-thrust belt
Arc collision   Miogeocline

C. Late Devonian–early Mississippian

Polarity reversal
Back-arc spreading   Miogeocline
Thrusts

D. Middle Mississippian-late Permian

Sonoma orogeny
Minor fold-thrust belt
Polarity reversal and collision   Miogeocline

E. Permian-Triassic

Polarity reversal; development of continental arc
Approach of arc (Stikinia)   Early Nevadan structures   Platform

F. Early Mesozoic

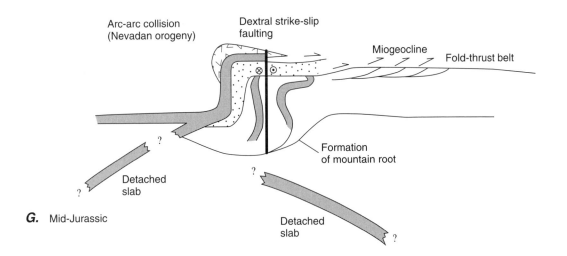

G. Mid-Jurassic

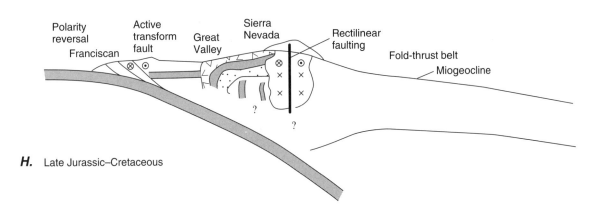

H. Late Jurassic–Cretaceous

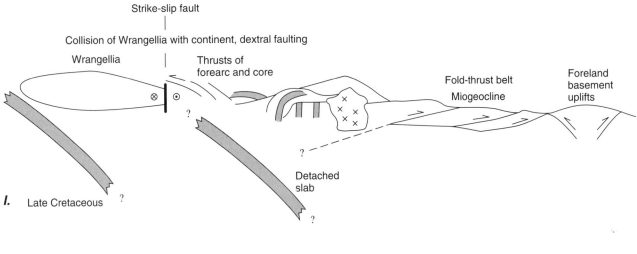

I. Late Cretaceous

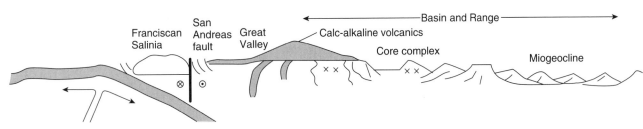

J. Tertiary-Quaternary

12.3 The Andes

The Andes are a 9000-km-long mountain chain along the western margin of South America. The predominant features of the modern mountain belt are the offshore presence of the active Peru-Chile Trench and subduction zone, the corresponding active volcanic arc on land, and an active fold-thrust belt along the eastern margin of the mountain chain. As shown in Figures 12.11 and 12.12, the Nazca Plate is actively subducting beneath South America along much of the length of the Andes. South of the Chile triple junction, a trench-trench-transform fault (TTF) junction, the Antarctic Plate is slowly subducting beneath South America.

The distribution of seismicity suggests that the Nazca Plate has broken into a number of segments of varying dip. Figure 12.11 shows the principal segments, with contour lines, which project onto the South American continent, drawn at 150 and 600 km. Two segments, north and south of the Nazca Ridge (which lies just north of the South American bight, also called the Arica bend), display earthquakes to depths of 600 km. Note that volcano-free segments exhibit relatively shallower subduction zone dips, as indicated by the width of the gap from the trench to the 150-km contour. The active volcanoes overlie segments that dip approximately 30° or more, whereas no volcanoes occur above segments that have a more gentle dip.

Like the North American Cordillera, the Andes have a long, complex history that extends back into the Paleozoic. Terrane tectonics are prominent in discussions of the pre-Mesozoic history of the belt. We focus here, however, on the Mesozoic-Recent, during which most workers believe that a mountain belt developed over an east-dipping subduction zone. For the purposes of this discussion, the Andes are divisible into seven segments, as indicated in Figure 12.12 (labeled A through G; note that the segments do not necessarily correspond to the segments of the Nazca Plate discussed above and shown in Fig. 12.11). We discuss each segment in Figure 12.12 briefly, starting in the north and moving south.

Segment A across Colombia and Ecuador is illustrated by the cross section in Figure 12.13A on page 335. The general features include a western region of collided oceanic blocks, a central region of Cenozoic volcanic and plutonic activity, and an eastern fold-and-thrust belt. Looking in more detail from west to east, we see mafic and ultramafic oceanic rocks mostly west of the Delores-Guayaquil Megashear or suture (DGM). These rocks include three island arc or ophi-olitic complexes; two were emplaced from west to east over the continental margin in Cretaceous time, followed by one that was thrust under the continental margin in late Cretaceous–early Tertiary time. Thus, the folds and thrusts in the western Cordillera show both east and west vergences. In the central Cordillera east of the Delores-Guayaquil Megashear, Mesozoic plutonic rocks intrude into Precambrian continental basement, and the vergence of folds and thrusts is toward the east. In the eastern Cordillera, thick sequences of Paleozoic-Mesozoic-Cenozoic rocks overlie Precambrian basement and are faulted in a foreland fold-and-thrust belt. The vergence of folds and thrusts is toward the continental interior in the east but toward the west near the Magdalena Valley.

Segment B across Peru and Bolivia is shown in cross section in Figure 12.13B. It is distinguished by Permian and Mesozoic batholithic rocks at the coast, which intrude a Precambrian unit, the Arequipa massif. This massif is allochthonous and is thrust eastward over Mesozoic rocks that are described as marginal basin type in the west and continental margin in the east. Most workers consider that the Arequipa massif has been an integral part of the South American Shield at least since middle Paleozoic time. All rocks are involved in east-vergent décollement-style folds and thrusts that affect rocks as young as late Cenozoic and that thrust rocks over the continental platform to the east.

Segment C includes northern Chile and Argentina and is shown schematically in Figure 12.13C. It includes a coastal batholith that is cut by a 1000-km-long strike-slip fault, the Atacama fault. To the east of the fault, but still in the Coastal Range, are allochthonous Paleozoic crystalline rocks and a Jurassic-Cretaceous basin of volcaniclastic deposits, interpreted as an aborted marginal basin. An east-vergent fold-and-thrust belt is present along the eastern Andes.

Segment D includes the region around Santiago, Aconcagua, and Mendoza (Fig. 12.13D). It is characterized by a coastal batholith, to the east of which is a Jurassic-Cretaceous aborted marginal basin similar to the structure in the Coastal Range in segment C. Folds and thrusts are east-vergent throughout the Cordillera Principal along the Chile-Argentina border, the Cordillera Frontal, and eastward into the Precordillera. Farther to the east are the Pampean Ranges, a region of basement uplift and faulting that is reminiscent of the eastern Rockies of Wyoming and Colorado.

Segment E includes the Nequen region of northern Patagonia (Fig. 12.13E). There, the Coastal Range consists of a late Paleozoic accretionary prism complex that is invaded by Paleozoic plutons. East of

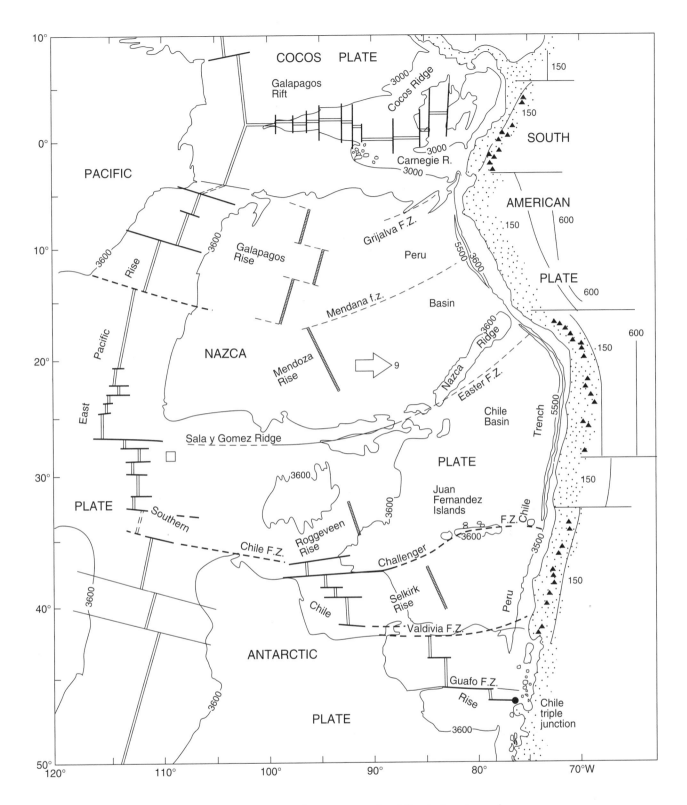

Figure 12.11 Map of western margin of South America and adjacent oceanic region showing principal plate boundaries and areas of active volcanism (solid triangles) that correspond to moderately dipping portions of the Nazca Plate. Lines with adjacent numbers 150 or 600 are contours of depth to seismic zone in kilometers. *(After Green and Wong, 1989)*

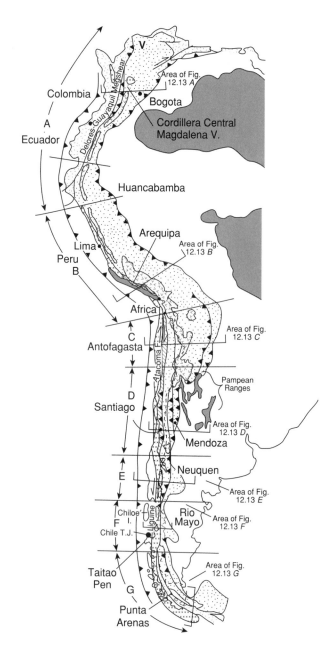

Quaternary volcanics
Precambrian rocks
Oceanic rocks
Marginal basin rocks
Miogeocline associated rift volcanics and sediments
Batholith and associated rocks
Metamorphic core complex

Figure 12.12 Map of Andes, South America, showing principal tectonic features. *(After Mégard, 1989; Mpodozis and Ramos, 1989)*

chain. To the east of this chain there is little or no deformation. As in the segment to the north, a region of presumed back-arc basalts occurs to the east of the mountain chain.

Segment G in southern Patagonia and Tierra del Fuego is represented by a north-south cross section (Fig. 12.13G). In the south, a coastal batholith is overlain by late Cretaceous volcanic rocks. Farther north, an ophiolite belt was thrust from southwest to northeast in middle Cretaceous time over a complex of metamorphic rocks; still farther north is a continent-vergent fold-and-thrust belt of late Cretaceous to early Tertiary age.

A comparison of the cross sections of the Andes (Fig. 12.13) with that of the model orogen of Chapter 10 (see Fig. 10.3) shows that a few of the features of the model are consistent with the actual Andean belt, but there are many incompatibilities as well. As in the model orogen, the continent-directed fold-and-thrust belt is present all along the length of the eastern margin of the Andes. The division between nonmetamorphic outer miogeoclinal belts and a central metamorphic core, however, does not hold up, since for most of its length the Andes belt lacks exposure of a central metamorphic core. Mesozoic ophiolitic rocks are present only in the extreme north and in the extreme south (an ophiolite belt of presumed Paleozoic age is present along the eastern margin of the Andes in Argentina and Bolivia), but the ophiolites in the model are not found consistently in the Andes. The thrust belts that are shown in the model on both sides of the orogenic core with vergence away from the core are present along only a part of the length of

the modern volcanic axis is the Agrio fold-and-thrust belt in which both east-and west-vergent folds and thrusts occur. Still farther east, along the eastern margin of the mountain range, is a region covered by Tertiary and Quaternary basalts that are interpreted to be back-arc basalts, although there is little or no actual spreading.

Segment F includes the Chiloé–Rio Mayo region of central Patagonia (Fig. 12.13F). It consists of a late Paleozoic accretionary complex intruded by Jurassic-Cretaceous plutons. A major left-lateral strike-slip fault, the Liquiñe-Ofqui fault, forms a tectonic boundary along the western edge of the modern volcanic

Figure 12.13 Cross sections of the Andes. *(A. after Vicente, 1989; B. after Roeder, 1988, and Vicente, 1989; C–G. after Mpodozis and Ramos, 1989)*

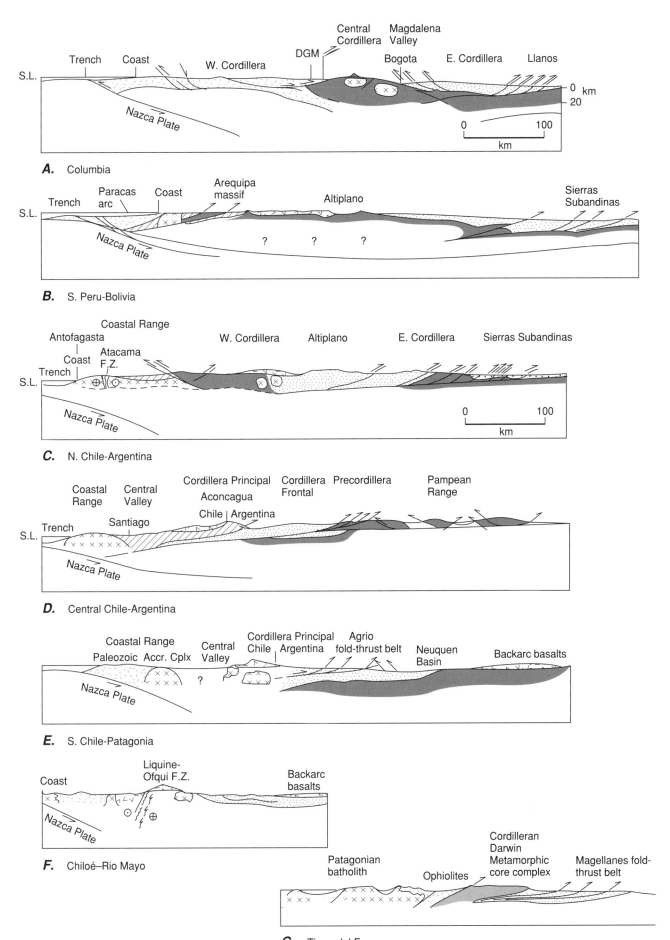

A. Columbia

B. S. Peru-Bolivia

C. N. Chile-Argentina

D. Central Chile-Argentina

E. S. Chile-Patagonia

F. Chiloé–Rio Mayo

G. Tierra del Fuego

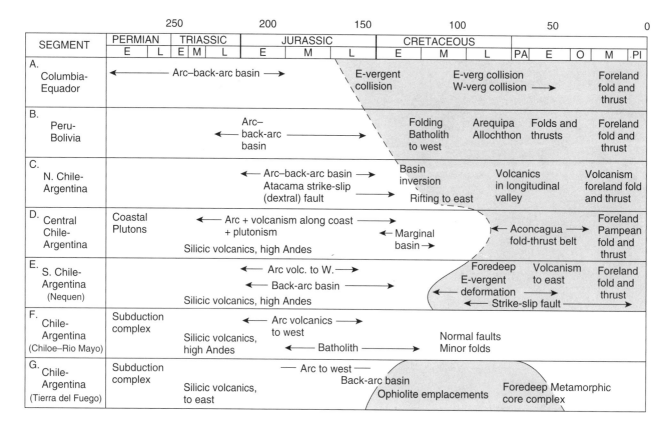

Figure 12.14 Time-space diagram showing principal tectonic events along length of Andes. Heavy dashed line is onset of main Andean deformation.

the Andes, if the active subduction zone along the coast is not included. The amount of crustal shortening, which by implication is substantial in the model, varies considerably along the Andean chain. As much as 250 km of shortening for the eastern fold-and-thrust belt alone has been proposed for segments B, C, and D, whereas in segment F little or no shortening is reported. Although extensional first motions have been recorded in areas of volcanic activity, no well-developed extensional regions have been documented, except for the long (1000 km or so) central valley of Chile.

The supposedly long-active subduction zone is a problem, too, because little accretionary prism is exposed or even developed. Only late Paleozoic subduction complexes are developed in any degree. Many workers attribute the lack of Mesozoic-Cenozoic accretionary deposits to the action of subduction erosion.

Most workers ascribe Andean orogenic evolution to noncollisional processes related to the absolute motion of the South American Plate relative to the hot-spot framework. Nevertheless, collisional processes probably emplaced the ophiolite complexes in the extreme north and south of the chain, and it is diffi-

cult to reconcile the presence of terranes and the thrust relationships with a noncollisional origin for the entire belt.

Figure 12.14 is a time-space diagram illustrating the main tectonic events along the Andes. From the diagram, it appears that volcanism began mostly in the Triassic. Marginal basin or back-arc basin development occurred in the Jurassic. Closure of the back-arc basin and deformation, as indicated by the heavy line, began in late Jurassic time in the south and north and moved toward the center. The onset of the main Andean deformation varies along strike, as well. Except for the undeformed segment F, it is earliest in the north and south and youngest in the center, as shown in Figure 12.14.

Figure 12.15 shows a cross section of segment B together with one for the Alps at the same scale. The width of the Andes cross section, compared with that of the Alps, is striking. As we see in the following section, a complex plate collisional scenario has been proposed for the Alps. For this reason, and because of the complexity of the Cordillera in the northern hemisphere, it seems difficult to us to ascribe the entire development of a much wider belt, the Andes, to interaction at a single subduction zone.

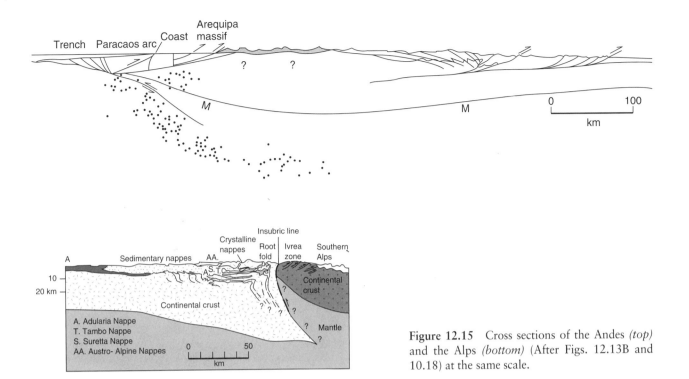

Figure 12.15 Cross sections of the Andes *(top)* and the Alps *(bottom)* (After Figs. 12.13B and 10.18) at the same scale.

12.4 The Alpine-Himalayan Orogenic Belt: Alpine-Iranian Sector

Regional Characteristics

The Alpine-Iranian orogenic belt constitutes the western part of the vast Alpine-Himalayan system that extends from Gibraltar through the Himalayas into China and Southeast Asia where it joins with the circum-Pacific orogenic belt. Figure 12.16 is a generalized map of the whole Alpine-Himalayan chain, and Figure 12.17 is a more detailed map of its western portion, the Alpine-Iranian segment. Although pre-Mesozoic rocks are involved, the Alpine-Himalayan belt developed principally during Mesozoic and Tertiary time and in fact is still in the process of formation. Its evolution, therefore, has occurred during ocean basin time and so can be correlated with plate motions as recorded in the oceanic spreading record.

The Alpine-Iranian belt was developed, however, on earlier orogenic belts, the late Paleozoic Hercynian and late Proterozoic–early Paleozoic Pan-African orogenic systems that developed in plate tectonic rather than ocean basin time. The pre-Alpine rocks of platforms and core zones west of Anatolia were deformed and metamorphosed in the late Paleozoic Hercynian orogeny (Fig. 12.18). In Turkey and Iran, pre-Alpine rocks were deformed and metamorphosed in the late Precambrian–Paleozoic Pan-African oro-

geny. Geologic relationships suggest that these regions of pre-Alpine orogeny are the remnants of microcontinents that had a rifting and subsequent deformational history somewhat different from that of the major continents.

In general, the history of this mountain belt is related to the opening of the Atlantic and Indian oceans, which drove the convergence and collision of the formerly separate continents of Eurasia, Africa (and its Neogene offshoot, Arabia), and India. Most important for the Alpine-Iranian segment of this system are the spreading histories of the northern and central Atlantic, which define the motion of Africa-Arabia relative to Europe. Figure 12.19 presents one interpretation of this relative motion as successive positions of Africa relative to Europe for the times indicated. This interpretation suggests that the principal motion was predominantly sinistral strike-slip from 190 to about 120 Ma (middle Jurassic to middle Cretaceous), followed by oblique convergence to the present.

Despite this relatively simple framework within which the tectonic history of the orogenic belt must have unfolded, the details are highly variable from place to place along the belt, and the motions within the chain are often difficult to link directly to the major plate motions. Thus, no single detailed tectonic history can apply along the whole length of the belt.

Tectonic activity, including rifting and subduction, began as long ago as the early Mesozoic, and in

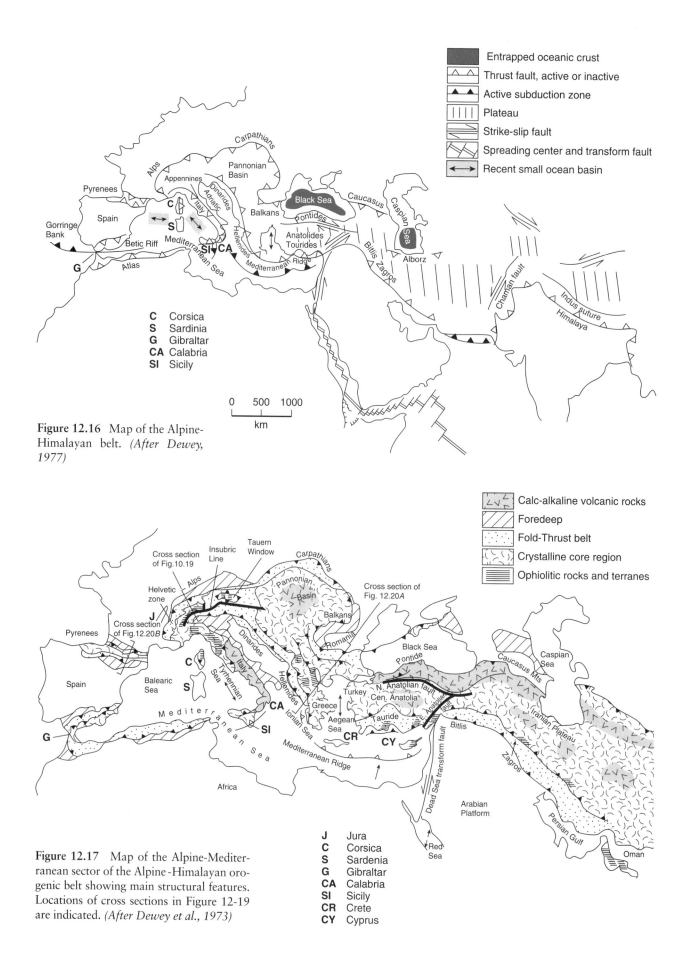

Figure 12.16 Map of the Alpine-Himalayan belt. *(After Dewey, 1977)*

Legend (Figure 12.16):
- Entrapped oceanic crust
- Thrust fault, active or inactive
- Active subduction zone
- Plateau
- Strike-slip fault
- Spreading center and transform fault
- Recent small ocean basin

C Corsica
S Sardinia
G Gibraltar
CA Calabria
SI Sicily

0 500 1000 km

Figure 12.17 Map of the Alpine-Mediterranean sector of the Alpine-Himalayan orogenic belt showing main structural features. Locations of cross sections in Figure 12-19 are indicated. *(After Dewey et al., 1973)*

Legend (Figure 12.17):
- Calc-alkaline volcanic rocks
- Foredeep
- Fold-Thrust belt
- Crystalline core region
- Ophiolitic rocks and terranes

J Jura
C Corsica
S Sardenia
G Gibraltar
CA Calabria
SI Sicily
CR Crete
CY Cyprus

places it continues today. Mesozoic oceanic crust underlies the eastern Mediterranean, Black, and southern Caspian seas. Neogene oceanic crust is present between Spain and Corsica-Sardinia. Subduction is currently active beneath the Aegean arc south of Crete and beneath Cyprus and Sicily (see Fig. 12.17). Back-arc spreading is taking place in the Tyrhennian Sea northwest of Calabria where new oceanic crust is forming and in the Aegean Sea north of Crete. In the Aegean Sea, back-arc spreading is still in a phase of continental extension, as true oceanic crust does not occur. Seismicity beneath Romania suggests ongoing convergence in this area.

Volcanic rocks related to subduction are rare in the western Mediterranean. East of the Alps, however, calc-alkaline volcanic rocks are present in the Pannonian Basin, the Dinarides, southern Italy, and the Aegean Sea. Large volcanic fields are present in eastern Anatolia and in Iran. The relative paucity of volcanics in the western part of the belt compared to the eastern part suggests that the original ocean that closed was narrow in the west and wider in the east, as indicated by the paleogeography shown in Figure 12.19.

In several places, the orogenic belt displays abrupt curves and large changes in the strike of fold-and-thrust belts. Bends of 90° to 180° characterize the Gibraltar region, the Alps, the Carpathians, and the Balkanides (see Fig. 12.17). These bends, together with the complex fold-and-thrust vergences, suggest that considerable rotation and/or strike-slip deformation must have occurred.

Thus, in general, the chain is more complex than is implied by the idealized bilaterally symmetrical model (see Fig. 10.3). Various parts of this chain, however, display features reminiscent of the model cross section.

Foredeep deposits are variable in extent along the margin of the belt. They are well-developed in the Pyrenees and Alps, but somewhat less so in such regions as the Dinarides, Hellenides (Fig. 12.20B), and the Turkish or Iranian ranges.

In several regions, thrust belts are directed away from a mountain core of highly deformed and metamorphosed rocks or from regions of volcanic rocks, consistent with the idealized model. The Alps themselves exhibit a typical cross section (see Fig. 10.18) marked by two outward-directed thrust sequences on either side of a metamorphic core, associated with an apparent offset in the Moho. The Pyrenees section (see Fig. 12.20A) also displays a similarity to the model section. The vergences of the Dinarides and the Carpathians (see Fig. 12.17) are in the opposite directions from each other and away from the volcanic-rich Pannonian Basin. In Turkey, the Tauride and Pontide thrust complexes bound on either side a central core of deformed and metamorphosed rocks and/ or young volcanic rocks that underly the central Anatolian

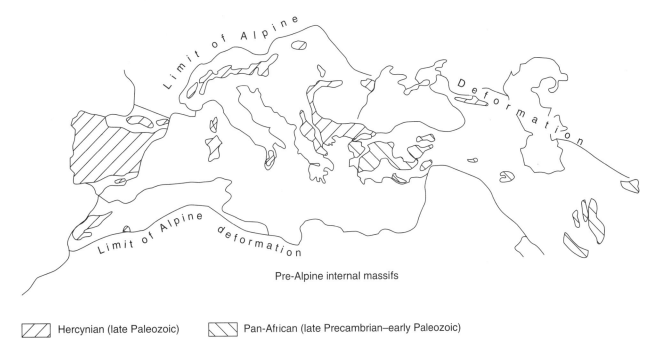

Pre-Alpine internal massifs

⬜ Hercynian (late Paleozoic)   ⬜ Pan-African (late Precambrian–early Paleozoic)

Figure 12.18 Map of the Alpine-Mediterranean region showing main regions of pre-Alpine crust involved in the later deformation. Twofold classification into late Paleozoic (Hercynian) and late Precambrian–early Paleozoic (Pan-African) ages is shown. *(After Dewey et al., 1973)*

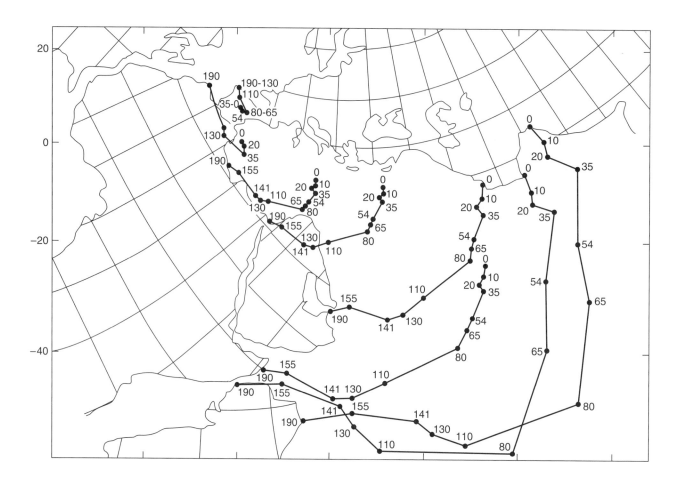

Figure 12.19 Motion of Africa with respect to Eurasia for various times in the Mesozoic-Cenozoic. Positions of the points on the African margin are shown for various times, indicated by numbers. Lines show the motion of points on the African margin relative to Eurasia. *(After Le Pichon et al., 1988)*

Plateau. In these regions, the entire orogenic belt displays a rough symmetry of the fold-and-thrust belts about the orogenic core (for example, see Fig. 12.20A). In other areas, however, the fold-and-thrust belts are directed toward each other and are separated by platforms or basins of minimally deformed rocks, such as the Ionian Sea and the Romanian Basin (see Fig. 12.17).

The character of the overthrust belts also varies considerably from place to place. Some belts are typical thin-skinned types, such as those of the Pyrenees, the southwestern Dinarides and Hellenides, the Taurides, and the main Zagros fold-and-thrust belt southwest of the Zagros Crush Zone (see Fig. 9.8). Others are quite different, such as the Helvetic zone of the Alps, which involves the external basement massifs in the overthrust belt (see Fig. 10.9).

Ophiolites are an important feature of the chain, and they are present from the western Alps and

Corsica in the west to Iran in the east (see Fig. 12.17). In most places, they occupy locations consistent with the model—that is, they are within or thrust over the miogeoclinal fold-thrust belt (for example, Italy, Greece, Turkey), or they form part of the crystalline core zone (for example, Alps, eastern Greece). In Cyprus and Oman, however, ophiolites are in positions at variance with the model. The Oman complex rests tectonically on the Arabian shelf, south of the Zagros fold-thrust belt and thus outside of the orogenic belt proper. The Troodos ophiolite in Cyprus is being emplaced today over the African margin along an active north-dipping subduction zone that is a continuation of the Mediterranean Ridge.

The dates of ophiolite emplacement are highly variable along the belt, indicating a long and complex history of collision. In the western Mediterranean and in central Iran-Afghanistan, the ophiolites were emplaced during the Jurassic and Cretaceous. Along the

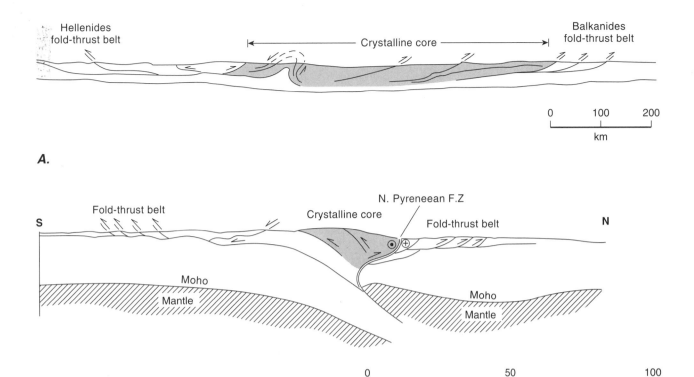

Figure 12.20 Cross sections through two segments of the Alpine-Mediterranean region. Section lines are shown in Figure 12.17. *A.* Hellenide-Balkan region. *B.* Pyrenees. *(A. after Burchfiel, 1983; B. after ECORS Pyrenees team, 1988)*

northern margin of Arabia, the Bitlis-Zagros-Oman zone, the ophiolites are latest Cretaceous in age and the Troodos complex is recent, as mentioned above. In Cyprus, however, the Troodos ophiolite is being emplaced today.

Melange terranes are extensive in parts of the Alpine-Iranian belt. They are present particularly in the Apennines of Italy, in the fold-and-thrust belt of the Dinarides and Hellenides, in the central core region of Anatolia, in the Caucasus, and in Iran where they also occupy an internal position. Melanges are also associated with all ophiolite complexes. Where subduction is still active, melanges are presumably associated with the accretionary prism. Subduction is still active, for example, under the Mediterranean Sea, where the Mediterranean Ridge, a topographic welt on the floor of the eastern Mediterranean, constitutes the currently active accretionary prism on the consuming margin in the eastern Mediterranean.

High-angle fault zones are present throughout the belt. Though not as spectacular as those in east and central Asia (see Fig. 9.11), they nevertheless are important. The principal ones illustrated in Fig. 12.17 are the Insubric Line in the Alps, which in part represents the remnant of the suture between the European

and Italian continents; the north Anatolian fault zone, an active dextral strike-slip zone located along a major melange zone called the Ankara melange; and the East Anatolian fault zone, an active sinistral fault zone.

High-grade metamorphic rocks are present along this belt in the Alps, in northern Italy, in the Hellenides, in the Aegean Sea, and in parts of central Anatolia. Extensive high-grade metamorphism seems to be lacking in the Carpathians and the Balkanides, and much of the central Iranian Plateau seems to be covered in volcanic or young sedimentary rock.

Extensional structures are present in the Tauern window in the eastern Alps. They are especially prominent in the Tyrhennian Sea, which is a back-arc basin beneath a subduction zone south of Sicily, and in the Aegean Sea, where a region of high-grade metamorphic rocks is undergoing extension directed north-south. Similar extension is also present along the eastern margin of the Aegean.

Models of Alpine-Himalayan Evolution

A number of attempts have been made to develop detailed evolutionary schemes for various parts of the

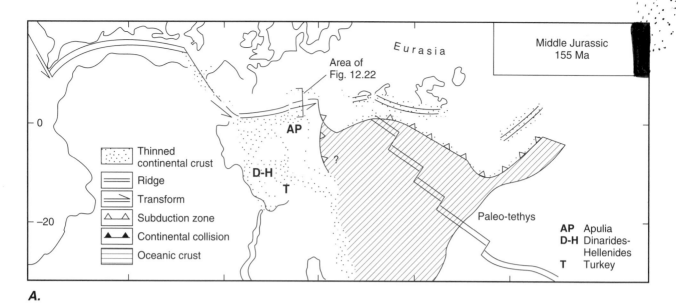

A.

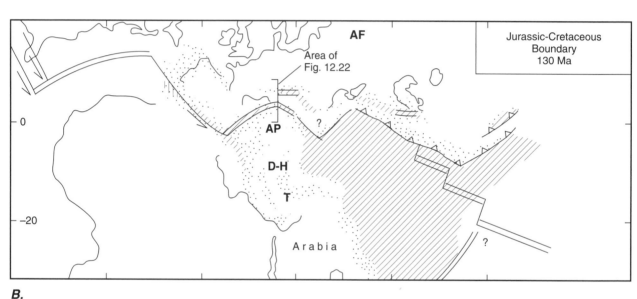

B.

Figure 12.21 Nine maps of the evolution of the Alpine-Himalayan region showing tectonic configurations that developed as the Atlantic and Indian oceans opened and the Tethys Ocean closed. *(After Le Pichon et al., 1988)*

orogenic belt. Figure 12.21 shows a model of the evolution of the Alpine-Himalayan belt based upon the reconstructions of Figure 12.19 as a framework. The figure shows evolution of the belt principally as a two-plate system, but with a number of smaller plates, such as Iberia and Apulia (eastern Italy, the Ionian Sea, Slovenia, Croatia, Bosnia, Albania, Montenegro, Greece, and western Turkey) in the intervening region. The illustrations are "snapshots" of a continuous evolution at various geologic times. The principal data used for the reconstructions are the Atlantic spreading

history and a tectonic interpretation of the geology in various regions.

As the central Atlantic first opened (Figure 12.21A), the relative motion between Africa and western Europe was predominantly strike-slip along some segments of the plate boundary and oblique to orthogonal opening along other segments. Subduction zones may have existed in eastern Europe and western Asia, based upon calc-alkaline volcanic rocks in these regions. Stratigraphic data suggest that rifted margins developed between Apulia and Europe.

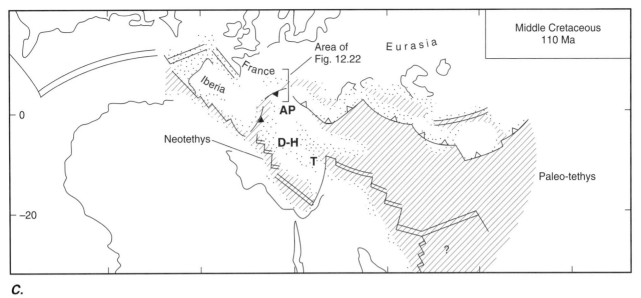

C.

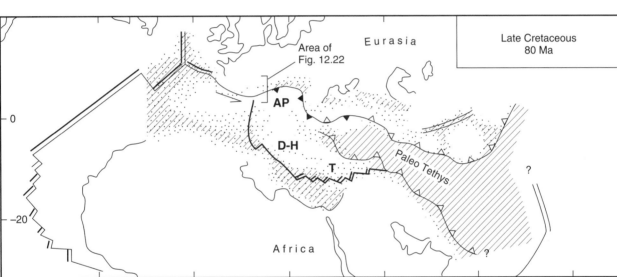

D.

Figure 12.21 continued.

Ophiolites in the Caucasus suggest opening in that region. A spreading ridge may have been present in the Paleotethys.

As the central Atlantic continued to open, strike-slip faulting in the Gibraltar region passed into continued opening between Apulia and Europe (Fig. 12.21B). Stratigraphic data indicate deep-water conditions between Apulia and Europe and the development of similar conditions between Apulia (Italy) and Africa. As the Atlantic continued to open, spreading developed between Iberia and France (Fig. 12.21.C).

Early thrust-faulting and ophiolite emplacement in the Alps suggest the development of a consuming margin in that region. Ophiolites developed along the southwestern margin of Apulia and the Dinarides-Hellenides, suggesting that the Neotethys began to develop. Ophiolites developed from Turkey to the Himalaya.

By late Cretaceous time (Fig. 12.21D), the equatorial and south Atlantic oceans had begun to open, and the Bay of Biscay had opened as Iberia rotated counterclockwise away from France. Blueschist

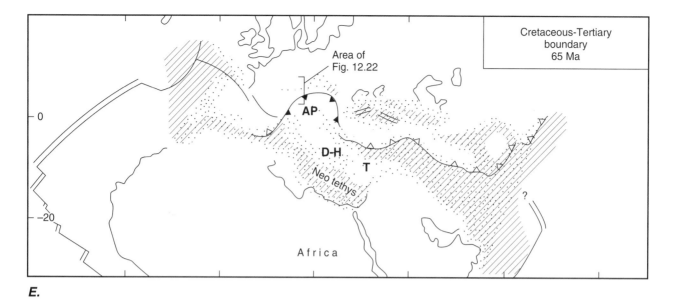

E.

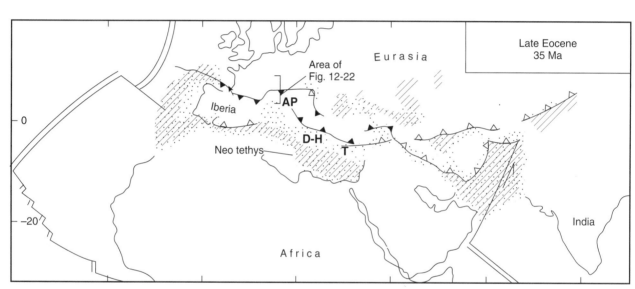

F.

Figure 12.21 continued.

metamorphism and ophiolite emplacement in the Alps suggest continuation of consuming margin activity, extending more or less continuously into western Asia. Ophiolite emplacement along the northeast margin of the African-Arabian continent suggests a consuming margin in that region as well.

By late Eocene time (Fig. 12.21*F*), convergent tectonics were present north and south of Iberia; in the Alps, Carpathians, and Dinarides-Helenides; and eastward into southwest Asia. India had become a part of the picture, with the early convergent structures in the Himalayas. The relative motion of Africa with respect to Europe had changed progressively

from oblique in the middle Jurassic to essentially orthogonal convergence since late Cretaceous time. Continued convergence through the Tertiary (Figure 12.21*G–I*) led to the present orogenic pattern.

No reconstruction such as that presented in Figure 12.21 is entirely satisfactory, however, because of the enormous complexity of the region's geology and the uneven detail of our current knowledge; the model can be considered only a working hypothesis. One important feature common to most models for this area, however, is that an ocean basin (often called Neotethys) opened between Africa and Apulia after the ocean basin that existed in Triassic time

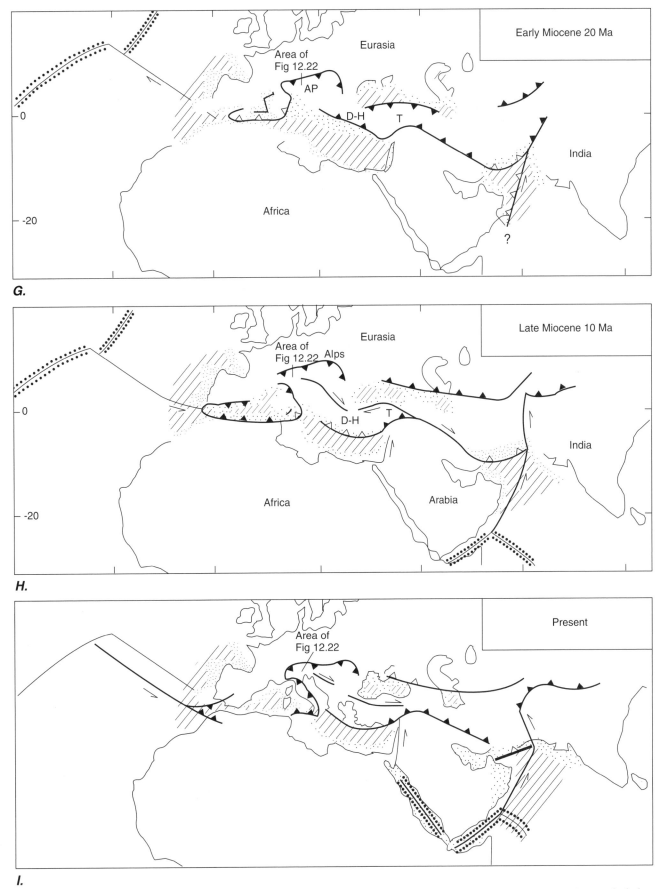

G.

H.

I.

Figure 12.21 concluded.

(Paleotethys) closed. This model also illustrates the potential origin of much of the confusing complexity of these orogenic zones, with the inclusion of collisional zones of various polarity, polarity reversals, and the development of strike-slip faulting as Africa and Europe approached.

An enormous amount of stratigraphic and tectonic information is available to constrain the possible plate tectonic models for such regions as the Alps, and we cannot even begin to review it here. Even with this large amount of data, however, a model for the plate tectonic evolution of even such a limited portion of the region as the Alps is still not settled. Figure 12.22 shows one proposed cross-sectional model of how such an orogen might fit in with the plate tectonic processes we have discussed in previous chapters, as well as with the model shown in Figure 12.21. This model also represents only a working hypothesis, and it will almost certainly be modified as critical questions posed by different models are answered by continued careful field work (see Box 12.1 for a discussion of slip-line models of the Alpine deformation).

In early to middle Jurassic time, the future realm of the Alps was characterized by sea floor spreading, deep-sea sedimentation, the formation of passive continental margins, and possibly strike-slip faulting, as indicated in Figures 12.21A and 12.22A. Between Apulia (1) and Europe (5) was a series of continental fragments or remnant arcs (3) separated by ocean basins (2 and 4) and possibly one or more subduction zones (suggested schematically in Fig. 12.21A), a situation that may have been comparable to the modern southwest Pacific (see Fig. 9.3). In the Cretaceous, development of a south-dipping subduction zone between the passive margin of Apulia (1) and Europe (5) (Figs. 12.21C and 12.22B) led to ophiolite emplacement on the Penninic microcontinent (3) (Fig. 12.22C). At this time, Apulia was separating from Africa to form the Neotethys (Fig. 12.21C). Renewed south-dipping subduction under the Apulia's continent led to the collapse of the intervening basin and to Apulia's eventually overriding the Penninic collision zone (1, 2, 3; Fig. 12.22D).

The cross-sectional model in Figure 12.22, more detailed than the maps of Figure 12.21, also shows subduction beneath the European passive margin (5) (Fig. 12.22E). By early Miocene time, 20 Ma, the intervening ocean basin (4) had been eliminated and the complex Apulian (1, 2, 3) and European (5) margins collided, with the European margin sliding under the Apulian margin (Fig. 12.22F; between Figs. 12.21F and 12.21G; compare Fig. 10.18). The geometry of the collision may account for the fact that the miogeoclinal sediments of the European margin never devel-

oped into a true fold-and-thrust belt, such as is forming in the Zagros area (see Figs. 9.8, 9.1E and 9.1F). With continued convergence, a large backfold developed (Fig. 12.22G), which formed the Alpine root zone and raised the Moho to relatively shallow levels to form the Ivrea body.

This cross-sectional model does not include possible strike-slip complexities, however. In particular, it does not portray such features as a possible collision of an irregular Apulian margin, as suggested by the strike-slip faulting in Figures 12.21H and 12.21I (see also Box 12.1).

The lack of calc-alkaline volcanic rocks in the western Mediterranean is a major exception to the composite orogenic model of Chapter 10 and to conventional ideas about mature subduction zones. It may indicate that the subducted ocean basins were never wide enough for the subducted slab to reach depths at which volcanics are formed and that much of the opening was transtensional, as implied in Figures 12.18, 12.21A, and 12.21B.

Figure 12.22 (*At right*) Plate tectonic cross-sectional model of the evolution of the alpine orogen. Black is crust, outlined shapes are lithosphere. A. Before 130 Ma. Preorogenic arrangement of continents and ocean basins, with Apulia (calcareous Alps) on the right, Europe of the Helvetic platform on the left, and a small continental (Penninic) terrane between the two ocean basins. B. About 120 Ma (middle Cretaceous). Subduction of the southern ocean basin to the south beneath an oceanic island are separated from the passive margin of the Apulian terrane by a marginal basin. C. About 110 Ma (middle Cretaceous). Collision of the Penninic terrane and deformation of the overriding arc and marginal basin. D. About 90 Ma (late Cretaceous). Overriding of the marginal basin and the collision zone by the Apulian continental margin (calcareous Alps). E. About 50 Ma (Eocene). Closing of the northern ocean basin by northward subduction beneath an arc separated from the Helvetic miogeocline by a back-arc basin. Detail of the earlier collision zone is omitted. F. About 20 Ma (early Miocene). Collision with the southern continental mass overriding the northern continent, and the deformation of the arc, back-arc basin, and Helvetic miogeocline. G. About 5–0 Ma (late Neogene to Recent). Continued convergence; nappes of the southern continent carried over the northern continent. A backfold, which carries the Ivrea body in its core, forms in the south. (*After Roeder and Bögel, 1978; Roeder, 1977*)

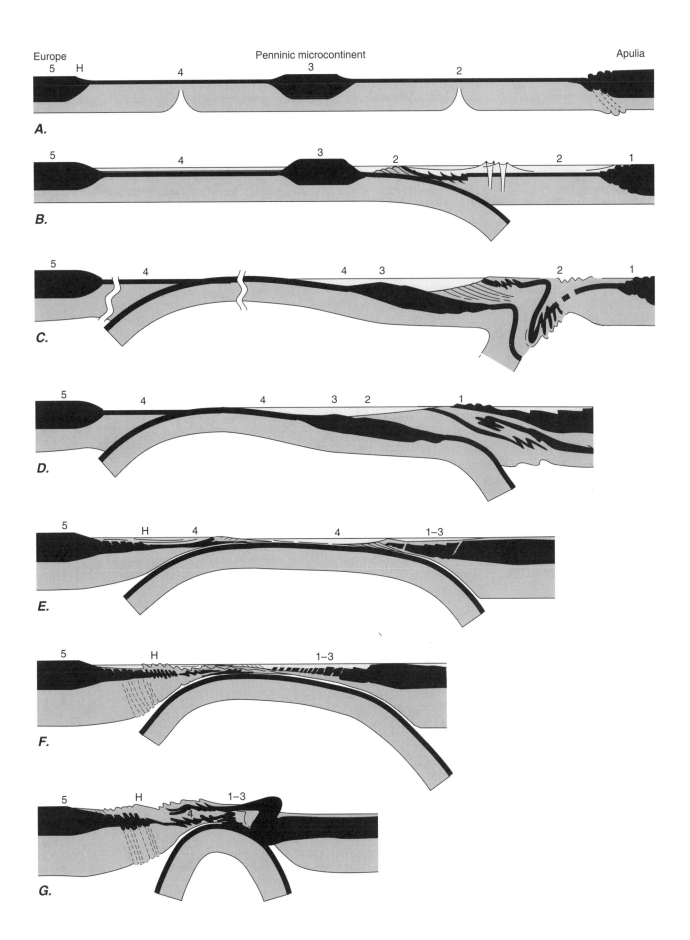

Box 12.1 **Slip-Line Models of the Alpine Collision**

An intriguing feature of the Alpine part of the Alpine-Himalayan system is the large lobes north of Italy that mark the arcs of the western Alps and the Carpathians. Figures 12.1.1 and 12.1.2 show two qualitative slip-line models that attempt to account for these features. Both models assume that a collision occurs between a promontory on the southern continent, representing Apulia, and a northern continent representing Europe.

The first model assumes a subduction zone dipping north under the northern continent so that the northern continent is hot and weak and the promontory is cold and stiff (Fig. 12.1.1; see Fig. 9.12). This geometry is similar to that used in

the plasticine model of the collision of India with Asia (see Fig. 9.17), except that the plastic northern plate does not have an unconstrained boundary on the east or a rigid boundary on the west. As the indenter intrudes, the plastic northern continent extrudes out from in front of the indenter along the slip lines, forming two lobes along either side of the indenter (Fig. 12.1.1B). The northern edge of the indenter is also modified by deformation. As the collision proceeds to completion, the lateral lobes of the northern continent collide with the edge of the southern continent, forming a large collision belt (Figure 12.1.1C). New faults may form in the previously undeformed region, and a

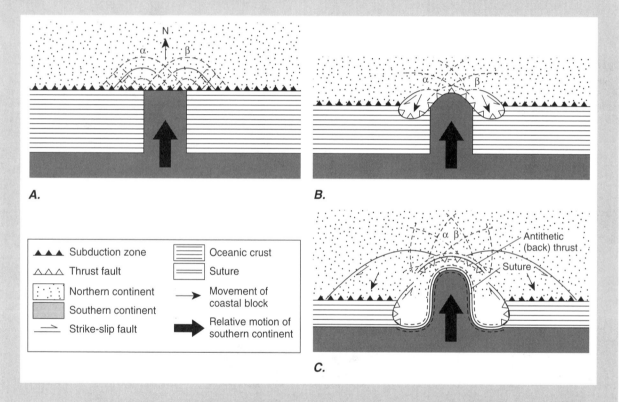

Figure 12.1.1 Schematic plastic slip-line model for the evolution of the Alps, assuming that a stiff promontory (Apulia) of a southern continent (Africa) on the downgoing plate collides with a subduction zone dipping north under a hotter and therefore more easily deformed northern continent (Europe) *A.* Configuration just at collision. *B.* Extrusion of northern crust around the indenter along slip lines. The end of the indenter becomes deformed, and the subduction zone conforms with the rounded end. *C.* Suturing of continental fragments and development of second phase of extrusion. *(After Tapponnier, 1977)*

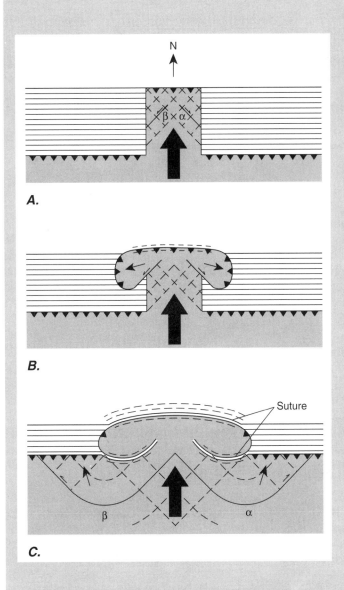

Figure 12.1.2 Schematic plastic slip-line model for the evolution of the Alps, assuming that a rigid northern continent on the downgoing plate collides with an easily deformed promontory (Apulia) on a southern continent (Africa). The subduction zone dips south under a hotter and therefore more easily deformed southern continent (Africa). *A.* Initial collision. *B.* The promontory deforms along plastic slip lines to form lobes on either side. *C.* The deformed promontory is caught between the two continental masses, and the weaker southern continent deforms along a set of slip lines. *(After Tapponnier, 1977)*

wide zone of intracontinental deformation may develop. Crustal thickening ahead of the indenter may lead to backfolding of the main thrust and a reversal of the vergence of thrusting (Fig. 12.1.1*C*).

The second model assumes that the consuming margin dips south under the southern continent and promontory (Fig. 12.1.2). In this case, the southern continent is the hotter and weaker plate, and plastic deformation is concentrated in the promontory. As convergence proceeds, a widening frontal zone of collision is formed by extrusion of the northern parts of the promontory out to the

sides to form lobes. Finally, the deformed promontory is caught between the edges of the relatively rigid northern continent and the plastic southern continent, and the main margin of the southern continent starts faulting along plastic slip lines.

The end results of the two models are distinctly different. In the rigid indenter model, the lobes are formed from the northern continent; in the plastic promontory model, they form from the promontory itself. Moreover, the zone of continental deformation is concentrated in the plastic plate, which is the northern continent in the first model

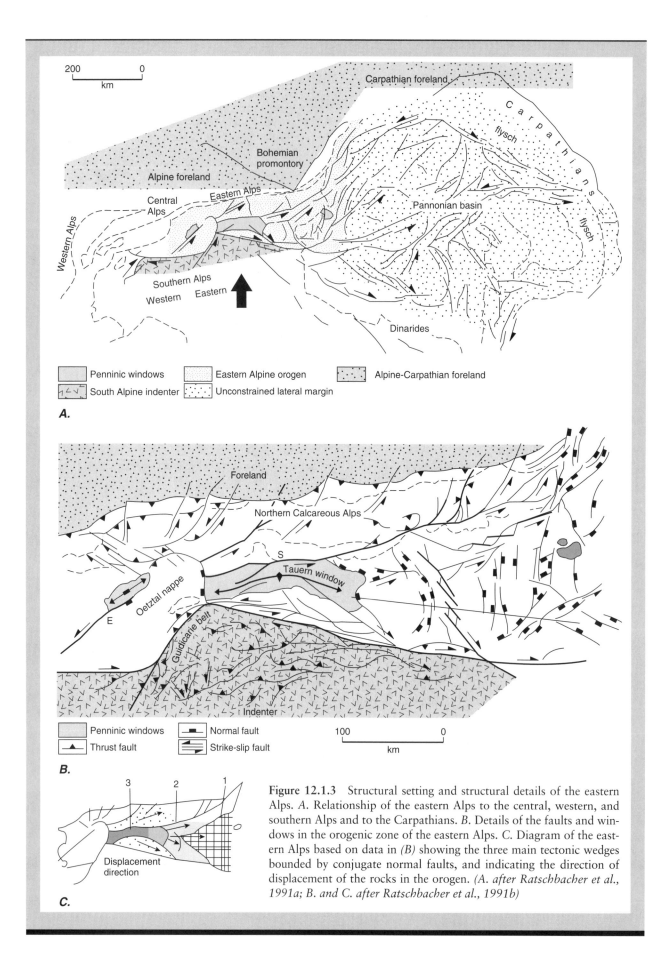

Figure 12.1.3 Structural setting and structural details of the eastern Alps. *A.* Relationship of the eastern Alps to the central, western, and southern Alps and to the Carpathians. *B.* Details of the faults and windows in the orogenic zone of the eastern Alps. *C.* Diagram of the eastern Alps based on data in *(B)* showing the three main tectonic wedges bounded by conjugate normal faults, and indicating the direction of displacement of the rocks in the orogen. *(A. after Ratschbacher et al., 1991a; B. and C. after Ratschbacher et al., 1991b)*

(Figure 12.1.1) and the southern continent in the second (Fig. 12.1.2). These models, however, can represent only the part of the Alpine collision since about 80 Ma (see Fig. 12.18), when the direction of convergence of Africa-Apulia and Europe was approximately perpendicular to the subduction zone. Thus, they cannot account for earlier thrusts and nappes associated with the highly oblique phase of convergence. A detailed analysis of the structures in the eastern Alps and a more carefully constructed set of model experiments reveal more about the tectonics of this part of the orogenic belt.

The structural data from the eastern Alps (Fig. 12.1.3A) indicate that late Oligocene to Miocene time (about 25–10 Ma) was dominated by north-south convergence along a south-dipping subduction zone. The Alpine-Carpathian foreland on the European continent to the north acted as a rigid buttress. The southern Alps acted as an indenter that was weaker than the foreland but stronger than the rocks of the orogenic zone that were caught in the middle. The orogenic zone (Fig.

12.1.3B) consisted of a tectonic pile of continental crust overthrust by Penninic and Austroalpine nappes, which were emplaced in Cretaceous time. As the isotherms that had been depressed by the nappe emplacement subsequently rose, the increasing temperatures weakened the rocks in the tectonic pile.

During this period, the southern Alps impinged on the orogenic zone, leading to shortening, uplift of the Tauern window, and eastward expulsion of the rocks along conjugate sets of strike-slip faults that are similar to slip lines in a plastic material (Fig. 12.1.4; see Fig. 9.13). Gravitational collapse of the uplift generated east-west extension, accommodated by a complex set of north-south normal faults, some of which exposed the Tauern window (see Figs. 12.1.3B and 12.1.3C; 12.1.4). The collapse contributed to the eastward expulsion of material. The ability of the orogenic rocks to move eastward implies that the eastern boundary, represented by the Carpathian basins, acted essentially as an unconstrained margin.

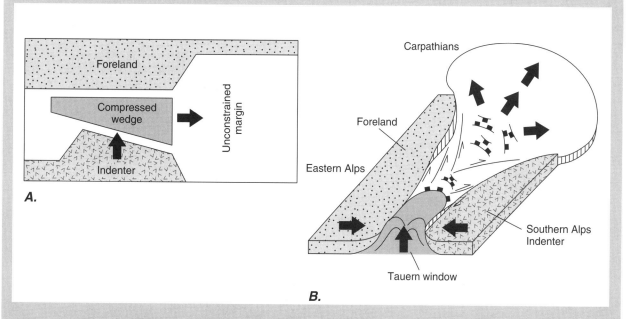

Figure 12.1.4 Models of the Oligocene-Miocene deformation in the eastern Alps. *A.* The compressed wedge represents the rocks in the eastern Alpine orogen caught between the southern Alpine indenter and the Alpine-Carpathian foreland. *B.* Three-dimensional sketch showing the uplift of the Tauern window, its exposure by low-angle normal faulting associated with gravitational collapse of the uplift, and the tectonic expulsion of rocks in the east on a conjugate set of strike-slip faults. *(A. after Ratschbacher et al., 1991a; B. after Ratschbacher et al., 1991b)*

Figure 12.1.5 shows the results of the most successful of numerous model experiments. This model was constructed from layers of sand, silicone putties, and syrup to approximate the scaled mechanical properties and density structure of the lithosphere. The model successfully reproduces the shortening and uplifts in the western areas, the wedges of material moving eastward and bounded by conjugate strike-slip faults, and the roughly north–south trending normal faults. Such experiments provide insight into the mechanical properties and boundary conditions that control the formation of an orogen.

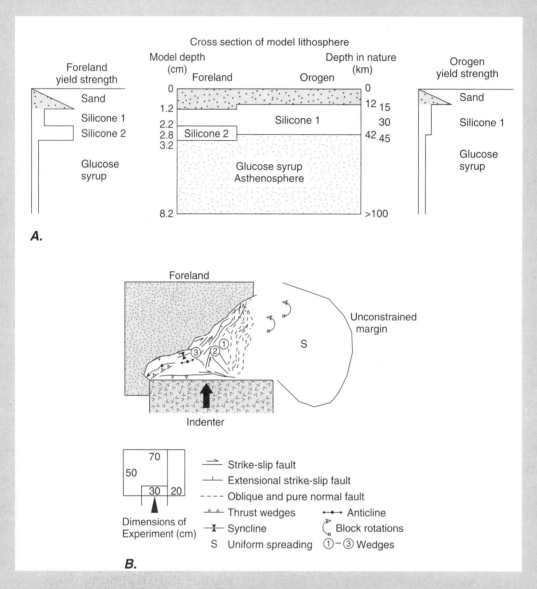

Figure 12.1.5 Scale Model experiment that reproduces the major features of the eastern Alpine orogen as shown in Figure 12.1.3. *A.* Vertical cross section of the model construction showing layers of sand, silicone, and syrup used to model the lithosphere of the foreland (*left*) and the orogen (*right*). The graphs on either side indicate the variation of yield strength with depth. *B.* Horizontal map showing the results of an indenter advancing from right to left in (*A.*) The resulting uplift, strike-slip faults, and normal faults show close similarities to the structure of the eastern Alps (see Fig. 12.1.3). (*After Ratschbacher et al., 1991a*)

12.5 The Alpine-Himalayan Orogenic Belt: Afghanistan-Himalayan-Tibetan Sector

The eastern sector of the Alpine-Himalayan orogenic belt extends from central Afghanistan in the west to southwestern China in the east. The sector is approximately 4000 km long and approximately 500 to 2000 km wide. The width is somewhat arbitrary, however, because (as discussed in Sec. 9.3) current seismicity in this region of active continent-continent collision is distributed over a width of approximately 3000 km.

Figure 12.23 shows a tectonic map of the region. This area includes five of the highest mountain ranges on Earth (Himalayas, Karakorum, Hindu Kush, Kün Lun, and Pamirs) and the highest plateau on Earth (Qinghai-Tibet). The topographic extremes of this region, coupled with the political boundaries and complexities that have affected this area during the nineteenth and twentieth centuries, have combined to make this region one of the least studied on Earth. This short discussion is therefore meant as a general introduction to the region.

Briefly, the region is a multiple collisional orogen in which two continents, Eurasia and India, enclose a number of distinct blocks (or terranes) separated from one another by sutures. West of India, oceanic crust of the northwest Indian Ocean is still actively subducting beneath the accretionary prism of the Makran region of southwestern Pakistan and Iran. The eastern boundary of the region is taken generally as the major bend or syntaxis in the thrust belt in Nagaland.

The major units of the region, from south to north, are as follows (see Figs. 12.23 and 12.24).

1. The southernmost belt of the orogen consists of the deformed rocks of the Himalayas and its continuations to the west, the Suleiman and Kirthar ranges of Pakistan, and the Burma Ranges to the east. In this region, marginal rocks of the Indian Shield are thrust south, southeast, or west, respectively. Thrusting commenced mostly in middle to late Tertiary time in these regions and continues today. Thrusts in the Himalayas include the main boundary thrust (MBT), along which late Precambrian and Paleozoic rocks are emplaced over Pleistocene rocks, and the main central thrust (MCT), which is a zone of ductile deformation several kilometers thick along which Precambrian crystalline rocks and overlying Paleozoic rocks are thrust over Paleozoic rocks above the MBT. In the Suleiman and Kirthar ranges, thrusts are principally of Jurassic and Cretaceous marginal rocks; in the Burma Ranges, they are mostly Tertiary rocks.

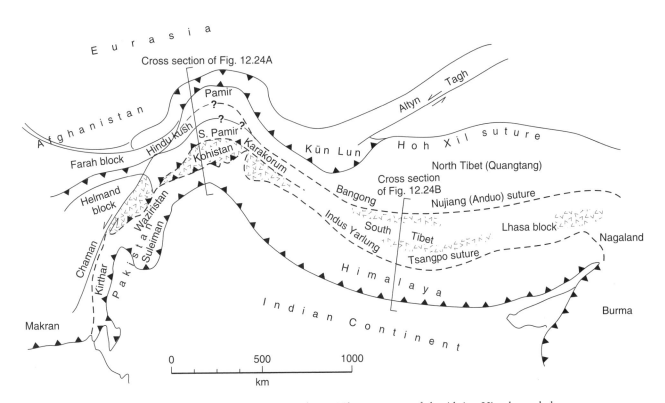

Figure 12.23 Map of Afghanistan-Himalayan-Tibetan sector of the Alpine-Himalayan belt. *(After Chang et al., 1989; Girardeau et al., 1989; Le Fort, 1989; Tapponnier et al., 1981)*

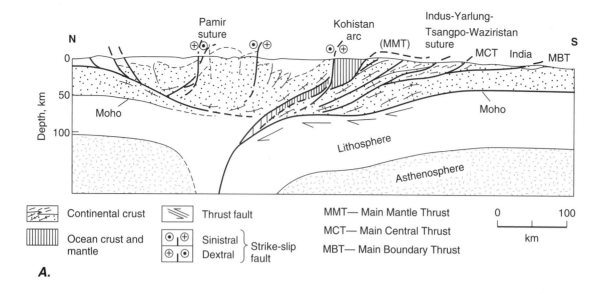

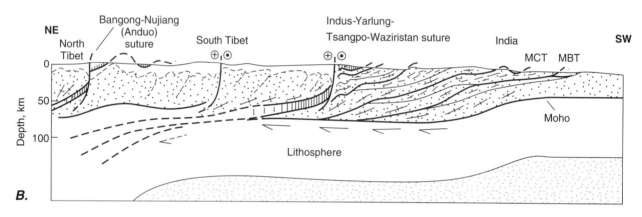

Figure 12.24 Cross sections of portions of the Afghanistan-Himalayan-Tibetan sectors. *(After Mattauer, 1986)*

2. The Indus-Yarlung-Tsangpo suture borders the Himalaya to the north. It consists of a narrow zone a few kilometers wide, chiefly of ophiolitic melange that dips steeply south. The suture separates India from the South Tibet–Lhasa block to the north. It continues westward as the main mantle Thrust (MMT) of Kohistan and the Waziristan suture in Pakistan. Rocks within the suture are mostly Cretaceous to early Tertiary in age.

3. Island arc rocks of the Kohistan arc (see Box 7.1) border the Yarlung-Indus-Tsangpo-Waziristan suture to the north. Generally, these rocks are Mesozoic to early Tertiary in age, ranging from 150 to 48 Ma. In Afghanistan and South Tibet, the volcanic rocks and associated plutonic rocks invade and overlie Precambrian–early Mesozoic continental crystalline and overlying rocks. The volcanic rocks in Kohistan appear to be part of an intraoceanic island arc. Perhaps taken together

they represent some sort of Japan-Marianas combination, although it is not clear that there was any original connection.

4. The Bangong-Nu Jiang (Anduo) suture zone is a zone of melange containing fragments of ophiolites that exhibit Jurassic ages. The ophiolites apparently were thrust from north to south in late Jurassic time. The extension of this zone to the west is problematic. It may connect with a suture dividing the southern and central Pamirs; alternatively, it may separate the central and northern Pamirs. West of the Pamirs, this suture continues as the Farah Basin or Farah block, a region of Triassic-Jurassic ophiolite and overlying sediments and thus an area where remnants of the original ocean basin still exist.

5. The North Tibet or Quangtang block consists of Precambrian basement overlain by a Silurian-Permian sequence of sediments of Gondwanan

affinity, extensive Cretaceous calc-alkaline volcanics, and a thick Tertiary sequence. It apparently is equivalent to the central Pamir block.

6. The available information on the Kün Lun mountains and the Hoh Xil suture zone suggest that together they represent an accretionary complex of late Paleozoic–early Mesozoic age superimposed upon an older continental platform. These mountains, and their continuation to the west, the North Pamirs, may represent a Paleozoic-Mesozoic consuming plate margin along the southern margin of Eurasia.

Figure 12.24 shows two cross sections across portions of the Afghanistan-Himalayan-Tibetan sector. Although the predominant thrust structures generally are portrayed as south-vergent, there are significant

regions of north-vergent structures as well, particularly along the Indus-Yarlung-Tsangpo suture and in the Pamirs. To our knowledge, however, no accurate true-scale cross section of the entire region exists.

Extensional structures have been described from the Himalayan region that are coeval in activity with the major south-vergent thrusts. One interpretation is that these extensional structures reflect collapse of an orogenic welt that had become too thick for its own strength to support.

The general evolution of the Himalayan-Tibetan sector apparently is the progressive northward migration of microcontinental blocks (or terranes) derived originally from the northern margin of Gondwanaland, the successive collision of these blocks, and the southward stepping of a north-dipping subduction zone. Figure 12.25 shows such an evolution. The

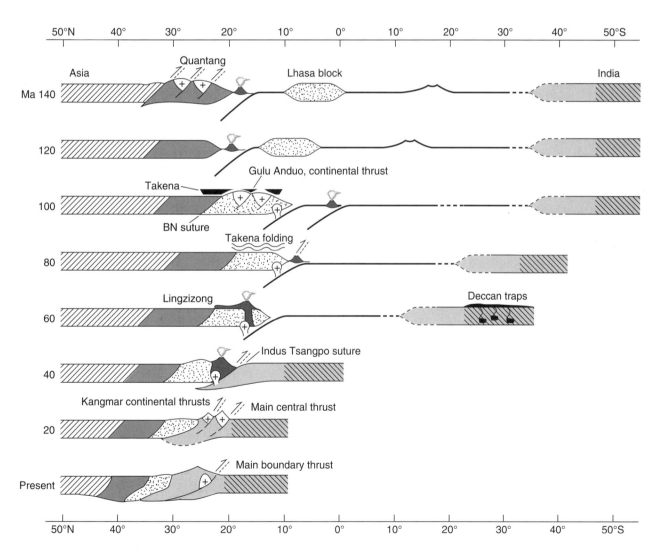

Figure 12.25 Evolutionary cross-sectional model of development of Kün Lun–Tibetan-Himalayan orogenic system. *(After Allegré et al., 1984)*

commencement of the continent-continent collision is thought by most workers to have been in late Eocene time (40 Ma). The convergence of India and Asia continues today; it is taken up by a combination of east-west extension, north-south compression, and a complex pattern of strike-slip faulting, as discussed in Section 9.3. The uplift of the Tibetan Plateau apparently has occurred since middle Miocene time.

12.6 The Appalachian-Caledonide System

Regional Characteristics

The Paleozoic Appalachian-Caledonide system is, on a reconstruction of Pangea, one of the world's great orogenic belts, having a total length of some 6000 km (Fig. 12.26). The system includes the Appalachians of the United States and Canada; the West African fold belt; and the Caledonides of Ireland, Britain, Scandinavia, east Greenland, and Svalbard. Branches off the Appalachian system continue southwest into the Ouachita Mountains in south-central North America, the Cordillera Oriental in Mexico, and the Merida or Venezuelan Andes in northwestern South America. Another branch continues east into the Variscan-Hercynian system of central Europe and central Asia. A branch off the Caledonian system extends around the northern margin of North America into the Innuitian mountain system in Arctic Greenland and Canada. Although it was once a continuous system, correlations along the belt have been complicated because it was broken up by the opening of the Atlantic and because much of it is now covered by Mesozoic continental margins and the Atlantic Ocean.

The Appalachian-Caledonide system and the Alps are the most intensively studied orogenic belts in the world, and information from these two belts has dominated our thinking about mountain belts and orogeny. The evolution of the Appalachian-Caledonide system, however, occurred almost entirely in plate tectonic time, before the advent of ocean basin time (see Fig. 12.2), and so reconstructing the plate tectonic history of the system must depend completely on evidence from continental rocks.

This orogenic belt is probably the best example of one that conforms to the composite orogen model shown in Figure 10.3, including the structural symmetry, the presence of foredeeps, thin-skinned fold-and-thrust belts, and a core zone with crystalline nappes. Perhaps this is not too surprising, because our prejudices about what constitutes an ideal orogenic belt have been molded by our detailed knowledge of this

particular one. Structurally, the vergence is away from the orogenic core and is bilaterally symmetric about the core (Figs. 12.26 and 12.27). Thus, in present coordinates the vergence is predominantly northwestward in North America but southeastward along the eastern margin of the chain in Nova Scotia and Newfoundland, as well as in the southeastern United States. Vergences in West Africa and Scandinavia are predominantly eastward, in Greenland westward, and in the British Isles and Svalbard in both directions.

Outer foredeeps are well developed west of the orogen in North America, east of the orogen in West Africa, and in parts of the British Isles and southern Scandinavia. The fold-and-thrust belts are well developed on the west side of the orogen in North America and Greenland and on the east side in West Africa and Scandinavia. In some parts of the fold-and-thrust belts, such as the central Appalachians, folding appears to be the predominant mode of deformation. In others, such as the southern Appalachians, imbricate thrust faulting is dominant. In the southern and central Appalachians, the fold-and-thrust belt forms the wide and well-developed Appalachian Valley and Ridge province. In the northern Appalachians, however, it narrows and in places is nonexistent. Fold-and-thrust belts are less pronounced in the British Isles and in Svalbard, but this may be because they are not exposed on land in this area.

Ophiolites are well preserved in Newfoundland (Fig. 12.26) and Scandinavia, but in the United States they occur chiefly as fragmented ultramafic bodies concordant with the local foliation (see Sec. 10.16). They are unknown in Greenland. Ultramafic rocks are present in Spain, but they are highly metamorphosed and their ophiolitic affinity is uncertain.

Crystalline core zones are well developed in North America, the British Isles, Scandinavia, Spain, and Greenland. They are less prominent in West Africa. Where present, they resemble the model core zone and, indeed, are the evidence that generated the model. Metamorphic grade is high, and the rock types affected include the entire range of lithologies mentioned in Section 10.4.

Metamorphism is chiefly Barrovian or Buchan (see Fig. 10.21), although blueschist metamorphism is preserved in the British Isles and eclogitic metamorphism is present in Spain and Scandinavia. The complexity of structure and/or the limited exposure of these rocks makes it difficult to confirm or deny the presence of any paired metamorphic belt, however.

High-angle faults are especially prominent in the southern Appalachians, in Nova Scotia, Newfoundland, the British isles, Spain, and North Africa. Faults in the Appalachians, Newfoundland, and Britain generally cut orogenic core rocks, whereas those in North

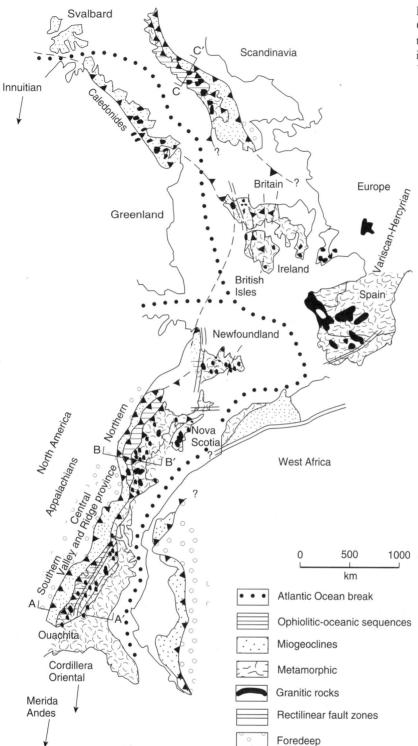

Figure 12.26 Map of Appalachian-Caledonide-West African mountain system. Locations of cross sections shown in Figure 12.27 are indicated. *(After Williams, 1984)*

Legend:
- • • • Atlantic Ocean break
- Ophiolitic-oceanic sequences
- Miogeoclines
- Metamorphic
- Granitic rocks
- Rectilinear fault zones
- Foredeep

Africa, Spain and Nova Scotia intersect the core and fold-thrust belts at high angles to their trends.

Extensional structures have been described in the metamorphic regions of Scandinavia. They may well be present elsewhere, however.

Granitic batholiths are variable in extent, being prominent in Spain and the United States, but less so elsewhere. These rocks may be part of the Appalachian-Variscan-Hercynean belt as opposed to the Appalachian-Caledonide belt. Where present, they

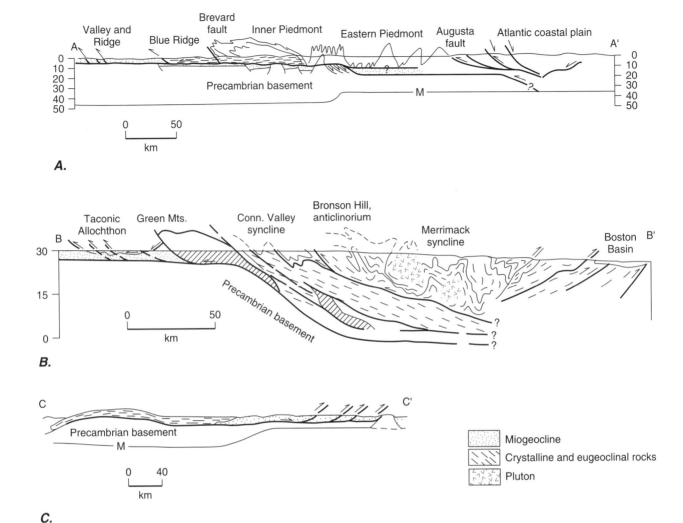

Figure 12.27 Representative cross sections of the Caledonide-Appalachian orogen. *A.* Cross section A–A′. The southern Appalachians. *B.* Cross section B–B′. Across the New England Appalachians. *C.* Cross section C–C′. Across the Scandinavian Caledonides. *(A. after Cook et al., 1981; Price and Hatcher, 1982; B. after Ando et al., 1982; Price and Hatcher, 1982; C. after Hossack and Cooper. 1986)*

tend to be localized within the metamorphic core. In the Canadian Appalachians, however, there is at least one large batholith in a slate belt.

Cross Sections of the Orogen

In cross section, the entire southern Appalachians are underlain by a subhorizontal series of seismic reflectors probably marking major décollements, and the entire chain therefore appears to be allochthonous (see Fig. 10.20). The structure above the décollements

from northwest to southeast (cross section A–A′, Fig. 12.27) apparently consists of a thin-skinned fold-and-thrust belt, some large northwest-vergent recumbent nappes in crystalline rocks, and then a series of more upright to southeast-vergent folds and faults. All structures are cut by the décollement and a set of west-vergent faults. The southeast end of cross section A–A′ is covered by the Atlantic coastal plain.

Similar structures are found by seismic reflection in the northern Appalachians of New England, although the surface geology indicates that they are older than those in the southern Appalachians (cross section B–B', Fig. 12.27), but it is unclear whether the subhorizontal reflectors extend beneath the entire exposed orogen. The structure in Scandinavia, much simplified (cross section C–C', Fig. 12.27), involves a highly metamorphosed core zone with east-vergent refolded recumbent folds to the west, which is in thrust contact with a thin-skinned fold-thrust belt to the east. All rocks overlie an undeformed Precambrian basement. Seismic reflection information from the British Isles, Scandinavia, and West Africa is not yet sufficient to discern whether large-scale subhorizontal reflectors are present there as they are in the Appalachians.

Exotic Terranes and Tectonic History

Most deformation in the Appalachian-Caledonide system took place in one of three major orogenies. An early Paleozoic orogeny that occurred chiefly during the middle to late Ordovician is known as the Taconian in the United States, the Grampian in Canada, and the Finnmarkian in Scandinavia. A second orogeny occurred in the middle Paleozoic, chiefly Silurian-Devonian, and is known as the Acadian orogeny in North America and the Caledonian orogeny in Great Britain and Scandinavia. A third orogeny in the late Paleozoic, chiefly late Carboniferous (Pennsylvanian)–Permian, is referred to as the Alleghanian orogeny in the United States and the Hercynian-Variscan orogeny in southern Europe. The timing of these orogenies is somewhat different at various places along strike in the chain (Fig. 12.28). Early Paleozoic deformation is present in North America, the British Isles, France, and Scandinavia, but not in Greenland, Spain, or Africa. Middle Paleozoic deformation is widespread in North America, the British Isles, Scandinavia, and Greenland, and it is also present in Africa. Late Paleozoic deformation is generally limited to the southern parts of the orogenic belt, in North America, Africa, and southwestern Europe.

Allochthonous terranes are apparently significant in the Appalachian-Caledonide system. Although controversial, these terranes have been recognized in all parts of the system except East Greenland. Figure 12.29 shows a map of the orogen indicating the terranes, and Figure 12.30 shows a diagram of the timing of accretion of the terranes and other continents to North America.

Five principal terranes have been proposed (Fig. 12.29).

Terrane I consists of Neoproterozoic–lower Paleozoic sedimentary rocks and Neoproterozoic crystalline basement. It is present along the entire length of the belt from the southern Appalachians to Svalbard.

This terrane is in thrust contact with the thin-skinned miogeoclinal fold-thrust belt and represents fragments derived from the Laurentian continent directly or from microcontinents near the continental margin.

Terrane II contains lower Paleozoic island arc and ophiolitic deposits with associated melange. It is present in North America, the British Isles, and Scandinavia. It includes the Taconic sequence of the United States and Canada, the Dunnage terrane of Canada, the Dundee terrane of the British Isles, and the Seve-Koli and Jotunheimen thrust sheets of Scandinavia.

Figure 12.28 Timing of deformation in the Appalachian-Caledonide orogen. *(After Williams, 1984)*

Early Paleozoic: Taconian, Grampian, Finnmarkian

Middle Paleozoic: Caledonian, Scandian, Acadian, Cymrian

Late Paleozoic: Alleghanian, Variscan, Hercynian

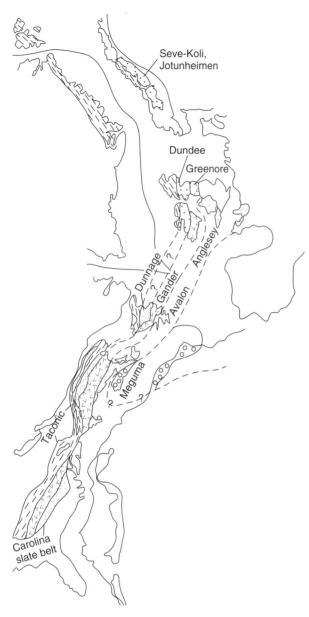

Seve-Koli, Jotunheimen

Dundee

Greenore

Dunnage ?

Gander ?

Avalon

Anglesey

Meguma

?

Taconic

Carolina slate belt

Terranes

I	Laurentian miogeocline or environs
II	Oceanic sequences
III	Lower Paleozoic arc volcanic sequences
IV	Late Precambrian, early Paleozoic sedimentary and volcanic rocks
V	Cambrian to Ordovician, off-continent sediments

Figure 12.29 Terrane map of the Appalachian-Caledonian orogen. See text for discussion. *(After Williams, 1984)*

Terrane IV is a thick sequence of Neoproterozoic diamictites and calc-alkaline volcanic rocks, overlain by Cambrian quartzite and shale with Atlantic or European faunal affinities. It includes the Carolina slate belt of the southern Appalachians and the Avalon sequence of the northern Appalachians. It also includes melange and the Welsh slate belt (see Fig. 10.10), which together constitute the Anglesey terrane in the British Isles, and similar rocks in northwestern France.

Terrane V is a sequence of Cambrian-Ordovician marine greywacke and shale, as much as 13 km thick, that resembles deposits formed off continental margins. No connection with any recognized continental margin is present, however. It includes the Meguma sequence of the northern Appalachians and similar rocks in northwestern Africa (Mauritanide orogen).

Because deformation of the Appalachian-Caledonide system ceased by the end of the Paleozoic, there is no ocean spreading history with which to compare it. The orogenic belt, however, does exhibit abundant evidence of the kinds of petrotectonic rock suites characteristic of modern plate tectonics. Thus, we infer that plate tectonics as currently operating also produced the Appalachian-Caledonide system. Paleomagnetic evidence for the movement of continents provides an approximate framework of plate motion that gave rise to the orogenic belt.

Inception of the pre-Appalachian-Caledonide ocean[2] began in late Precambrian time. Paleomagnetic and stratigraphic evidence suggests the existence of a late Precambrian supercontinent that fragmented before the beginning of Cambrian time and gave rise to a number of continents. Figure 12.31*A* shows a possible reconstruction of the early Cambrian Earth showing the location of known rifted margins.[3] Identifiable separate continents include Laurentia or proto-North America; Baltica or proto-Europe; Siberia; Gondwana; and south China. Consuming

Terrane III consists of lower Paleozoic volcanic and sedimentary rocks, including greywackes and quartzites. It is called the Gander terrane in the United States and Canada and the Greenore terrane in the British Isles.

---

[2] Commonly called the Iapetus Ocean after the Greek mythical god Iapetus, the father of Atlantis. The pre-Appalachian-Hercynian ocean is commonly called the Rheic Ocean, after the Greek mythical goddess Rhea, the daughter of Uranus and Gaea and mother of Zeus.

---

[3] In this and following reconstructions, it is important to keep in mind that only latitude and orientation are given by paleomagnetic measurements. Longitude is arbitrary.

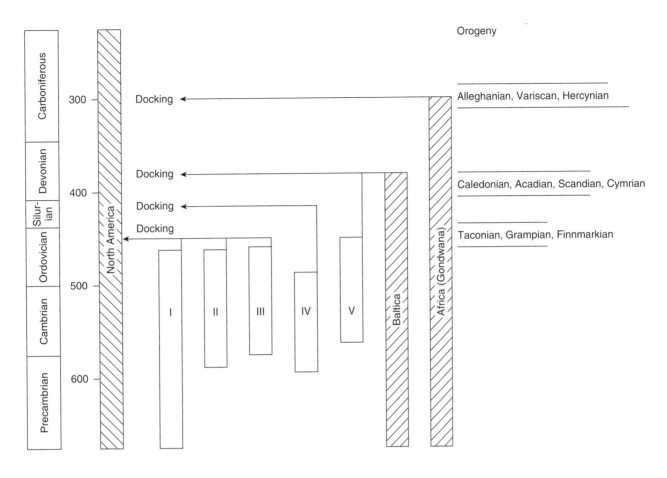

Figure 12.30 Terrane amalgamation diagram of the Appalachian-Caledonian orogen showing timing of terrane accretion and continental collisions. *(After Williams, 1984)*

margins are shown along the margin of Gondwana, within Gondwana in Africa-Arabia, and along possible terranes III and IV.

By late Ordovician time, the world may have looked as portrayed in Figure 12.31*B*. The illustration also shows a possible subduction zone off the margin of Laurentia and Baltica to account for the Taconian, Grampian, and Finnmarkian orogenies.

Figure 12.31*C* shows a possible reconstruction for early Devonian time. The middle Paleozoic orogenies may have included the collision of Baltica with Laurentia to produce the Caledonian and Acadian orogenies, which resulted in a larger continent, sometimes called Laurussia. A small continent, Kazakhstania, formed by amalgamation of several oceanic island arcs.

Figure 12.31*D* shows a possible reconstruction for continents in late Carboniferous time, indicating the development of the Alleghanian-Hercynian-Variscan orogeny by the collision of Gondwana and Laurussia. The major west-vergent nappes of the

Appalachians imply the existence of a subduction zone with the polarity as shown. Much of the evidence, however, may well be covered by the Atlantic Ocean and its margins.

A comparable reconstruction is shown in Figure 12.32 for a cross section through the southern Appalachians. The slight differences between this reconstruction and that of Figure 12.31 indicate the uncertainty of interpretation of the rock record, such as the relationship between thrust-belt polarity and subduction and the significance of ophiolite emplacement.

Although these reconstructions must be considered to be only working hypotheses, they indicate clearly that we cannot expect a simple sequence of events or a uniform history along an entire orogen. The orogenic belt was produced by various events along different parts at various times over a time span on the order of 100 million years. Thus, the simple Wilson cycle concept, despite its original brilliance, does not reflect the enormous complexity characteristic of the orogen.

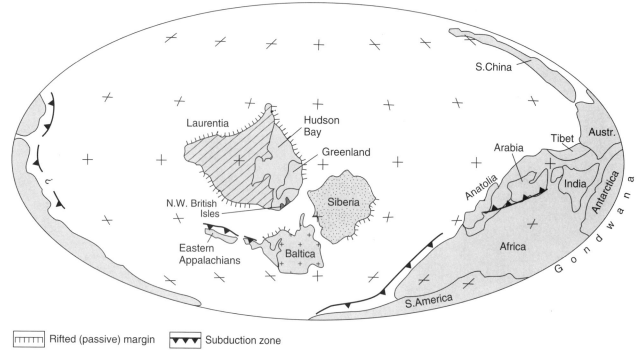

Rifted (passive) margin ⊔⊔⊔⊔⊔   Subduction zone ▼▼▼

A. Eocambrian-Cambrian

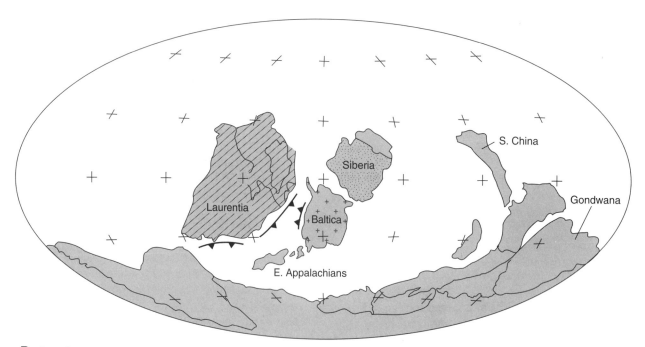

B. Late Ordovician

Figure 12.31 Paleogeographic maps illustrating possible continental configurations and a few selected plate margins. *(After Scotese, 1984; rifted margins on Part A after Bond et al., 1984)*

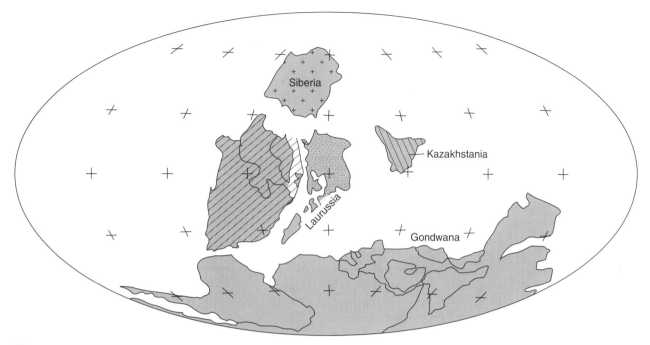

C. Early Devonian

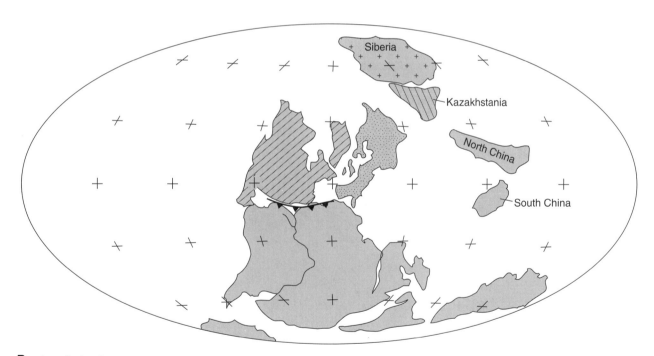

D. Late Carboniferous

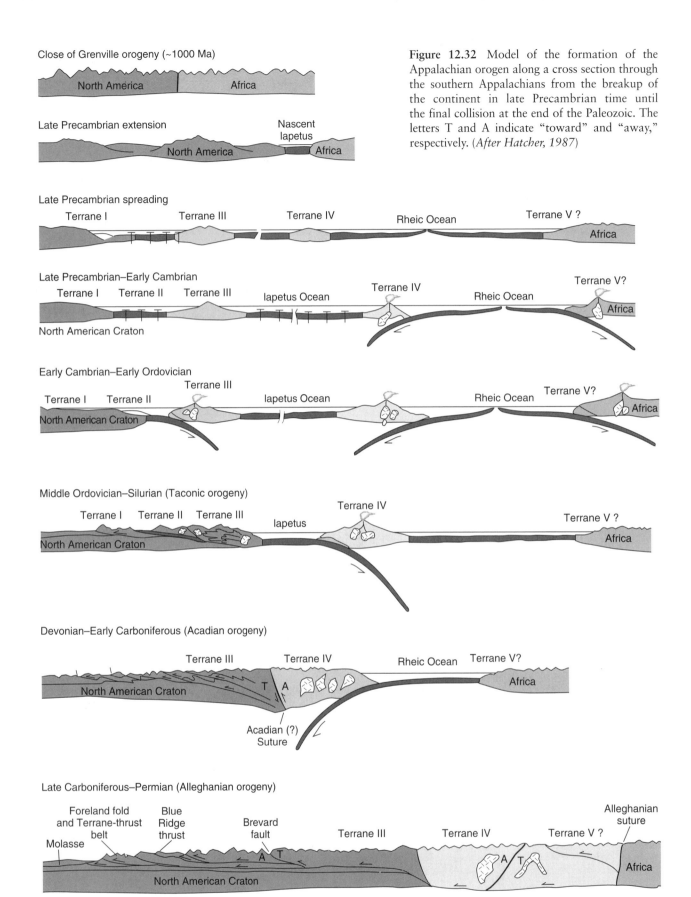

Figure 12.32 Model of the formation of the Appalachian orogen along a cross section through the southern Appalachians from the breakup of the continent in late Precambrian time until the final collision at the end of the Paleozoic. The letters T and A indicate "toward" and "away," respectively. (*After Hatcher, 1987*)

Close of Grenville orogeny (~1000 Ma)

Late Precambrian extension

Late Precambrian spreading

Late Precambrian–Early Cambrian

Early Cambrian–Early Ordovician

Middle Ordovician–Silurian (Taconic orogeny)

Devonian–Early Carboniferous (Acadian orogeny)

Late Carboniferous–Permian (Alleghanian orogeny)

The comparative discussion of mountain belts shows that each has similarities to and differences from the model outlined in Chapter 10. With this perspective, we now are in a position to consider these and other orogenic belts in light of their time of formation and to speculate on possible changes in the plate tectonic process over Earth's history.

The Alpine-Himalayan system developed during ocean basin time (see Fig. 12.2), and so the orogenic events can be compared directly with the oceanic spreading history. Even with the plate tectonic control of the motion and good exposure in many places, however, it has been remarkably difficult to reconstruct the details of the plate tectonic geometry that resulted in the orogenic belt and to make a unique correlation between deformational events and plate tectonic motions. This system provides the only modern example of a continent-continent collision.

Orogenic activity in the North American Cordillera and the Andes spanned plate tectonic and ocean basin time (see Figs. 12.1 and 12.2), with much of the activity occurring early in ocean basin time. Thus, only the younger events can be correlated with plate motions defined by oceanic magnetic anomalies. Development of each of these orogen is dominated by the interaction of a continent with an oceanic basin and the exotic terranes within it. Despite the lack of independent control on the plate motions, the structures that we observe are comparable to younger structures that we can connect directly with plate tectonics, and so the interpretation of the orogenic events in plate tectonic terms seems reasonably secure.

The Appalachian-Caledonide belt is an example of an orogen that developed entirely during plate tectonic time and was terminated by a continent-continent collision. Its structures also are comparable to those we can associate with younger plate tectonic motions, so it seems likely that the same processes we know today were operating in the Paleozoic. The Gondwanide system may be another example of an orogen of comparable age (see Fig. 12.2).

The oldest orogenic system that originated in plate tectonic time is the Pan-African–Baikalian–Braziliano system present in east Asia and throughout much of Africa and South America. It dates back to the Neoproterozoic (see Fig. 12.2) and has many features that also resemble those of Mesozoic-Cenozoic orogenic regions. Long-distance correlations are hampered by the lack of knowledge of the relative positions of continents at that time. Indications are, however, that around approximately 1000 Ma, a

supercontinent formed by the convergence and collision of formerly separate continental masses along orogenic zones known as the Grenville belt and its correlatives (Fig. 12.33B). At this time, North America and East Gondwanaland (Australia and Antarctica) may have been joined (Fig. 12.33A). In latest Precambrian time, this supercontinent broke up; Laurentia separated; and another supercontinent formed, probably as a result of the Pan-African–Baikalian–Braziliano orogeny, which may have had a configuration like that shown in Figure 12.33B. The Canadian geologist Paul Hoffman suggests that one supercontinent (Gondwanaland) may have developed by turning inside out or "extraversion" of a previous supercontinent. Some workers call this first supercontinent Rodinia.[4] The rifting of western (in present coordinates) North America away from Antarctica-Australia and the opening of the proto-Pacific (or Panthalassa) ocean resulted from this extraversion.

During Meso-Paleoproterozoic time, several features that we ascribe to modern plate tectonic processes are present, including continental platforms; evidence of crustal-scale brittle behavior such as dike swarms; rifted continental margins; linear deformed belts, such as the Wopmay orogen in northwestern Canada; and aulacogens. Other features are generally lacking, including blueschist belts and ophiolites as we know them from the Phanerozoic. Still other features, such as batholith-sized anorthosites, large mafic-ultramafic plutonic complexes, Rapakivi granites, and associated volcanic rocks (see Chapter 3, Sec. 3.3), are not characteristic of plate tectonic time. These factors attest to significant differences in mantle processes, the physical conditions accompanying those processes, or the composition of the mantle from which igneous melts are derived.

Paleomagnetic evidence is contradictory. Some workers argue that all continental regions were joined or nearly so throughout much of the time in question, and they ascribe a complex polar wander path to this supercontinent. Others suggest that large-scale relative movement did take place, reminiscent of Phanerozoic patterns. The latter hypothesis argues that most Proterozoic rocks have been remagnetized one or more times and that polar wander paths based upon these rocks are possibly in error.

The lack of blueschists and ophiolites also presents problems in any attempt to apply modern plate tectonic models to middle to early Proterozoic terranes. Blueschists are ubiquitous characteristics of consuming margins, and ophiolites are typical in collisional belts. Their absence in Mesoproterozoic and

---

[4] After the Russian *rodit*, to beget.

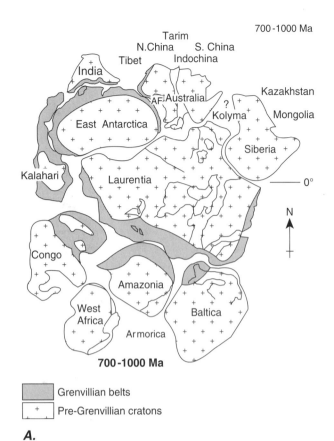

700-1000 Ma

700-1000 Ma

Grenvillian belts
+ Pre-Grenvillian cratons

A.

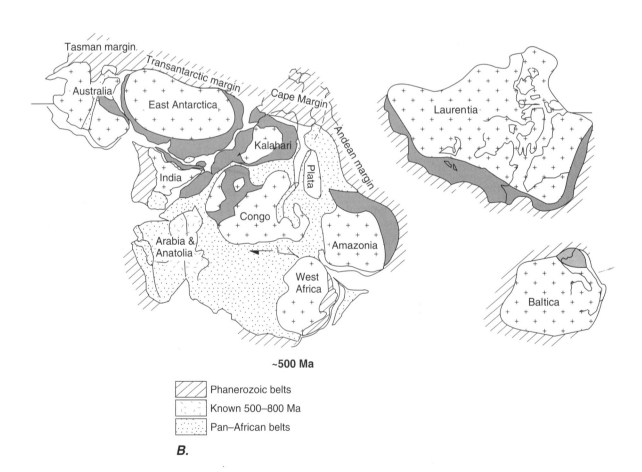

~500 Ma

Phanerozoic belts
Known 500–800 Ma
Pan–African belts

B.

Figure 12.33 Possible late Precambrian continental configurations. *A.* Possible prebreakup configurations of Precambrian cratonic region in supercontinent Rodinia; areas of Grenvillian (800–1500 Ma) orogenic belts. *B.* Possible configuration of Gondwanaland formed by breakup and rearrangement of fragments shown in *(A). (After Hoffman, 1991)*

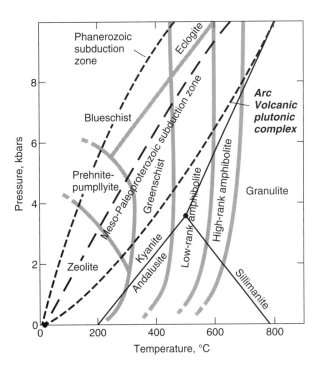

Figure 12.34 Pressure-temperature regimes of metamorphism, showing possible relationship between Proterozoic subduction zones and metamorphic facies. The lack of blueschists may result from higher geothermal gradients in Proterozoic times. Thus, the high-pressure member of paired metamorphic belts would be composed of kyanite-bearing rocks rather than blueschist assemblages. *(After Ernst, 1974)*

Paleoproterozoic belts means either that plate tectonics as we know it was not operating or that some other petrotectonic suites have taken the place of blueschists and ophiolites. If the geothermal gradient had been higher in this era, the temperatures in a subduction zone would have been higher, and the metamorphic conditions characteristic of a subduction zone would have been a high-pressure, medium-temperature suite, as illustrated in Figure 12.34. In such a belt, greenschist or kyanite-bearing rocks would occupy the tectonic position of blueschist in late Neoproterozoic or Phanerozoic deformed belts.

The ophiolite problem stems from the fact that the rock sequences observed in areas where ophiolites would be expected are not ophiolites. Instead, one finds thick deformed stratiform mafic-ultramafic complexes and associated shallow intrusive and extrusive rocks. Although ophiolites have been reported from middle to lower Proterozoic and even Archean regions, so far few, if any, possess a magmatic-tectonite textural boundary in the ultramafic rocks that characterizes Phanerozoic ophiolite complexes. All ultramafic rocks in Proterozoic complexes are deformed igneous rocks. One hypothesis to account for this fact is

that the magmatic oceanic crust may have been thicker, possibly owing to higher temperatures in the Earth and/or greater partial melting of a more fertile (that is, less depleted) mantle. Figure 12.35 shows a plot of the thickness of magmatic rocks versus time for ophiolites and their Proterozoic equivalents that seems to support this hypothesis. Figure 12.36 shows the implications of this hypothesis. This illustration is identical to Figure 9.28, except that it shows the effect of a magmatic crust two and three times thicker. If the oceanic crust were thicker, emplacement of the oceanic lithosphere over continental crust might fail to expose the mantle, leaving only deformed magmatic rocks as seen in the middle to lower Proterozoic complexes.

In Archean time, from 2.5 to 3.8 billion years B.P. (see Sec. 3.3), the characteristic rock sequences differ substantially from those of the Phanerozoic. Continental platforms and associated sediments are lacking. Greenstone belts resemble modern island arc or marginal basin sequences, and a number of writers have used this resemblance to invoke such an origin for these belts. These simplistic applications of modern plate tectonic models to the Archean are unconvincing, however, for several reasons.

Extensive differences between modern and Proterozoic tectonics indicate that we probably need to modify the modern model substantially when applying it to Mesoproterozoic and Paleoproterozoic terranes. It seems only reasonable that even more modification will be necessary for Archean times.

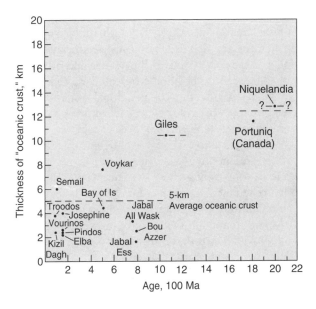

Figure 12.35 Plot of thickness of mafic (oceanic) crust of Phanerozoic and latest Precambrian ophiolites and possible middle to early Proterozoic equivalents. Compare with Figures 3.3 and 5.27 *(After Moores, 1986)*

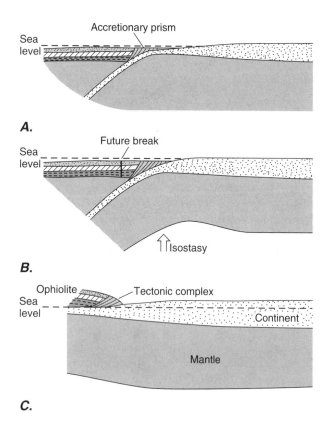

A.

B.

C.

Figure 12.36 Model of emplacement of oceanic crust and mantle with crust two and three times thicker than modern oceanic crust, so that the mantle is not exposed. This model may explain possible deformed mafic-ultramafic intrusive and extrusive sequences along Proterozoic sutures. *(After Moores, 1986)*

Although greenstone belts bear some resemblance to modern island arc or marginal basin sequences, significant differences cast doubt on a simple correlation. Ultramafic lavas (komatiites), abundant in the Archean, are almost entirely lacking in modern marginal basins or island arcs. If the greenstone belts represent marginal basins, they lack the accompanying mature arc that is ubiquitously associated with modern marginal basins. If they are island arcs, they both differ in composition from modern island arcs and seem to lack the compositional zonation possessed by modern arcs that may reflect the polarity of the subducting slab.

The quartzite-carbonate–mafic-ultramafic association present in gneissic terranes has no obvious modern equivalent. Many workers have suggested that the sedimentary rocks represent shelf sequences. The ubiquitous association of the quartzite-carbonate rocks with mafic-ultramafic complexes prompts us to wonder whether the latter represent the Archean equivalents of oceanic crust.

The possibility of a higher geothermal gradient in Archean times also implies a tectonic regime that is significantly different from that of today. A number of workers have argued that the increased heat flux out of the Earth during Archean times must have resulted in many smaller plates with shorter residence times at the surface. Several recent studies have provided evidence, however, that the gradient actually present during crystallization of some metamorphic assemblages was approximately the modern average.

Continental regions where the geothermal gradient is two to three times normal may provide useful tectonic insight into the Archean. In these regions, principally such areas of active continental rifting as the Basin and Range province, a progression in mode of deformation is present that ranges from brittle at the surface to ductile at depth. In places, the transition is marked by a detachment surface of mylonitic rocks (for example, see Sec. 3.3). Similar mylonitic structures are also present in Archean terranes and may have the same origin. For example, the tectonic boundaries between greenstone and gneiss terranes, in regions where the latter include highly metamorphosed greenstone belt equivalents (mafic amphibolites or granulites), may be a similar detachment feature.

Thus, the degree of correspondence between modern plate tectonic processes and those that must have been active in Earth's history before plate tectonic time seems to become more and more problematical. Despite these differences, however, mantle convection almost certainly did occur. This fact, coupled with crusts and/or lithospheres of contrasting densities and thicknesses, argues for some form of plate tectonics in early Proterozoic and Archean times.

Additional Readings

Allegré, C. J., et al. 1984. Structure and evolution of the Himalaya-Tibet orogenic belt. *Nature* 307:17–22.

Ando, C. J., F. A. Cook, J. E. Oliver, L. D. Brown, and S. Kaufman. 1982. Crustal geometry of the Appalachian orogen from seismic reflection studies. *GSA Memoir* 158:83–103.

Burchfiel, B. C. 1983. The continental crust. In *The Dynamic Earth*. R. Siever, ed. Special issue of *Scientific American*. September.

Clark, S. P. Jr., B. C. Burchfiel, and J. Suppe, eds. *Processes in Continental Lithospheric Deformation*. GSA Special Paper 218.

Closs, H., D. Roeder, and K. Schmidt, eds. *Alps, Apennines, Hellenides*. Interunion Commission on Geodynamics, Scientific Report No. 38.

Coward, M. P., and A. C. Ries, eds. *Collision Tectonics*. Geological Society London Special Publication 19.

Davies, G. F. 1992. On the emergence of plate tectonics. *Geology* 20: 963–966.

Dewey, J. F., W. C. Pitman III, W. B. F. Ryan, and J. Bonnin. 1973. Plate tectonics and the evolution of the Alpine system. *GSA Bull.* 84:3137–3180.

Erickson, G. E., P. M. T. Cañas, and J. A. Reinemund, eds. *Geology of the Andes and Its Relation to Hydrocarbon and Mineral Resources*. Houston: Circum-Pacific Council for Energy and Mineral Resources Earth Science Series, vol. 11.

Ernst, W. G., ed. 1974. *Geotectonic Development of California*. Rubey Vol. 1. Englewood Cliffs, N.J.: Prentice Hall.

Hoffman, P. F. 1991. Did the break-out of Laurentia turn Gondwanaland inside-out? *Science* 252:1409–1412.

King, P. B. 1977. *Evolution of North America*. Princeton, N.J.: Princeton University Press.

Le Fort, P. 1989. The Himalayan orogenic segment. 1989. In *Tectonic Evolution of the Tethyan Region*. A. M. C. Sengör, ed. Boston: Kluwer.

Monger, J. W. H. 1984. Cordilleran tectonics: A Canadian perspective. *Bull. Soc. Geol.* France (no. 26):255–278.

Oldow, J. S., A. W. Bally, and H. G. Ave Lallemant. 1990. Transpression, orogenic float, and lithospheric balance. *Geology* 18:991–994.

Sengör, A. M. C., ed. 1989. *Tectonic Evolution of the Tethyan Region*. Boston: Kluwer.

Trompette, R. 1994. *Geology of Western Gondwana (2000–500 Ma): Pan-African-Braziliano Aggregation of South America and Africa*. Translated by A. V. Carozzi. Vermont: Balkema.

Williams, H. 1984. Miogeoclines and suspect terranes of the Caledonian-Appalachian orogen: Tectonic patterns in the North Atlantic region. *Can. J. Earth Sci.* 21:887–901.

13 Tectonics of Terrestrial Planets

13.1 Introduction

Although comparative planetary tectonics is still a young field of study, an enormous amount of information is already available about the tectonic features exhibited by other terrestrial bodies. These bodies include Mercury; Mars; the Moon; Venus; and several moons of Jupiter, principally Io and Europa, and of Saturn, principally Titan. Most of our information comes from the various planetary missions that have been sent by the United States and the former Soviet Union, and much of the information is indirect, based upon the interpretation of surface features as revealed by remote sensing. So far, samples have been retrieved only from the Moon, although some chemical information is available from spacecraft that made soft landings on Venus and Mars. Some meteorites found on Earth are in addition inferred to have come from Mars, ejected by large, high-velocity meteorite impacts on its surface.

The terrestrial bodies all appear to resemble Earth in composition, although their size and the proportions of their constituents vary. The three terrestrial planets (Mercury, Venus, and Mars), as well as the Moon all apparently have a core, a man-

tle, and a crust, but the relative sizes of these features differ.

The surface features of the terrestrial planets also differ from those of Earth. The most obvious tectonic difference between the Earth and several neighboring bodies, including the Moon, is the relative lack of impact structures on Earth. Mercury, Mars, and the Moon all display ubiquitous and abundant crater structures, and there is also evidence of craters on Venus. Data on the timing of these impact events are available only from the Moon. Correlation between the density of impact features and the ages of samples retrieved from the Moon by the Apollo expeditions indicates that most craters are very old. Figure 13.1 shows a plot of age versus density for major cratering events on the Moon. It shows that most craters apparently formed by 3500 Ma, and it provides a way to infer the approximate ages of cratered surfaces that have not yet been dated directly. Presumably, the other terrestrial planets also experienced meteorite bombardment at approximately the same time as the Moon, or earlier.

On Earth, however, evidence of abundant cratering events before 3500 Ma is rare because of the scarcity of crust older than 3500 Ma. Impact structures are found on Earth (Fig. 13.2A, p. 372), but the

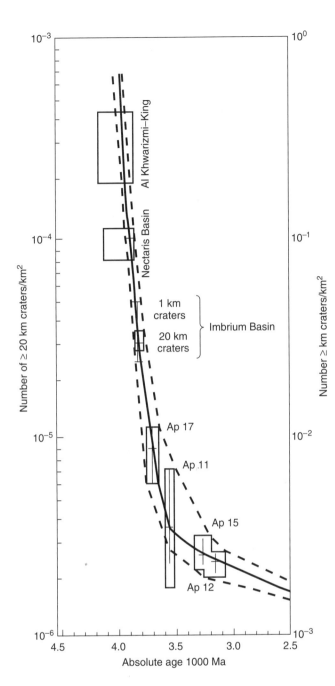

Figure 13.1 Estimate of frequency of large craters on the Moon versus time, based upon available age data from *Apollo* samples and regional surveys of crater density. Upper part of curve is based upon large craters 20–40 km in diameter; lower part of curve based upon small craters ≥1 km in diameter. Tie-in between two parts of curve made by counts of both small and large craters in Imbrium Basin (see Fig. 13.4). *(After Wilhelms, 1984)*

13.2 Mercury

Observations of the surface of Mercury are limited to one flyby made by the *Mariner 10* spacecraft. The craters on Mercury's surface are similar in density, and presumably in age, to those of the Moon. In addition to heavily cratered regions, the surface shows features attributed to volcanic activity (Fig. 13.3, p. 373). Mercury's surface also exhibits a widespread system of curved topographic ridges and scarps that reach lengths up to several hundred kilometers and heights up to 2 km. These features are interpreted as more or less randomly oriented thrust faults and/or folds in the planetary surface that may be the result of a planetwide contraction associated with cooling. Mercury possesses a weak magnetic field, suggesting that the core is liquid iron.

13.3 Earth's Moon

The Moon displays two principal types of surface: lighter colored highlands and darker colored plains or **maria** (Fig. 13.4, p. 373). The heavily cratered highlands appear to be composed primarily of anorthosite, a rock made up mostly of plagioclase, whereas the maria are chiefly basaltic. The distribution of these two physiographic types is markedly asymmetric. The Moon's visible side contains far more maria than the hidden side. The maria are associated with concentrations of mass (so-called mascons) that are thought to represent remnants of large impacts late in the cratering epoch. Data from crater counts and from *Apollo* lunar samples suggest that the highlands are older, around 4000 Ma or more, whereas the maria are younger, 3000 to 3500 Ma.

Tectonic features on the Moon seem to be limited to faults that are associated with impact features. There appear to be no major structural features of internal origin.

ages of all such structures that have been preserved are much younger than those produced by the major lunar bombardments. Figure 13.2B, p. 372, shows a plot of the ages of impact structures versus their diameters.

In addition to impact structures, each terrestrial planet reveals other features that are apparently of tectonic origin. These structures imply that there are considerable differences between the tectonic processes and histories of these planets and those of Earth. We describe the terrestrial planets and the Moon in the following four sections.

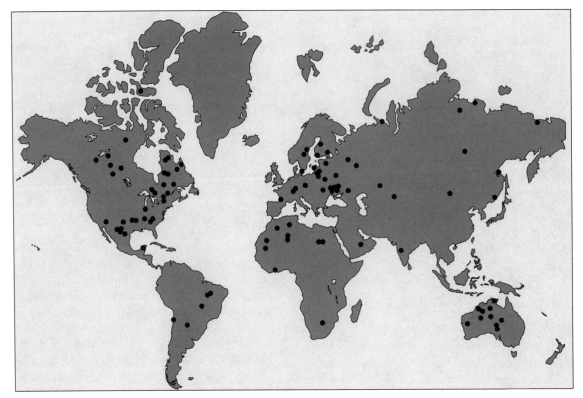

A.

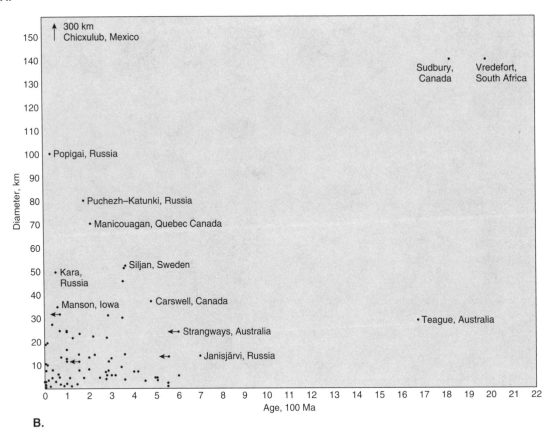

B.

Figure 13.2 Impact structures on Earth. *A.* Geographic distribution of known impact structures on Earth. *B.* Graph showing diameter versus age of known impact structures on Earth. Dots with arrows show maximum size of craters. *(After Grieve, 1982)*

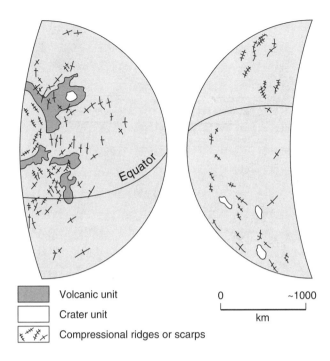

Volcanic unit

Crater unit

Compressional ridges or scarps

0   ~1000
km

Figure 13.3 Schematic map of known area of Mercury showing volcanic units, cratered units, and scarps or ridges inferred to be of compressional origin. *(After Strom, 1984; Head and Solomon, 1981)*

13.4 Mars

Mars is the most intensively studied planet other than Earth, having been the subject of seven spacecraft missions, as of 1995, including two landers. It is enveloped by only a very thin atmosphere, and it has no magnetic field. In general, the older regions of the Martian surface appear to be higher. The Tharsis bulge (Fig. 13.5, p. 374) stands approximately 10 km above its surroundings.

The complex Martian surface exhibits several general physiographic units (see Fig. 13.5).

1. Polar features include permanent ice, layered features thought to be wind-deposited sediments, and etched plains thought to be areas that were once subjected to wind erosion.
2. Ancient features consist of highly cratered upland regions.
3. Volcanic features include volcanoes and sparsely to moderately cratered plains.
4. Complex features include a number of diverse physiographic types such as channels and water-laid deposits, which indicate that running water once existed on Mars.

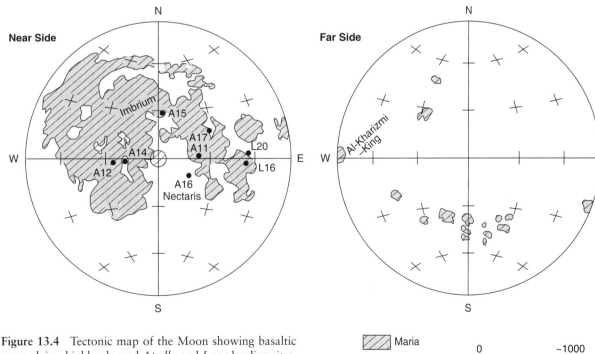

Figure 13.4 Tectonic map of the Moon showing basaltic mare plains, highlands, and *Apollo* and *Luna* landing sites. *(After Wilhelms, 1984)*

Maria

Highlands

0   ~1000
km

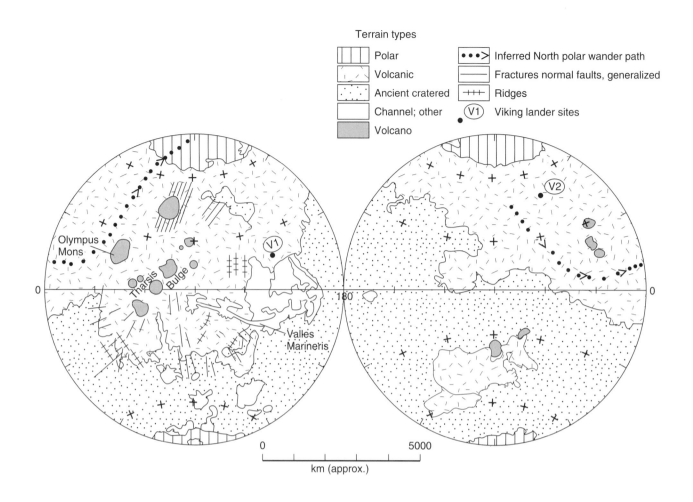

Terrain types

⊞ Polar		•••>	Inferred North polar wander path
⊡ Volcanic		⊟	Fractures normal faults, generalized
⊡ Ancient cratered		⊞	Ridges
☐ Channel; other		(V1)	Viking lander sites
▨ Volcano			

Figure 13.5 Tectonic map of Mars showing four different terrain types: polar, volcanic, cratered, and channel or other. The Tharsis bulge is indicated, together with inferred radial fractures and concentric compressional ridges. Dotted line is the inferred polar wander path. *Viking* lander sites indicated. *(After Carr, 1984; Head and Solomon, 1981; Schulz, 1985)*

Structural features on the Martian surface are of two principal types: (1) normal faults and fractures that radiate from the center of the Tharsis bulge, and (2) concentric ridges interpreted as fold structures that are arrayed around the bulge (see Fig. 13.5). On the eastern flanks of the Tharsis bulge, a huge network of canyons, the Valles Marineris, extends for 4500 km and is up to 600 km wide and 7 km deep. This enormous feature has been interpreted as a rift valley, and it is similar in size to the East African Rift Valley.

Several large shield volcanoes sit atop the Tharsis bulge. Because they are similar in geometry to Earth's basaltic shield volcanoes, they are inferred to be basaltic in composition. They are the largest volcanoes so far identified in the solar system. Olympus Mons, the largest, is 25 km high and 550 km in diam-

eter. In contrast, Mauna Loa in Hawaii, the largest volcano on Earth, rises approximately 9 km from the ocean floor and is approximately 120 km in diameter at its base. In other words, Olympus Mons is about 10 times the volume of Mauna Loa.

Observations of Mars suggest that it possesses a thick, rigid lithosphere. There is no evidence of features ascribable to Earthlike plate tectonics. The Tharsis Mons region, however, shows a strong correlation of observed gravity and topography, suggesting that it is not isostatically compensated.

Mars contains a number of intriguing features that suggest possible large-scale shifts of the poles, or true polar wander. For example, ice-related surface deposits, which should have formed in polar regions, have been found at low latitudes, and asymmetric craters have been found at high latitudes. The craters are thought to have been formed by grazing impacts of meteorites near the equator, with their long axes oriented parallel to the equator. These craters now have different orientations in higher latitudes. These features lead to a speculative polar wander path for Mars, as indicated by the dotted line in Figure 13.5.

The timing of any proposed events is not yet known. By assuming that the cratering on the Moon and Mars originated during the same event, it is possible to estimate times by comparing specific crater densities on the two bodies. Such comparisons suggest that the Martian surface preserves a record of the formation of crustal features that spans the entire length of Martian time since 3800 Ma. They also suggest that polar wander occurred throughout this time, with a significant amount taking place in the past 1000 m.y.[1]

13.5 Venus

Observations of the surface of Venus have always been hampered by the thick cover of clouds that envelops the planet. The earliest information was obtained with Earth-based radar, which provided a resolution of only 20 to 70 km. Later, an order of magnitude improvement in resolution (to about 1 to 2 km) was obtained with a Venus-orbiting radar. Beginning in 1972, we obtained our first images of the Venusian surface from the former Soviet Union's six successful *Venera* surface landers, but they had only a limited view and a short life on the extremely hot surface. The U.S. *Magellan* spacecraft was launched in April 1989 and went into orbit around Venus on August 10, 1990. Its principal scientific instrument was a synthetic aperture radar (SAR), which is capable of seeing through the cloud cover and returning radar images with a resolution of 120–300 m, another order of magnitude improvement over earlier radar images. By 1992, some 95 percent of Venus's surface had been radar mapped, providing a more detailed database than is available for Earth's surface. (On Earth, our view is obscured not by clouds but by the liquid water that covers more than 70 percent of the surface. Because water is opaque to radar, imaging this part of Earth's surface depends largely on ship-based sonar.) Analysis of the rapidly arriving data from Venus has

proved to be a daunting task. Therefore, the following discussion is only a snapshot of a constantly changing kaleidoscope of Earth's nearest and most similar neighbor.

Although the data obtained before the *Magellan* mission displayed little detail of the planet's surface features, the generalized topography was known with considerable precision (Fig. 13.6). *Magellan* has confirmed this generalized view and added a rich store of new information. The resulting map of Venus shows areas of highlands (terrae) and lowlands (planitiae), defined relative to the mean surface radius. The lowland areas make up some 80 percent of the surface. Observations from the *Venera* landers are consistent with lowland regions composed of basalts. *Venera 8* indicated the possible presence of material similar in composition to Earth's continents.

A synthesis of the *Magellan* observations of tectonics on Venus up until 1992 (Solomon et al., 1992) discussed the highlands, the lowlands, the intermediate ridged regions or *tesserae*, and the large circular features called *coronae*. These authors divided the highlands into three types, based partly upon tectonic style, regional morphology, and apparent depth to the level of isostatic compensation.

The first type of highland, designated Beta after the region named Beta Regio, is characterized by broad topographic rises, abundant shield volcanoes, evidence for extensional tectonics, and an apparent depth of isostatic compensation of more than 100 km. (Gravity data collected by *Magellan* will add considerably to this existing data.) These regions are thought to be similar to the East African Rift (Fig. 13.7), as was initially inferred from analyses of pre-*Magellan* data.

The second type of highland, called Ovda after the region of the same name, is characterized by an upland plateau exhibiting complex tectonic structures. Some areas appear to be similar to fold-thrust belts on Earth. Limited volcanism occurs in some areas, and this type of highland is characterized by a depth of isostatic compensation of less than 100 km.

The third type of highland, called Lakshmi after the region of that name, is a plateau with peripheral mountain belts, a moderate amount of volcanism, and considerable evidence of crustal shortening (Fig. 13.8). Fine-scale linear topographic variations in the Ishtar Terra region may represent folds and associated faults (Figs. 13.6 and 13.8*A*). Figure 13.8*A* shows an interpretive map of these structures suggesting folds, strike-slip faults, and thrust faults. Figure 13.8*B* shows a generalized tectonic map of the northwestern corner of India for comparison. The Ishtar Terra region stands some 10 km above the

[1] The precision of the crater density versus time plot is extremely low in the last 3000 m.y.; thus, for the most recent times, it is possible only to estimate ages to plus or minus 500 m.y.

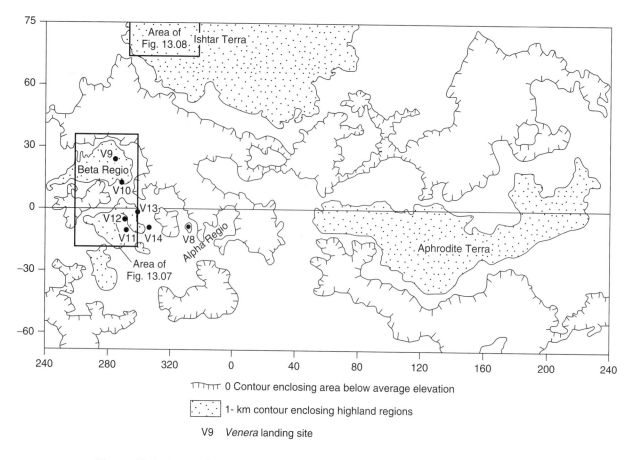

Figure 13.6 Map of Venus surface based on radar observations. *(After Head and Crumpler, 1987)*

surrounding plains. Given Venus's high surface temperature (approximately 450°C), rocks must be ductile at relatively shallow depths. This implies that the topographic elevation is supported by dynamic processes such as convection rather than by the static strength of the rock. One image from *Magellan* (Fig. 13.8C) shows a highly fractured dome structure surrounded by folds and faults, suggesting a close correlation between extensional fractures and compressional folds and faults.

Tesserae are regions of complex topography and evident structure that may represent multiply deformed terranes (Fig. 13.9).

The plains regions, making up 80 percent of Venus's surface, display a low-relief volcanogenic surface that shows evidence of considerable interplay between volcanism and tectonism. Numerous graben structures and related straight features interpreted to be fractures often coexist with an orthogonal set of comparable features (Fig. 13.10*A*, p. 379). In some cases, these fractures intersect young-looking volcanic features that appear to be steep-sided, flat-topped circular mounds, suggesting extrusion of highly viscous (siliceous?) lavas (Fig. 13.10*B*). Elsewhere, the plains

exhibit belts of parallel ridges that have been interpreted as fold belts.

Corona structures are another feature that appears unique to be to Venus (Fig. 13.11, p. 380), and about 21 of them have been recognized so far. They are ring-shaped structures, 150–1000 km in diameter, with a slightly uplifted central volcanic-rich region surrounded by a lower annulus that exhibits evidence of crustal shortening. Three hypotheses have been proposed to explain coronae: A rising diapir or hot spot, a sinking diapir, or gravitational relaxation of topography. Current thinking favors the first hypothesis.

These topographic features and the similarity in size of Venus and Earth make it tempting to speculate that Earth-like plate tectonics may be operating there. The high surface temperature, however, seems to argue against the existence of thick, relatively dense lithospheric slabs, and because the surface temperature is at or above the Curie temperature for most minerals, no magnetic anomalies are expected to exist.

Despite the presence of highlands and lowlands, as indicated on Figure 13.6, Venus's elevation shows a

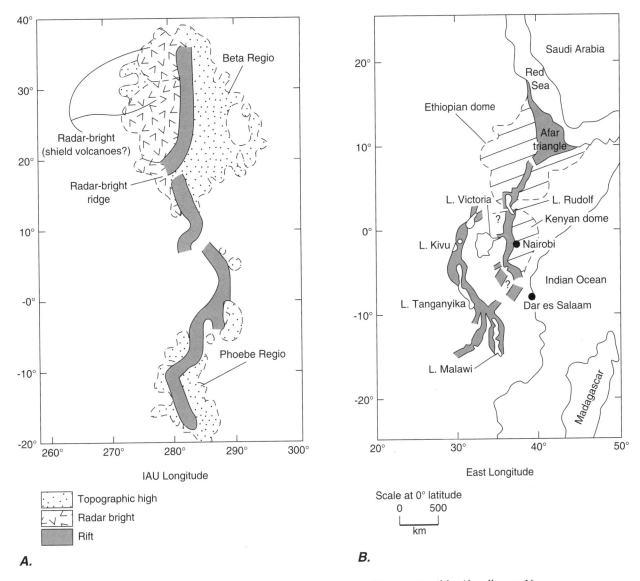

Figure 13.7 *A.* The Beta Regio and adjacent regions of Venus. Possible rift valley on Venus. *B.* The East African Rift Valley for comparison. *(After Phillips and Malin, 1984)*

unimodal distribution, or hypsometric curve, in contrast to the bimodal distribution exhibited on Earth (Fig. 13.12, p. 380; compare Fig. 1.4). Theoretical models of the structures on Venus argue for a thinner boundary layer (lithosphere) than that common on Earth. Crater counts on Venus imply that the entire surface is less than 500 million years old.

Some writers believe that features related to subduction exist on Venus, whereas others disagree and think that the main tectonic style is hot-spot volcanism, with alternating cold spots or mantle downwellings.

13.6 Discussion

This brief overview of Mercury, Venus, Mars, and the Moon reveals a number of similarities and differences, especially in comparison with Earth. All the planets show evidence of volcanism at some stage in their history; all show evidence of a history of meteorite impact; and at least two (Earth and Mars) may have undergone a shift in the position of the rotation poles. Only on Venus is there a suggestion of anything like plate tectonic processes reminiscent of those on Earth,

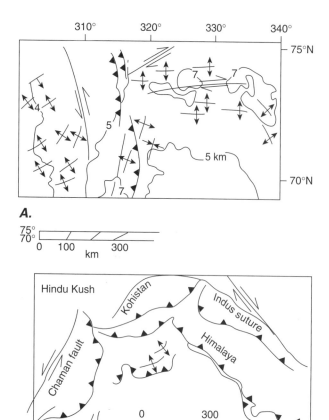

A.

B.

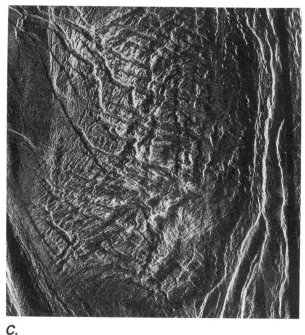

C.

Figure 13.8 Possible compressional features on Venus. *A.* Interpretive map of radar images with ridges of inferred compressive origin, scarps of inferred thrust fault origin, and offset features of inferred strike-slip origin; Ishtar Terra, Maxwell Montes. *B.* Comparison, at approximately the same scale, of the Kashmir syntaxis. *C. Magellan* image showing apparent extension fractures across a dome, suggesting that the dome has been pulled apart. Image 125 by 125 km. Located on eastern flanks of Freyja Montes in Ishtar Terra. (*A. and B. after Crumpler et al., 1986; C. NASA image P-37138*)

and these are apparently subordinate, or at least episodic. Mars, the Moon, and Mercury appear to possess a single global lithospheric plate. Transfer of heat from the interior to the surface is either by conduction or by hot-spot volcanism, the latter being especially important on Mars.

Is it possible that Proterozoic or Archean Earth may be a more suitable analogue for Mars or Venus than the present Earth? Or conversely, can we gain insights into Proterozoic or Archean processes by looking at present-day Venus? A feature of Proterozoic Earth discussed earlier is large mafic-ultramafic plutonic complexes, such as the Bushveld

Figure 13.9 *Magellan* image of tessera terrain. Image 125 by 150 km. Ridges and valleys thought to reflect intense multiple folding, faulting, compression, and extension. (*NASA image P-37322*)

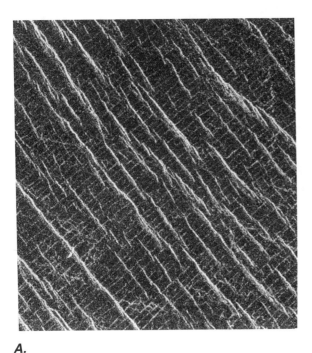

A.

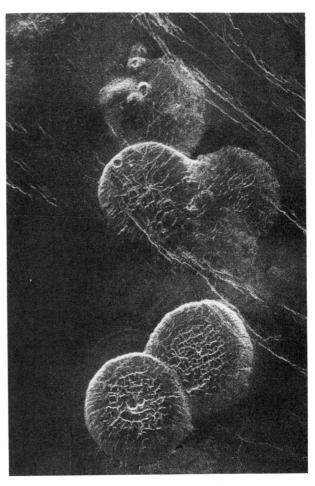

B.

Figure 13.10 *Magellan* images of structures on Venus. *A.* Intersecting lineations interpreted as fracture sets. Area 37 by 80 km. Lakshmi region (30°N, 333°E). *B.* Image from eastern edge of Alpha Regio showing fractured volcanic domes, averaging 25 km in diameter. *(A. NASA image P-36699; B. NASA image P-37125)*

complex (see Figs. 3.11 and 3.12). The size of this complex is comparable to that of the Martian volcanoes. Thus, we might imagine the structure at depth beneath Olympus Mons, for example, to be equivalent.

The widespread existence of komatiites and deformed mafic-ultramafic complexes in Archean terranes argues for large amounts of partial melting of the mantle during this Archean time. Some workers have proposed magma oceans and whole-mantle melting during early stages of Earth's history. Thus, it is easy to imagine that these complexes, which formed early in Earth's history, could represent a counterpart to the large volcanic features so prominent on neighboring bodies.

The tectonic features exhibited by neighboring planets raise fascinating questions not only about the planets themselves, but also about the tectonic pro-

cesses operating on planet-sized bodies in general and on Earth in particular. A general model for the tectonic evolution of planetary bodies should be able to account not only for what we observe on Earth but also for what we observe on other planets. The similarities and differences in tectonic features that we observe ultimately must be explainable in terms of the fundamental characteristics of the different bodies, because it is the similarities and differences among the planets that must give rise to the similarities and differences in their tectonic evolution. To the extent that our observations of other planetary bodies constrain the models we use to understand Earth, we can learn about our own planet from studying those around us. Perhaps the next few decades will see major advances in the formulation of a unified theory of planetary tectonics and evolution.

Figure 13.11 Corona structures on Venus. *A. Sketch map of idealized corona structure showing main features. B. B. Magellan image of corona structure, 375 km in diameter. It is located approximately 1000 km south of Sphrodite Terra. The black vertical strip at the right represents an area for which there is no data. (A. after Stofan et al., 1991; B. Image courtesy of J. J. van der Woude at the Jet Propulsion Laboratory, no. P37946 MGN 39)*

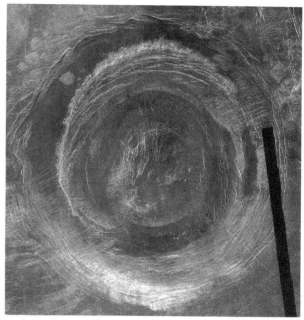

B.

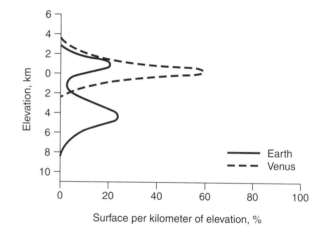

Figure 13.12 Hypsographic curves for Earth and Venus, showing bimodal and unimodal distributions, respectively. *(After Saunders and Carr, 1984; Moores, 1986)*

Features	Characteristics
	Annulus of ridges; spaced 5–15 km apart– possibly compressional and/or extensional in origin.
	Interior ridges; < 100 km long–may be either compressional or extensional.
	Interior grooves; < 100 km long–may be 5-20 km wide grahens.
	"Chaotic terrain"; Ridges and grooves.
•	Domes; 5-15 km in diameter–may be volcanic.
	Radar-bright and-dark regions; up to hundreds of kilometers long– may be lara flows
	Volcanic (?) features with summit craters or pits; radial lines may be faults or fissures.

A.

Additional Readings

Carr, M. H. 1984. Mars. In *The Geology of the Terrestrial Planets*. M. H. Carr, ed. Washington, D.C.: NASA Special Paper 469.

Head, J. W., and R. S. Saunders. 1991. Geology of Venus: A perspective from early *Magellan* mission results. *GSA Today* 1:49.

Phillips, R. J., and M. C. Malin. 1984. Tectonics of Venus. *Ann. Rev. Earth Planet. Sci.* 12:411–443.

Saunders, R. S., et al., Magellan at Venus. *J. Geophys. Res.* (Part 1) 97:13,059–13,690; (Part 2) 97:10, 921–16,382.

Solomon, S. C., et al. 1992. Venus tectonics: An overview of *Magellan* observations. *J. Geophys. Res.* 97: 13,199–13,255.

Stofan, E. R., D. L. Bindschadler, J. W. Head, and E. M. Parmentier. 1991. Corona structures on Venus: Models of origin. *J. Geophys. Res.* 96:20,933–20,947.

Turcotte, D. L. 1993. An episodic hypothesis for Venusian tectonics. *J. Geophys. Res.* 98:17,061–17,068.

Wilhelms, D. E. 1984. Moon. In *The Geology of the Terrestrial Planets*. M. H. Carr, ed. Washington, D.C.: NASA Special Paper 469.

APPENDIX

Map Projections

A.1 Introduction

Maps and other projections are an integral part of any study in structural geology and tectonics. They serve two purposes: (1) to represent the three-dimensional geometry of structures in two dimensions, and (2) to show the relationships between the geometry and other structural or tectonic information. Although the purposes are different, the principles involved are the same, and we will discuss them together in general terms.

The Earth is spherical; maps are flat. It is impossible to transform the surface of the Earth into a flat surface without introducing some sort of distortion. A number of different systematic distortion schemes may be used to flatten a spherical surface, and these schemes are called **projections** of the spherical surface onto a plane or onto a curved surface that can be simply flattened. Each type of projection distorts lines, areas, and angular relationships in different ways, and thus, each provides a different compromise for depicting the spherical surface. The type of projection used depends upon what one is trying to show or to analyze.

A.2 Types of Projections

The basic process used to construct any map projection includes selecting a type of surface and projecting information from the globe onto that surface. The procedure symbolized in Figure A1.1 is followed to construct a map. We begin with a spheroid approximating the Earth, reduce it to a globe, and then project features of the globe upon the surface. The most common projections are to a cylindrical surface, a conical surface, or a plane. The resulting projections are called cylindrical, conic, or azimuthal projections, respectively. It is also possible to calculate a map **graticule** directly, without going through a projection. (A graticule is a distribution of points of latitude or longitude.) Sinusoidal projections are constructed in this fashion by using a specific trigonometric formula to produce a desired projection.

The scale of a map is not uniform throughout. A projection generally contains one or more **standard lines**, which are lines whose length on the map equals that on the globe. The lengths of all other lines do not equal their lengths on the globe. The **scale factor** is the map distance divided by the globe distance. Thus, for scale factors greater or less than 1, the lines are increased or decreased in length on the map relative to the globe.

Maps of all projection types are constructed to be equidistant, equal area, conformal, or equal angle. On equidistant maps, the distance between lines of latitude is the same everywhere on the map. Equal-area maps preserve the relative areas of sectors on the globe bounded, for example, by specific lines of latitude and longitude. On conformal maps, distances are distorted the same amount in all directions from a given point. On equal-angle projections, the angles between tangents to great circles at their intersection is the same as the angle between two real planes in space; small circles plot as circles. Table A.1 on pages 384–385 summarizes the map projections most often used in the study of structural geology and tectonics. These projections and their characteristics are illustrated in Figures A1.2 through A1.12. In particular, Figure A1.9 shows the western hemisphere on various azimuthal projections. Note the difference in the shapes and sizes of land surfaces and the shapes of parallels and meridians. Figure A1.10, the distortion of a face, shows a comparison of several azimuthal projections with that of the Mercator projection. Figure A1.12 shows continental configurations on four different projections for 220 Ma (early Triassic), when the land masses were assembled into Pangea.

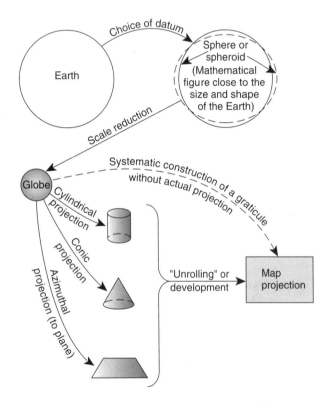

A1.1 The evolution of a map projection showing choice of a reference sphere or spheroid, scale reduction, choice of projection, and development. Development of the actual map projection involves "unrolling" of the projection surface. As indicated, one also can construct a map without actual projection by calculation of a graticule. *(After McDonnell, 1979)*

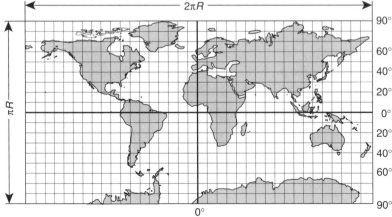

A1.2 Cylindrical equidistant projection. Dimensions of map relative to reference globe are shown. *(After McDonnell, 1979)*

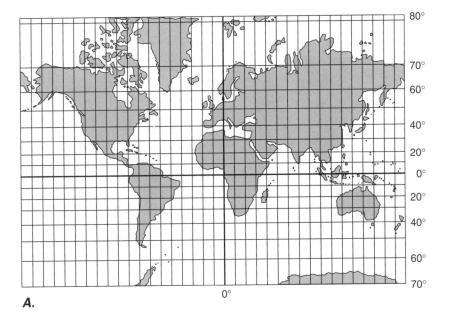

A.

A1.3 Two cylindrical projections showing effects of different requirements. *A.* Mercator projection, in which N-S exaggeration = E-W exaggeration = secant (latitude). The map is conformal; that is, angular relationships are preserved, but exaggeration is very great in high latitudes (for example, Greenland is actually 1/10 the area of South America). *B.* Cylindrical equal-area map in which the N-S distance is inversely proportional to the E-W distance. In other words, N-S is proportional to cosine (latitude), whereas E-W distance is proportional to secant (latitude). *(A. after Deetz and Adams, 1928; B. after McDonnell, 1979; Deetz and Adams, 1928)*

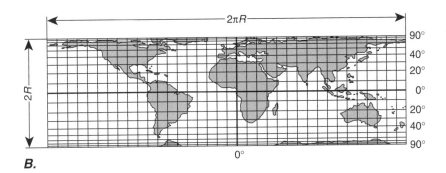

B.

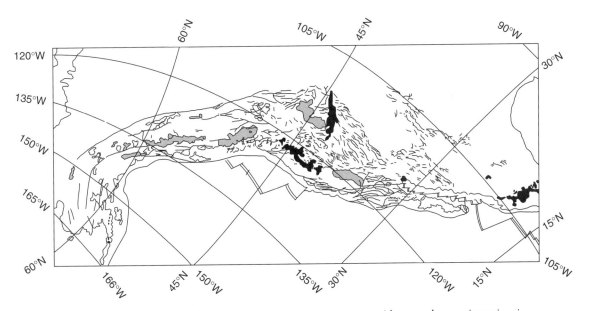

A1.4 Geologic example of the use of a transverse Mercator grid, one whose orientation is not parallel to latitude and longitude. Tectonic map of western North America, with plate boundary features, faults, recent volcanics (black) and Mesozoic batholiths (gray). The map is oriented so that the relative motion vector between the Pacific and North American plates is horizontal. Standard line approximately along San Andreas fault. *(After Atwater, 1970)*

Table A.1 Map Projections Used in Structural Geology and Tectonics

Projection Class	Projection Type	Figure	Projection Principle	Scale Factor	Standard Lines	Comments
Cylindrical						
1	Cylindrical equidistant	A1.2 A1.12B	Equal longitudinal distance on globe and map	North-South = 1 East-West = secant (latitude)	Meridians Equator	Used for world, equator-centered maps on which polar regions must also be depicted.
2	Mercator	A1.3A A1.10D A1.12A	Conformal	secant (latitude) = 1 at equator 2 at 60° ∞ at poles	Equator	North-South exaggeration = East-West exaggeration = secant (latitude). Latitude and longitude are straight and mutually perpendicular. No angular distortion, but much areal distortion away from equator. Used in navigation for 400 years because line of constant bearing is a straight line. Transverse Mercator used on new tectonic map of North America.
	Oblique (transverse) Mercator	A1.4			Any other great circle	
3	Cylindrical equal area	A1.3B	Equal Area	East-West = secant (latitude) North-South times east-west = 1.	Equator	Used for world, equator-centered maps on which straight latitude and longitude lines and area conservation are desired.
Conic Projections		A1.6 A1.7A,B				
4	Albers	A1.5	Equal area	North-South > 1 East-West < 1 Product of E-W and N-S scale factors = 1	Two specified lines of latitude	Useful for maps that extend large distances in the E-W direction. Used for the 1975 Geologic Map of the United States.
5	Lambert conformal	A1.6	Conformal	East-West = North-South	Two specified lines of latitude	Similar to Albers, except latitude lines are slightly closer together. Used on recent U.S. Geological Survey maps: the 1977 Geological Map of California.
6	Polyconic	A1.7C	Equidistant	Approximately 1 for small areas	Every latitude line is a standard line.	Used on older U.S. Geological Survey topographic maps. Bipolar polyconic projection used on older tectonic maps of North and South America.

Azimuthal						
			A1.8			
	7	Gnomonic	A1.9H	Projection from center of sphere onto a plane tangent at the pole		Pole plots in center of projection. All great circles are straight lines. Useful for maps near the pole. Very great distortion at large distances away from center of map (that is, away from the pole).
	8	Globular	A1.9G A1.10A	Distances between latitude and longitude remain equal.		All latitude lines are arcs of circles.
	9	Stereographic or Equal-angle	A1.9A A1.9D A1.10C	Projection from pole onto plane tangent at opposite pole		Angle between tangents to two great circles at their intersection on projection is the same as between two real planes in space. Small circles on the sphere plot as small circles. Can show nearly entire sphere, but the areal distortion of the back hemisphere is very large. The Wulff net (or stereographic or equal-angle net) is an equatorial stereographic projection. A diagram constructed on the Wulff net is a polar stereographic projection.
	10	Orthographic	A1.9F A1.10B	Projection along lines perpendicular to the projection plane		Equivalent to viewing a hemisphere from infinite distance. Can show only one hemisphere.
	11	Lambert Equal-Area	A1.9C A1.10E A1.12C	Point on projection plane tangent to the sphere is specified by a chord that is drawn from the tangent point to the point on the sphere, and then rotated to the projection plane about the tangent point in a plane normal to the projection plane.		Relative areas are preserved, but angles become distorted. Entire sphere can be shown, but back hemisphere is greatly distorted. Point antipodal to the tangent point of the projection plane is represented as a circle. Schmidt or equal-area net is an equatorial Lambert equal-area projection.
	12	Equidistant	A1.9B	Projection from point, distances from point are the same, azimuths are preserved.		Useful for specific stations, e.g. airports, seismographic stations, etc., where distance from station is important.
Sinusoidal						
		Mollweide	A1.11			
	13		A1.12D	Entire globe is projected onto an ellipse whose major axis is the equator. Meridians are half of an ellipse. Primary meridian is a straight line. The two meridians 90° from the primary form a circle. Meridians are equally spaced along lines of latitude. Latitude lines are spaced so that area is preserved.	Major axis of ellipse = 4 R √2 Minor axis of ellipse = 2 R √2	Equal-area property is preserved over the entire projection. Useful for projecting world-wide features (for example, see Fig. 12.31). Scale factor is inconsistent throughout the projection.

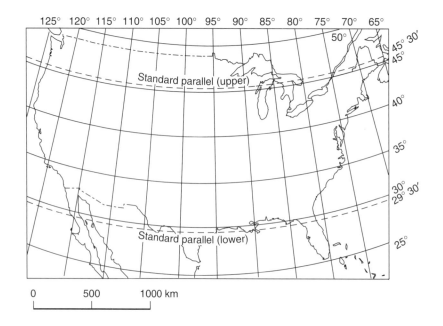

A1.5 Albers equal-area projection of coterminous United States, a model for that used on the Geologic Map of the United States. Standard parallels are shown. *(After Deetz and Adams, 1928)*

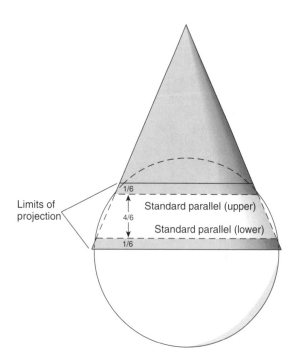

A1.6 Principle of conic projection with two standard parallels. Basis of Albers and Lambert conformal conic projections.

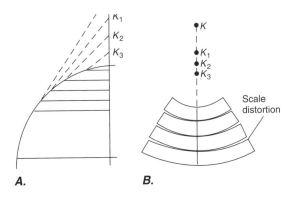

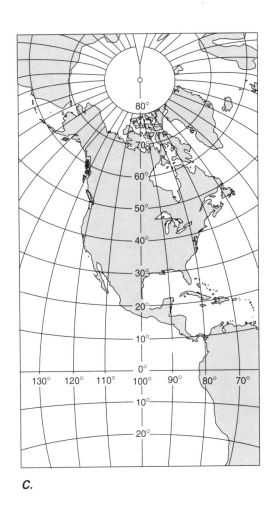

A1.7 Polyconic projection. *A.* Principle of the polyconic projection, with each parallel serving as a standard parallel, with its own cone defined by tangent lines. *B.* Unfolding of resultant surface, with scale distortion increasing away from central meridian. *C.* Polyconic projection of region around North America. *(B. and C. after Deetz and Adams, 1928)*

C.

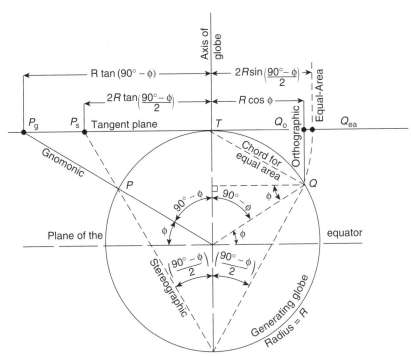

A1.8 Projection principal for four azimuthal projections showing generating globe with radius *R*; two points on globe surface, *P* and *Q*, at equivalent latitudes; and their positions on the tangent planes in various projections. Gnomonic point (P_g) results from projection from center of sphere; stereographic point P_s results from projection from the point opposite the point of tangency, orthographic point Q_o results from projection perpendicular to tangent plane, and equal area point Q_{ea} is located at a distance from the tangent point equal to the chord between the point of tangency *T* and the point in question *Q*. *(After McDonnell, 1979)*

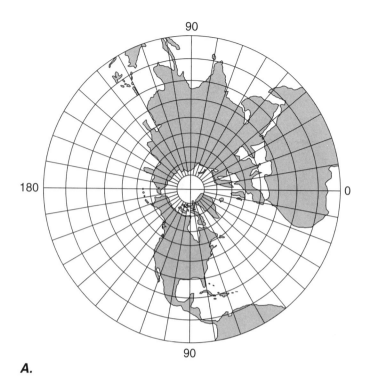

A.

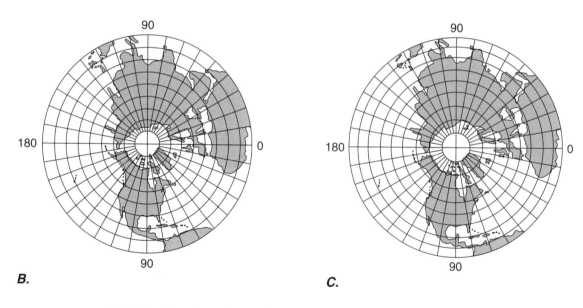

B.

C.

A1.9 Configurations of parts of the Earth's surface on various projections. *A.* Stereographic projection of northern hemisphere. *B.* Azimuthal equidistant projection of northern hemisphere. *C.* Lambert equal-area projection of northern hemisphere. *D.* Stereographic projection of western hemisphere. *E.* Lambert equal-area projection of western hemisphere. *F.* Orthographic projection of western hemisphere. *G.* Globular projection of western hemisphere. *H.* Gnomonic projection of western hemisphere. The gnomonic projection is only part of the hemisphere because of the character of the projection. *(After Deetz and Adams, 1928)*

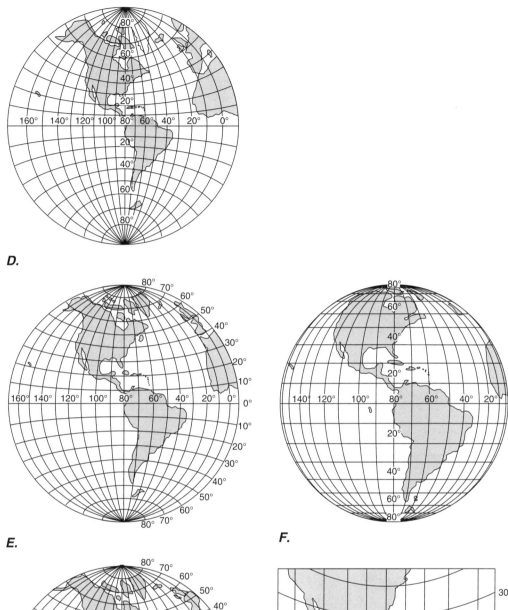

D.

E.

F.

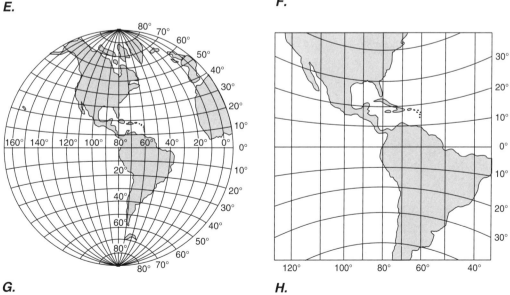

G.

H.

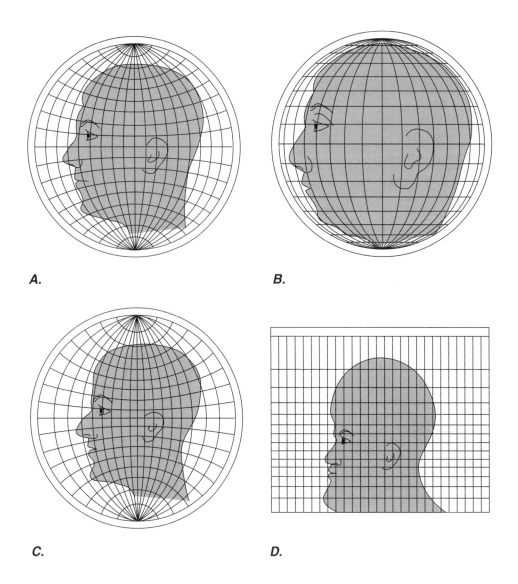

A1.10 Illustration of distortion from different projections. A. A man's head drawn on globular projection and then redrawn on *B.* orthographic, *C.* stereographic, and *D.* Mercator projections using sam latitude and longitude values obtained from *A.*

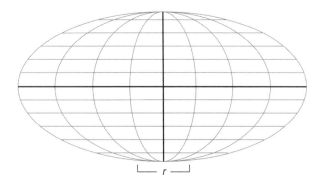

A1.11 Mollweide projection showing parallels at 15° intervals and meridians at 45° intervals.

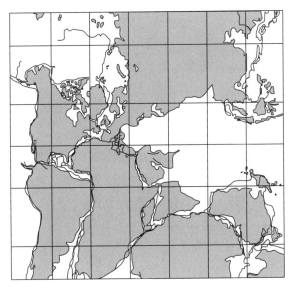

A.

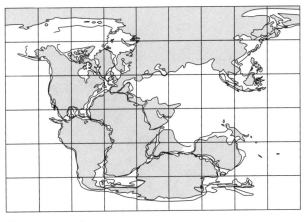

B.

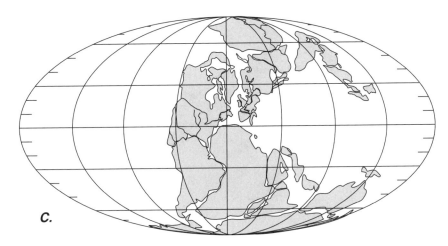

C.

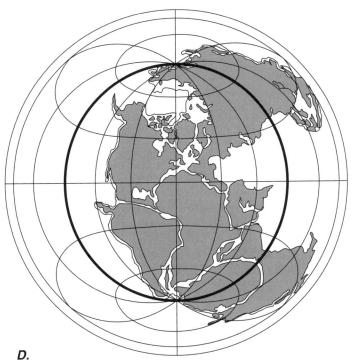

D.

A1.12 Continental configurations at time of Pangea (late Permian–early Triassic, 220–240 Ma). *A.* Mercator projection of Pangea (compare with Fig. A1.3). *B.* Cylindrical equidistant diagram of Pangea (compare with Fig. A1.2). *C.* Lambert equal-area map of Pangea. *D.* Mollweide projection of Pangea, using different reconstruction and earlier (late Permian) configuration. The heavy line encloses the front hemisphere. *(A. after Smith and Briden, 1977; B. after Smith et al., 1980; C. after Smith and Briden, 1977; D. after Scotese et al., 1979)*

References and Illustration Sources

Chapter 1

Baily, E. B. 1968. *Tectonic Essays, Mainly Alpine*. London: Oxford, University Press.

Bally, A. W. 1980. Basins and subsidence: A summary. In *Dynamics of Plate Interiors*. A. W. Bally, P. L. Bender, T. R. McGetchin, and R. I. Walcott, eds. Geodynamics Series, vol. 1. Washington, D.C.: American Geophysical Union; Boulder, Colo.: Geological Society of America.

Goguel, Jean. 1962. *Tectonics*. From the French edition of 1952. H. E. Thalman, trans. San Francisco: W. H. Freeman and Company.

Siever, R., ed. 1983. *The Dynamic Earth*. Special issue of *Scientific American*. September.

Twiss, R. J., and E. M. Moores. 1992. *Structural Geology*. New York: W. H. Freeman and Company.

Uyeda, S. 1978. *The New View of the Earth*. San Francisco: W. H. Freeman and Company.

Wyllie, P. J. 1975. The earth's mantle. *Scientific American* 232(3):50–63. Reprinted in *Continents Adrift and Continents Aground*. J. T. Wilson, ed. 1976. San Francisco: W. H. Freeman and Company.

Chapter 2

Fowler, C. M. R. 1990. *The Solid Earth: An Introduction to Global Geophysics*. Cambridge (England); New York: Cambridge University Press.

Gurnis, M. Long-term controls on eustatic and epeirogenic motions by mantle convection. *GSA Today* 2(7):141.

Lindseth, R. O. 1982. *Digital Processing of Geophysical Data. A Review*. Continuing Education Program, Society of Exploitation Geophysicists. Calgary, Alberta, Canada: Teknica Resource Development Ltd.

McElhinny, M. W. 1973. *Paleomagnetism and Plate Tectonics*. London: Cambridge University Press.

Press, F., and R. Siever. 1986. *Earth*. New York: W. H. Freeman and Company.

Sheriff, R. E. 1978. *A First Course in Geophysical Exploration and Interpretation*. Boston: International Human Resources Development Corporation.

Stacey, F. D. 1969. Physics of the earth. In space text series. New York: Wiley.

Turcotte, D. L., and G. Schubert. 1982. *Geodynamics: Applications of Continuum Physics to Geological Problems*. New York: Wiley.

Chapter 3

Anhaeusser, K. 1984. In *Precambrian Tectonics Illustrated*. Kröner, A., and R. Greiling, eds. Stuttgart: Schweizerbartsche.

Anonymous. 1950. Der Bau der Erde, Gotha, Justus Perthes, 1:140,000,000.

Bally, A. W., et al. 1979. *Continental Margins: Geological and geophysical research needs and problems*. Washington, D.C.: National Academy of Sciences.

Bally, A. W., et al. 1980. Geodynamics Series, vol. 1. Washington, D.C.: American Geophysical Union. Boulder, Colo.: Geological Society of America.

Burchfiel, B. C. 1983. The continental crust. In *The Dynamic Earth*. R. Siever, ed. Special issue of *Scientific American*. September.

Burke, K. 1980. Intracontinental rifts and aulacogens. In Geophysics Study Committee, Continental Tectonics, National Academy of Sciences.

Cloetingh, S. 1986. Intraplate stresses: A new tectonic mechanism for fluctuations of relative sealevel. *Geology* 14:617–620.

Francheteau, J. 1983. The oceanic crust. In *The Dynamic Earth*. R. Siever, ed. Special issue of *Scientific American*. September.

Hoffman, P. 1988, United Plates of America. *Ann. Rev. Earth and Plan Sci.* 16:543–603.

Jackson, M. P. A. 1984. Archaean structural styles in ancient gneiss complex of Swaziland, southern Africa.

In A. Kröner and R. Greiling, eds. *Precambrian Tectonics Illustrated*. Stuttgart: Schweizerbartsche.

Kay, M. 1981. North American geosynclines. *Geological Society of America Memoir*, p. 48.

King, P. B. 1977. *The Evolution of North America*, rev. ed. Princeton: Princeton University Press.

Kröner, A., and R. Greiling, eds. *Precambrian Tectonics Illustrated*. Stuttgart: Schweizerbartsche.

National Academy of Sciences–National Research Council. 1980. *Continental Tectonics*. Washington, D.C.

Nisbet, E. G. 1987. *The Young Earth: An Introduction to Archaean Geology*. Boston: Allen and Unwin.

Pakiser, L. C., and W. D. Mooney, eds. 1989. *Geophysical Framework of the Continental United States*. Geological Society of America Memoir 172. Boulder: Geological Society of America.

Press, F., and R. Siever. 1986. *Earth,* 4th edition. New York: W. H. Freeman and Company.

Sclater, J. G., C. Jaupart, and D. Galson. 1980. The heat flow through oceanic and continental crust and the heat loss of the Earth. *Rev. Geophys. and Space Phys.* 18:269–311.

Smithson, S. B., P. N. Shive, and S. K. Brown. 1977. Seismic velocity, reflection, and structure of the crystalline crust. In *The Earth's Crust, its Nature and Physical Properties.* J. G. Heacock, ed. American Geophysical Union Monograph 20:254–270.

Stanton, R. L. 1972. *Ore Petrology*. New York: McGraw-Hill.

Stanley, S. 1986. *Earth and Life Through Time*. New York: W. H. Freeman and Company.

Trendall, A. P. 1968. Geological Society of America Bulletin 79, 1527–1544.

Uyeda, S. 1978. *The New View of the Earth*. San Francisco: W. H. Freeman and Company.

Vail, P. R., R. M. Michum, Jr., and S. Thompson. *AAPG Memoir* 19:84.

Windley, B. F. 1984. *The Evolving Continents*. New York: Wiley.

Wright, L. A., B. W. Troxel, E. G. Williams, M. T. Roberts, and P. E. Diehl. 1971. *Precambrian sedimentary environments of the Death Valley region, eastern California; G.S.A. Cordilleran Section, Field Trip No. 1, Guidebook, Death Valley Region, California and Nevada.* Shoshone, Calif.: Death Valley Publishing Company.

Zonenshain, Lev. P., M. I. Kuzmin, and L. M. Natapov. 1990. Geology of the USSR: A plate-tectonic synthesis. American Geophysical Union Geodynamic Series, vol. 21. B. M. Page, ed.

Chapter 4

Chase, C. G. 1978. Plate kinematics: The Americas, east Africa, and the rest of the world. *Earth. Plan. Sci. Lett.* 37:355–368.

Cox, A. 1972. *Plate Tectonics and Geomagnetic Reversals*. San Francisco: W. H. Freeman and Company.

Cox, A., and R. B. Hart. 1986. *Plate Tectonics: How It Works*. Palo Alto: Blackwell's, pp. 138 and 362. Reprinted by permission of Blackwell Science Publications, Inc.

Cronin, V. S. 1987. Cycloid kinematics of relative plate motion. *Geology* 15:1006–1009.

Cronin, V. S. 1991. The cycloid relative-motion model and the kinematics of transform faulting. *Tectonophysics* 187:215–249.

Cronin, V. S. 1992. Types of kinematic stability of triple junctions. *Tectonophysics* 207:287–301.

DeMets, C. 1993. Earthquake slip vectors and estimates of present-day plate motions. *J. Geophys. Research* 98: 6703–6714.

Demets, C., R. G. Gordon, D. F. Argus, and S. Stein. 1990. Current plate motions. *Geophys. J. International* 101: 425–478.

Dewey, J. F. 1975. Finite plate implications: Some implications for the evolution of rock masses at plate margins. *Am. J. Sci.* 275A:260–284.

Dziewonski, A., et al. 1987. A window into the lower mantle. *Science* 236:41–45.

Engebretson, D. C., A. Cox, and R. G. Gordon. 1985. Relative motions between oceanic and continental plates in the Pacific basin. GSA Special Paper 206.

Engebretson, D. C., K. P. Kelley, H. J. Cashman, and M. A. Richards. 1992. 180 Million Years of Subduction. *GSA Today* 2(6):93–95, 100.

Hilde, T. W., S. Uyeda, and L. Kroenke. 1977. Evolution of the western Pacific and its margin. *Tectonophysics* 38: 145–165.

Isacks, B., and P. Molnar. 1972. Mantle earthquake mechanisms and the sinking of the lithosphere. From *Nature* 223:1121–1124, reprinted in A. Cox, ed., *Plate Tectonics and Geomagnetic Reversals*. San Francisco: W. H. Freeman and Company.

Jeanloz, R., and F. Richter. 1979. Convection, composition and the thermal state of the lower mantle. *J. Geophys. Res.* 81:5497–5504.

Larson, R. L., and W. C. Pitman III. 1972. World-wide correlation of Mesozoic magnetic anomalies and its implications. *GSA Bull.* 83:3645–3662.

Le Pichon, X., and J. Francheteau. 1976. *Plate Tectonics*. New York: Elsevier.

McKenzie, D. P., and F. Richter. 1976. Convection currents in the Earth's mantle. *Scientific American* 235(5): 72–84.

McKenzie, D. P. 1983. The Earth's mantle. In *The Dynamic Earth*. R. Siever, ed. *Special Issue of Scientific American*. September.

Minster, J. B., and T. H. Jordan. 1978. Present-day plate motions. *J. Geophys. Res.* 83:5331–5354.

Minster, J. B., T. H. Jordan, P. Molnar, and E. Haines. 1974. Numerical modeling of instantaneous plate tectonics. *Geophys. J. Roy. Astron. Soc.* 36:541–576.

Morgan, W. J. 1968. Rises, trenches, great faults, and crustal blocks. *J. Geophys. Res.* 73:1959–1982.

Morgan, W. J. 1972. Plate motion and deep mantle convection. *GSA Mem.* 132:7–22.

Morgan, W. J. 1981. Hotspot tracks and the opening of the Atlantic and Indian Oceans. In *The Oceanic Lithosphere: The Sea*. C. Emiliani, ed. New York: Wiley.

Patriat, P., and V. Courtillot. 1984. On the stability of triple junctions and its relationship to episodicity in spreading. *Tectonics* 3:317–332.

Press, F., and R. Siever. 1986. *Earth*. 4th ed. New York: W. H. Freeman and Company.

Richardson, R. M., S. C. Solomon, and N. H. Sleep. 1979. Tectonic stress in the plates. *Rev. Geophys. and Sp. Phys.* 17:981–1019.

Smith, A. G., A. M. Hurley, and J. C. Brideu. *Phanerozoic Paleocontinental World Maps*. New York: Cambridge University Press.

Uyeda, S. 1978. *The New View of the Earth*. San Francisco: W. H. Freeman and Company.

Zoback, M. L., and M. D. Zoback. 1980. State of stress in the conterminus United States. *J. Geophys. Research* 85: 6113–6156.

Chapter 5

Allmendinger, R. W., T. A. Hauge, E. C. Hauser, C. J. Potter, S. L. Klemperer, K. D. Nelson, P. Knuepfer, and J. Oliver. 1987. Overview of the COCORP 40°N transect, western United States: The fabric of an orogenic belt. *GSA Bull.* 98:364–372.

Burk, C. A. 1968. Buried ridges within continental margins. *N.Y. Acad. Sci. Trans.* 30:135–160.

Burke, K. 1980, Intracontinental rifts and aulacogens. In *Continental Tectonics*. Washington, D.C.: National Academy of Sciences.

Burke, K. and A. J. Whiteman. 1973. Uplift, rifting, and the breakup of Africa. In *Implications of Continental Drift to the Earth Sciences*, vol. 2. D. H. Tarling and S. K. Runcorn, eds. New York: Academic Press.

Burke, K., and J. T. Wilson. 1976. Hot Spots on the Earth's Surface. In *Continents Adrift and Continents Aground*. Readings from Scientific American. J. T. Wilson, ed. San Francisco: W. H. Freeman and Company.

Christensen, N. I., and M. H. Salisbury. 1975. Structure and constitution of the lower oceanic crust. *Rev. Geophys. Space Phys.* 13:57–86.

Clifford, T. N., and I. G. Gass. 1970, *African Magmatism and Tectonics*. Edinburgh: Oliver & Boyd.

Coleman, R. G. 1974. Geologic background of the Red Sea. In *Continental Margins*. C. A. Burk and C. L. Drake, eds. New York: Springer.

Coward, M. D., J. F. Dewey, and P. L. Hancock. 1987. Continental extension tectonics. Geol. Society of London. Special Publication 28.

Deffeyes, K. S. 1970. The axial valley, a steady state feature of the terrain. In *The Megatectonics of Continents and Oceans*. H. Johnson and B. L. Smith, eds. New Brunswick, N.J.: Rutgers Univ. Press.

Detrick, R. S., P. Buhl, E. Vera, J. Mutter, J. Orcutt, J. Madsen, and T. Brocher. 1987. Multi-channel seismic imaging of a crustal magma chamber along the East Pacific Rise. *Nature* 326:35–41.

Dixon, T. H., E. R. Ivins, and B. J. Franklin. 1989. Topographic and volcanic asymmetry around the Red Sea: Constrains on rift models. *Tectonics* 8:1193–1216.

Eaton, G. P. 1980. Geophysical and geological characteristics of the crust of the Basin and Range Province. In *Continental Tectonics*. Washington, D.C.: National Academy of Sciences.

Fairhead, J. D. 1986. Geophysical controls on sedimentation within the African rift systems. In *Sedimentation in the African Rifts*. L. E. Frostick et al., eds. Geol. Society of London. Special Publication 25.

Hamilton, W. 1978. Mesozoic tectonics of the western United States. In Soc. Econ. Paleontol. Mineralog. Pacific Coast Paleogeogr. Symp. 2. Los Angeles.

Heezen, B., and M. Ewing. 1963. *The Mid-Oceanic Ridge in the Sea*, vol. 3. M. N. Hill, ed. New York: Wiley-Interscience.

Hey, R. N. 1977. A new class of pseudofaults and their bearing on plate tectonics: A propagating rift model. *Earth Planet Sci. Lett.* 37:321–325.

JOIDES. 1991. North Atlantic Rifted Margins Detailed Planning Group Report.

King, P. B. 1977. *Evolution of North America*. Rev. ed. Princeton: Princeton University Press.

Lachenbruch, A. 1976. Dynamics of a passive spreading center. *J. Geophys. Res.* 81:1883–1902.

Lister, G., M. A. Etheridge, and P. A. Symonds. 1986. Detachment faulting and the evolution of passive continental margins. *Geology* 14:246–250.

Lowell, J. D., G. J. Genik, T. H. Nelson, and P. M. Tucker. 1975. Petroleum and plate tectonics of the southern Red Sea. In *Petroleum and Plate Tectonics*. A. G. Fischer and S. Judson, eds. Princeton: Princeton University Press.

McClain, J. S., and C. Atallah. 1986. Thickening of the oceanic crust with age. *Geology.* 14:574–576.

Macdonald, K. C. 1982. Mid-ocean ridges: Fine scale tectonic, volcanic and hydrothermal processes within the plate boundary zone. *Ann. Rev. Earth Planet. Sci.* 10: 155–190.

Macdonald, K. C. 1989a. Tectonic and magmatic processes on the East Pacific Rise. In *Geology of North America*. vol. N, East Pacific Ocean and Hawaii. E. L. Winterer, D. M. Hussong, and R. W. Decker, eds. Boulder, Colo.: Geological Society of America.

Macdonald, K. C. 1989b. Anatomy of the magma reservoir. *Nature* 339:178–179.

Macdonald, K. C., and P. J. Fox. 1990. The mid-ocean ridge. *Scientific American* 262(6):72–79.

Macdonald, K. C., J.-C. Sempere, P. J. Fox, and R. Tyce. 1987. Tectonic evolution of ridge-axis discontinuities by the meeting, linking, or self-decapitation of neighboring ridge segments. *Geology* 15:993–997.

May, P. R. 1971. Pattern of Triassic-Jurassic diabase dikes around the North Atlantic in the context of predrift position of the continents. *GSA Bull.* 82:1285–1292.

Mayer, L. 1986. Topographic constraints on models of lithospheric stretching of the Basin and Range province, western United States. GSA Special Paper 208.

Menard, H. W., and T. E. Chase. 1970. Fracture zones. In *The Sea*, vol. 4, part 1. New concepts of sea floor evolution. A. E. Maxwell, ed. New York: Wiley.

Montadert, L., D. G. Roberts, O. de Charpal, and P. Guennoc. 1979. Rifting and subsidence of the northern continental margin of the Bay of Biscay. Initial Reports Deep-Sea Drilling Project 48: 1205–1260. San Diego: Scripps Institute, Oceanography.

Moores, E. M. 1982. Origin and emplacement of ophiolites. *Rev. Geophys. and Space Physics* 20:735–760.

Moores, E. M., and F. J. Vine. 1971. The Troodos complex, Cyprus, and other ophiolites as oceanic crust: Evaluation and implications. *Trans. Roy. Soc.* 286A:144–166.

Nur, A., and A. Ben-Avrahem. 1982. Oceanic plateaus, the fragmentation of continents, and mountain building. *J. Geophys. Res.* 87:3644–3662.

Rabinowitz, P. D. 1974. The boundary between oceanic and continental crust in the western North Atlantic. In *Continental Margins.* C. A. Burk and C. L. Drake, eds. New York: Springer.

Richardson, R. M., S. C. Solomon, and N. H. Sleep. 1979. Tectonic stress in the plates. *Rev. Geophys. and Space Phys.* 17:981–1019.

Rosendahl, B., D. J. Reynolds, P. M. Lorber, C. F. Burgess, J. McGill, D. Scott, J. J. Lambiase, and S. J. Derksen. 1986. Structural expressions of rifting: Lessons from Lake Tanganyika, Africa. In *Sedimentation in the African Rifts.* L. E. Frostick et al., eds. Geological Society of America, Special Publication 25.

Schlee, J. 1980. A comparison of two Atlantic-type continental margins. USGS Professional Paper 1187.

Sclater, J. G., and J. Francheteau. 1970. The implications of terrestrial heat flow observations on current tectonic and geochemical models of the crust and upper mantle of the Earth. *Geophys. J., Roy. Astron. Soc.* 20:509–542.

Sclater, J. G., C. Jaupart, and D. Galson. 1980. The heat flow through oceanic and continental crust and the heat loss of the Earth. *Rev. Geophys. and Space Phys.* 18:269–311.

Sheridan, R. E., J. A. Grow, and K. D. Klitgord. 1988. Geophysical data. In *Geology of North America,* vol. I-2. The Atlantic continental margin: U.S. R. E. Sheridan and J. A. Grow, eds. Geological society of America. Boulder, Colo.: 177–196.

Smith, R. B. 1978. Seismicity, crustal structure, and intraplate tectonics of the interior of the western Cordillera. *GSA Memoir* 152:111–144.

Stewart, J. H. 1972. Initial deposits in the Cordilleran geosyncline: Evidence of a late Precambrian (<850 m.y.) continental separation. *GSA Bull.* 83:1345–1360.

Stewart, J. H. 1978. Basin-range structure in western North America: A review. *GSA Memoir* 152:1–31.

Stewart, J. H. 1980. *Geology of Nevada.* Nevada Bureau of Mines & Geology, Special Publication 4.

Talwani, M. 1965. Crustal structure of the mid-ocean ridges. 2. Computed model from gravity and seismic refraction data. *J. Geophys. Res.* 70:341–452.

Tapponnier, P., and J. Francheteau. 1978. Necking of the lithosphere and the mechanics of slowly accreting plate boundaries. *J. Geophys. Res.* 83:3955–3970.

Varga, R. J., and E. M. Moores. 1985. Spreading structure of the Troodos ophiolite, Cyprus. *Geology* 13:846–850.

Varga, R. J., and E. M. Moores. 1990. Intermittent magmatic spreading and tectonic extension in the Troodos ophiolite: Implications for exploration for black smoker-type ore deposits. In *Ophiolites: Oceanic Crustal Analogues.* Nicosia: Geological Survey of Cyprus. J. Malpas, E. Moores, A. Panayitou, and C. Kenophontos, eds.

Vine, F. J. 1968. Magnetic anomalies associated with mid-ocean ridges. In *The History of the Earth's Crust.* R. A. Phinney, ed. Princeton: Princeton University Press.

Williams, H., and R. K. Stevens. 1974. The ancient continental margin of eastern North America. In *The Geology of Continental Margins.* C. A. Burk and C. L. Drake, eds. New York: Springer.

Wilson, R. A. M. 1959. The geology and mineral deposits of the Xeros-Troodos area. *Cyprus Geological Survey Memoir 1.*

Chapter 6

Abbate, E., V. Bortolotti, and G. Principi. 1980. Appenine ophiolites: A peculiar oceanic crust. *Ofioliti.* Special issue of *Tethyan Ophiolites* 1:59–97.

Anderson, D. L. 1971. The San Andreas fault. *Scientific American 225:* 52–68.

Argus, D. K., and R. G. Gordon. 1991. Current Sierra Nevada–North America motion from very long baseline interferometry: Implications for the kinematics of the western United States. *Geology* 19:1085–1088.

Atwater, T., and P. Molnar. 1973. Relative motion of the Pacific and North American plates deduced from seafloor spreading in the Atlantic, Indian, and south Pacific oceans. In *Proceedings of the Conference on Tectonic Problems of the San Andreas Fault System.* R. L. Kovach and A. Nur, eds. Stanford: Stanford Univ. Publications.

Carter, R. M., and R. H. Norris. 1976. Cainozoic history of southern New Zealand: An accord between geological observations and plate-tectonic predictions. *Earth and Planetary Sci. Let.* 31:85–94.

Casey, J. F., and J. F. Dewey. Initiation of subduction zones along transform and accreting plate boundaries, triple junction evolution, and forearc spreading centres: implications for ophiolite geology and subduction. In *Ophiolites and Oceanic Lithosphere.* I. G. Gass et al., eds. Geological Society Special Publication 13:269–290. London: Blackwell's.

DeMets, C., R. G. Gordon, D. F. Argus, and S. Stein. 1990. Current plate motions. *Geophys. J. International* 101: 425–478.

Ernst, W. G., Y. Seki, H. Onuki, and M. C. Gilbert. 1970. Comparative study of low-grade metamorphism in the California Coast Ranges and the Outer Metamorphic Belt of Japan. *GSA Mem.* 124.

Fox, P. J., R. S. Detrick, and G. M. Purdy. 1980. Evidence for crustal thinning near fracture zones: Implications for ophiolites. In *Proceedings of the International*

Ophiolite Symposium. A. Panayiotou, ed. Cyprus Geol. Surv. Dept.

Fox, P. J., and D. G. Gallo. 1984. A Tectonic model for ridge-transform-ridge plate boundaries: Implications for the structure of oceanic lithosphere. *Tectonophysics* 104:205–242.

Fox, P. J., E. Schreiber, H. Rowlett, and K. McCamy. 1976. The geology of the Oceanographer fracture zone: A model for fracture zones. *J. Geophys. Res.* 81:4117–4128.

Grindley, G. W. 1974. New Zealand. In *Mesozoic-Cenozoic Orogenic Belts.* A. M. Spencer, ed. Geological Society of London, Special Publication No. 4.

Karson, J. 1986a. Lithosphere age, depth and structural complications from migrating transform faults. *J. of the Geological Society, London* 153:785–788.

Karson, J. 1986b. Variations in structure and petrology in the Coastal complex, Newfoundland: Anatomy of an oceanic fracture zone. Geol. Soc. London Spec. Pub. 13.

Macdonald, K. C., K. Kastens, S. Miller, and F. N. Spiess. 1979. Deep-tow studies of the Tamayo transform fault. *Mar. Geophys. Res.* 4:37–70. Reprinted by permission of Kluwer Academic Publishers.

Menard, H. W., and T. E. Chase. 1970. Fracture zones. In *The Sea.* Vol. 4. A. E. Maxwell, ed. New York: Wiley.

Moores, E. M. 1982. Origin and emplacement of ophiolites. *Rev. Geophys. and Space Phys.* 20:735–760.

Moores, E. M., and F. J. Vine. 1971. The Troodos massif, Cyprus, and other ophiolites as oceanic crust: Evaluation and implications. *Philosophical Trans. Roy. Soc. London* 278A:443–466.

Robinson, P. T., B. T. R. Lewis, M. F. Flower, M. H. Salisbury, and H.-U. Schmincke. 1983. Crustal accretion in the Gulf of California, an intermediate-rate spreading axis. In *Initial Reports of the Deep Sea Drilling Project* 65:739–752.

Simonian, K., and I. G. Gass. 1978. Arakapas fault belt, Cyprus, a fossil transform fault. *GSA Bull.* 89:1220–1230.

Wallace, R. E., ed. 1990. The San Andreas Fault System, California. *U.S.G.S. Prof. Paper.* 1515:283.

Chapter 7

Abbate, E., V. Bortolotti, and P. Passerini. 1970. Olisto-stromes and olistoliths. *Sedimentary Geology.* 4: 521–557.

Abbate, E., V. Bortolotti, and G. Principi. 1980. Appenine ophiolites: A peculiar oceanic crust. *Ofioliti.* Special issue of *Tethyan Ophiolites,* vol. 1. Western Area. G. Rocci, ed.

Bard, J. P. 1983. Metamorphism of an obducted island arc: Example of the Kohistan sequence (Pakistan) in the Himalayan collided range. *Earth and Planet. Sci. Lett.* 65: 133–144.

Byrne, T. 1982. Structural evolution of coherent terranes in the Ghost Rocks Formation, Kodiak Island, Alaska. Geological Society of London. Special Paper 10.

Cowan, D. G. 1985. Structural styles in Mesozoic and Cenozoic melanges in the western Cordillera of North America. *GSA Bull.* 96:451–462.

Coward, M. P., B. F. Windley, R. D. Broughton, I. W. Luff, M. G. Petterson, C. J. Pudsey, D. C. Rex, and M. Asif Khan. 1986. Collision tectonics in the NW Himalayas. In *Collision Tectonics.* M. P. Coward and A. C. Ries, eds. Geological Society, London. Special Publication 19.

Curray, J., and D. G. Moore. 1974. Sedimentary and tectonic processes in the Bengal deep-sea fan and geosyncline. In *The Geology of Continental Margins.* C. A. Burk and C. L. Drake, eds. New York: Springer.

Davis, D., J. Suppe, and F. A. Dahlen. 1983. Mechanics of fold-and-thrust belts and accretionary wedges. *Geophys. Res.* 88(B2):1153–1172.

Dewey, J. F. 1976. Plate tectonics. In *Continents Adrift and Continents Aground.* J. T. Wilson, ed. New York: *Scientific American.*

Dickinson, W. R. 1970. Relations of andesites, granites, and derivative sandstones to arc-trench tectonics. *Rev. Geophys. and Space Phys.* 8:813–860.

Ernst, W. G. 1974. Arcs and subduction zones. In *Geological Interpretations from Global Tectonics, with Applications for California Geology and Petroleum Exploration.* W. R. Dickinson, ed. Bakersfield, Calif.: San Joaquin Geological Society.

Ernst, W. G., Y. Seki, H. Onuki, and M. C. Gilbert. 1970. Comparative study of low-grade metamorphism in the California Coast Ranges and outer metamorphic belt of Japan. *GSA Memoir* 124.

Hayes, D. E. 1974. Continental margin of western South America. In *Geology of Continental Margins.* C. A. Burk and C. L. Drake, eds. New York: Springer.

Isacks, B., and P. Molnar. 1971. Distribution of stresses in the descending lithosphere from a global survey of focal-mechanism solutions of mantle earthquakes. *Rev. Geophys. and Space Phys.* 9:103–174.

Karig, D. E. 1972. Remnant arcs. *GSA Bull.* 83:1057–1068.

Karig, D. E. 1974. Evolution of arc systems in the western Pacific. *Ann. Rev. Earth Planet Sci.* 2:51–76.

Kennett, J. P., A. R. McBirney, and R. C. Thunell. 1977. Episodes of Cenozoic volcanism in the circum-Pacific region. *J. Volcanol. and Geothermal Res.* 2:145–163.

Lawver, L., and J. W. Hawkins. 1978. Diffuse magnetic anomalies in marginal basins: Their possible tectonic and petrologic significance. *Tectonophysics* 45:323–339.

Lehner, F., H. Douslt, G. Bakker, P. Allenbach, and T. Gueneau. 1983. Active margins, part 3. In *Seismic Expressions of Structural Styles.* A. W. Ball, ed. Tulsa, Okla.: American Association of Petrology Geologists.

Marsh, B. D. 1979. Island-arc volcanism. *Am. Sci.* 67:161–172.

Marsh, B. D., and I. S. E. Carmichael. 1974. Benioff zone magmatism. *J. Geophys. Res.* 79:1196–1206.

Maxwell, J. C. 1974. Anatomy of an orogen. *GSA Bull.* 85:1195–1204.

Meissner, R. O., E. R. Flueh, F. Stibane, and E. Bekrg. 1976. Dynamics of the active plate boundary in southwest Colombia according to recent geophysical measurements. *Tectonophysics* 35:115–137.

Moore, G. F., J. R. Curray, and D. G. Moore, and D. E. Karig. 1980. Variations in geologic structure along the Sunda forearc, NE Indian Ocean. In *The Tectonic and Geologic Evolution of Southeast Asian Seas and Islands*. D. E. Hayes, ed. AGU Monograph 23.

Mwrozoski, C. L., and D. E. Hayes. 1980. A seismic reflection study of faulting in the Mariana forearc. In *The Tectonic and Geologic Evolution of Southeast Asian Seas and Islands*. D. E. Hayes, ed. AGU Monograph 23.

Oxburgh, E. R., and D. L. Turcotte. 1970. Thermal structure of island arcs. *GSA Bull.* 82:1665–1688.

Petterson, M. G., and B. F. Windley. 1985. Rb-Sr dating of the Kohistan arc-batholith in the Trans-Himalaya of north Pakistan, and tectonic implications. *Earth Planet. Sci. Lett.* 74:45–57.

Platt, J. P., ed. 1986. Dynamics of orogenic wedges and the uplift of high-pressure metamorphic rocks. *GSA Bull.* 97:1106–1121.

Raymond, L. A., ed. 1984. *Melanges: Their Nature, Origin, and Significance*. GSA Special Paper 198.

Scott, R., and L. Kroenke. 1980. Evolution of back arc spreading and arc volcanism in the Philippine Sea: Interpretation of Leg 59 DSDP results. In *The Tectonic and Geologic Evolution of Southeast Asian Seas and Islands*. D. E. Hayes, ed. AGU Monograph 23.

Shipley, T. H., K. D. McIntosh, E. A. Silver, and P. L. Stoffa. 1992. Three-dimensional seismic imaging of the Costa Rica Accretionary Prism: Structural diversity in a small volume of the lower slope. *J. Geophys. Res.* 97:4439–4459.

Shreve, R. L., and M. Cloos. 1986. Dynamics of sediment subduction, melange formation and prism accretion. *J. Geophys. Res.* 91:10,229–10,245.

Silver, E. A., M. J. Ellis, N. A. Breen, and T. H. Shipley. 1985. Comments on the growth of accretionary wedges. *Geology* 13:6-9.

Toksoz, M. N. 1976. Subduction of the lithosphere. In *Continents Adrift and Continents Aground*. J. T. Wilson, ed. New York: W. H. Freeman and Company.

Turcotte, D. L., and G. Shubert. 1982. *Geodynamics*. New York: Wiley.

Twiss, R. J., and E. M. Moores. 1992. *Structural Geology*. New York: W. H. Freeman and Company.

Uyeda, S. 1976. Some basic problems in the trench-arc-bac arc system. In *Island Arcs, Deep Sea Trenches, and Back-arc Basins*. M. Talwani and W. Pitman, eds. Maurice Ewing Series 1.

Uyeda, S. 1978. *The New View of the Earth*. San Francisco: W. H. Freeman and Company.

von Huene, R., and D. W. Scholl. 1991. Observations at convergent margins concerning sediment subduction. Subduction erosion, and the growth of continental crust. *Rev. Geophys.* 29:279-316.

Weissel, J. K. 1981. Magnetic lineations in marginal basins of the western Pacific. *Phil. Trans. Roy. Soc. Lond.* 300A:223–247.

Zhao, W.-L., D. M. Davis, F. A. Dahlen, and J. Suppe. 1986. Origin of convex accretionary wedges: Evidence from Barbados. *J. Geophys. Res.* 91:10,246–10,258.

Chapter 8

Dickinson, W. R., W. S. Snyder. 1979a. Geometry of triple junctions related to San Andreas transform. *J. Geophys. Res.* 84:561–572.

Dickinson, W. R., and W. S. Snyder. 1979b. Geometry of subducted slabs related to San Andreas transform. *J. Geol.* 87:609–627.

Griscome, A., and R. C. Jachens. 1989. Tectonic history of the north portion of the San Andreas fault system, California, inferred from gravity and magnetic anomalies. *J. Geophys. Res.* 94:3089–3099.

Huchon, P., and P. Labaume. 1989. Central Japan triple junctions: A three-dimensional compression model. *Tectonophysics* 160:117–133.

Merrits. D., and W. B. Bull. 1989. Interpreting Quaternary uplift rates at the Mendocino triple junction, northern California, from uplifted marine terraces. *Geology* 17: 1020–1024.

Ogawa, X., T. Seno, H. Akiyoshi, H. Tokuyama, K. Fujioka, and H. Taniguchi. 1989. Structure and development of the Sagami Trough and the Boso triple junction. *Tectonophysics* 160:135–150.

Patriat, P., and V. Courtillot. 1984. On the stability of triple junctions and its relation to episodicity in spreading. *Tectonics* 3:317–332.

Sclater, J. D., C. Bowin, R. Hey, H. Hoskins, J. Pierce, J. Phillips, and C. Tapscott. 1976. The Bouvet triple junction. *J. Geophys. Res.* 81:1857–1869.

Searle, R. C., and H. Francheteau. 1986. Morphology and tectonics of the Galapagos triple junction. *Marine Geophys. Res.* 8:95–129.

Silver, E. A. 1971. Tectonics of the Mendocino triple junction. *GSA Bull.* 82:2965–2978.

Yamazaki, T., and Y. Okamura. 1989. Subducting seamounts and deformation of overriding forearc wedges around Japan. *Tectonophysics* 160:207–229.

Chapter 9

Addicott, W., and P. W. Richards. 1982. Plate tectonic map of the circum-Pacific region, Pacific basin sheet. Tulsa, Okla.: American Association of Petrological Geologists.

Backofen, W. A. 1972. *Deformation Processing*. Reading, Mass.: Addison Wesley.

Bain, J. H. C. 1973. A summary of the main structural elements of Papua New Guinea. In *Western Pacific Island Arcs*. P. J. Coleman, ed. Nedlands (Australia): University of Western Australia Press.

Baranowski, J., J. Armbruster, L. Seeber, and P. Molnar. 1984. Focal depths and fault plane solutions of earthquakes and active tectonics in the Himalaya. *J. Geophys. Res.* 89:6918–6928.

Bird, P. 1978. Finite element modelling of lithosphere deformation: The Zagros collision orogeny. *Tectonophysics* 50:307–336.

Cardwell, R. K., B. L. Isacks, and D. E. Karig. 1980. The spatial distribution of earthquakes, focal mechanism solutions, and subducted lithosphere in the Philippine and northeastern Indonesian islands. AGU Monograph 23.

Carney, J. N., and A. MacFarlane. 1982. Geological evidence bearing on the Miocene to recent structural evolution of the New Hebrides arc. *Tectonophysics* 87: 147–175.

Casey, J. F., and J. F. Dewey. 1984. Initiation of subduction zones along transform and accreting plate boundaries, triple-junction evolution, and forearc spreading centres: Implications for ophiolitic geology and obduction. In *Ophiolites and Oceanic Lithosphere*. I. G. Gass, S. Lippard, and A. Shelton, eds. Geological Society of London. Special Publication 13.

Couch, R., R. Whitsett, B. Huehn, and L. Briceno-Guarupe. 1981. Structures of the continental margin of Peru and Chile. *GSA Memoir* 154:703–728.

Coward, M. P., and A. C. Ries, eds. 1986. *Collision Tectonics* GSA Special Publication 19.

Dewey, J. F. 1977. Suture zone complexities: A review. *Tectonophysics* 40:53–67.

Dewey, J. F. 1980. Episodicity, sequence, and style at convergent plate boundaries. In *The Continental Crust and its Mineral Deposits*. D. W. Strangway, ed. Geological Association of Canada Special Paper 20.

Dewey, J. F., and J. M. Bird. 1970. Mountain belts and the new global tectonics. *Geophys. Res.* 75:2625–2647.

Eaton, J. P. 1966. Crustal structure in northern and central California from seismic evidence. *Calif. Div. Mines Geol. Bull.* 190:419–426.

England, P. 1982. Some numerical investigations of large-scale continental deformation. In *Mountain Building Processes*. K. J. Hsü, ed. New York: Academic Press.

Gansser, A. 1980. The significance of the Himalayan suture zone. *Tectonophysics* 62:37–53.

Gass, I. G., et al. "Origin and emplacement of ophiolites." July 1975. *Geodynamics Today; A Review of the Earth's Dynamic Processes*. London: Royal Society, pp. 54–65.

Gass, I. G. 1982. Ophiolites. *Scientific American*: August.

Helwig, J. 1976. Shortening of continental crust in orogenic belts and plate tectonics. *Nature* 260:768–780.

Hilde, T. W. C., S. Uyeda, and L. Kroenke. 1977. Evolution of the western Pacific and its margin. *Tectonophysics* 38:145–166.

Krogstad, E. J., S. Balakrishnan, D. K. Mukhopadhyay, V. Rajamani, and G. N. Hanson. 1989. Plate tectonics 2.5 billion years ago: Evidence at Kolar, south India. *Science*: 243:1337–1340.

Liou, J. G. C.-Y. Lan, J. Suppe, and W. G. Ernst. 1977. The East Taiwan ophiolite, its occurrence, petrology, metamorphism, and tectonic setting. Taipei (Taiwan): Mining Research and Service Organization.

McCaffrey, R., E. A. Silver, and R. W. Raitt. 1980. Crustal structure of the Molucca Sea collision zone, Indonesia. *AGU Monograph* 23.

McCaffrey, R., and G. A. Abers. 1991. Orogeny in arc-continent collision: The Banda arc and western New Guinea. *Geology* 19:563–566.

McKenzie, D. P. 1972. Active tectonics in the Mediterranean region. *Geophys. J. Roy. Astron. Soc.* 30:109–185.

Monger, J. W. H., and E. Irving. 1978. Northward displacement of north-central British Columbia. *Nature* 285: 289–294.

Moores, E. M. 1982. Origin and emplacement of ophiolites. *Rev. Geophys. and Space Phys.* 20:735–760.

Ni, J., and M. Barazangi. 1984. Seismotectonics of the Himalayan collision zone: Geometry of the underthrusting Indian plate beneath the Himalaya. *J. Geophys. Res.* 89:1145–1163.

Nur, A., and Z. Ben-Avrahem. 1982. Oceanic plateaus, the fragmentation of continents and mountain building. *J. Geophys. Res.* 87:3644–3661.

Pigram, C. J., and H. L. Davies. 1987. Terranes and the accretion history of the New Guinea orogen. *BMR J. Australian Geol. and Geophys.* 10:193–211.

Silver, E. A., L. D. Abbott, K. S. Kirchoff-Stein, D. L. Reed, B. Bernstein-Taylor, and D. Hilyard. 1991. Collision propagation in Papua New Guinea and the Solomon Sea. *Tectonics* 10:863–874.

Stöcklin, J. 1974. Possible ancient continental margins of Iran. In *Geology of Continental Margins*. C. Burk and C. L. Drake, eds. New York: Springer.

Tapponnier, P. 1977. Evolution tectonique du systeme alpine en mediterranee: Poinçoinnement et ecrasement rigid-plastique. *Bull. Soc. Geol. France* 7 (xxix):437–460.

Tapponnier, P., and P. Molnar. 1976. Slip-line field theory and large-scale continental tectonics. *Nature* 264:319.

Tapponnier, P., et al. 1982. Propagating extrusion tectonics in Asia: New insights from simple experiments with plasticine. *Geology* 10(12):614–616.

Vink, G. E., W. J. Morgan, and W.-L. Zhao. 1984. Preferential rifting of continents: A source of displaced terranes. *J. Geophys. Res.* 89:10,072–10,076.

Whittington, H. B. 1973. Ordovician trilobites. In *Atlas of Paleobiogeography*. A. Hallam, ed. New York: Elsevier.

Chapter 10

Bally, A. W., et al. 1979. Continental margins: Geological and geophysical research needs and problems. Washington, D.C.: National Academy of Sciences.

Bally, A. W. 1980. *Basins and Subsidence: A Summary*. AGU-GSA Geodynamic Series 1.

Bally, A. W., P. L. Gordy, and G. A. Stewart. 1986. Structure seismic data and orogenic evolution of the southern Canadian Rockies. *Canadian Petroleum Geology Bulletin* 14:337–381.

Bateman, P. C. 1981. In *The Geotectonic Evolution of California*. W. G. Ernst, ed. Englewood Cliffs, N.J.: Prentice-Hall.

Burchfiel, B. C. 1983. The continental crust. In *The Dynamic Earth*. R. Siever, ed. Special issue of *Scientific American*. September.

Burchfiel, B. C., C. Zhiliang, K. V. Hodges, L. Yuping, L. H. Royden, D. Changrong, and X. Jiene. 1991. The South Tibetan detachment system, Himalayan orogen: Extension contemporaneous with and parallel to shortening in a collisional mountain belt. GSA Special Paper 269.

Clark, S. P., Jr., B. C. Burchfiel, and J. Suppe, Eds. 1988. Processes in continental lithospheric deformation. GSA Special Paper 218.

Cook, F. A., et al. 1981. COCORP seismic profiling of the Appalachian orogen beneath the coastal plain of Georgia. *GSA Bull.* 1:92, 739.

Coward, M. P., and A. W. B. Siddans. 1979. The tectonic evolution of the Welsh Caledonides. In *The Caledonides of the British Isles Reviewed*. A. L. Harris, C. H. Holland, and G. E. Leake, eds. Geol. Soc. of London Special Publication 8.

Dalziel, I. W. D., and R. D. Forsythe. 1985. *Andean Evolution and the Terrane Concept*. Circum-Pacific Council for Energy and Mineral Resources, Earth Science Series, No. 1.

Davis, D., J. Suppe, and F. A. Dahlen. 1983. Mechanics of fold-and-thrust belts and accretionary wedges. *J. Geophys. Res.* 88:1153–1172.

Dewey, J. F. 1977. Suture zone complexities: A review. *Tectonophysics.* 40:53–67.

Durney, D. W., and G. Ramsay. 1973. Incremental strain measured by syntectonic crystal growths. In *Gravity and Tectonics*. K. A. De Jong and R. Scholten, eds. New York: Wiley.

England, P. C., and A. B. Thompson. 1984. Pressure-temperature-time paths of regional metamorphism. I. Heat transfer during the evolution of regions of thickened continental crust. *J. Petrol.* 25:894–928.

Eskola, P. 1949. The problem of mantled gneiss domes. *Quatern. J. Geol. Soc. London* 104:461–476.

Francheteau, J. 1983. The oceanic crust. In *The Dynamic Earth*. R. Siever, ed. Special issue of *Scientific American*. September.

Frey, M. et al. 1974. Alpine metamorphism of the Alps: A review. *Schweizerische Mineralogische und Petrografische Mitteilung* 54:247–290.

Gansser, A. 1974. *Geology of the Himalayas*. New York: Wiley-Interscience.

Hansen, E. C. 1971. *Strain Facies*. New York: Springer Verlag.

Harris, L. D., and K. C. Bayer. 1979. Sequential development of the Appalachian orogen above a master décollement: A hypothesis. *Geology* 7:568–572.

Hatcher, R., and R. T. Williams. 1986. Mechanical model for single thrust sheets. *GSA Bull.* 97:975–985.

Hoffman, P. 1988. United Plates of America. *Ann. Rev. Earth and Plan. Sci.* 16:543–603.

Hoffman, P., J. F. Dewey, and K. Burke. 1974. SEPM Special Publication 19.

Jackson, M. P. A. 1984. Archean structural styles in ancient gneiss complex of Swaziland, southern Africa. In *Precambrian Tectonics Illustrated*. A. Kröner and R. Greiling, eds. Stuttgart: Schweizerbartsche.

Jordan, T. E., B. L. Isacks, R. W. Allmendinger, J. A. Brewer, V. A. Ramos, and C. J. Ando. 1983. Andean tectonics related to geometry of subducted Nazca plate. *GSA Bull.* 94:341–361.

King, P. B. 1977. *Evolution of North America*. 2nd ed. Princeton: Princeton University Press.

Kröner, A., and Greiling, eds. *Precambrian Tectonics Illustrated*. Stuttgart: Schweizerbartsche.

Laubscher, H. P. 1982. Detactment, shear, and compression in the central Alps. *GSA Memoir* 158:191–211.

LeForte, P. 1975. Himalayas: The collided range. Present knowledge of the continental arc. *Amer. J. Sci.* 275A:1–44.

Merle, O. 1986. Patterns of stretch trajectories and strain rates in spreading-gliding nappes. *Tectonophysics* 124:211.

Miller, H., S. Mueller, and G. Perrier. 1983. Structure and dynamics of the Alps: A geophysical inventory. In *Alpine Mediterranean Geodynamics*. H. Berckhemer and K. Hsü, eds. GSA-AGU Geodynamics Series 7.

Misra, K. C., and F. B. Keller. 1978. Ultramafic bodies in the southern Appalachians: A review. *Amer. J. Sci.* 278:389–418.

Moore, J. C. 1978. Orientation of underthrusting during latest Cretaceous and earliest Tertiary time, Kodiak Islands, Alaska. *Geology* 6:196–213.

National Academy of Sciences-National Research Council. 1980. *Continental Tectonics*. Washington, D.C.

Nisbet, E. G. 1987. *The Young Earth: An Introduction to Archaean Geology*. Boston: Allen and Unwin.

Press, F., and R. Siever. 1986. *Earth,* 4th ed. New York: W. H. Freeman and Company.

Price, R. A., and R. D. Hatcher, Jr. 1983. Tectonic significance of similarities in the evolution of the Alabama-Pennsylvania Appalachians and the Alberta-British Columbia Canadian Cordillera. *GSA Memoir* 158:149–160.

Ramsay, J. G. 1963. Stratigraphy, structure and metamorphism in the western Alps. *Geologists' Assoc. Proc. London* 74:357–392.

Roeder, D. H. 1973. Subduction and orogeny. *J. Geophys. Res.* 78:5005.

Roeder, D. H., O. E. Gilbert, Jr., and W. D. Witherspoon. 1978. Evolution and macroscopic structure of Valley and Ridge thrust belt, Tennessee and Virginia. *Studies in Geology,* vol. 2. Chattanooga: Dept. Geological Sci. Univ. of Tenn.

Sanderson, D. J. 1982. Models of strain variation in nappes and thrust sheets: A review. *Tectonophysics* 88:201–233.

Schaer, J.-P., and J. Rodgers, eds. 1987. *The Anatomy of Mountain Ranges*. Princeton: Princeton University Press.

Seeber, L., J. G. Armbruster, and R. C. Quittmeyer. 1981. Seismicity and continental subduction in the Himalayan arc. In *Zagros-Hindu Kush-Himalaya Geodynamic Evolution*. H. K. Gupta and F. M. Delany, eds. AGU-GSA Geodynamics Series Volume 3.

Siddans, A. W. B. 1983. Finite strain patterns in some Alpine nappes. *J. Struc. Geol.* 5(no. 3/4):441.

Smithson, S. B., J. Brewer, S. Kaufman, J. Oliver, and C. Hurich. 1979. Nature of the Wind River thrust, Wyoming, from COCORP deep-reflection data and from gravity data. *Geology* 6:648–652.

Smithson, S. B., P. N. Shive, and S. K. Brown. 1977. Seismic velocity, reflection, and structure of the crystalline crust. In *The Earth's Crust: Its Nature and Physical Properties*. J. G. Heacock, ed. *AGU Monograph* 20: 254–270.

Spicher, A. 1980. *Tektonische Karte der Schweiz.* Bern: Schweizer Geologisches Kommission.

Thompson, J. B., et al. 1968. Nappes and gneiss domes in west-central New England. In *Structural studies of the Northern Appalachians*. E.-A. Zen, et al., eds. New York: Wiley.

Turner, F. J. 1968. *Metamorphic Petrology, Mineralogical and Field Aspects*. New York: McGraw Hill.

Twiss, R. J., and E. M. Moores, 1992. *Structural Geology*. New York: W. H. Freeman and Company.

Uyeda, S. 1978. *The New View of the Earth*. New York: W. H. Freeman and Company.

Vail, P. R., R. M. Michum, Jr., and S. Thompson. 1979. *AAPG Memoirs* 19:84.

Williams, H. 1984. Miogeoclines and suspect terranes of the Caledonian-Appalachian orogen: Tectonic patterns in the North Atlantic region. *Canadian J. of Earth Sci.* 21: 887–901.

Windley, B. F. 1984. *The Evolving Continents*. New York: Wiley.

—— 1995. *The Evolving Continents*, 3d ed. New York: Wiley.

Windley, B. F., F. C. Bishop, and J. V. Smith. 1981. *Ann. Rev. Earth and Planetary Sci.* 10:179.

Zingg, A., and R. Schmid. 1979. Multidisciplinary research on the Ivrea zone. *Schweische Mineralogische und Petrografische Mitteilung* 59:189–97.

Chapter 11

Allen, C. A. 1986. Seismological and paleoseismological techniques of research in active tectonics. In *Active Tectonics*. Geophysics Study Committee. Washington, D.C.: National Academy Press.

Bull, W. B., and L. D. McFadden. 1977. Tectonic geomorphology north and south of the Garlock fault, California. In *Geomorphology in Arid Regions*. D. O. Doehring, ed. Proceedings of 8th Annual Geomorphology Symposium, State University of New York at Binghamton.

Carter, W. E., and D. S. Robertson. 1986. Studying the earth by very-long-baseline interferometry. *Scientific American* 255(5):46–54.

Geophysics Study Committee. 1986. *Active Tectonics*. Washington, D.C.: *National Academy Press*.

Lajoie, K. R. 1986. Coastal tectonics. In *Active Tectonics*. Geophysics Study Committee. Washington, D.C.: National Academy Press.

Lisowski, M., J. C. Savage, and W. H. Prescott. 1991. The velocity field along the San Andreas fault in central and southern California. *J. Geophys. Res.* 96:8369–8389.

Minster, J. B., and T. Jordan. 1978. Present-day plate motions. *J. Geophys. Res.* 83:5331–5354.

Nash, D. B. 1986. Morphologic dating and modeling degradation of fault scarps. In *Active Tectonics*. Geophysics Study Committee. Washington, D.C.: National Academy Press.

Press, F., and R. Siever. 1986. *Earth*. 4th ed. New York: W. H. Freeman and Company.

Schumm, S. A. 1986. Alluvial river response to active tectonics. In *Active Tectonics*. Geophysics Study Committee. Washington, D.C.: National Academy Press.

Schwartz, D. P., and K. J. Coppersmith. 1986. Seismic hazards: new trends in analysis using geologic data. In *Active Tectonics*. Geophysics Study Committee. Washington, D.C.: National Academy Press.

Sieh, K. E. 1978. Prehistoric large earthquakes produced by slip on the San Andreas fault at Pallett Creek, California. *J. Geophys. Res.* 83:3907–3939.

Sieh, K. E., M. Stiover, and D. Brillinger. 1989. A more precise chronology of earthquakes produced by the San Andreas fault in southern California. *J. Geophys. Res.* 94:603–623.

Slemmons, D. B., and C. M. Depolo. 1986. Evaluation of active faulting and associated hazards. In *Active Tectonics*. Geophysics Study Committee. Washington, D.C.:National Academy Press.

Chapter 12

Allegré, C. J., et al. 1984. Structure and evolution of the Himalaya-Tibet orogenic belt. *Nature* 307:17–22.

Allmendinger, R. W., T. A. Hauge, E. C. Hauser, C. J. Potter, S. L. Klemperer, K. D. Nelson, P. Knuepfer, and J. Oliver. 1987. Overview of the COCORP 40°N transect, western USA. *GSA Bull.* 98:364–372.

Ando, C. J., F. A. Cook, J. E. Oliver, L. D. Brown, and S. Kaufman. 1982. Crustal geometry of the Appalachian orogen from seismic reflection studies. *GSA Memoir* 158:83–103.

Bond, G. C., P. A. Nickeson, and M. A. Kominz. 1984. Breakup of a supercontinent between 625 Ma and 55 Ma: New evidence and implications for continental histories. *Earth and Planet. Sci. Lett.* 70:325–345.

Burchfiel, B. C. 1983. The continental crust. In *The Dynamic Earth*. R. Siever, ed. Special issue of *Scientific American*. September.

Chang, C.-F., Y.-S. Pan, and Y.-Y. Sun. 1989. The tectonic evolution of Qinghai-Tibet Plateau: A review. In *Tectonic Evolution of the Tethyan Region*. A. M. C. Sengor, ed. Boston: Kluwer.

Cloos, H., D. Roeder, and K. Schmidt, eds. 1978. *Alps, Apennines, Hellenides*. Interunion Commission on Geodynamics, Scientific Report No. 38.

Coney, P. J., D. L. Jones, and J. W. H. Monger. 1980. Cordilleran suspect terranes. *Nature* 288:329–333.

Cook, F. A., et al. 1981. COCORP seismic profiling of the Appalachian orogen beneath the coastal plain of Georgia. *GSA Bull.* 92:738–748.

Csejtey, B., D. P. Cox, R. C. Evarts, G. D. Stricker, and H. L. Foster. 1982. The Cenozoic Denali fault system and the Cretaceous accretionary development of southern Alaska. *J. Geophys. Res.* 87:3742–3754.

Debiche, M. G., A. Cox, and D. Engebretson. 1987. The motion of allochthonous terranes across the North Pacific Basin. GSA Special Paper 207.

Dercourt, J., et al. 1985. Presentation de 9 cartes paleogeographique au 1/20.000.000 s'entendant de l'Atlantic au Pamir pour la periode du Lias a l'Actuel. *Bull. Soc. Geol. France.* 8(no. I):637–652.

Dewey, J. F. 1977. Suture zone complexities: A review. *Tectonophysics* 40:69–100.

Dewey, J. F., W. C. Pitman III, W. B. F. Ryan, and J. Bonnin. 1973. Plate tectonics and the evolution of the Alpine system. *GSA Bull.* 84:3137–3180.

ECORS Pyrenees team. 1988. The ECORS deep reflection seismic survey across the Pyrenees. *Nature* 33:508–510.

Erickson, G. E., P. M. T. Cañas, and J. A. Reinemund, eds. *Geology of the Andes and Its Relation to Hydrocarbon and Mineral Resources.* Houston: Circum-Pacific Council for Energy and Mineral Resources Earth Science Series, vol. 11.

Ernst, W. G., ed. 1974. *Geotectonic Development of California.* Rubey Vol. 1. Englewood Cliffs, N.J.: Prentice Hall.

Girardeau, J., J. Marcoux, and C. Montenat. 1989. The Neo-Cimmerian ophiolite belt in Afghanistan and Tibet: Comparison and evolution. In *Tectonic Evolution of the Tethyan Region.* A. M. C. Sengör, ed. Boston:Kluwer.

Green, H. G., and F. L. Wong. 1989. Ridge collisions along the plate margins of South America compared with those in the southwest Pacific. In *Geology of the Andes and Its Relation to Hydrocarbon and Mineral Resources.* G. E. Erickson, P. M. T. Cañas, and J. A. Reinemund, eds. Houston: Circum-Pacific Council for Energy and Mineral Resources Earth Science Series, vol. 11.

Hatcher, R. D., Jr. 1987. Tectonics of the southern and central Appalachian internides. *Ann. Rev. Earth Planet. Sci.* 15:337–362.

Hoffman, P. F. 1991. Did the break-out of Laurentia turn Gondwanaland inside-out? *Science* 252:1409–1412.

Hossack, J. R., and M. A. Cooper. 1986. Collision tectonics in the Scandinavian Caledonides. In *Collision Tectonics.* M. P. Coward and A. C. Ries, eds. Geological Society London Special Publication 19.

King, P. B. 1977. *Evolution of North America.* Princeton, N.J.: Princeton University Press.

Le Fort, P. 1989. The Himalayan orogenic segment. 1989. In *Tectonic evolution of the Tethyan Region.* A. M. C. Sengör, ed. Boston: Kluwer.

Le Pichon, X., F. Bergerat, and M. J. Roulet. 1988. Plate kinematics and tectonics leading to the Alpine belt formation: A new analysis. In S. P. Clark, Jr., B. C. Burchfiel, and J. Suppe, eds. *Processes in Continental Lithospheric Deformation.* GSA Special Paper 218.

Mattauer, M. 1986. Intracontinental subduction, crust-mantle décollement and crustal-stacking wedge in the Himalayas and other collision belts. In Coward, M. P. and A. C. Ries, eds. *Collision Tectonics.* Geological Society of London. Special Publication 19.

Maxwell, J. C. 1974. Anatomy of an orogen. *GSA Bull.* 85:1195–1204.

Megard, F. 1989. The evolution of the Pacific Ocean margin in South America north of the Africa elbow (18°S). In *The Evolution of the Pacific Ocean Margins.* Z. Ben-Avrahem, ed. New York: Oxford University Press.

Monger, J. W. H. 1984. Cordilleran tectonics: A Canadian perspective. *Bull. Soc. Geol. France* (no. 26):255–278.

Moores, E. M. 1970. Ultramafics and orogeny, with models of the U.S. Cordillera and the Tethys. *Nature* 228:837–842.

Moores, E. M. 1986. The Proterozoic ophiolite problem, continental emergence, and the Venus connection. *Science* 234:65–68.

Mpodozis, C., and V. Ramos. 1989. The Andes of Chile and Argentina. In *Geology of the Andes and Its Relation to Hydrocarbon and Mineral Resources.* G. E. Erickson, P. M. T. Cañas, and J. A. Reinemund, eds. Houston: Circum-Pacific Council for Energy and Mineral Resources Earth Science Series, vol. 11.

Piper, J. D. A. 1983. *Dynamics of continental crust.* GSA Memoir 161:14.

Potter, C. J., et al. 1986. COCORP transect of the Washington-Idaho Cordillera. *Tectonics* 5:1007–1026.

Price, R. A., and R. D. Hatcher, Jr. 1982. Tectonic significance of similarities in the evolution of the Alabama-Pennsylvania Appalachians and the Alberta-British Columbia Canadian Cordillera. *GSA Memoir* 158:149–161.

Ratschbacher, L., O. Merle, P. Davy, and P. Cobbold. 1991a. Lateral extrusion in the Eastern Alps, Part I: Boundary conditions and experiments scaled for gravity. *Tectonics* 10(no. 2):245–256.

Ratschbacher, L., W. Frisch, H.-G. Linzer, and O. Merle. 1991b. Lateral extrusion in the Eastern Alps, Part II: Structural analysis. *Tectonics* 10:256–272.

Roeder, D. H. 1977. Continental convergence in the Alps. *Tectonophysics* 40:339–350.

—— 1988. Andean-age structure of eastern Cordillera (Province of La Paz, Bolivia). *Tectonics* 7: 23–40.

Roeder, D. H., and H. Bögel. 1978. Geodynamic interpretation of the Alps. In *Alps, Apennines, Hellenides.* H. Cloos, D. Roeder, and K. Schmidt, eds. Interunion Commission on Geodynamics, Scientific Report No. 38.

Roeder, D. H., and C. G. Mull. 1978. Ophiolites of the Brooks Range. *AAPG Bull.* 62:1692–1702.

Schweickert, R. A., and W. S. Snyder. 1980. Paleozoic plate tectonics of the Sierra Nevada and adjacent region. In *The Geotectonic Development of California*. Rubey, vol. 1. W. G. Ernst, ed. N.J.: Prentice-Hall.

Scotese, C. R. 1984. Paleozoic paleomagnetism and the assembly of Pangea. In *Plate Reconstruction from Paleozoic Paleomagnetism*. R. Van der Voo, C. R. Scotese, and N. Bonhommet, eds. GSA-AGU Geodynamics series vol. 12.

Sengör, A. M. C., ed. *Tectonic Evolution of the Tethyan Region*. Boston: Kluwer.

Tapponnier, P. 1977. Evolution tectonique du system alpin in Mediterranee, poinçonnement et ecrasement rigidplastique. *Bull. Soc. Geol. France*. 7(xxix):437–460.

Tapponnier, P., M. Mattauer, F. Proust, and C. Cassaigneau. 1981. Mesozoic ophiolites, sutures, and large scale tectonic movements in Afghanistan. *Earth Planet Sci. Lett.* 52:355–371.

Vicente, J.-C. 1989. Early late Cretaceous overthrusting in the western Cordillera of southern Peru. In *Geology of the Andes and Its Relation to Hydrocarbon and Mineral Resources*. Erickson, G. E., Cañas, P. M. T. and J. A. Reinemund, eds. Houston: Circum-Pacific Council for Energy and Mineral Resources Earth Science Series, vol. 11.

Williams, H. 1984. Miogeoclines and suspect terranes of the Caledonian-Appalachian orogen: Tectonic patterns in the North Atlantic region. *Can. J. Earth Sci.* 21:887–901.

Chapter 13

Bindschadler, D. L., G. Schubert, and W. M. Kaula. 1992. Coldspots and hotspots: Global tectonics and mantle dynamics of Venus. *J. Geophys. Res.* 97:13,495–13,532.

Carr, M. H. 1984. Mars. In *The Geology of the Terrestrial Planets*. M. H. Carr, ed. Washington, D.C.: NASA Special Paper 469.

Crumpler, L. S., and J. W. Head. 1988. Bilateral topographic symmetry patterns across Aphrodite Terra, Venus. *J. Geophys. Res.* 93:301–312.

Crumpler, L. S., J. W. Head, and D. B. Campbell. 1986. Orogenic belts on Venus. *Geology* 14:977–1092.

Grieve, R. A. F. 1982. Impact structures on Earth. GSA Special Paper 190.

Head, J. W., and L. S. Crumpler. 1987. Evidence for divergent plate-boundary characteristics and crustal spreading on Venus. *Science* 238:1380–1385.

Head, J. W., and R. S. Saunders. 1991. Geology of Venus: A perspective from early *Magellan* mission results. *GSA Today* 1:49.

Head, J. W., and S. C. Solomon. 1981. Tectonic evolution of the terrestrial planets. *Science* 213:62–76.

Herrick, R. R. 1994. Resurfacing history of Venus. *Geology* 22:703–706.

McKenzie, D., et al. 1992. Features on Venus generated by plate boundary processes. *J. Geophys. Res.* 97:13,533–13,544.

Moore, E. M. 1986. The Proterozoic ophiolite problem, continental emergence, and the Venus connection. *Science* 234:65–68.

Phillips, R. J., and M. C. Malin. 1984. Tectonics of Venus. *Ann. Rev. Earth Planet. Sci.* 12:411–443.

Saunders, R. S., and M. H. Carr. 1984. Venus. In *The Geology of the Terrestrial Planets*. M. H. Carr, ed. Washington, D.C.: NASA Special Paper 469.

Saunders, R. S., et al. 1992. Magellan at Venus. *J. Geophys. Res.* (Part 1) 97:13,059–13,690; (Part 2) 97:10,921–16,382.

Schulz, P. H. 1985. Polar wandering on Mars. *Scientific American* 253(6):94–102.

Solomon, S. C., et al. 1992. Venus tectonics: An overview of *Magellan* observations. *J. Geophys. Res.* 97:13,199–13,255.

Squyres, S. W., et al. 1992. The morphology and evolution of coronae on Venus. *J. Geophys. Res.* 97:13,611–13,634.

Stofan, E. R., D. L. Bindschadler, J. W. Head, and E. M. Parmentier. 1991. Corona structures on Venus: Models of origin. *J. Geophys. Res.* 96:20,933–20,947.

Strom, R. G. 1984. Mercury. In *The Geology of the Terrestrial Planets*. M. H. Carr, ed. Washington, D.C.: NASA Special Paper 469.

Suppe, J., and C. Connors. 1992. Critical taper wedge mechanics of fold and thrust belts on Venus: Initial results from *Magellan*. *J. Geophys. Res.* 97:13,545–13,561.

Turcotte, D. L. 1993. An episodic hypothesis for Venusian tectonics. *J. Geophys. Res.* 98:17,061–17,068.

Wilhelms, D. E. 1984. Moon. In *The Geology of the Terrestrial Planets*. M. H. Carr, ed. Washington, D.C.: NASA Special Paper 469.

Appendix

Atwater, T. 1970. Implications of plate tectonics for the Cenozoic tectonic evolution of western North America. *GSA Bull.* 81:1,3513–3536.

Deetz and Adams, 1928. Elements of map projections. U.S. Coast and Geodetic Survey, Serial 146. 2d ed. Special Publication 68.

McDonnell, P. W., Jr. 1979. *Introduction to Map Projections*. New York: M. Dekker.

Smith, A. G., and J. C. Briden. 1977. *Mesozoic-Cenozoic Paleocontinental Maps*. New York: Cambridge University Press.

Smith, A. G., A. M. Hurley, and J. C. Briden. 1981. *Phanerozoic World Maps*. New York: Cambridge University Press.

Snyder, J. P. 1987. Map projections: A working manual. Washington, D.C.: U.S. Geol. Surv. Prof. Pap. No. 1395.

Index

Note: Page numbers in *italics* indicate illustrations; those followed by *n* indicate footnotes.

Coláiste na hOllscoile Gaillimh

3 1111 30115 7465